HYDRAULIC FRACTURING IN UNCONVENTIONAL RESERVOIRS

Theories, Operations, and Economic Analysis

HYDRAULIC FRACTURING IN UNCONVENTIONAL RESERVOIRS

Theories, Operations, and Economic Analysis

HOSS BELYADI
EBRAHIM FATHI
FATEMEH BELYADI

Gulf Professional Publishing is an imprint of Elsevier
50 Hampshire Street, 5th Floor, Cambridge, MA 02139, United States
The Boulevard, Langford Lane, Kidlington, Oxford, OX5 1GB, United Kingdom

© 2019 Elsevier Inc. All rights reserved.

No part of this publication may be reproduced or transmitted in any form or by any means, electronic or mechanical, including photocopying, recording, or any information storage and retrieval system, without permission in writing from the publisher. Details on how to seek permission, further information about the Publisher's permissions policies and our arrangements with organizations such as the Copyright Clearance Center and the Copyright Licensing Agency, can be found at our website: www.elsevier.com/permissions.

This book and the individual contributions contained in it are protected under copyright by the Publisher (other than as may be noted herein).

Notices
Knowledge and best practice in this field are constantly changing. As new research and experience broaden our understanding, changes in research methods, professional practices, or medical treatment may become necessary.

Practitioners and researchers must always rely on their own experience and knowledge in evaluating and using any information, methods, compounds, or experiments described herein. In using such information or methods they should be mindful of their own safety and the safety of others, including parties for whom they have a professional responsibility.

To the fullest extent of the law, neither the Publisher nor the authors, contributors, or editors, assume any liability for any injury and/or damage to persons or property as a matter of products liability, negligence or otherwise, or from any use or operation of any methods, products, instructions, or ideas contained in the material herein.

Library of Congress Cataloging-in-Publication Data
A catalog record for this book is available from the Library of Congress

British Library Cataloguing-in-Publication Data
A catalogue record for this book is available from the British Library

ISBN: 978-0-12-817665-8

For information on all Gulf Professional publications
visit our website at https://www.elsevier.com/books-and-journals

Publisher: Brian Romer
Senior Acquisition Editor: Katie Hammon
Editorial Project Manager: Joanna Collett
Production Project Manager: Bharatwaj Varatharajan
Cover Designer: Matthew Limbert

Typeset by SPi Global, India

Contents

Author biography	xv
Introduction	xvii

1. Introduction to unconventional reservoirs — 1
- Introduction — 1
- Different types of natural gas — 4
- Natural gas transport — 5
- Unconventional reservoirs — 5

2. Advanced shale reservoir characterization — 13
- Introduction — 13
- Pore-size distribution measurement of shale — 13
- Shale sorption measurement techniques — 15
- Shale porosity measurements — 20
- Pore compressibility measurements of shale — 21
- Shale permeability measurement techniques — 22

3. Shale initial gas-in-place calculation — 27
- Introduction — 27
- Total gas-in-place calculation — 27
- Density of adsorbed gas — 31
- Recovery factor — 33

4. Multiscale fluid flow and transport in organic-rich shale — 35
- Introduction — 35
- Multicontinuum modelling of shale reservoirs — 36
- Interfacial tension and capillary pressure — 39
- Wettability effects on shale recovery — 44

5. Hydraulic fracturing fluid systems — 47
- Introduction — 47
- Slick water fluid system — 49
- Cross-linked gel fluid system — 52
- Hybrid fluid system — 54
- Foam fracturing — 55
- Foam quality — 57
- Foam stability — 58

Tortuosity	59
Typical slick water frac steps	61
Acidization stage	61
Pad stage	63
Proppant stage	65
Flush stage	66
Frac fluid selection summary	69

6. Proppant characteristics and application design — 71

Introduction	71
Sand	71
Precured resin-coated sand	72
Curable resin-coated sand	72
Intermediate-strength ceramic proppant	74
Lightweight ceramic proppant	74
High-strength proppant	74
Proppant size	76
Proppant characteristics	79
Proppant particle-size distributions	82
Proppant transport and distribution in hydraulic fracture	82
Fracture conductivity	85
Dimensionless fracture conductivity	86
International Organization for Standardization conductivity test	86
Non-Darcy flow	88
Multiphase flow	88
Reduced proppant concentration	89
Gel damage	89
Cyclic stress	90
Fines migration	91
Time degradation	93
Finite vs infinite conductivity	93

7. Unconventional reservoir development footprints — 97

Introduction	97
Casing selection	98
Conductor casing	98
Surface casing	99
Intermediate casing	100

Contents   vii

Production casing	100
Hydraulic fracturing and aquifer interaction	101
Hydraulic fracturing and fault reactivation	102
Hydraulic fracturing and low-magnitude earthquakes	106

8. Hydraulic fracturing chemical selection and design   **107**

Introduction	107
Friction reducer	108
FR flow loop test	108
Pipe friction pressure	111
Reynolds number	112
Relative roughness of pipe	112
FR breaker	114
Biocide	114
Scale inhibitor	115
Linear gel	116
Gel breaker	117
Buffer	118
Cross-linker	118
Surfactant	119
Iron control	120

9. Fracture pressure analysis and perforation design   **121**

Introduction	121
Pressure (psi)	121
Hydrostatic pressure (psi)	122
Hydrostatic pressure gradient (psi/ft)	122
Instantaneous shut-in pressure (psi)	123
Fracture gradient (psi/ft)	124
Bottom-hole treating pressure (psi)	125
Total friction pressure (psi)	126
Pipe friction pressure (psi)	126
Perforation friction pressure (psi)	127
Open perforations	128
Perforation efficiency	128
Perforation design	129
Numbers of holes (perfs) and limited entry technique	130
Perforation diameter and penetration	131
Perforation erosion	131

Near-wellbore friction pressure	131
Fracture extension pressure	132
Closure pressure	132
Net pressure	135
Surface-treating pressure (psi)	137
Production casing design	139
Rate step-down test workflow	141
10. Fracture treatment design	**149**
Introduction	149
Absolute volume factor (AVF, gal/lbs)	149
Dirty (slurry) vs clean frac fluid	150
Slurry (dirty) density (ppg)	151
Stage fluid clean volume (BBLs)	152
Stage fluid slurry (dirty) volume (BBLs)	153
Stage proppant (lbs)	153
Sand per foot (lb/ft)	154
Water per foot	154
Sand-to-water ratio (SWR, lb/gal)	155
Slick water frac schedule	155
Foam frac schedule and calculations	161
Foam volume	161
Nitrogen volume	162
Blender sand concentration	163
Slurry factor (SF)	163
Clean rate (no proppant)	164
Clean rate (with proppant)	165
Slurry rate (with proppant)	165
Nitrogen rate (with and without proppant)	166
11. Horizontal well multistage completion techniques	**177**
Introduction	177
Conventional plug and perf	178
Composite bridge (frac) plug	179
Sliding sleeve	182
Sliding sleeve advantages	182
Sliding sleeve disadvantages	183
Frac stage spacing (plug-to-plug spacing)	184
Shorter stage length	184
Cluster spacing	186
Refrac overview	186

12. Completions and flowback design evaluation in relation to production — 191

- Introduction — 191
- Landing zone — 192
- Stage spacing — 193
- Cluster spacing — 193
- Number of perforations, entry-hole diameter, and perforation phasing — 194
- Sand and water per foot — 194
- Proppant size and type — 195
- Bounded vs unbounded (inner vs outer) — 196
- Up dip vs down dip — 198
- Well spacing — 199
- Water quality — 200
- Flowback design — 201
- Flowback equipment — 203
- Flowback equipment spacing guidelines — 212
- Tubing analysis — 213

13. Rock mechanical properties and in situ stresses — 215

- Introduction — 215
- Young's modulus (psi) — 215
- Poisson's ratio (ν) — 217
- Fracture toughness (psi/$\sqrt{}$ in) — 220
- Brittleness and fracability ratios — 220
- Vertical, minimum horizontal, and maximum horizontal stresses — 222
- Vertical stress — 223
- Minimum horizontal stress — 224
- Biot's constant (poroelastic constant) — 225
- Biot's constant estimation — 225
- Maximum horizontal stress — 226
- Various stress states — 228
- Fracture orientation — 228
- Transverse fractures — 229
- Longitudinal fractures — 229

14. Diagnostic fracture injection test — 233

- Introduction — 233
- Typical DFIT procedure — 234
- DFIT data recording and reporting — 234
- Before-closure analysis — 235
- After-closure analysis — 252

15. Numerical simulation of hydraulic fracturing propagation — 257
Introduction — 257
Stratigraphic and geological structure modeling — 258
Development of hydraulic fracturing simulators — 259
2D hydraulic fracturing models — 262
Fluid flow in hydraulic fractures — 265
Solid elastic response — 266
Pseudo-3D hydraulic fracturing models — 266
3D hydraulic fracturing models — 268
Hydraulic and natural fracture interactions — 270
Hydraulic fracture stage merging and stress shadow effects — 271

16. Operations and execution — 273
Introduction — 273
Water sources — 273
Hydration unit ("hydro") — 277
Blender — 279
Sand master (sand mover, sand king, or sand castle) — 280
T-belt — 282
Missile — 290
Frac manifold (isolation manifold) — 291
Frac van (control room) — 292
Overpressuring safety devices — 294
Water coordination — 299
Sand coordination — 300
Chemical coordination — 300
Stage treatment — 301
Tips for flowback after screening out — 301
Frac wellhead — 304

17. Decline curve analysis — 311
Introduction — 311
Decline curve analysis — 311
Anatomy of decline curve analysis — 312
Primary types of decline curves — 315
Arps decline curve equations for estimating future volumes — 319
Multisegment decline — 330
Pressure normalized rate — 333

18. Economic evaluation — 341

Introduction — 341
Net cash-flow model — 341
Royalty — 342
Working interest — 343
Net revenue interest — 345
British thermal unit content — 346
Shrinkage factor — 346
Operating expense — 348
Total Opex per month — 351
Severance tax — 354
Ad valorem tax — 355
Net Opex — 356
Revenue — 357
New York Mercantile Exchange (NYMEX) — 359
Henry Hub and basis price — 360
Cushing Hub and West Texas Intermediate (WTI) — 361
Mont Belvieu and Oil Price Information Services (OPIS) — 362
Capital expenditure (CAPEX) — 362
Opex, Capex, and pricing escalations — 364
Profit or net cash flow (NCF) — 365
Before federal income tax monthly undiscounted net cash flow — 366
Discount rate — 366
Weight of debt and equity — 368
Cost of debt — 369
Cost of equity — 370
Capital budgeting — 372
Net present value — 372
BTAX and ATAX monthly discounted net cash flow — 376
Internal rate of return (IRR) — 377
NPV vs IRR — 382
Modified internal rate of return (MIRR) — 382
Payback method — 384
Profitability index (PI) — 386
Tax model (ATAX calculation) — 387

19. The role of federal reserve — 405

Introduction — 405
Interest rate management — 405
Money supply management — 408

Federal funds rate, discount rate, and prime rate — 410
Historically important financial and political events — 410

20. NGL and condensate handling, calculations, and breakeven analysis — 415
Introduction — 415
NGL yield calculation — 418
Blending vs processing gas step-by-step guide and calculation — 423

21. Well spacing and completions optimization — 443
Introduction — 443
Gas pricing and CAPEX influence — 444
Lateral length influence — 444
Inventory influence — 445
Simultaneous optimization of well spacing and completions design — 445

22. Stacked pay development — 467
Introduction — 467
Step-by-step workflow for codevelopment analysis — 468

23. Production analysis and wellhead design — 475
Introduction — 475
Nodal analysis — 475
Pseudo-pressure concept and calculation — 476
Rawlins and Schellhardt (back pressure model) — 479
Tubing performance relationship — 485
Erosional velocity calculation — 488
Gas production operation issues — 491
Turner rate — 492
Coleman rate — 493

24. Application of machine learning in hydraulic fracture optimization — 499
Introduction — 499
Theory — 500
ML case usages in various industries — 502
Knowledge of data science and domain expertise — 504
Methodology — 505
Artificial neural network (ANN) — 508
Example #1: Marcellus shale completion and stimulation optimization using supervised ML — 517

	K-means clustering	530
	Other application of unsupervised ML algorithms	537

25. Numerical simulation of real field Marcellus shale reservoir development and stimulation — **541**

- Introduction — 541
- Problem statement — 541
- Phase I: Petrophysics and well log analysis — 545
- Phase II: Shale gas reservoir base model development — 554
- Phase III: Sensitivity analysis and history-matching — 556
- Phase IV: Well performance analysis — 565
- Phase V: Frac job sand schedule optimization — 577
- Phase VI: Modified hyperbolic decline curve — 584
- Conclusion — 585

References — *587*
Index — *595*

Author biography

Hoss Belyadi

Hoss Belyadi is currently a senior subsurface engineer at EQT Corporations specializing in production and completions optimization, completions and reservoir modeling, machine learning, and project evaluation. Mr. Belyadi is also an adjunct faculty member at Marietta College and Saint Francis University teaching Natural Gas Engineering, Enhanced Oil Recovery, and Hydraulic Fracture Stimulation Design. Previously, he was an adjunct professor at West Virginia University teaching Hydraulic Frac Stimulation Design and a senior reservoir engineer at CNX Resources Corporation. Mr. Belyadi has been a member of the Society of Petroleum Engineers (SPE) since 2006 and has authored and coauthored several SPE papers on reservoir/completions optimization, production enhancement, and machine learning. Hoss earned his BS and MS, both in Petroleum and Natural Gas Engineering from West Virginia University.
Affiliations and Expertise
Senior Subsurface Engineer, EQT Corporations, USA

Ebrahim Fathi

Ebrahim Fathi is currently an associate professor of Petroleum and Natural Gas Engineering at West Virginia University. He has received the 2014 (AIME) Rossiter W. Raymond Memorial Award and SPE outstanding Technical Editor Award in 2013 and 2018. Ebrahim has authored several peer-reviewed journal publications and conference papers. He has been heavily involved in research on various aspects of unconventional reservoir developments, including the application of artificial intelligence, multistage hydraulic fracturing of horizontal wells, innovative research developing a new generation of flow simulators for unconventional reservoirs, and the area of computational fluid dynamics. He earned a BS in the Exploration of Mining Engineering and an MS in the Exploration of Petroleum Engineering, both from Tehran University, a PhD and postdoc in Petroleum Engineering from the University of Oklahoma.
Affiliations and Expertise
Associate Professor, Petroleum and Natural Gas Engineering, West Virginia University, USA

Fatemeh Belyadi

Fatemeh Belyadi is currently the founder and CEO of Integrated Smart Data Solutions, LLC. Dr. Belyadi was formerly an assistant professor of Petroleum and Natural Gas Engineering at West Virginia University, process engineer at Exterran Energy Solution Company, and petroleum engineer at Kish Oil and Gas Company (KOGC) where she specialized in process equipment, gas plant, refinery reactor design, reservoir management, and machine learning. SPE publishes a large portion of her professional activities, where she regularly attends and presents papers in SPE conferences. Her research interests include multiphase fluid flow, wellbore integrity, designing and managing drilling fluids, machine learning, and enhanced oil recovery, especially from stripper wells. Dr. Belyadi received her BS, MS, and PhD in Petroleum and Natural Gas Engineering from West Virginia University.

Affiliations and Expertise

Founder and CEO of Integrated Smart Data Solutions, LLC, Morgantown, West Virginia, USA

Introduction

We live in a unique time in the oil and gas industry, a time that automation and machine learning (ML) is becoming more and more common amongst various companies across the world. The industry has gone through rapid changes to adapt, survive, improve processes, create automation, and generate value. Disruptive thinking and ideas will overtake the industry as companies look for ways and opportunities to increase value for their shareholders. Therefore, it is important to adapt to new changes and embrace these changes as new and exciting opportunities. Just think about the industry before the unconventional oil and gas revolution. The industry has adapted to so many exciting and new changes that will shape its future. We are indeed very fortunate to live in such an exciting era—an era of unconventional oil and gas, an era of big data and data analytics, an era of fast adaptability and new changes, an era if companies do not adapt, a new start-up with disruptive thinking could blow the rest out of the park, an era that has generated massive wealth for ordinary people that are excited and optimistic about what lies ahead. This is the reality of the oil and gas industry and the purpose of this book is to provide the latest and greatest information about hydraulic fracturing in unconventional reservoirs directly impacting production performance, well spacing optimization, as well as many other aspects of the process. We have added multiple exciting new chapters on well spacing and completions design optimization, application of machine learning for completions design optimization, NGL, CND processing, handling and step-by-step workflow and calculations, production analysis and wellhead design, stacked pay development and analysis, and the role of the federal reserve and economic impact. The second edition also includes a Marcellus shale field case study example based on a data source available to the public and following step-by-step methodology that can be applied to any unconventional reservoir and used for senior design project for undergraduate students. In addition, we have added multiple sections to the already existing decline curve analysis and perforation design chapters.

Conventional vs unconventional

We live in an important era of the oil and gas industry—an era when unconventional reservoirs created tremendous value across various basins and countries. We are essentially moving from resources that were hard to find but easy to produce (conventional reservoirs) to resources that are easy to find but hard to produce (unconventional reservoirs). It is harder to produce from unconventional reservoirs since hydraulic fracturing would have to be applied, or otherwise, the resource could not be produced at an economically feasible rate. The permeability of reservoirs that are classified as unconventional is <0.1 md vs permeability of conventional reservoirs that by text book definition is more than 0.1 md. In conventional reservoirs, the hydrocarbon gets produced in the source rock and gets migrated until it is faced with some sort of geologic and structural traps; however, in unconventional reservoirs, we tap right into the source rock. Another primary difference between conventional and unconventional reservoirs is the availability of production history. There are conventional wells with longer than a century of production data; however, production history from unconventional wells in some of the oldest basins is only two or three decades. Some basins have less than a decade of production data. The heterogeneity and complexity of unconventional reservoirs due to the existence of natural fractures and faults are evident based on micro-seismic and diagnostic tests. Geologic complexity has proven to have a direct impact on production performance of wells, and therefore, it requires engineers and geologists to do their due diligence to make sure that wells are prohibited from drilling in complex geologic areas where the production performance could be severely jeopardized and hindered. The heterogeneous nature of unconventional reservoirs has made production forecasting much more challenging as compared to conventional reservoirs where the volumetric calculation was used to calculate the base reserve. Use of analogous wells for creating production type curve is essential; however, stochastic analysis is another important piece of the puzzle for accurate calculations and estimations of reserves in unconventional reservoirs. In conventional reservoirs, transient flow (pressure pulse that moves into an infinite or semiinfinite acting reservoir) lasts for days; however, unconventional reservoirs are known for having a long history of transient flow. In conventional reservoirs, fewer wells can achieve commerciality; however, in unconventional reservoirs, many wells will be needed to make a field commercially viable and successful. One of the main advantages of unconventional reservoirs is having the optionality of changing the hydraulic frac stimulations jobs to increase the net asset value of the field. Just comparing frac design from 2005 to today, reducing clustering spacing and increasing proppant job sizes have brought massive change and value to many companies across various basins. This shows the uplift potential and uplift that exists within the stimulation (completions) optimization of unconventional reservoirs. With the rise in big data and automation, the future of the oil and gas industry and people who adapt to new changes is bright. What an exciting journey to have ahead of all of us, and we hope we can satisfy your needs by providing you the most accurate and up-to-date information about the hydraulic fracturing and unconventional reservoirs. As each author has various academic and practical industry experience, our goal is to provide the most useful step-by-step workflow that can be applied to both industry and practical applications. We hope you enjoy reading this book, and we look forward to many more editions to come.

CHAPTER ONE

Introduction to unconventional reservoirs

Introduction

Oil and natural gas are extremely important. Our society is dependent on fossil fuels. They alone afford many of our greatest everyday comforts and conveniences. From the packaging used for our foods to the way we heat our homes, to all of our various transportation needs, without fossil fuels our way of life would come to a screeching halt. In light of current technological advancements, oil and natural gas will be the major player in the energy industry for years to come. Other sources of energy, such as wind, solar, electricity, biofuel, and so forth will eventually contribute along with fossil fuels to meet the growing global energy demand. When compared to different fossil fuels, natural gas is the cleanest because it emits much smaller quantities of CO_2 when burnt. Natural gas is a hydrocarbon mixture that primarily consists of methane (CH_4). It also includes varying amounts of heavier hydrocarbons and some nonhydrocarbons (as presented in Table 1.1). General usages of natural gas components are also presented in Table 1.2.

Natural gas can be found in pockets as structural or stratigraphic gas reservoirs or in oil deposits as a gas cap. Gas hydrates and coalbed methane are considered as a major source of natural gas. Natural gas is measured by MSCF, which is 1000 standard cubic feet (SCF) of gas. Combustion of 1 ft^3 of natural gas produces an equal amount of 1000 British thermal units (BTUs), the traditional unit for energy. One BTU by definition is the amount of energy needed to cool or heat one pound of water by 1°F. Each hydrocarbon has a different BTU and the heavier the hydrocarbon the higher the BTU becomes. Table 1.3 shows the BTU/SCF and BTU factor for each natural gas component. As can be seen below, methane has a BTU of 1012. If the price of gas is assumed to be $4/MMBTU, 1 MSCF of pure methane would be valued at $4.048/MSCF. To measure the actual BTU of natural gas, a gas sample is taken from a producing well. This sample is

Table 1.1 Typical natural gas components

Natural gas components	Chemical formula	Short formula	
Methane	CH_4	C_1	Light ends
Ethane	C_2H_6	C_2	
Propane	C_3H_8	C_3	Heavier hydrocarbons
i-Butane	C_4H_{10}	$i\text{-}C_4$	
n-Butane	C_4H_{10}	$n\text{-}C_4$	
i-Pentane	C_5H_{12}	$i\text{-}C_5$	
n-Pentane	C_5H_{12}	$n\text{-}C_5$	
Hexane$^+$	C_6H_{14}	C_6^+	
Nitrogen	N_2	N_2	Inert/no heat content
Carbon dioxide	CO_2	CO_2	
Oxygen	O_2	O_2	

Table 1.2 General uses for natural gas components

General uses for natural gas components

Methane	Cooking, heating, fuel, hydrogen gas production for oil refining, and ammonia production
Ethane	Ethylene for plastics, petrochemical feedstock
Propane	Residential and commercial heating, cooking fuel, petrochemical feedstock
i-Butane	Refinery feedstock, blend in gasoline, petrochemical feedstock
n-Butane	Petrochemical feedstock, gasoline blend stock
i-Pentane	"Natural gasoline" blended into gasoline, jet fuel, naphtha cracking
n-Pentane	"Natural gasoline" blended into gasoline, jet fuel, naphtha cracking
Hexane$^+$	"Natural gasoline" blended into gasoline, jet fuel, naphtha cracking
Nitrogen	Air is 78% N_2
Carbon dioxide	Air is 0.04% CO_2
Oxygen	Air is 21% O_2

Table 1.3 BTU of each natural gas component

Natural gas components	BTU/SCF	MMBTU per MSCF (BTU factor)
Methane	1012	1.012
Ethane	1774	1.774
Propane	2522	2.522
i-Butane	3259	3.259
n-Butane	3270	3.27
i-Pentane	4010	4.01
n-Pentane	4018	4.018
Hexane$^+$	4767	4.767
Nitrogen	–	–
Carbon dioxide	–	–
Oxygen	–	–

Fig. 1.1 Gas chromatograph.

then taken to the lab, and by using a device called a gas chromatograph the natural gas composition (mol%) can be measured by component. After measuring the gas composition of the natural gas sample, the approximate weighted average BTU of the gas can be calculated. It is important to note that natural gas is sold by volume and heat content. Therefore, the heat content (weighted average BTU) of natural gas must be measured and calculated for sales purposes. Fig. 1.1 shows the gas chromatograph instrument.

Example

A gas sample was taken from a producing well site and transferred to the lab. Using a gas chromatograph, the composition of the natural gas sample was measured. The result is reported in Table 1.4 as mol% for each component. Calculate the approximate BTU of the gas sample, discarding compressibility factor because the compressibility factor will slightly change the BTU.

To calculate the weighted average BTU of gas, take the mol% (measured from the gas chromatograph) and multiply it by the BTU factor of each component. The summation of the product of mol% and BTU factor will yield the weighted average BTU factor. The BTU of the gas sample is 1113 (not corrected for compressibility), but the BTU factor is 1.113. If the price of gas is \$4/MMBTU, the value of the gas based on the heat content is actually $4 \times 1.113 = \$4.452/MSCF$.

Table 1.4 Weighted average BTU factor example

Natural gas components	BTU/SCF	MMBTU per MSCF (BTU factor)	mol%	Product BTU factor and mol%
		Known	Chromatograph	Simple product
Methane	1012	1.012×	88.2187	=0.8928
Ethane	1774	1.774×	9.3453	=0.1658
Propane	2522	2.522×	1.4754	=0.0372
i-Butane	3259	3.259×	0.1768	=0.0058
n-Butane	3270	3.27×	0.2125	=0.0069
i-Pentane	4010	4.01×	0.0586	=0.0023
n-Pentane	4018	4.018×	0.0236	=0.0009
Hexane$^+$	4767	4.767×	0.0313	=0.0015
Nitrogen	–	–	0.3323	–
Carbon dioxide	–	–	0.0932	–
Oxygen	–	–	0.0323	–
Total (weighted average BTU factor)				1.113

Different types of natural gas

Natural gas can be found in different forms, such as natural gas liquid (NGL), compressed natural gas (CNG), liquefied natural gas (LNG), and liquefied petroleum gas (LPG). NGLs refer to the components of natural gas that are liquid at surface facilities or gas processing plants. For the purpose of this book, NGLs consist of ethane, propane, butane, pentane, and hexane$^+$, but do not include methane. Iso-pentane, n-pentane, and hexane$^+$ are also called "natural gasoline." CNG is the compression of natural gas to <1% of the volume occupied in standard atmospheric pressure. CNG is stored and transported in cylindrical and spherical high-pressure containers. LPG consists of only propane and butane and has been liquefied at low temperatures and moderate pressures. LPG has many uses including heating, cooking, refrigeration, motor fuel, and so on. A simple example of LPG is a propane tank used for grilling. In addition to the aforementioned types of natural gas, terms like associated or nonassociated gas are also used in the oil and gas industry. Associated gas refers to the gas associated with oil deposits either as free gas or dissolved in solution. Nonassociated gas is not in contact with significant quantities of liquid petroleum. Nonassociated gas is sometimes referred to as dry gas.

Natural gas transport

Natural gas can be transported using three different methods. The first method is via pipeline, which is currently used across the United States. The second method is by liquefying natural gas, and the third method is by converting natural gas to hydrates and transporting the hydrates. In the case of LNG, natural gas is cooled to $-260°F$ at atmospheric pressure to condense. The main purpose of LNG is the ease of storage and transportation. LNG occupies approximately 1/600th of the volume of gaseous natural gas. LNG is transported through ocean tankers. Another advantage of liquefying natural gas is the removal of oxygen, sulfur, carbon dioxide (CO_2), hydrogen sulfide (H_2S), and water from natural gas.

One of the main disadvantages of converting natural gas to LNG is the cost. However, technological advancements can decrease the cost and make the process economically feasible. In some places, the construction of pipeline facilities could be more expensive because of the lack of infrastructure. A disadvantage or risk of LNG is when cooled natural gas comes in contact with water it can result in a rapid phase transition explosion. In this type of explosion, a massive amount of energy is exchanged between water at a normal temperature and LNG at $-260°F$. This transfer of energy causes rapid phase transition, which is also known as cold explosion. When the gas reservoir is far from pipelines, the third method of gas transport, which is converting gas to gas hydrates, can be used. The economy plays a major role in choosing the gas-transport technique. In some cases, as studied by Gudmundsson et al. (1995), it is economically more viable to convert gas to gas hydrates, and then transport natural gas as frozen hydrate. One major concern in gas hydrate transport is the hydrate stability. Mid-refrigeration at $-20°F$ prevents gas dehydration. This is due to the generation of an ice shell around the hydrate that prevents early gas dehydration. There are several centers around the world working on the pilot and laboratory-scale experimental studies of gas hydrate transport, including British Gas, Ltd., and the Japanese National Marine Research Institute.

Unconventional reservoirs

As time passes, more technological advancements will result in more commercial production of oil and natural gas. For example, shale was a known resource decades before it could be exploited in an economically feasible process to produce significant amounts of oil and natural gas. The development of drilling horizontal wells and using multistage hydraulic fracturing have made the exploitation of previously untapped resources not only possible but also profitable reserves for small and big operators. These new extraction methods have led to the shale reservoirs playing a major role in the oil and gas industry. These burgeoning technologies will enable us to extend the life of Earth's finite natural gas resources. Therefore, in 50 years, if the question of "how much oil and gas is left on this earth" is proposed, the answer would be another 50 years. Technology continuously advances and improves as such, that they will cause oil and gas to be recovered more efficiently and economically. For example, the development of unconventional shale reservoirs has added a tremendous amount of reserves and value to the oil and gas industry.

Unconventional oil and gas reservoirs are playing an important role in providing clean energy, environmental sustainability, and increased security. The US Energy Information Administration (EIA) predicted that shale gas production would increase by 23% in 2010 and 49% by 2035. The US Geological Survey in 2008 estimates the mean undiscovered volume of hydrocarbon in only the Bakken formation in the United States portion of the Williston Basin of Montana and North Dakota to be 3.65 billion barrels of oil, 148 million barrels of NGL, and 1.85 trillion ft^3 of associated/dissolved natural gas. The United States will play a critical role in changing the global energy landscape because of production from these resources. The potential for transferring the production and development technologies has led to growing interest in unconventional oil/gas resources all over the world as reflected in the World Shale Map published by the Society of Petroleum Engineers (SPE) in the Journal of Petroleum Technology (JPT, March 2014).

Due to the tight and multiscale nature of shale structures, knowledge of shale characteristics is limited and there are difficulties associated with stimulation and production strategies causing diminished production from these substantial resources (between 5% and 10% with current technology from shale oil resources) (Hoffman, 2012). A conventional enhanced oil recovery technique, such as water flooding, is also a suboptimal method for stimulation because of the ultralow permeability. The current industry standard practice is to decrease the well spacing and increase the number of stages in hydraulic fracturing treatments to increase production. This approach raises serious environmental concerns for governmental entities. There is a critical need to develop new technologies that improve recovery and minimize the environmental impact associated with these activities. In the absence of such technology, our prediction and optimization of field-scale production in this new generation of clean energy will likely remain limited.

Unconventional gas resources are different than conventional resources in that they are technically difficult to produce because of low permeability or poorly understood production mechanisms. There are also challenges associated with the risk analysis and economics of these resources. Fig. 1.2 shows the resource pyramid where gas resources are divided into three categories of "good," "average," and "poor" based on their formation permeability. The majority of the "good" resources have already been produced and we are now looking into "average" and "poor" resources. As the oil and gas industry moves to produce from "average" and "poor" resources, more advanced technology, time, and research must be devoted to producing from these resources.

Unconventional gas reservoirs fall into the "poor" resource category and are comprised of mainly tight gas sands, coalbed methane, shale gas, and gas hydrates. Gas sands, coalbed methane, and shale gas are currently being produced. Natural gas hydrates, with perhaps the largest volume of gas in place, pose the greatest future challenges with respect to technology, economics, and environment. Tight gas sand, shale gas, and coalbed methane can be distinguished based on their total organic contents (TOCs). TOC is represented by weight percent of organic matter. Shale gas reservoirs require a value of at least 2% to be economically feasible for investment. Shale reservoirs with a TOC of more than 12% are considered to be excellent.

Tight gas sands have a minimum amount of TOC—<0.5%. Most of the gas presented in tight gas sands is free gas. Shale gas reservoirs have a TOC of between 0.5% and 40% and coalbed methane reservoirs are mainly made of organic matter (more than 40%). Among these unconventional gas

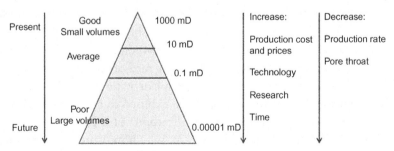

Fig. 1.2 Gas resources pyramid.

resources, coalbed methane, and shale gas reservoirs are very similar. They are both sedimentary rock with organic materials having low to ultralow permeability and a multiscale pore structure. Coal is a mixture of various minerals and organic materials exhibiting an intricate pore network. Coalification is defined as the process of gradual change in the physical and chemical properties of coal as pressure and temperature increase during geological time. Coalification, also known as metamorphism, delineates different ranks of coal. As coal reaches a higher rank, it contains more carbon content and volatile components, and less moisture.

Shale is the most common sedimentary rock and is composed of fine-grained and clay-sized particles. The more quartz in the matrix of a shale sample, as compared to clay minerals, leads to a more brittle or fracable shale formation. Shale sediments with potential for natural oil and gas production are generally rich in a type of organic matter known as kerogen (Kang et al., 2010). The color of shale ranges from gray to black depending on the organic content. Oftentimes, as shale gets darker, more organic material will be present. Shale can be presented as a source rock or cap rock in unconventional and conventional reservoirs. Source rock is what generates oil and gas; it is known as black shale when it has a high TOC. Often organic-rich black shale has a high TOC and gas content, and low water saturation. During digenesis, most of the organic content of shale and coal is transformed into large molecules known as kerogen. Increasing the temperature and reducing the microbial activities transform kerogen to bitumen, which has smaller and more mobile molecules. Kerogen is made of maceral, which is equivalent to the minerals in the inorganic material. Of the four different Kerogen types, type I is simultaneously the most valuable and vulnerable because it has the highest capacity to produce liquid. Type II is also a good source for hydrocarbon liquid production. However, kerogen type III produces mainly gas, except when it is mixed with type II. Kerogen type IV is highly oxidized and has no hydrocarbon generation potential. Waples (1985) categorized different kerogen types based on their original organic matter and maceral (as illustrated in Table 1.5). In addition to kerogen type and TOC, the thermal maturity (TM) of shale is also a key parameter in shale reservoir evaluation. TM is a measure of the heat-induced process of converting organic matter to oil or natural gas. TM measures the degree to which a formation has been exposed to the high heat needed to break down organic matter into hydrocarbon. This parameter is quantified based on vitrinite reflectance (% Ro), which measures the maturity of the organic matter. Vitrinite reflectance varies from 0.7% to 2.5+%. A

Table 1.5 Different Types of Kerogen

Kerogen type	Maceral	Origin of maceral
I	Alginite	Freshwater algae
II	Exinite; cutinite	Pollen, spores; land-plant cuticle; land-plant resins
	Resinite; liptinite	All land-plant lipids, marine algae
III	Vitrinite	Woody and cellulosic material from land plants
IV	Inertinite	Charcoal; highly oxidized or reworked material of any origin

Table 1.6 Vitrinite reflectance values and reservoir relationship

Reservoir	Vitrinite reflectance values (%)
Immature	<0.60
Oil window	0.60–1.10
Condensate/wet gas	1.10–1.40
Dry gas window	>1.40

vitrinite reflectance of >1.4% indicates that the hydrocarbon is dry. A TM closer to 3% indicates overmaturation resulting in gas evaporation. Table 1.6 summarizes vitrinite reflectance and its significance in various reservoir fluid windows. The range of vitrinite reflectance for different reservoir fluid windows (oil, gas, and condensate) may vary depending on the kerogen type.

Both shale and coal have multiscale pore structures important for gas transport and production that consist of primary pores (inorganic materials with free and adsorbed gas) and secondary pores (in inorganic materials). Fig. 1.3 shows schematics and sample pictures of coalbed methane and shale from the Black Warrior Basin and Marcellus. Fig. 1.3 illustrates that the coalbed methane matrix consists of mainly organic materials, whereas the shale matrix organic materials are represented as islands inside of the inorganic matrix. Table 1.7 shows the typical TOC of North American shale gas plays.

It is important to examine the different natural fracture systems present in coalbed methane and shale reservoirs. Coalbed methane has a uniform fracture network making it easy to model using dual-porosity and dual-permeability models, conventionally called "cubic sugar" models. In contrast, shale matrixes possess a nonuniform fracture system that requires sophisticated numerical models, such as quad-porosity and double-permeability models. Natural fractures are very important in economically producing coalbed and shale formations. The connection of hydraulic

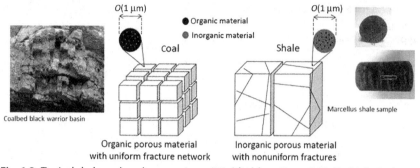

Fig. 1.3 Typical shale and coal comparison. *(Modified from Kang, S.M., Fathi, E., Ambrose, R.J., Akkutlu, I.Y., Sigal, R.F. 2010. CO_2 applications. Carbon dioxide storage capacity of organic-rich shales. SPE J. 16(4), 842-855.)*

Table 1.7 Typical TOC of North American shale plays

Shale or play	Average TOC (weight %)
Barnett	4
Marcellus	1–10
Haynesville	0–8
Horn river	3
Woodford	5

fractures (created during a frac job) with natural fractures in the reservoir creates the necessary channels for optimum production. Therefore, a moderate presence of natural fractures is necessary to economically produce from shale reservoirs.

In addition to the amount and quality of shale organic content, water saturation must also be <45% for production to be economically feasible. Water saturation of Marcellus Shale is typically <25% while Bakken Shale in North Dakota has a varying water saturation of 25%–60%. The clay content of shale is another important parameter to investigate for shale reservoir evaluation. Clays are soft and loose materials formed as a result of weathering and erosion over time. The clay minerals most often found in shale gas reservoirs are illite, chlorite, montmorillonite, kaolinite, and smectite. Some clay swells when in direct contact with water, and this can cause a reduction in the efficiency of hydraulic fracturing. A moderate clay content (of <40%) is needed for a marketable production in shale reservoirs. Rock mechanical properties, such as brittleness, Young's modulus, and Poisson's ratio also play an important role when designing a fracturing job. A high Young's modulus and a low Poisson's ratio is the goal in hydraulically fracturing a zone. Rock brittleness is often used as an indication of a formation fracability. Formation density must be determined to decide where to land the horizontal well. For this purpose, a density log is commonly used to determine the density of the formation. The lower the density of the formation, the better suited the zone is for landing the well. In addition, a lower density is typically indicative of higher organic content.

A gamma ray log is one of the most common logs used in drilling operations. It can detect the presence of shale inside the tubing or casing, and it can be run in salt-mud or nonconductive mud, such as oil or synthetic-based mud. A gamma-ray log measures the natural radiations in the formation. Sandstone and limestone have a lower gamma ray, and shale has a higher gamma ray. In a gamma-ray log light emissions are counted and ultimately displayed as counts per second (CPS) vs depth on a graph. The unit for a gamma ray is converted from CPS to gamma ray, American Petroleum Industry unit (GAPI) and is shown as GAPI on the log. When uranium is the driver in Marcellus Shale, a higher gamma ray is often associated with a higher TOC and organic content in the rock. When uranium is not the driver, density logs can be used to determine the zones with higher organic contents. Fig. 1.4 shows a gamma-ray log and interpretation.

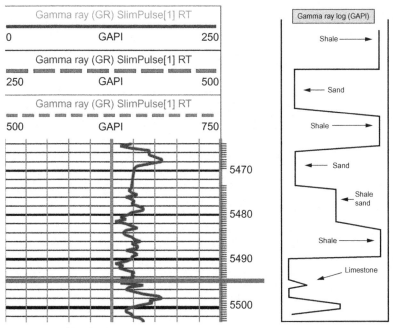

Fig. 1.4 Gamma-ray log.

Reservoir pressure, also known as pore pressure, is another important parameter in commercial production from shale gas reservoirs. Reservoir pressure needs to be above normal, which is defined as any reservoir with a pressure gradient >0.465 psi/ft. Areas that have above normal reservoir pressure gradients are considered optimal for production enhancements. The highest ultimate recoveries will be from abnormal reservoir pressures. Reservoir pressure can be calculated using build-up tests or more often calculated using diagnostic fracture injection tests (DFITs), which will be discussed in detail in the DFIT chapter 14.

This book will focus on many critical considerations regarding shale development, namely shale reservoir characterization, modeling, hydraulic fracturing, enhanced shale oil and gas recovery, and economic analysis. In addition, optimum field development/strategy and various case studies of applying various machine learning algorithms to optimize production and generate value will be discussed.

CHAPTER TWO

Advanced shale reservoir characterization

Introduction

Unconventional shale reservoir characterization is important for accurate estimation of original oil- and gas-in-place (OOIP and OGIP) and production rates. Production from unconventional reservoirs is a function of reservoir matrix porosity, permeability, hydrocarbon saturation, pore pressure, contact area, and conductivity provided by hydraulic fracturing and effective enhanced oil recovery techniques (Rylander et al., 2013). Characterization often includes laboratory measurements of pore volume, permeability, molecular diffusivities, saturations, and sorption capacity of selected shale samples. Conventional methods of sampling and measuring these properties have limited success due to the tight and multiscale nature of the core samples. Therefore, new experimental techniques are needed to analyze shale samples.

Pore-size distribution measurement of shale

As shale oil and gas resources gain popularity it is critical to search for more information regarding their rock and fluid characteristics. One such piece of critical information is the porosity of shale rocks. Knowing the total and effective porosity of shale resources is crucial to determine OOIP and OGIP and gas storage capacity. In addition to shale matrix porosity, understanding pore shapes and connectivity can provide information about how fast oil and gas can be produced and how oil and gas flow will be impacted as reservoir pressure changes. Therefore, to retrieve the most accurate storage capacity of a reservoir, the pore-size distribution must be analyzed and interpreted.

Pore sizes are classified in four main categories by the International Union of Pure and Applied Chemistry (IUPAC) and are defined as macropores, mesopores, micropores, and ultramicropores. These have diameters of >50 nm, between 2 and 50 nm, between 0.7 and 2 nm, and <0.7 nm, respectively. One of the main characteristics of organic-rich shale is the matrix micropore structure that controls the oil and gas storage and transport in these tight formations. Using focused ion beam scanning electron microscopy (FIB/SEM), Ambrose et al. (2010) showed that a significant portion of the pores associated with gas storage are found within shale organic material known as kerogen. Kerogen has a pore-size distribution between 2 and 50 nm, with the average kerogen pore-size typically below 10 nm (Akkutlu and Fathi, 2012; Adesida et al., 2011). The range of pore sizes shows that the organic-rich shale can also be considered as organic nanoporous material.

There are different pore-size distribution measurement techniques, each capable of capturing different ranges of pore sizes. To capture the whole range of pore-size distribution, a combination of different pore-size measurement techniques are required. The earliest work on pore-size distribution measurements goes back to 1945 by Drake and Ritter. They injected mercury into the porous material and used the intrusion pressure and volume of mercury displaced to obtain the pore-size distribution. A high-pressure mercury injection in shale samples is a common technique to find the pore-size distribution. In this technique, the pressure profile is collected during mercury injection, and will be used in Washburn equation (Eq. 2.1) (Washburn, 1921) to obtain the pore diameter.

$$\text{Washburn}: D = \frac{-4\sigma \cos\theta}{P} \quad (2.1)$$

D is the pore diameter, σ is surface tension, and θ is contact angle. In the case when mercury is used for the experiment, a contact angle of 130 degrees and surface tension of 485 dyn/cm are commonly used.

Nuclear magnetic resonance (NMR) is used in the industry to estimate pore-size distribution and rock matrix grain sorting. In this technique, a sample saturated with brine is exposed to NMR where collecting the single fluid relaxation time reflects the pore-size distribution and grain sorting of the sample matrix. The assumption is that water molecules inside pores excited by an NMR pulse will diffuse, hiding in the pore walls much like Knudsen diffusion. Given enough time, these fluid-rock molecular collisions lead to the relaxation of the NMR signal, which can be modeled with exponential function such as Eq. (2.2).

$$\text{NMR exponential function}: N(t) = \omega_0 e^{-1/T} \quad (2.2)$$

In this equation ω_0 is the total relaxation time and T is a function of total bulk relaxation, surface relaxation, and molecular diffusion gradient effect. For simplicity, T is considered a function of surface relaxation that is related to fluid-rock molecular collision. Fluid-rock molecular collision is a function of pore radius, pressure, temperature, and fluid type. In the case of a sample saturated with water, a linear relationship between relaxation time and pore diameter can be developed and used for pore-diameter estimation. Micropores are detected in an NMR signal by the shortest T value while mesopores have a middling length, and macropores the longest T value.

Recent advancements in imaging techniques and the availability of three dimensional images of organic-rich shales in different scales have made it possible to investigate the fundamental physics governing fluid flow, storage, and phase coexistence in organic nanopores. These advanced technologies offer new opportunities to unlock this abundant source of oil and natural gas. FIB/SEM is used to image the microstructure of shale samples (Ambrose et al., 2010). FIB/SEM is also used to provide detailed information on microstructure, rock, and fluid characteristics of organic-rich shale samples. The FIB system is used to remove very thin slices of material from shale rock samples, while the SEM provides high-resolution images of the rock's structure, distinguishing voids, and minerals. Curtis et al. (2010) used the FIB/SEM technique and measured pore-size distribution of different shale samples. They concluded that small pores were dominant based on their number; however, large pores still provided the major pore volume in the samples investigated. Scanning transmission electron microscopy imaging (STEM) is also used to image and measure pore-size distribution of shale samples. STEM has similar resolution as FIB/SEM.

It is also possible to use adsorption-desorption data to characterize the pore structure of different materials. Homfray and Physik (1910) were the pioneers using sorption behavior of different gases to characterize the charcoal pore structure. Currently, low-temperature nitrogen adsorption techniques are widely used to determine the pore-size distribution of shale samples, estimate an effective pore size, and determine sorption behavior of shale samples.

Shale sorption measurement techniques

Sorption is a physical or chemical process in which gas molecules become attached or detached from the solid surface of a material. There are physical and chemical sorption processes. Physical sorption is caused by electrostatic and van der Waals forces, while chemical sorption (high-heat sorption) is the result of a strong chemical bond (Ruthven, 1984). As free gas pressure increases, the amount of the sorbed gas will increase. This is referred to as the adsorption process. Desorption is the process that occurs when free gas pressure drops and adsorbed gas molecules start desorbing from a solid surface. Sorption isotherms are often used to determine maximum adsorption capacity and the amount of adsorbed gas at different pore pressures. Here we are concerned with sorption behavior of clay minerals and organic materials such as coal and shale.

Among several models describing equilibrium sorption behavior, the Henry's law isotherm is the simplest. It considers the linear relationship between adsorbed and free gas. That is, $C_\mu = KC$ where C_μ is the adsorbed gas concentration, K is the Henry's constant, and C is the free gas concentration. Even though the relationship between adsorbed and free gas concentrations is not linear, Henry's law has been used extensively because of its simplicity. There are other isotherm models presented, including Gibbs, potential theory, and Langmuir. The Gibbs model defines the sorption process by the equation of state in terms of two-dimensional films. Several authors including Saunders et al. (1985) and Stevenson et al. (1991) have used this model for the gas sorption measurement in the coal. The potential theory model defines sorbed volume as the thermodynamic sorption potential. The Gibbs and potential theory models were largely implemented for coal gas sorption measurements. The Langmuir model is defined as the equilibrium between condensation and evaporation. The Langmuir model consists of three different types of isotherms including Langmuir, Freundlich, and the combination of both (Langmuir and Freundlich) isotherms (Yang, 1987).

Irvin Langmuir (1916) developed the theory of Langmuir isotherm, which is the most common model used in the oil and gas industry describing the sorption relationship. The main assumptions for deriving the Langmuir equation are as follows:
- In each adsorption site, one gas molecule is adsorbed.
- There is no interaction between adsorbed gas molecules at the neighboring site.
- The energy at the adsorption site is equal (homogenous adsorbent).

The Langmuir isotherm has been extensively considered as $C_\mu = abC/(1 + aC)$. In this case, a is the Langmuir equilibrium constant, and b represents complete monolayer coverage of the open surface by the gas molecules. The Langmuir equilibrium equation is a special form of the multilayer

Brunauer, Emmett, and Teller adsorption equation, $C_\mu = abC/(1-/(1+b(C-1)))$. The Langmuir equation is rearranged as Eq. (2.3).

$$\text{Langmuir isotherm (gas content)}: V = V_L \frac{P}{P + P_L} \quad (2.3)$$

V is the adsorbed gas volume (gas content) in scf/ton at pore pressure P (psi). V_L is the maximum monolayer adsorption capacity of the sample in scf/ton. P_L is the Langmuir pressure (psi), which is the pore pressure at which half of the adsorbed sites are taken (Fig. 2.1). The Langmuir model could be presented in a linear form by taking the reciprocal of the terms on both sides of the above equation (Mavor et al., 1990; Santos and Akkutlu, 2012; Fathi and Akkutlu, 2014).

$$\text{Linearized form of Langmuir equation}: \frac{1}{V} = \frac{1}{V_L} + \left(\frac{P_L}{V_L}\right)\frac{1}{P} \quad (2.4)$$

The Freundlich isotherm is given by Eq. (2.5).

$$\text{Freundlich equation}: V = KP^n \quad (2.5)$$

The combined Langmuir/Freundlich isotherm is presented as follows:

$$\text{Combined Langmuir equation}: V = V_L \frac{KP^n}{1 + KP^n} \quad (2.6)$$

The relationship between adsorbed gas volume and free gas pressure is nonlinear at equilibrium conditions, homogeneous conditions, and isotropic media. Experimental studies on the sorption behavior of different materials show six different adsorption isotherm types as illustrated in Fig. 2.2 (Sing, 1985).

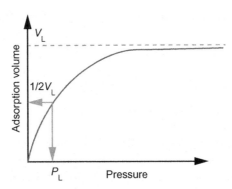

Fig. 2.1 Schematic of typical Langmuir isotherm.

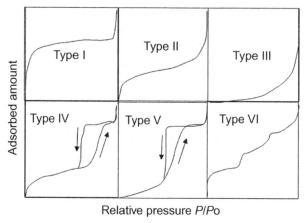

Fig. 2.2 Different adsorption isotherm types.

In Fig. 2.2, the adsorption amount is plotted versus relative pressure, which is the ratio of absolute pressure to the saturation pressure. Saturation pressures have been found empirically for many gases and can be found by increasing the pressure of a gas until it condenses. As absolute pressure approaches saturation pressure the adsorption is maximized. The desorption isotherm can also be obtained by reaching the maximum adsorption, and then systematically reducing the pressure and plotting the quantity of molecules desorbed versus the pressure. The shape of sorption isotherms can also be used to characterize the pore structure of the material. Type I isotherms are typically representative of microporous materials with monolayer adsorption as discussed in Langmuir-type adsorption. The adsorption isotherm of natural gas in organic-rich shale typically follows the type I adsorption isotherm. Types II and IV adsorption isotherms are very similar except type IV experiences hysteresis, or a deviated curve on the desorption isotherm that could be related to condensation and type II has a larger saturation pressure. These are often indicative of nonporous or macroporous materials. Type II adsorption isotherms can be seen when monolayer and multilayer adsorption exist on solid surfaces. Types III and V are also very similar in shape. Type V shows hysteresis on the desorption curve unlike type III. Type III isotherms are usually representative of large pores while type V is representative of mesopores. Type VI adsorption isotherm corresponds to multilayer adsorption on a completely uniform surface without pores. IUPAC has introduced four different types of hysteresis as shown in Fig. 2.3. Type I represents uniform distribution of pores with no

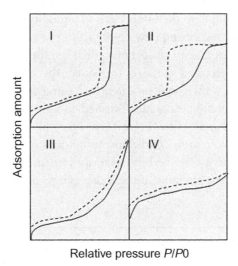

Fig. 2.3 Different hysteresis types.

interconnecting channels. Type II shows interconnecting channels, and types III and IV mainly represent slit-like pores. Types II and IV are different in the sense that the former does not show adsorption reaching the plateau while Type IV shows limited adsorption even at a very high pressure (Sing, 1985).

There are different experimental techniques available to measure the adsorption capacity of the sample including volumetric, gravimetric, chromatographic, pulse, and dynamic adsorption methods. The last two techniques are an extension to the traditional volumetric technique. Among all of these different techniques, the volumetric technique is the most commonly used in the oil and gas industry to measure the sorption capacity of shale. The low-temperature nitrogen adsorption technique is one of the volumetric techniques used to measure the sorption capacity of the sample. This technique as discussed earlier can also be used to determine the pore-size distribution of shale and to characterize the effective porosity of the shale sample. Volumetric sorption measurement techniques usually consist of a double-cell gas expansion porosimeter in a constant temperature unit. The experiment is performed at multiple stages and involves the following steps:

- accurate measurement of sample and reference cell pressures at initial condition (in the case where the adsorption measurement reference cell has higher pressure)

- bringing the sample in the sample cell to equilibrium pressure with reference cell and measuring new equilibrium pressure
- charging the reference cell to new initial condition and repeating the experiment in elevated pressures to recover the whole isotherm curve

Sorbed gas quantity then will be calculated using material balance and a compressibility equation of state. Crushed samples are usually used for adsorption measurements. However, it is not possible to perform the experiment under reservoir stress conditions, using crushed samples. Kang et al. (2010) used a new five-stage adsorption measurement technique where they performed the measurements using core plugs under actual reservoir conditions.

Shale porosity measurements

Porosity is defined as total pore volume over bulk volume, and effective porosity is effective pore volume divided by bulk volume. Effective pore volume is defined as interconnected pore volumes. Effective pore volumes of the samples can be obtained using the difference between bulk and grain densities. One needs to first obtain the bulk density, and then measure the grain density of the sample. Sample bulk volume can be obtained by immersing the sample in a mercury bath and measuring the mercury displacement. Bulk density of the sample can then be obtained by measuring the dry sample weight. To obtain the grain density the sample will be crushed and low-pressure gas pycnometry is used to measure the grain density. For this purpose gas, typically helium, is introduced to the gas pycnometer and a change in pressure with and without the sample is used to obtain the grain density. This calculation is based on Boyle's law and the real gas compressibility equation of state. If shale samples are used, this method might overestimate the effective pore volume since helium has a much smaller molecular size when compared to methane molecules. Helium can access pores that are inaccessible to methane molecules. Since we are interested in finding the porosity of the shale sample to methane (methane is the dominant gas component) an experiment needs to be performed using methane gas that requires additional safety considerations. In addition to molecular size, methane has a much higher adsorption capacity that leads to a reduction in the pore-size diameter and therefore the effective pore volume. Organic-rich shale samples act as a molecular sieve for the gas measurement.

The high-pressure mercury injection is also conventionally used to measure the sample's effective pore volume. In this case, mercury is injected into penetrometer and pressure increases until mercury invades and fills all connected pore volumes. The sample effective pore volume is then equal to the volume of displaced mercury. If a shale sample is used then the mercury intrusion will start at high pressure, typically 10,000 psi, due to very small pore-size distribution. To invade all of the interconnected pores pressure has to rise to more than 60,000 psi. At this pressure the instrument cannot detect the contribution of micropores and some of the mesopores on pore volume.

Several other techniques are available for the measurement of total and effective pore volumes based on different principles such as thermogravimetry, NMR spectrometry, SEM, and low-temperature adsorption. When the sample under investigation is shale all of these techniques have their own limitations. These measurements have limitations due to the fact that they are not performed under reservoir conditions (effective in situ stress and temperature) (Akkutlu and Fathi, 2012).

It is believed that pore volume associated with organic matter is linked to the thermal maturity of the shale. Therefore, thermal maturity can impact both storage (porosity) and transport (permeability) potential of the organic-rich shales (Curtis et al., 2013).

Pore compressibility measurements of shale

Pore compressibility is defined as the change in pore volume of a sample with respect to pressure at a constant temperature and denoted by C_p:

$$\text{Pore compressibility}: \quad C_p = \frac{-1}{V_o} \times \frac{dV}{dP} \tag{2.7}$$

In this equation the relative change of sample pore volume with respect to some referenced volume (usually at standard conditions) is measured. The relationship is inverse since increasing the pressure at a constant temperature will result in a reduction in volume. Pore compressibility can also be used as an indication of rock mechanical properties such as bulk modulus. Therefore, accurate measurement of heterogeneous rock pore compressibility is not an easy task. The problem becomes more complicated considering the change in pore pressure and overburden pressure during oil and gas production. This will result in dynamic pore compressibility. There have been several studies on the relationship between pore compressibility and mineralogy of different consolidated and unconsolidated formations such as Newman (1973), Anderson (1988), Zimmerman (1991), and Cronquist (2001).

However, except for special cases, no distinct and universal relation is found. Therefore, most of the relationships developed are used for qualitative and comparison studies. If shale samples are used, finding the correlation is harder due to the quasibrittle/ductile characteristics of shale samples. For this purpose, a special experimental setup for shale samples was designed by Kang et al. (2010). Later, Santos and Akkutlu (2012) used modified pulse-decay permeameter and measured the pore compressibility of shale samples using the two-stage gas expansion technique. The details of the experimental setup are shown in Fig. 2.5.

Shale permeability measurement techniques

Shale reservoirs are known to have ultralow matrix permeability. Permeability is defined as the ability of rock to transmit fluid and is measured based on Darcy's unit (1 darcy is equivalent to 9.869233×10^{-13} m^2). The shale body can be divided into the shale matrix and fractures. Effective permeability of shale, which is the combination of matrix and natural fracture permeability, can be measured using well test analysis, diagnostic fracture injection tests, advanced steady state, or pressure pulse-decay permeability measurement techniques. Darcy's law describes the fluid flow through porous media, which is a proportional relationship between the discharge rate through a porous medium, geometry of the media, length and cross section, viscosity of the fluid, and pressure gradient over the length of media. The negative sign in the equation is necessary since fluid always flows from high pressure to low pressure. Eq. (2.8) shows that Darcy's law can be rearranged to find the absolute permeability K using Eq. (2.9).

$$\text{Darcy's law}: Q = \frac{-kA \times (\Delta P)}{\mu \times L} \quad (2.8)$$

where Q is the flow rate (m^3/s), K is absolute permeability (m^2), ΔP is the pressure gradient across the core sample (Pascal), μ is the fluid viscosity (Pa.s), and L is the sample length (m).

Darcy's equation can be rearranged to solve for permeability:

$$\text{Permeability}: K = \frac{-Q \times \mu \times L}{A \times (\Delta P)} \quad (2.9)$$

Fig. 2.4 shows the schematic of Darcy's experiment, in which the sample with a cross section of A and a length of L is exposed to a pressure gradient of $\Delta P (P_1 > P_2)$ between points 1 and 2. An incompressible fluid such as

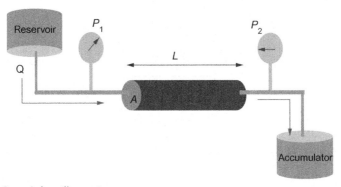

Fig. 2.4 Darcy's law illustration.

water will be injected with the constant flow rate of Q until a steady-state condition is reached. At the steady-state condition, K (sample absolute permeability) can be obtained using Eq. (2.9). If the sample is shale, conventional steady-state methods of permeability measurements are not practical because of very low flow rates and the extremely long time needed to reach the steady-state condition. Therefore, unsteady state methods based on pressure pulse-decay measurement have been extensively used to estimate permeability of the shale samples (Brace, 1968; Ning, 1992; Finsterle and Persoff, 1997). The unsteady state methods are faster and can be used to measure permeabilities as low as 10E-9 md (Ning, 1992). It is crucial to perform the experiment under reservoir conditions since pore pressure, temperature, and confining stress conditions could lead to changes in shale rock characterization.

New pulse-decay permeameters perform the experiment under high pressure and high temperature. They are designed and assembled to precisely measure shale matrix, fracture, and effective permeability using a pulse-decay technique. This technique is used at pressure and temperatures up to 10,000 psia and 340°F under different effective stress conditions. In the pulse-decay permeability measurement technique, the shale core plug (after preparation) will be placed in a core holder and brought to equilibrium pressure conditions. Different pulses will then be applied to the system and pressure decay upstream and pressure buildup downstream will be recorded with high accuracy as illustrated in Fig. 2.5. In Fig. 2.5 temperature is kept constant to reservoir temperature, and confining pressure is applied to resemble reservoir overburden stresses. Different history-matching algorithms can then be used to match the pressure profiles and extract the permeability values. Depending on the magnitude of the pressure pulse, the fracture,

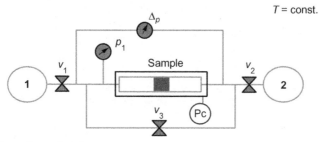

Fig. 2.5 Core plug pulse-decay permeameter.

matrix, or effective permeability can be obtained. Small pulses likely carry the impact of fractures and are used to measure fracture permeability. On the other hand, large pulses are affected by both fracture and matrix and can be used to extract effective matrix permeability. Fig. 2.6 shows the schematic of the setup for performing the experiment with different pulse magnitudes.

Fig. 2.6 Automated high-pressure, high-temperature (HPHT) pulse-decay permeameter.

Generally the slope of pressure versus time in a semilog plot is used to estimate the matrix permeability. Yamada (1980) developed analytical solutions for transient behavior of pressure during pulse decay. However, his solution was only valid under very specific and simplified conditions. The most common method used by the industry is the technique that was introduced by Jones (1997). He modified the conventional pulse-decay setup by using equal upstream and downstream volumes and added two large dead volumes to reduce the time required to reach equilibrium pressure. In Jones' technique, the adsorption capacity of shale and the possibilities of solid or surface transport were neglected. Akkutlu and Fathi (2012) and Fathi and Akkutlu (2013) introduced new sets of governing equations to simulate gas transport and adsorption in shale gas reservoirs and used that in a nonlinear history-matching algorithm to obtain unique shale rock properties.

To use the pressure decline curves obtained from pressure pulse-decay techniques, the sample pore volume and porosity at different pressures are needed. A double-cell Boyle's law porosimeter can be used to provide a precise estimation of these quantities as discussed earlier. Interpretations of the data obtained from transient techniques introduce a large margin of uncertainty due to the nonuniqueness of the results and reproducibility (except if more advanced techniques are used). To avoid the complications in interpretation of the pulse-decay techniques and to perform the experiment in a much shorter time, most of the commercial laboratories are using a crushed sample permeability measurement known as the Gas Research Institute (GRI) technique presented by Luffel (1993). In this case, a double-cell porosimeter is used to provide the crushed sample permeability. It is believed that by crushing the sample, the effect of natural fractures will be removed and the permeability measured by this technique can be a good representation of the shale matrix permeability. Fig. 2.7 illustrates the GRI technique used for crushed sample permeability measurements. The GRI permeability measurement technique on crushed rock is highly affected by particle size, average pressure of the experiment, and gas type (Tinni et al., 2012; Fathi et al., 2012). This results in a discrepancy in permeability values of up to three orders of magnitude measured by different commercial labs using the same samples (Miller, 2010; Passey et al., 2010). In addition, recently injecting mercury in different sizes of crushed shale samples and imaging using microcomputed tomography, Tinni et al. (2012) showed that crushing the shale samples does not remove the microcracks from the matrix. Therefore, the permeability measured by the GRI technique is *not* shale matrix

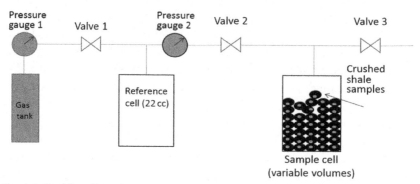

Fig. 2.7 Double-cell Boyle's law porosimeter.

permeability, but rather a combination of matrix and fracture permeability without the impact of confining stresses.

Recently, Zamirian et al. (2014a,b) designed a new pseudo-steady-state permeability measurement experiment to overcome difficulties presented with conventional steady-state permeability measurements, such as the extremely long time required to reach a steady state, and the inability to measure extremely low flow rates. The laboratory system is referred to as Precision Petrophysical Analysis Laboratory (PPAL). The experimental setup is very similar to the pulse-decay permeameter. In this setup, a pressure gradient will be applied between the upstream and downstream after the initial equilibrium pressure has been reached. As pressure builds up downstream, an ultraprecise pressure differential gauge measures the pressure difference between the sample and downstream. As pressure in downstream builds up by 0.5 psi, a bypass valve opens, discharging the gas to keep the downstream pressure constant. Rate of pressure buildup in downstream versus time is then used in Darcy's law to calculate matrix effective permeability. To reach a steady-state condition, usually more than 50 cycles of 0.5 psi pressure buildup in the downstream is required. This technique enables us to measure flow rates as small as 10^{-6} cm^3/s.

CHAPTER THREE

Shale initial gas-in-place calculation

Introduction

Initial hydrocarbon-in-place calculation is crucial in determining the economic feasibility of shale oil and gas reservoirs and reserve estimation. In ultralow permeability reservoirs, transient flow regime can last for a long period of time. Therefore, having a good understanding of original oil- and gas-in-place helps in determining the long-term production forecast and it will decrease the uncertainty when performing the reserve estimation. There are different techniques developed for original oil- and gas-in-place calculations in unconventional reservoirs. These are either numerical or analytical techniques based on volumetric or material balance calculations. The volumetric method is the most common technique, requiring detailed information regarding reservoir rock and fluid properties such as porosity, compressibility, saturations, and formation volume factor. This information is mostly extracted from well logs or obtained using experimental techniques as discussed earlier in chapter two. In the volumetric approach, the shale matrix will be divided into grain volume (e.g., clay minerals, nonclay inorganic minerals, and organic materials), volume occupied by water, oil and free gas, volume occupied by clay-bound water, some dead ends, and isolated pores (Hartman et al., 2011). Fig. 3.1 shows the schematic of bulk volume of shale sample.

Total gas-in-place calculation

In the case of shale gas reservoirs, the gas storage can be divided into free, adsorbed, absorbed, and dissolved gas. Free gas is stored in natural and induced fractures, in inorganic macropores, and organic meso- and micropores, while adsorbed gas is stored mainly at the solid surface of organic

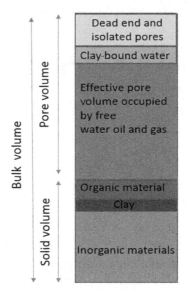

Fig. 3.1 Shale matrix bulk volume.

materials and some by clay minerals. The amount of absorbed gas as discussed earlier is assumed to be negligible. However, there are new studies investigating the impact of gas absorption in original gas-in-place (OGIP) calculation (Ambrose et al., 2012). Gas dissolved in water and hydrocarbon cannot be distinguished from adsorbed gas with current experimental techniques and is usually considered as part of adsorbed gas in gas-in-place calculations. Therefore, OGIP embraces free G_{free} and adsorbed gas G_{ads} scf/ton.

$$\text{Total OGIP}: \text{OGIP} = G_{free} + G_{ads} \qquad (3.1)$$

The volumetric approach is used to measure organic pore volumes, and therefore, the free gas amount. However, due to adsorption in organic nanopores, part of the pore volume will be occupied by adsorbed gas and is not available for free gas storage. In other words, free gas calculation overestimates on the gas storage capacity due to porosity including pore space that is already occupied by adsorbed gas. Therefore, Ambrose et al. (2010) proposed a model where free gas pore volume is corrected to include the adsorption layer thickness as follows:

$$\text{Free OGIP}: G_{free} = \frac{32.0368}{B_g}\left[\frac{\varphi(1 - S_W - S_O)}{\rho_b} - \Psi\right] \qquad (3.2)$$

In this equation, B_g represents the gas formation volume factor, φ is the effective rock porosity, S_w is the water saturation, and S_o is oil saturation, ρ_b is the bulk density of shale in g/cc, and Ψ is the correction factor for adsorbed layer thickness defined as follows:

$$\text{Adsorbed layer correction:} \quad \Psi = \frac{1.318 \times 10^{-6} M \rho_b}{\rho_S}\left[V_L \frac{P}{P+P_L}\right] \quad (3.3)$$

M is the molecular weight of the single-component gas or apparent molecular weight of the gas mixture, ρ_s is adsorbed gas density, P is the pore pressure, P_L and V_L are Langmuir pressure and volume, respectively. Adsorbed gas density is a parameter that requires more detailed studies to obtain. Different analytical and numerical techniques are suggested to obtain adsorbed gas density including the application of the Van der Waals equation of state or molecular dynamic technique. G_{ads} as discussed earlier assumes monolayer adsorption. The Langmuir equation is as follows:

$$G_{ads} = V = V_L \frac{P}{P+P_L}$$

Ambrose et al. (2010) also extended the calculation to a multicomponent single-phase case. In general, ignoring the adsorbed layer effect in OGIP calculations might result in more than 30% overestimation. Therefore, reservoir engineers and petrophysicists typically calculate free GIP using both techniques which include conventional old methodology and new approach proposed by Ambrose et al. (2010).

Example
Calculate **adsorbed gas-in-place (MSCF)** with the following properties for a **half-foot** section of the reservoir. The following data were obtained from core and log analyses:

$A = 640$ acres $= 640$ acres is also referred to one section
$h = 0.5'$
Bulk density $= 2.6$ g/cc (obtained from log)
$V_L = 60$ SCF/ton (obtained from core analysis)
$P_L = 800$ psia (obtained from core analysis)
$P = 4400$ psia (obtained from diagnostic fracture injection tests or DFIT)

$$G_{ads} = V = V_L \frac{P}{P+P_L} = \frac{60 \times 4400}{800+4400} = 50.77 \, \text{SCF/ton}$$

The amount of adsorbed gas for a half-foot of reservoir section given can then be obtained as follows:

$$\text{Adsorbed gas} = A \times h \times \rho_b \times G_{ads}$$

$$= 640(\text{acres} \times 43{,}560\,\text{ft}^2/\text{acres}) \times 0.5(\text{ft}) \times 2.6\left(\text{g/cc} \times \frac{1}{3.531e^{-5}\text{ft}^3/\text{cc}}\right)$$

$$\times 50.77 \left(\frac{\text{SCF}}{907{,}185\,\text{g}}\right) = 1359 \times A \times h \times \rho_b \times G_{ads} = 57{,}405\,\text{MSCF}$$

This indicates that a half-foot section of the reservoir contains 57.4 MMSCF of adsorbed gas-in-place in addition to the free gas-in-place.

Example

Calculate free gas and adsorbed gas-in-place (BCF) given the following information (assume 100% methane gas):

$A = 640$ acres, $h = 100'$, bulk density $= 2.35\,\text{g/cc}$, adsorbed gas density $= 0.37\,\text{g/cc}$, $V_L = 60\,\text{SCF/ton}$, $P_L = 700\,\text{psia}$, $P_R = 4800\,\text{psia}$, $S_w = 20\%$, $S_O = 0\%$, porosity $= 10\%$, $B_{gi} = 0.0038$

$$\Psi = \frac{1.318 \times 10^{-6} M \rho_b}{\rho_s}\left[V_L \frac{P}{P+P_L}\right]$$

$$= \frac{1.318 \times 10^{-6} \times 16.04 \times 2.35}{0.37}\left[\frac{60 \times 4800}{700 + 4800}\right] = 0.00703$$

$$G_{free} = \frac{32.0368}{B_g}\left[\frac{\varphi(1 - S_w - S_o)}{\rho_b} - \Psi\right] = \frac{32.0368}{0.0038}\left[\frac{0.1 \times (1-0.2)}{2.35} - 0.00703\right]$$

$$= 227.7357\,\text{SCF/ton}$$

$$\text{Free gas in place} = 43{,}560 \times A \times h \times \rho b \times G_{free}$$

$$= 43{,}560 \times 640 \times 100 \times 2.35 \times \left(\text{g/cc} \times \frac{1}{3.531e^{-5}\,\text{ft}^3/\text{cc}}\right)$$

$$\times 227.7357\left(\frac{\text{SCF}}{907{,}185\,\text{g}}\right)$$

$$= 4.66 \times 10^{10}\,\text{SCF} = 46.6\,\text{BCF}$$

$$G_{ads} = V = \frac{V_L \times P_R}{P_L + P_R} = \frac{60 \times 4800}{700 + 4800} = 52.36\,\text{SCF/ton}$$

$$\text{Adsorbed gas} = 1359 \times A \times h \times \rho_b \times G_{ads}$$

$$= 1359 \times 640 \times 100 \times 2.35 \times 52.36 = 10.7\,\text{BCF}$$

Total GIP = 46.6 + 10.7 = 57.3 BCF

This example indicates that 57.3 BCF of the total gas is present in one section (640 acres) of the reservoir. This does not mean that the entire amount of gas can be recovered. In unconventional shale reservoirs, the recovery factor (RF) can range anywhere from 10% to 80% depending on the reservoir properties and the completions design. For example, if RF is assumed to be 25% for this particular reservoir, only 25% of 57.3 BCF can be recovered per section. Therefore, 14.325 BCF of gas can be recovered from this reservoir. Finally, 14.325 BCF is also called estimated ultimate recovery (EUR) per 640 acres.

Density of adsorbed gas

As discussed earlier, adsorbed gas density is required to calculate the OGIP in shale gas reservoirs. However, this is not an easy quantity to measure in the laboratory. Dubinin in 1960 proposed to use the Van der Waals equation of state as a means to calculate the adsorbed gas density. The Van der Waals equation of state relates the density of gases to pressure, temperature, and volume and is one of the earliest attempts to modify the ideal gas equation of state.

$$\text{Van der Waals equation of state}: \left(P + \frac{a}{V^2}\right)(V - b) = RT \quad (3.4)$$

In the Van der Waals equation of state, Eq. (3.4), the interaction forces and volume of gas molecules that are neglected in the ideal gas equation of state are considered using the correction factors a and b, respectively. Dubinin (1960) suggested that in the cases where adsorption is of importance, the b constant in the Van der Waals equation of state is equal to the volume taken by the adsorbed phase v divided by the actual adsorbed gas amount μ as follows:

$$v/\mu = b$$

Therefore, the adsorbed gas density can be written as

$$\rho_s = \frac{M\mu}{v} = \frac{M}{b}$$

where M is the molecular weight of the gas and b is the Van der Waals coefficient. The coefficient b can be obtained using first and second derivative of pressure in the Van der Waals equation of state with respect to volume at critical temperature as follows:

Van der Waals equation of state correction factor for gas molecules volumes:

$$b = \frac{RT_C}{8P_C} \qquad (3.5)$$

In this equation, R is the universal gas constant and T_c and P_c are critical temperature and pressure of the gas. Therefore, the adsorbed gas density using Eq. (3.5) can be obtained as follows:

$$\text{Adsorbed gas density}: \ \rho_S = \frac{8P_C M}{RT_C} \qquad (3.6)$$

The critical properties of the pure components are constant values that can be obtained from the physical properties table. In the case of gas mixtures, the apparent molecular weight of gas mixture and the pseudocritical pressure and temperature can be used in Eq. (3.6). Pseudocritical properties of gas mixture are defined as the weighted average of the critical properties of pure components in the mixture. Ideal adsorption solution (IAS) theory in this case can also be used to calculate the gas mixture adsorption if the individual adsorption properties of the pure components are known (Myers and Prausnitz, 1965). Recently, molecular dynamic simulations have been extensively used to investigate the gas mixture adsorption and adsorbed-phase density and thickness when multicomponent gas mixtures are considered (see Kim et al., 2003; Rahmani Didar and Akkutlu, 2013, for more detailed discussions).

Overall, there is a substantial understanding of the application of different equations of state to investigate the phase transitions in bulk fluids where the system size is not of importance. However, as the volume of the system reduces to the meso- and microscales, the phase equilibriums become size dependent, where the wall confinement effects significantly change the thermodynamic properties of the fluids. Experimental and numerical investigations on equilibrium and nonequilibrium thermodynamical properties of fluids in nanoporous materials show dramatic deviations from their bulk values obtained using pressure-volume-temperature (PVT) measurements. Results of recent studies show that as pore size decreases to the nanoscale, critical temperature, freezing, and melting points decrease. It has also been observed that water viscosity is significantly reduced with critical pressure and interfacial tensions.

Recovery factor

The RF equation is as follows:

$$RF = \frac{EUR}{IGIP} \quad (3.7)$$

RF is the recovery factor (%), EUR is the estimated ultimate recovery (BCF), and IGIP is the initial gas-in-place (BCF).

In old OGIP calculation techniques, the pore volume occupied by adsorbed gas was neglected, which could lead to up to 30% overestimation of the OGIP depending on the amount of total organic contents (TOCs) and nanoorganic pore-size distribution (Ambrose et al., 2012). Belyadi (2014) studied the impact of adsorbed gas on OGIP, total gas production, and RF using information from the Marcellus Shale in West Virginia and Pennsylvania. Using compositional reservoir simulation, she showed that an increase in adsorbed gas amount increases the initial gas-in-place, and therefore, total gas production. However, total gas recovery decreased by increasing the amount of adsorbed gas during a specific time period of production. An increase in the adsorbed gas amount leads to a longer transient regime, while a decrease in adsorbed gas amount leads to a faster boundary dominated regime. Table 3.1 shows the details of the calculations and Fig. 3.2 compares total gas recovery from a single horizontal well with 13 hydraulic fracture stages assuming different Langmuir volumes (m). It can be observed that during the early production period, Langmuir volume has a minor impact on the total gas recovery because during this period the production is controlled by the hydraulic fractures and mainly free gas is produced. At middle and late time production periods, higher Langmuir volume leads to lower RF. This is due to fact that even though more gas-in-place and gas

Table 3.1 Cumulative gas productions, initial gas in place, and gas recovery factor obtained for different Langmuir volume conditions

Langmuir volume (m) MSCF/ton	Total gas production (G_p) (BCF)	Initial gas in place (IGIP) (BCF)	Total gas recovery factor Fraction
0	4.21	5.77	0.73
0.05	4.65	7.53	0.62
0.089	4.98	8.89	0.56
0.1	5.06	9.28	0.55

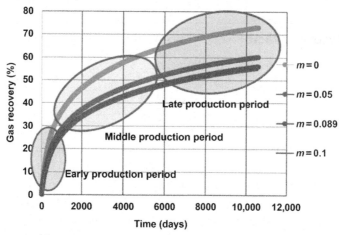

Fig. 3.2 Impact of Langmuir volume on the total gas recovery from horizontal shale gas.

production can be obtained by having more adsorption, most of the adsorbed gas is not available for production due to ultralow matrix permeability resulting in lower RF. In the following chapter, we will discuss the effect of adsorption and organic nanopore size distribution on fluid flow and transport in shale reservoirs.

CHAPTER FOUR

Multiscale fluid flow and transport in organic-rich shale

Introduction

Gas transport and storage in organic-rich shale reservoirs are not very well understood. Previously dual-porosity single-permeability models were used to model fluid flow and transport in shale reservoirs. This approach follows conventional reservoir simulators developed for naturally fractured carbonate or coalbed methane reservoirs. To include the sorption rate and mass exchange between matrix and fractures, bidisperse models were developed with diffusion rates introduced as a controlling factor for sorption rates (Gan et al., 1972; Shi and Durucan, 2003). In these models, instantaneous adsorption/desorption of gas to and from the organic materials are assumed. The resistance to flow is considered to be governed by transfer function, which is a function of pressure gradient between two media, matrix transport, and a shape factor. This basically follows the approach introduced earlier by Warren and Root (1963) to model mass exchange between matrix and fractures. In this approach, two main assumptions have been made. First, the presence of a uniformly distributed fracture network with a known fracture matrix interface that clearly confines the matrix block is assumed. Second, the application of a matching parameter called the shape factor, which controls the mass transfer between matrix and fracture, is used. However, in the organic-rich shale reservoirs, multiscale pore structure presents nonuniformly distributed natural fractures with different dimensions through the reservoir matrix. The magnitude of the contribution of these multidimensional natural fractures in the oil and gas storage and transport is a function of reservoir effective stress, which is defined as follows:

$$\text{Effective stress}: \sigma_e = \sigma_n - \alpha P \qquad (4.1)$$

In Eq. (4.1), σ_e is the effective stress; σ_n is the normal stress applied to the rock, or overburden pressure; and α is the Biot's poroelastic coefficient.

Several authors applied discrete fracture network modeling technique to investigate fluid flow, transport, and storage in these formations. However, their approach does not only require detailed information on the explicit distribution of fractures, but also requires an accurate approximation of pressure distribution in the matrix. In some cases, a general parabolic equation to describe the pressure distribution in the matrix is used.

Multicontinuum modelling of shale reservoirs

Recently multicontinuum models have been used to simulate fluid flow and transport in shale gas reservoirs that can bypass some major difficulties associated with the application of discrete fracture network modeling. In this approach, first the number of components in the reservoir will be identified. Next, the hydraulic properties such as nature of fluid flow and transport in each component will be recognized. Finally, different coupling scenarios describing the mass exchange between components are investigated. Unlike discrete models, this approach does not need to explicitly define the spatial distribution of each component. At any location in the space, all of the components of the multicontinuum model are present and their contribution to the flow will be identified though coupling and mass exchange terms between each component. Fig. 4.1 shows how the conceptual shale matrix model is separated to three continua (inorganic, organic, and fractures) and then uniformly combined to generate the multicontinuum structure.

There are three different types of coupling introduced in the literature to generate the multicontinuum structure, including series, parallel, and selective couplings. Series coupling is defined where the coupling is in the order of hydraulic conductivity. Kang et al. (2010) showed that in samples under investigation in the Barnett Shale, the coupling between organic–inorganic and fractures mostly occurred in a series fashion. After initial gas production from fractures, the organic materials supply gas to inorganic materials, which exchange mass with the system of natural fractures. This behavior can be seen by two distinct changes in the slope of the pressure decay test. In the case of parallel coupling, both organic and inorganic materials are in hydraulic communication and both will supply gas to the system of natural fractures. Selective coupling occurs when two continuums are not hydraulically connected but exchange mass with the third continuum. Fig. 4.2 shows schematics of different possible hydraulic couplings in shale reservoirs.

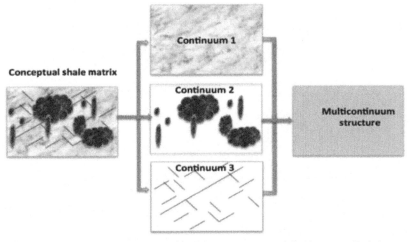

Fig. 4.1 Schematic representation of multicontinuum modeling approach: *light gray*, inorganic matrix; *brown islands* (black in print versions), organic materials (kerogen); *Continuum 3*, discrete natural fracture system.

Recent experimental and numerical studies have shown that the shale matrix can be divided in organic, inorganic, and fractures (Kang et al., 2010; Akkutlu and Fathi, 2012). The transport in organic materials can be represented by free and solid (surface) diffusions, while the transport in inorganic materials is mostly governed by free gas diffusion and convection (Darcy flow). Eq. (4.2) show the mass balance in organic micropores, and Eq. (4.3) illustrate material balance in inorganic macropores of shale matrix, respectively.

Material balance in organic matters

$$: \frac{\partial(\varepsilon_{kp}\phi C_k)}{\partial t} + \frac{\partial[\varepsilon_{ks}(1-\phi-\phi_f)C_\mu]}{\partial t}$$
$$= \frac{\partial}{\partial x}\left(\varepsilon_{kp}\phi D_k \frac{\partial C_k}{\partial x}\right) + \frac{\partial}{\partial x}\left(\varepsilon_{ks}(1-\phi-\phi_f)D_S \frac{\partial C_\mu}{\partial x}\right) \quad (4.2)$$

Material balance in inorganic matters : $\dfrac{\partial\bigl[(1-\varepsilon_{kp})\phi C\bigr]}{\partial t}$

$$= \frac{\partial}{\partial x}\left[(1-\varepsilon_{kp})\phi D \frac{\partial C}{\partial x}\right] + \frac{\partial}{\partial x}\left((1-\varepsilon_{kp})\phi C \frac{k_m}{\mu}\frac{\partial p}{\partial x}\right) - W_{km} \quad (4.3)$$

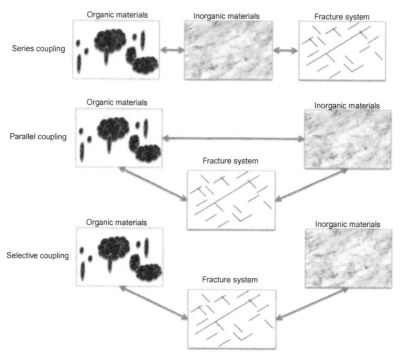

Fig. 4.2 Different hydraulic coupling used in multicontinuum approach.

Free gas mass balance in fracture networks can also be shown as:

$$\text{Material balance in fracture system}: \frac{\partial(\phi_f C_f)}{\partial t}$$

$$= \frac{\partial}{\partial X}\left(\phi_f K_L \frac{\partial C_f}{\partial X}\right) + \frac{\partial}{\partial X}\left(C_f \frac{k_f}{\mu}\frac{\partial p_f}{\partial X}\right) - W_{mf} \qquad (4.4)$$

where W_{km} and W_{mf} are mass exchange terms between different continua and defined as follows:

$$W_{kmi} = \Omega_m \Psi_{ki}(C_i - C_{ki})$$
$$W_{mfi} = \Omega_f \Psi_{mi}(C_{fi} - \overline{C}_i)$$

In the above equations, the subscripts k, m, and f refer to quantities related to the organic (kerogen), inorganic matrix, and fracture, respectively. The variables x and t are the space and time coordinates. $C(x, t)$ and $C_\mu(x, t)$ represent the amounts of free and adsorbed gas in terms of moles per pore volume and moles per organic solid volume, respectively. P is the pore pressure and ϕ and ϕ_f are the total matrix and fracture porosity, respectively. ε_{ks} is the total organic content in terms of organic grain volume per total grain volume and

ε_{kp} is organic pore volume per total matrix pore volume. D represents total free gas diffusion such as bulk plus Knudsen diffusion. D_s is solid or surface diffusion coefficient. k is the absolute permeability, KL represents the macrodispersion coefficient in the fracture network, and μ the dynamic gas viscosity. Ω is the shape factor and Ψ is the transport function in source media, and $\overline{C}$ is referred to as average free gas concentration (Akkutlu and Fathi, 2012).

To describe the gas sorption behavior in organic-rich shale reservoirs, Fathi and Akkutlu (2009) suggested nonlinear adsorption kinetics as shown in Eq. (4.5). This approach was also suggested earlier by Srinivasan et al. (1995) and Schlebaum et al. (1999) to study the carbon molecular sieve and organic contaminant fraction in soil. They argued that the nonlinear sorption kinetic could significantly impact the diffusion processes. The general nonlinear sorption kinetics model in shale can be presented as follows:

$$\text{Gas sorption kinetics}: \frac{\partial C_\mu}{\partial t} = k_{\text{desorp}} \left[K \left(C_{\mu s} - C_\mu \right) C_k - C_\mu \right] \quad (4.5)$$

In this equation, $C_{\mu s}$ represents the maximum monolayer gas adsorption, K is the ratio of adsorption to desorption rate that is known as equilibrium coefficient, and k_{desorp} is the gas desorption rate. In a limiting case where the system reaches equilibrium conditions, that is, $\partial C_\mu / \partial t = 0$, Eq. (4.5) will simplify to single-component monolayer Langmuir isotherm as described earlier.

Interfacial tension and capillary pressure

Interfacial tension (IFT) is the enhancement in intermolecular attraction forces of one fluid facing another and has dimension of force per unit length. IFT is responsible for many fluid behaviors such as interfacial behaviors of vapor-liquid and liquid-liquid. Laplace's law indicates that there is a linear relationship between the pressure difference between two phases and the radius of interface curvature. Laplace's law has been used to study both interfaces between a liquid and its own vapor (surface tension) and between different fluids (IFT) as presented in Eq. (4.6):

$$\text{Interfacial tension}: \Delta P = \frac{\sigma}{r} \quad (4.6)$$

where σ is the surface tension, r is the radius of curvature, and ΔP is the pressure difference between the inside and outside of the interface. This linear relationship is used to obtain the surface tension by simulating series of bubbles with various sizes, measuring their radius, and inside and outside densities (liquid and gas). Different experimental techniques have been used in the oil and gas industry to measure the IFT including the capillary rise and du Noüy ring technique. In the du Noüy ring method, the IFT can be obtained as follows:

$$\sigma = \delta \times \frac{g_c}{2\pi d}$$

where σ is the IFT in dynes per centimeter, g_c is the gravitational constant (980 cm/s^2), d is the ring diameter in cm, and δ is grams-force measured with analytic balance.

In the capillary rise method, the height of the liquid rise in a capillary can be obtained as follows:

$$h = \frac{2\sigma \times \cos \theta}{r \rho g_c}$$

In this equation, r is the radius of capillary in cm, ρ is the density of the denser fluid in g/cc, and $\cos \theta$ is the cosine of the angle between the surface inside the capillary and the capillary wall.

The IFT measurements are also a function of temperature of the experiment. As the temperature increases, the IFT drops. As discussed earlier, phase behavior and phase coexistence properties of fluids under confinement are different from those in the bulk system, especially in organic-rich shales. Recent studies using molecular dynamic simulations revealed that IFTs under pore-wall confinements decrease manyfold and the results are highly sensitive to the temperature (Singh et al., 2009). Bui and Akkutlu (2015) also showed that the surface tension of methane is smaller under confinement using molecular dynamic simulation and is a function of liquid saturation and pore width.

Example

IFT measurement of synthetic oil is done using du Noüy ring where the ring diameter was 1.55 cm and the force measured by analytic balance was 0.38 g-force. Calculate the IFT:

$$\sigma = \delta \times \frac{g_c}{2\pi d} = 0.38 \times \frac{980}{2 \times 3.1416 \times 1.55} = 38$$

The experiment is repeated this time with a capillary rise experiment. The height of the rise in the capillary with a radius of 0.2 cm was 0.36 cm and the density of the synthetic oil is measured to be 0.99 g/cc. Calculate the IFT using capillary rise technique. Assume $\cos\theta = 1.0$.

$$\sigma = \frac{hr\rho g_c}{2\cos\theta} = \frac{0.36 \times 0.2 \times 0.99 \times 980}{2 \times 1} = 35$$

Most often, we are interested in phase coexistence in porous media where two immiscible fluids are in contact. In this case, depending on the chemical properties of each fluid and formation solid surface, one fluid tends to have a higher affinity to wet the formation solid surface. The fluid that adheres to the solid surface is called the wetting phase and the other fluid is called the nonwetting phase. The Young-Dupré equation describes the relationship between imbalance forces of fluid-fluid and fluid-solid interactions as follows:

$$\text{Young} - \text{Dupré equation}: A_T = \sigma_{nw-s} - \sigma_{w-s} = \sigma_{nw-w} \cos\theta_{eq} \quad (4.7)$$

where σ_{nw-s}, σ_{w-s}, σ_{nw-w} are IFTs between nonwetting phase and solid, wetting phase and solid, and nonwetting phase and wetting phase, respectively, and θ_{eq} is the equilibrium contact angle between solid surface line and liquid. Fig. 4.3 shows the schematic of three-phase gas (air), liquid (water), and solid interactions and quantities of the Young-Dupré equation. The contact angle between liquid and solid is not constant and will change as a function of liquid volume. If we were to inject a small amount of liquid into the droplet using a needle, the contact line between liquid and solid can stay constant; however, the contact angle will increase (maximum angle θ_{max}). On the other hand, removing the liquid from the droplet with constant contact line between the solid and liquid will result in a decrease in the

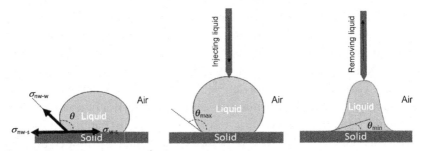

Fig. 4.3 Dynamic contact angle measurement.

contact angle (minimum angle θ_{min}) as shown in Fig. 4.3. The equilibrium contact angle can be calculated using the Tadmor (2004) equation as follows:

Equilibrium contact angle:
$$\theta_{eq} = \arccos\left(\frac{r_{max}\cos\theta_{max} + r_{min}\cos\theta_{min}}{r_{max} + r_{min}}\right) \quad (4.8)$$

where the r_{max} and r_{min} are defined as follows:

$$r_{max} = \left(\frac{\sin^3\theta_{max}}{2 - 3\cos\theta_{max} + \cos^3\theta_{max}}\right)^{1/3} \text{ and}$$

$$r_{min} = \left(\frac{\sin^3\theta_{min}}{2 - 3\cos\theta_{min} + \cos^3\theta_{min}}\right)^{1/3}$$

is defined as the difference between the pressure in the nonwetting phase and the wetting phase, and can be obtained as follows:

$$P_c = P_{nw} - P_w$$

Considering the presence of water and oil in the porous media, if the contact angle between solid and water phase falls between 0 and 70 degrees, we call the formation water wet. If the contact angle is between 70 and 110 degrees, the formation is called neutrally wet, and if the contact angle is greater than 110 degrees, the formation is called oil wet. Capillary pressure in the formation is mainly a function of formation wettability, saturation of different phases, and pore geometry. The Young-Laplace equation describes this relationship at equilibrium condition where there is no flow as follows:

$$\text{Capillary pressure}: P_c = \frac{2\sigma\cos\theta}{r} \quad (4.9)$$

where r is the capillary radius. There are different techniques available to measure the capillary pressure including porous diaphragm method, mercury injection method, centrifuge method, and dynamic method.

The mercury injection method is the most common and rapid technique to measure capillary pressure. In this technique, mercury as a nonwetting fluid is forced to the sample, and the pressure required to get excess mercury volume into the core sample is recorded. The mercury saturation is calculated from a known injection and pore volumes of a sample. This technique has a major disadvantage because the sample cannot be used for further analysis after it is exposed to the mercury. Special considerations are also needed to convert the capillary data obtained from the mercury/air system to the reservoir fluid system. The mercury injection technique has been conventionally used to measure the capillary pressure of shale samples. In this case, mercury is injected to the crushed shale samples with increasing pressure of up to 60,000 psi. Three different volumes will be invaded by mercury including closure or conformance volume, that is, the volume that the mercury needs to fill to overcome the sample surface roughness, pore volume of the sample, and the volume caused by relative change in the sample volume due to compression exposed by mercury. Crushing the shale samples introduces an artificial interparticle volume that will be occupied at low pressure by mercury. This volume is considered as the closure volume and needs to be corrected. Actual intrusion of mercury in shale pore volume occurs after injection pressure exceeds the capillary pressure required for mercury to enter large pores. This will continue until all possible pores of the sample are invaded by mercury. Mercury can enter pores as small as 3 nm, however, there are shale samples with a significant amount of pores of less than 1 nm that cannot be invaded by mercury even at 60,000 psi injection pressure. To accurately measure the capillary pressure, it is crucial to be able to distinguish between the end of closure and start of intrusion, that is, the pressure at which mercury starts invading the larger pores in the sample. This can be identified by the rapid change in the slope of mercury injection pressure vs. mercury saturation. In extremely tight shale samples with the majority of pores in the order of nanometer, identifying this point is extremely difficult and leads to a significant error in capillary pressure measurements. As mercury is injected to the sample and before it reaches the minimum pressure required to invade the larger pores, the mercury pressure applies external stress on both pore and bulk volumes of the sample. Depending on the difference between pore and grain compressibility and pore-throat volume, this external stress on the sample increases the intraparticle volumes due to sample compression and can impair the actual intrusion pressure. Detailed study on pore compressibility of shale as a function of pressure is required to advance the understanding of capillary measurement using mercury injection in ultratight samples (Bailey, 2009).

As previously discussed, the mercury injection technique uses crushed samples. Therefore, it is not performed at in situ reservoir conditions. To be able to perform the capillary pressure measurement at the reservoir condition, the advanced high-pressure/high-temperature porous diaphragm method is proposed in which a core plug is used in a resistivity core holder. The core sample is then exposed to confining and pore pressure, and reservoir temperature is achieved through the application of a heating jacket. In this technique, a low-permeability porous plate saturated with core sample fluid will be set at the downstream of the core holder and a precise pump is used to inject different fluids to perform imbibition or drainage testing. The average fluid saturation of the core sample is determined volumetrically from the displaced fluid received from downstream and fluid volume injected upstream. The equilibrium condition is tested through resistivity measurements performed in axial and radial directions of the core sample. Equilibrium in electrical resistance is assumed when the variation is less than 0.5% in 1 h. The experiment will be repeated at different differential pressures between upstream and downstream and at elevated temperatures. Even though the porous diaphragm method can be applied at reservoir condition and provides more accurate results, calibration, and preparation of the test; also, performing the experiment is very time consuming.

Recently, there have been several studies using nonequilibrium molecular dynamics on the flow of hydrocarbons in organic nanocapillaries to understand the physics of capillary pressure and IFT in organic-rich shale reservoirs. The main drive behind these studies was the difficulty associated with direct measurements of these properties in shale samples and huge uncertainty associated with these techniques (Feng and Akkutlu, 2015).

Wettability effects on shale recovery

As discussed earlier, wettability is defined as the relative adhesion of fluid to the solid surface. Wettability is conventionally measured using three different techniques including contact angle, Amott wettability index, and US Bureau of Mines (USBM) wettability index measurements. In the Amott wettability index test, the sample is imbibed with water to its residual oil saturation first and then immersed in oil for 20 h. The amount of water displaced by spontaneous drainage of water is then measured as a volume of water (V_{wsp}). Next, water is drained to its residual water saturation and the maximum amount of water recovered is recorded as the total water volume (V_{wt}). The sample is then immersed in brine for 20 h and oil volume displaced by natural water imbibition is measured as V_{osp}. The remaining oil in the sample is then forced out by injecting brine to the sample to its residual oil saturation to measure the total oil volume V_{ot}. The Amott wettability index can then be obtained using Eq. (4.10).

$$\text{Amott wettability index}: I_w = \frac{V_{osp}}{V_{ot}} - \frac{V_{wsp}}{V_{wt}} \qquad (4.10)$$

In this equation, I_w is the Amott wettability index and ranges between -1 and 1, in which -1 stands for oil-wet, 0 stands for neutral-wet, and $+1$ indicates the water-wet formation. The wettability characteristic of the formation highly impacts the hydrocarbon recovery and multiphase flow in porous media and is a function of the solid surface chemistry and microscale roughness of the surface. Due to the wide range of applications of wettability characteristics of the material, there have been several studies on either alteration or restoration of the solid surface wettability. Wettability of the solid surface can be changed by altering the solid surface chemistry or by changing the microscale surface roughness. The chemistry of the solid surface can be altered using different techniques including oxidation of the solid surface, deposition of nonwetted material at the solid surface, or application of the electric fields. However, they are poorly developed techniques to change the microscale roughness of the solid surfaces (Aria and Gharib, 2011). In the oil and gas industry, to change the wettability of the formation, different techniques such as treating the solid surface with a coating agent (e.g., organosilanes), using naphthenic acid or asphaltenes, and adding surfactants to the injected fluids are used. In the case of restoration of the wettability in the core samples, toluene followed by ethanol to extract the toluene is also used. In some cases, drying the core sample or aging in crude oil for 100h at 65°C is also recommended. In organic-rich shale reservoirs, these techniques are not practical. This is due to the complex pore structure, extremely low permeability, and heterogeneity in mineral compositions of organic-rich shales. Recently, nondestructive techniques such as nuclear magnetic resonance (NMR) are used to study the wettability characteristics of the shale reservoirs and to monitor sequential imbibition and drainage processes in shale samples. Ousina et al. (2011) used a total of 50 samples from different US shale basins and measured their wettability using the NMR technique. In this approach, they first performed NMR study on the samples as received. Next, they immersed the samples in brine at room-temperature conditions and ran NMR after 48h of spontaneous imbibition. The samples were then immersed for 48h in dodecane and NMR analysis was done on the drained samples. They found that shale samples in general show mixed wettability, with organic material contributing to oil-wetting characteristics. To have a better understanding of the wettability characteristics in organic-rich shale reservoirs, a combination of different direct and nondestructive approaches are required.

CHAPTER FIVE

Hydraulic fracturing fluid systems

Introduction

Hydraulic fracturing or fracing has become one of the most important parts of completing a well. Hydraulic fracturing is essentially the act of pumping sand, water, and specific chemicals at a very high rate and pressure to break the rock and release the hydrocarbon. Hydraulic fracturing stimulation is used to increase the permeability and reduce the skin damage caused by drilling. Unconventional shale reservoirs are known for having an extremely low permeability. In an attempt to increase the production volumes of unconventional shale reservoirs, hydraulic fracturing is performed on every well. Without hydraulic fracturing, reservoirs with low permeability will never produce at an economically feasible rate.

The first use of hydraulic fracturing was in 1947; however, modern hydraulic fracturing referred to as "slick water multistage horizontal stimulation" or "slick water frac" was first performed in the Barnett Shale, located in Texas, in 1998 using more water and higher pump rate than previously attempted techniques. The introduction of slick water horizontal frac made the production of low-permeability reservoirs promising. This is when the industry started looking at various shale plays across the United States and the world. The industry is moving from conventional resources with high permeability, which is hard to find, but easy to produce, to resources such as shale that are much easier to find, but more difficult to produce. Conventional resources are hard to find, but once the appropriate reservoirs are found no hydraulic fracturing is typically necessary to increase the permeability. The permeability of conventional resources is usually high enough that the hydrocarbon trapped in the reservoir will automatically flow into the wellbore right after perforation. In contrast, unconventional resources would not be economically feasible to produce without hydraulic fracturing. There are many different applications for hydraulic fracturing and they are as follows:

Hydraulic Fracturing in Unconventional Reservoirs
https://doi.org/10.1016/B978-0-12-817665-8.00005-9

1. Increase the flow rate from low-permeability reservoirs such as shale formation in general.
2. Increase the surface area or the amount of formation contact with the wellbore.
3. Reduce the number of infill wells with horizontal hydraulic fracturing stimulation.
4. Connect hydraulic fractures with existing natural fractures.
5. Increase the flow rates from wells that have been damaged (near wellbore skin damage) because of drilling.
6. Decrease the pressure drop around the well, which will cause a reduction in sand production.

The first application listed above (the most important application) is the essence of hydraulic fracturing since the main reason behind hydraulic fracturing is to increase the permeability of the reservoir. Not only does the flow rate increase in a naturally fractured formation with low permeability but also hydraulic fracturing will connect the natural fractures and faults (if present) in the formation. When the formation is hydraulically fractured, the amount of the formation in contact with the wellbore will increase, and as a result, the flow rate will also increase. Porous and permeable reservoir rocks filled with lots of hydrocarbons are any company's dream to obtain. However, poorly cemented sandstone formations with high permeability can cause lots of issues when it comes to production. In this type of formation, sand grains flow into the wellbore with the produced hydrocarbon and cause various issues. These issues can lead to severe pipe erosion/damage, flow line blockage, and finally a reduction in production. Various completion techniques, such as gravel packing, frac packing, and expandable sand screens can be used to fight this problem. Gravel packing is essentially the placement of a steel screen and packing the surrounding annulus with specific and designed size gravel. The designed size gravel prevents the passage of formation sand into the wellbore. Hydraulic fracturing can be used in conjunction with the conventional gravel-packing technique in a process called *frac packing*. In the frac-packing process, hydraulic fracturing occurs after the placement of a gravel pack to create a good conduit for the flow of hydrocarbon at some distance from the wellbore. Therefore, hydraulic fracturing can also have a positive impact on conventional high-permeability sandstone formations with sand production issues. There are various types of hydraulic fracture fluid systems in the industry and every formation requires a specific system. The most commonly used frac fluid systems are described in the following sections.

Slick water fluid system

This type of fluid system is well known in the industry and is currently being used in the Marcellus Shale, Barnett Shale, Eagle Ford, Hayesville, Utica/Point Pleasant, and many other low-permeability reservoirs. In this technique, water, sand, and specific chemicals are pumped downhole to create a complex fracture system within the reservoir. The main goal in low-permeability reservoirs using water frac is to create a complex fracture system and maximum surface area. When there is not enough surface area created in low-permeability rocks, the well productivity is not maximized. This is why this system uses a huge amount of water to create the maximum possible surface area. Additionally, the *rate* is the drive needed to create a complex fracture system within the formation. Using a higher rate yields more surface area which could potentially result in better production. Some operators limit their rate to prevent fracture height growth and paying less for hydraulic horsepower if and only if the partial cost of the job or the contract depends on the horsepower. More rates require more pumps and sometimes the size of the pad (well site) and many other factors do not allow the operator to have as many pumps as necessary for the job.

Another limiting factor in achieving the necessary rate is *pressure*. There are various limitations on pressure such as surface equipment and casing burst pressure ratings. For example, the maximum allowable surface-treating pressure when fracturing in Marcellus Shale is usually 9500 psi based on 5½ in., 20 lb/ft, P-110 production casing. This pressure is determined from the casing, surface, and wellhead pressure ratings used for the job. For example, if during a frac stage treatment the surface-treating pressure is about 9500 psi at 60 bpm (BBLs of frac fluid per minute) the rate will be limited to 60 bpm. The rate can only be increased if pressure decreases below the maximum allowable treating pressure during the frac stage. Rate is basically the most important parameter in water frac; however, sometimes the size of the pad, cost, and pressure limitations can prevent achieving the designed rate during frac jobs.

One important lesson that is crucial to emphasize is that a higher rate will create more surface area. George Mitchel is the pioneer of the slick water frac and he spent several years designing the best practice to economically produce from Barnett Shale. The introduction of high-rate slick water frac was the key to his success. Many shale plays across the United States are full of natural fractures, which are one of the main sources of transferring fluid into the wellbore. As more natural fractures and surface areas are contacted by hydraulic fracturing stimulation, better productivity will be attained. Low-viscosity fluid, such as water frac, tends to follow natural fractures, contact more surface area, and create a complex fracture system within the reservoir. The reason a water frac is called "slick water frac" is because of a chemical additive called friction reducer (FR). Without FR, a slick water frac cannot be pumped at a high rate. The addition of the FR to water reduces friction and makes the water very slick.

The best type of frac fluid is not necessarily freshwater. Formation water (flowback water) is actually believed to be a better frac fluid in some areas since it contains the Earth's minerals. Using freshwater for fracing could cause a filter cake along the created fractures, which causes a reduction in permeability and conductivity. The majority of companies use a mixture of treated or untreated flowback and freshwater to obtain the volume of water needed for the job. Some companies have even tested 100% produced water for frac with significant advancements in proper FR selection that can handle high total dissolved solids (TDS), irons, etc. Each slick water hydraulic fracturing stage typically uses about 4000–11,000 BBLs of water (168,000–462,000 gallons of water) depending on the size of the job (sand volume), treatment complexity, production results, etc. For example, if the designed sand volume in a stage is 200,000 lb, it will take less volume of water to pump the stage compared to the designed sand volume of 500,000 lb. Higher sand volume basically requires more water to be placed into the formation. Some stages could take more water because the stage is very difficult to treat when higher sand concentrations are run. For example, some stages do not like higher sand concentration, such as 3 ppg (3 lb of sand per gallon of water); therefore, the stage is treated at a lower sand concentration to put all the designed sand into the formation. In a water frac, if the stage is hard to treat, it is more important to place the designed sand into the formation at a lower concentration (to achieve more surface area) compared to running higher sand concentration and cutting the stage short of design.

Another important candidate for water frac is formations with high brittleness. When a rock is brittle, it helps to keep the fractures open after breaking down. For example, glass is a brittle material and when a glass is broken, it is scattered. The main application for a water frac is in formations with high Young's modulus and low Poisson's ratio. High Young's modulus and low Poisson's ratio is basically an indication that the rock is brittle and slick water frac can be used to break the rock. In a water frac, a maximum sand concentration of 3–3.5 ppg (lb per gallon) can be pumped in the best case scenario. Given a healthy, high rate and ease of the formation, 4 ppg sand concentrations can also be achieved with slick water (very rare). Pumping higher sand concentrations (more than 4 ppg) is not possible with a slick water frac and can lead to sanding off the wellbore (screening out). Achieving higher sand concentration can be performed using other frac fluid systems, such as cross-linked gel, which will be discussed.

As previously mentioned, the main objective of a water frac treatment is to create a complex, but not a dominant fracture network. In general, in a slick water frac, low-viscosity fluid leads to a complex fracture network, while converting to higher-viscosity frac fluid (e.g., linear gel and cross-linked gel) tends to create dominant hydraulic fractures. The essence of a water frac in naturally fractured reservoirs is to follow natural fractures while creating multiple flow paths as a result of applying high energy to the rock. This energy is only achieved at a higher rate. The combination of a higher rate and low-viscosity fluid slick water will cause the sand to be placed farther into the formation and result in better long-term productivity. One of the major issues with our industry is being so dependent on short-term production. Some companies only look at the initial production of the well and ignore the long-term production. This is another recipe for failure when designing and comparing the performance of a well. It is also crucial not to make decisions based on a single well's production data. Instead, a field of production data can be used to make critical economic decisions. In the machine learning (ML) chapter of this book, we will review how various statistical techniques can be used to find hidden patterns in a field of interest and use them for important economic decisions.

Average treating rate in slick water frac varies from stage to stage based on the pressure limitations discussed. The goal is to achieve the maximum designed rate, which is typically 70–100 bpm. In Barnett Shale, the average surface-treating rate is even higher and pump rates as high as 130 bpm have been achieved. Rate also overcomes leak-off and fracture-width problems during the frac stage. Leak-off refers to the fracturing fluid getting lost in the formation. Having a high rate during the frac stage eliminates the concern of having high leak-off. Having high leak-off could lead to sanding off the well. One of the main reasons rate should not be sacrificed in low-permeability unconventional reservoirs is because of not creating the surface area needed for long-term production. When hydraulic fracturing is performed on a low-permeability reservoir with a limited rate, only limited surface area is achieved. Once the reservoir is drained in that particular surface area, the production will be decreased significantly. In high-permeability conventional reservoirs, the surface area is not the only deciding factor, because even after creating so much surface area the permeability beyond the stimulated reservoir volume region to transmit fluid into the created hydraulic fractures is still high.

Over time, naturally fractured reservoirs become the best refrac candidates to enhance oil and gas recovery. For example, if after 20 years of production, the production rate declines below the economic limit, it is highly recommended to restimulate the well due to natural fractures that exist in the reservoir to enhance the recovery. In addition to having a natural fracture system, reservoirs should have high initial gas in place (IGIP), high pore pressure, and superior reservoir properties to be the most successful candidates for refrac. It is also very important to select candidates based on their original completion design. Having all of the conditions discussed above, poor initial completion design wells are better candidates for refrac. Fig. 5.1 shows the schematic of typical hydraulic fracture and complex fracture network interactions in slick water frac.

Cross-linked gel fluid system

This type of fluid system is used in conventional and unconventional reservoirs to achieve the so-called bi-wing fracture system. Cross-linked gel is a heavy viscous fluid. In this type of frac fluid system, viscosity (not velocity) is used to place proppant into the formation. Cross-linked gel is typically used in ductile formations with higher permeability (e.g., oil windows of Eagle Ford and Bakken Shales). Cross-linked gel is also heavily used in

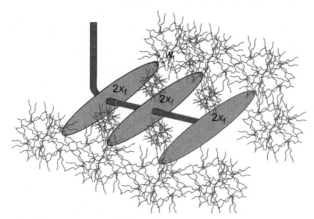

Fig. 5.1 Complex fracture system illustration.

oil windows of various shale plays to be able to obtain the necessary fracture width for optimum oil production. The goal in this type of frac is to achieve the maximum sand concentration near the wellbore (higher conductivity) through the use of a viscous fluid. As opposed to water frac that uses velocity to carry proppant, cross-linked gel uses heavy viscous fluid (cross-linked gel) to place the proppant into the formation. High rate is not required in this type of frac fluid system and usually 25–70 bpm is used to place the proppant into the formation. Higher sand concentration up to 10 ppg can be obtained if and only if a great cross-linked gel is obtained. If for any reason throughout the stage cross-linked gel is cut (not pumped) due to an equipment malfunction, the first thing to do is to cut sand and flush the well to prevent sanding off the well. The reason being is that viscosity carries the high sand concentrations into the formation and without the heavy viscous fluid the well could be sanded off easily at such high sand concentrations. One of the most common mistakes with using cross-linked fluid is pumping the job at lower sand concentrations (<6 ppg). The advantage of using a cross-linked fluid system is the utilization of heavy, viscous fluid to pump very high sand concentrations, which will create a dominant fracture with a large proppant pack near the wellbore.

Another advantage of the cross-linked fluid system is fluid leak-off reduction. In high-permeability reservoirs where fluid leak-off is significant, the cross-linked fluid system is known to reduce fluid leak-off and keep the proppant suspended until closure. In addition, very high viscosities (thousands of centipoise) can be reached using the cross-linked fluid system.

Another main criterion for choosing cross-linked gel is ductility. Formations with very low Young's modulus and very high Poisson's ratio that have higher permeability are the best candidates for cross-linked frac. It is extremely important to use production data to come up with the best frac fluid system in any formation. Sometimes after taking all the necessary parameters into account, one type of frac does not yield the best production results. For example, if after hydraulically fracturing a formation using cross-linked gel, the well produces below expectation, a different technique ought to be utilized to maximize production. Theories are good to know and understand, however, the main deciding factor in choosing the frac fluid system is production data. In both Marcellus and Barnett Shale, the reason a water frac is chosen as the main frac fluid system by the majority of operating companies is because of successful production results. If production using slick water frac was not promising, many companies would have tried a different frac fluid system.

As previously mentioned, in a cross-linked frac, viscosity (as opposed to rate or velocity) is used to place high proppant concentrations into the formation. One of the biggest concerns with cross-linked fluid is the gel residue that this type of fluid system leaves in the formation. The cross-linked residue, if not broken properly at reservoir conditions, can cause serious damage to the created fractures by reducing the permeability and fracture conductivity (to be discussed). Fig. 5.2 illustrates the schematic of a bi-wing fracture system using cross-linked gel.

Hybrid fluid system

In this type of fluid system, slick water is used to pump at a lower sand concentration followed by cross-linked or linear gel to pump at a higher sand concentration to maximize near wellbore conductivity. Some companies use this type of frac in unconventional reservoirs in the event there are severe issues with placing higher sand concentrations into the formation. This is because some formations do not like higher sand concentrations and the only way to put all the designed sand away is either using linear gel (less viscous compared to cross-linked gel) or cross-linked gel at higher sand concentrations. For example, in a Marcellus Shale operation, if all the designed sand is not able to be placed into the formation by using slick water frac fluid, the linear gel is used to increase the viscosity of the fluid and as a result, increase the fracture width and overcome near wellbore tortuosity. Some stages in a Marcellus Shale formation need to have some type of viscous fluid such as linear gel for all the designed sand to be successfully placed into the formation. Typically if there are issues with establishing a good rate during

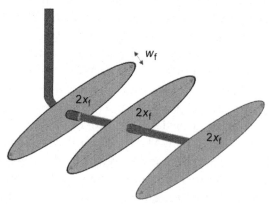

Fig. 5.2 Bi-wing fracture system illustration.

higher sand concentrations, 5–10 lb linear gel is used at the start to provide more fracture width and better proppant transport. In some stages, if a baseline flow rate (bpm) is not established to start even lower sand concentrations, 15–20 lb linear gel could be used to carry the proppant into the formation without screening out at a lower rate. Once the necessary baseline flow rate is established, the linear gel can be reduced or completely eliminated throughout the stage. For example, if only 25 bpm is reached at the maximum allowable surface pressure, a 15–20 lb gel system can be used and 0.1–0.25 ppg (lb of sand per gallon of water) is started. Once a base rate is established, gel concentration can be lowered or cut throughout the stage. Gel concentrations are typically provided in lb of polymer per 1000 gallons of base fluid (water). For example, a 20 lb ABC gel system is prepared with 20 lbs of ABC per 1000 gallons of base fluid. Gel typically comes in two forms: dry and wet liquid gel concentrate (LGC). The LGC is made by mixing high concentrations of dry gel in a solvent to make an LGC. The concentration of dry gel used varies but it is usually 4 lb of gel per gallon of solvent (water). The main equation for linear gel usage is shown in Eq. (5.1).

$$\text{Linear gel (gpt)} = \frac{\text{Linear gel system in lb}}{4} \qquad (5.1)$$

For example, 5 lb linear gel system is equal to 5 divided by 4, which yields 1.25 gpt (gallons of gel per thousand gallons of water). In other words, LGC is made using 4 lbs of gel in a gallon of solvent. To make a 5 lb/1000 gallon gel system, 1.25 gpt of LGC will be needed. The 5 lb gel system is not a superviscous fluid and this type of gel system has just enough viscosity to overcome the friction pressure and provide more fracture width to be able to place the sand into the formation. When tortuosity is severe, higher gel concentrations, such as 15–20 lb gel systems will be needed. Fig. 5.3 shows a sample of a 20 lb linear gel system that was used during a Marcellus stage treatment.

Foam fracturing

Foam fracturing is not a common frac fluid system in the majority of unconventional shale plays, but this type of fracturing fluid system provides some attributes that others do not provide. Foams are made up of two parts. The first part is gas bubbles, referred to as the internal phase, and the second part is liquids, referred to as the external phase. Nitrogen foam frac is the

Fig. 5.3 20 lb linear gel system.

most commonly used form of the foam-fracturing fluid system. In this type of fluid system, nitrogen is typically pumped with water and other additives to form a foam-like fluid. The nitrogen foam-fracturing fluid system is common in coalbed methane, tight sands, and some low-permeability shale reservoirs that are normally <5000 feet deep.

Foam-fracturing fluid, just like other types of fracturing fluid systems, has advantages and disadvantages. Since a nitrogen foam frac has less fluid in the system and a big percentage of the fluid system is composed of nitrogen, it is ideal for water-sensitive formations (e.g., clay-containing formations). Due to the fact that less liquid is pumped in a nitrogen foam frac, clay swelling and formation damage are minimized in water-sensitive formations. A nitrogen foam frac is ideal for low-pressure and depleted formations in which the energy of nitrogen is used to help the well cleanup and flowback after the frac job is completed. As foam frac fluid is mostly composed of >60% gas, recovery of fracturing fluid in low-pressure reservoirs is more efficient as compared to nonfoam-fracturing fluid systems. The compressible nature of the foam frac fluid will help recover the liquid due to gas expansion as the fracturing fluid travels to the wellbore. Due to the fast well cleanup after frac, cleanup time will be minimized in a nitrogen foam frac. Without this type of fluid system in depleted and low-pressure formations, the reservoir does not have the energy to recover the frac fluid pumped downhole. Since the foam-fracturing fluid system only contains 5%–35% liquid, low liquid percentage foam will have less hydrostatic pressure acting on the formation.

Another advantage of foam-fracturing fluid system is the fluid-loss capability. As previously discussed, less fluid is pumped downhole in this type of fluid system and as a result, a foam frac provides better fluid efficiency, which in turn yields low fluid loss. This fluid-loss capability can be demonstrated by putting some shaving cream on your hand and flipping your hand upside down. The shaving cream does not readily fall off of the hand. This is indicative of the fluid-loss capability of the foam-fracturing fluid system. When fluid-loss additives are not required, any detrimental damage to fracture permeability and conductivity can be reduced. Note that sometimes fluid-loss capability might be required in highly naturally fractured formations with higher permeability. When nitrogen is injected into a liquid such as water, some foaming will occur. However, due to water being thin, some bubbles will rupture. Adding a foaming agent, such as soap will cause the bubbles to become more stable. Soap is known to be a type of surfactant that will stabilize the foam when injecting nitrogen. The general rule of thumb is that in formations with permeability >1 md, fluid-loss additives could be beneficial.

Another important advantage of foam-fracturing fluid is proppant transport. As opposed to a slick water fluid system, foam allows proppant transport into the formation without settling out. This will allow for uniform distribution of proppant particles throughout the fractures. The amount of proppant that foaming frac fluid can suspend will depend on the foam quality (to be discussed). When regular sand (SG = 2.65) is used with a foaming agent and 3 lb of sand per gallon of foam is desired, the sand concentration at the blender will be 9 ppg for 67% quality foam, 12 ppg for 75% quality foam, and 15 ppg for 80% quality foam. Fig. 5.4 illustrates this concept for regular sand with a specific gravity of 2.65.

Foam quality

Foam quality is the ratio of gas volume to foam volume (gas + liquid) over a given pressure and temperature. Nitrogen or CO_2 can be used to create foam in liquid status, but nitrogen is typically preferred because CO_2 can be extremely harsh and eroding when water is existent.

$$FQ = \frac{\text{Gas volume}}{\text{Gas volume} + \text{liquid volume}} \quad (5.2)$$

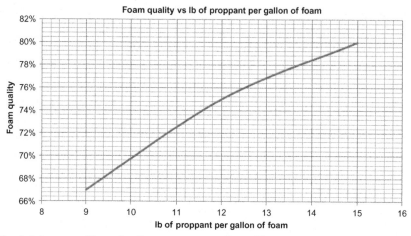

Fig. 5.4 Foam quality vs lb of proppant per gallon of foam.

where FQ is the foam quality, %; gas volume = BBLs or gallons; and liquid volume = BBLs or gallons.

When foam quality is between 0% and 52%, gas bubbles do not contact each other and are spherical. Foam viscosity is also low because there is a lot of free fluid in the system, which in turn will affect the fluid-loss capability. When foam quality is between 52% and 96%, the gas bubbles are in contact with each other and as a result, an increase in viscosity will occur. Foam qualities of 52% and 60% do not have the proppant-suspension capability. Finally, when foam quality is more than 96%, the foam will degenerate into mist and as a result, there will be a loss in viscosity. Note that higher foam quality has a higher viscosity and is better able to suspend proppant. As foam quality increases, more hydraulic horsepower will be needed. This is because an increase in foam quality will decrease the hydrostatic pressure and in turn will increase the surface-treating pressure. An increase in surface-treating pressure will cause an increase in hydraulic horsepower. The most frequently used foam quality is typically 70%–75%.

Foam stability

There are several factors that affect foam stability. Foam quality, surfactant type/concentration, and polymer type/concentration are some examples. One of the most important aspects of foam fracturing is to keep the foam in motion. If the foam is not in motion, it will be unstable. When foam stops moving, gravity will cause the free fluid in the foam to drain. This drainage can cause foam instability issues. The rate of foam drainage will depend on many factors, such as temperature, viscosity of the liquid phase, and foaming-agent concentration. An increase in temperature can potentially cause a reduction in the viscosity of the fluid. As temperature increases, more foaming agents must be used. Gelling agents are also very important because they can be used to add stability to the fluid. Gelling agents will increase the viscosity (not considerably), but will improve proppant transport and fluid-loss control. Gelling agents must be used in moderation because higher fluid viscosity will be harder to foam and pump, and as a result, will require more hydraulic horsepower.

Tortuosity

Tortuosity refers to the pressure loss by fracturing fluid between the perforations and main fracture(s). It is basically the restricted pathways between the perforations and main fractures. Tortuosity can be justified as one of the main causes for the majority of screen-outs. Tortuosity was not an issue in vertical wells; however, tortuosity seems to be very severe in horizontal wells, wells with moderate-to-severe inclinations, hard rock reservoirs, and wells with dispersed perforations. Pumping conditions and rock properties have a direct impact on tortuosity. Tortuosity can be severe in some stages and this is why viscous fluid, such as linear gel is run to fight the problem and be able to successfully put all the designed sand in the formation. In general, not being able to obtain sufficient rate during a frac stage could potentially be due to severe tortuosity issues. This problem can be easily solved by pumping higher viscous fluid such as linear gel, which will cause the surface-treating pressure to drop as soon as the linear gel hits the perforations. A drop in the surface-treating pressure is an indication of overcoming the tortuosity issues between the perforations and main fractures. There are various ways to figure out if tortuosity is the problem. The first and the most commonly used method for identifying the tortuosity is pumping a sand slug at low concentrations after the pad. If the sand slug hits the formation and pressure rises, it is an indication of tortuosity. If the sand slug causes an increase in pressure followed by a considerable break in pressure, it means the removal of tortuosity. Finally, if there is no impact when sand hits the perfs, then there are no issues with tortuosity.

When sufficient rate is not established during the pad stage in a slick water fluid system, a low-concentration sand slug (typically 0.1–0.25 ppg) is run to see the impact on pressure and figure out whether severe tortuosity exists in the formation or not. Another option is to run the sand slug for a second time to attack the tortuosity if pressure permits. Another process for determining if tortuosity exists or not during a frac stage is by subtracting the closure pressure from the instantaneous shut-in pressure (ISIP). If the difference between ISIP and closure pressure is more than 400 psi, there is a high possibility of tortuosity. Some of the most commonly used techniques (as discussed) to combat severe tortuosity are as follows:

1. pump low-concentration proppant slugs,
2. use high gel loading (>15 lb system), and
3. increase rate (if possible).

It is challenging to determine whether tortuosity or high perforation friction is the cause of not being able to pump into a zone. When dealing with high perforation friction pressure, the following techniques can be used to overcome the issue:

1. pump low-concentration proppant slugs,
2. spot acid (for the second time), and
3. reperforate.

Sometimes spotting acid for the second time might help resolve the issue in the event all the perforations were not cleaned of cement and debris the first time. When the first five options listed above fail to deal with tortuosity and high perforation friction pressure, reperforating is used to overcome the problem and get into the stage. Some companies do not even try any of the techniques listed above and simply reperforate since the cost of reperforating could possibly be cheaper than trying various techniques. Fig. 5.5 shows the schematic of possible tortuosity between perforations and hydraulic fracture. Fig. 5.6 illustrates the basics of frac fluid design

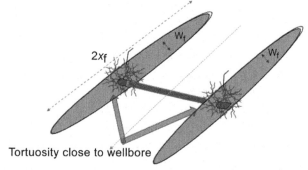

Fig. 5.5 Tortuosity.

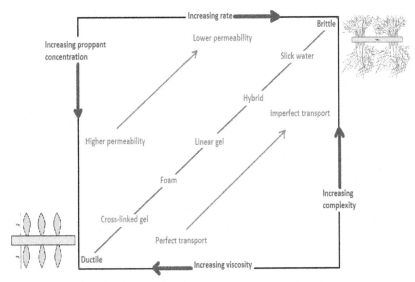

Fig. 5.6 Frac design basis (Britt, 2011).

selection based on rock mechanical properties from brittle rocks with high Young's modulus and low Poisson's ratio to ductile rocks. As Fig. 5.6 illustrates, moving from brittle rocks to ductile rocks will need a change in the frac fluid system (slick water to cross-linked gel). This will lead to an increase in viscosity of the fluid, better proppant deliverability, lower fracture complexity, and lower flow rate.

Typical slick water frac steps

As previously discussed, slick water is the most commonly used frac fluid system in shale plays. There are four main steps during each slick water hydraulic frac stage that are followed in sequence; they are described in the following sections.

Acidization stage

In this stage, various concentrations of HCl (hydrochloric) or HF (hydrofluoric) acid is pumped downhole to clean the perforations (holes) of any type of debris or cement. The purpose of acid is to clean the perforations of any cement or debris and is not meant to acidize the formation. Acid does help when the formation has limestone streaks or calcite. Different companies have various theories regarding the volume and concentration of acid. Depending on the operating region and formation type, typically 500–4000 gallons of 3%–15% acid is pumped downhole to clean the perforations. The acid stage can easily be seen on the surface-treating pressure chart because as soon as the acid hits the perforations (holes in the casing), the surface-treating pressure decreases and a higher rate can be established. Table 5.1 shows specific gravities of HCl acid at various HCl acid concentrations.

Contact time is very important when pumping acid. In general, contact time can be reached by pumping lower acid concentrations. For example, instead of pumping 3000 gallons of 15%, 3000 gallons of 3%–7% can be used in an attempt to have longer contact time with the perforations. This could possibly enhance the cleaning process compared to pumping higher acid concentrations. Typically service companies have acid-blending plants and bring 31.45% acid into their acid-blending plant. It is then mixed as needed in the yard and sampled to confirm the proper mix. Typically, the maximum concentration is 28% acid that can be hauled to a location in normal oilfield trucks. Since the acid pumped downhole is usually 5%–15%, the desired percentage is mixed on the fly. It is a lot easier operationally to haul the acid out hot and mix it on the fly. One of the advantages of mixing the acid on the fly is that more stages can be pumped out of one acid tube. This will be helpful because fewer complications will occur while moving acid in and out after every stage. The acid concentration achieved from mixing on the fly will be close enough to the desired designed concentration. It is the company representative's responsibility to calculate the dilution rates to ensure the proper rates for getting the desired acid percentage

Table 5.1 Specific gravity (SG) of HCl acid

%	SG
3	1.015
5	1.025
10	1.048
15	1.075
20	1.100
25	1.126
28	1.141
30	1.151
31.45	1.160
36	1.179

are used. Eq. (5.3) can be used to acquire gallons of acid required to convert to the desired acid concentration:

$$\text{Original acid volume} = \frac{\%\text{of acid desired} \times \text{SG of desired acid}}{\%\text{of original acid} \times \text{SG of original acid}} \times \text{gallons of desired acid} \quad (5.3)$$

Example

How much acid and water are needed given 28% hydrochloric acid (hauled to location) in order to obtain 3000 gallons of 5% acid?

$$\text{HCl acid volume} = \frac{5 \times 1.025}{28 \times 1.141} \times 3000 = 481.3 \text{ gallons of } 28\% \text{ HCl acid}$$
$$\text{Water volume} = 3000 - 481.3 = 2518.7 \text{ gallons of water}$$

As can be seen in this example, to make 3000 gallons of 5% hydrochloric acid (28% original trucked to location), only 481 gallons would be acid and the rest would be water. Now, let's do one more calculation: how much acid is required to make 3000 gallons of 15% instead of 5%?

$$\text{HCl acid volume} = \frac{15 \times 1.075}{28 \times 1.141} \times 3000 = 1514.2 \text{ gallons of } 28\% \text{ HCl acid}$$
$$\text{Water volume} = 3000 - 1514.2 = 1485.8 \text{ gallons of water}$$

Therefore, more than 28% of acid is needed to achieve higher acid concentration.

Pad stage

After pumping the designed volume and concentration of acid, pad (which is a combination of only water and chemicals) is pumped downhole to initiate the hydraulic fractures by creating fracture length, height, and width before starting the main proppant stage. In other words, pad is the volume of fluid pumped downhole to create a sufficient fracture network before pumping the proppant stage. It is very crucial to obtain as much rate as possible during the pad stage for a bigger fracture network. Pad volume is extremely important to determine in order to prevent premature sand-off (tip screen-out). Engineers strongly believe that if a sufficient fracture network is not created during this stage, a premature screen-out can be the consequence. The hydraulic fracture network is created throughout the entire treatment; however, the majority of the fracture network is created during pad injection. If not enough pad is pumped, at some point during the treatment the sand will reach the tip of the created fractures causing them to bridge with sand and eventually pack off all the fractures. This will result in sanding off the wellbore if the stage is not ended early by cutting sand. On the other hand, too much pad can be harmful as well. If too much pad is pumped, fracture tips continue to propagate after pumping is stopped resulting in a large unpropped (unpropped means no proppant) region near the tip of the fracture. Propped fracture regions can move toward the unpropped region and essentially leave a poor final proppant distribution inside the main body of the fractures. This underlines the importance of calculating and understanding the pad volume before the main treatment. Pad volume is calculated using Eqs. (5.4)–(5.6), which are functions of fluid efficiency in the formation.

$$\text{Nolte method} = \text{Pad volume}\% = (1 - \text{FE})^2 + 5\% \qquad (5.4)$$

$$\text{Shell method} = \text{Pad volume}\% = \frac{1 - \text{FE}}{1 + \text{FE}} \qquad (5.5)$$

$$\text{Kane method} = \text{Pad volume}\% = (1 - \text{FE})^2 \qquad (5.6)$$

Fluid efficiency is the ratio of stored volume within the fracture to the total fluid injected. Fluid efficiency is inversely related to fluid leak-off. Higher fluid efficiency means lower fluid leak-off and lower fluid efficiency means higher fluid leak-off. Leak-off is the amount of fracturing fluid lost to the formation during or after treatment. Unconventional shale reservoirs, in general, have lower leak-off because the permeability is very low. Low fluid leak-off in unconventional shale reservoirs indicates that the fluid does not get lost in the formation as much as it would in high-permeability formations. The fluid pumped in low-permeability formations will effectively create fractures because of low fluid leak-off. Generally, shale has a high fluid efficiency (low leak-off); therefore, it requires less pad to be pumped. Fluid efficiency can be calculated using the diagnostic fracture injection test, which will be discussed.

Sanding off the wellbore or screening out can be a costly issue when hydraulically fracturing high-permeability formations because the fluid gets lost very quickly to the formation and the amount of pad that was originally pumped is lost in the formation as well. This is the main reason that high-permeability reservoirs, which have higher fluid leak-off, need more pad volume to effectively place all the designed sand into the formation.

Example
Calculate the pad volume % needed for a frac stage with 70% fluid efficiency if 7000 BBLs of frac fluid is designed for the stage:

$$\text{Nolte method} = \text{pad volume}\% = \left((1 - 0.7)^2 + 0.05\right) \times 100$$
$$= 14\% \text{pad volume}$$
$$\text{Pad volume} = 7000 \times 14\% = 980 \text{ barrels}$$
$$\text{Kane method} = \text{pad volume}\% = (1 - 0.7)^2 \times 100 = 9\% \text{pad volume}$$
$$\text{Min pad volume} = 7000 \times 9\% = 630 \text{ barrels}$$
$$\text{Shell method} = \text{pad volume}\% = \frac{1 - 0.7}{1 + 0.7} \times 100 = 17.6\% \text{pad volume}$$
$$\text{Max pad volume} = 7000 \times 17.6\% = 1232 \text{ barrels}$$

In the example above, a minimum of 9% pad volume will be needed for the job. Another important parameter to keep in mind during the frac job treatment is the pressure chart. If halfway or closer to the end of the frac job, surface-treating pressure starts rising, it could be an indication of high fluid leak-off and losing the pad. In this particular scenario, a minipad or extended sweep can be pumped in the middle of the stage to clear the near wellbore of sand accumulation, create more room for getting back into the stage by pumping more sand, and place the existing sand farther into the formation.

A sweep is essentially when sand is cut and only water and chemicals are pumped downhole as the sand starts packing off. Usually, after a hole casing volume sweep, surface-treating pressure begins to decline, which can be an indication of sand accumulation being swept away from the near wellbore area. On the other hand, an *extended sweep*, also referred to as a *minipad*, is when more than one hole casing volume is pumped until the surface-treating pressure shows some relief by having a downward pressure trend. Sweeps can be very common in some areas, especially when pumping high volumes of sand. Some sweeps are scheduled in the design, while others are only run as needed. Cutting sand on time and running sweeps as needed is strongly recommended to be able to get back to the stage and put all the designed sand away. If the formation gives out (this can be easily seen on the surface-treating pressure chart) and sweeps are not run, screening out can be the consequence. Experienced frac engineers and consultants are not afraid to cut sand and run a sweep as needed throughout any frac stage.

Proppant stage

After pumping the calculated pad volume the proppant stage can be started. The proppant stage is the stage during which combinations of proppant, water, and chemicals (called slurry) are pumped downhole. In a slick water frac, it is very important to establish enough flow rate before starting sand. As previously discussed, when using slick water fluid system, rate is the primary mechanism for placing the sand into the formation. If enough rate is not achieved (at least 35 bpm), the proppant stage should not be started because it might result in sanding off the wellbore. Sometimes small concentrations of proppant slugs (such as 500–1000 lb of 0.1–0.25 ppg) are pumped downhole to make sure the formation is able to take in the introduced sand slug before starting the actual pump schedule. The sand stage typically starts with 0.1–0.25 ppg and is gradually increased to higher sand concentrations in a slick water frac. It is important to make sure the current sand concentration hits the perforations before staging up to the next sand concentration to make sure the formation tolerates the amount of sand concentration. For example, if 1.5 ppg of the sand stage is being pumped downhole it is crucial to let it reach the perforations before staging up to 1.75 or 2.00 ppg of the sand stage.

The entire casing volume capacity (of slurry fluid) must be used for the sand to hit the perforations. This maximum casing volume is calculated to discern when the sand will hit the perforations. In the field operation, there is a common saying: "did sand hit the bottom?" This question is asking whether the sand has reached the perforations. In a slick water frac, typically 0.25 ppg jumps are taken to increase sand concentration. However, with more aggressive schedules, 0.5 ppg jumps are attempted as well. It is very crucial to start up sand at very low concentration, such as 0.1 or 0.25 ppg, to erode the perforations in a slick water frac. Starting with higher sand concentrations such as 1 ppg can cause the packing off of all of the perforations and as a result screening out in a slick water frac. A frac stage is very similar to an individual starting to run. Typically, the individual stretches before running and starts at very low speed and gradually gets up to speed. Frac stage follows the same pattern in that it starts with low sand concentrations and gradually stages up throughout the stage.

Flush stage

After pumping the designed pump schedule, proppant is cut and the well is flushed. Flushing means water and chemicals are only pumped downhole to clear the inside of the production casing of sand until all of the remaining proppant in the casing has been removed/flushed to the formation. Flush volume can be calculated given the casing size, grade, weight, and bottom perforation. The rule of thumb is to pump at least one hole casing volume of water and chemicals to the bottom perforation depth *after* all the surface lines are cleared of sand. There is a densometer (reads sand concentration) at the end of the surface lines and before the entrance to the wellhead. This densometer indicates 0 ppg when all the surface lines are clear of sand. As soon as the densometer shows 0 ppg, the flush count starts. The flush stage is very important to pay attention to because after cutting sand, the hydrostatic pressure increases (due to losing slurry hydrostatic pressure) and pressure needs to be monitored to make sure the maximum allowable pressure is not exceeded. The flush volume is calculated using Eq. (5.7).

$$\text{Flush volume} = \text{Casing capacity} \times \text{bottom perforation measured depth} \quad (5.7)$$

where casing capacity is BBL/ft and bottom perforation MD is ft.

Casing capacity can also be calculated using Eq. (5.8).

$$\text{Casing capacity} = \frac{ID^2}{1029.4} \quad (5.8)$$

where ID is inside diameter of production casing, ft.

Example

Calculate the flush volume if a 5½ in., 20 lb/ft, P-110 (ID = 4.778 in.) production casing is used and the bottom perforation of the stage is located at 12,650 ft.

$$\text{Casing capacity} = \frac{4.778^2}{1029.4} = 0.0222 \, \text{BBL/ft}$$

Please note that casing capacity can be found from any casing table, which can be found in any service company's standard handbook.

$$\text{Flush volume} = 0.0222 \times 12650 = 280 \, \text{barrels}$$

Therefore, 280 BBLs are needed to flush the well to the bottom perf after the densometer reads 0 ppg on the surface lines.

Some operators flush 10–40 BBLs over the bottom perf flush volume (overflush) just to make sure the wellbore is completely clear of any sand. This is just a safety precaution taken by some of the operators to make sure that during a plug-and-perf completion technique, the composite bridge plug and perforation guns can be pumped downhole without any issues. If the wellbore is not fully clear of proppant near the perforations, the settled proppant can sand off some of the perforations, and pressuring out while pumping down composite bridge plug and peroration guns can be the

Fig. 5.7 Densometer.

consequence. Overflushing is basically a taboo in vertical wells because by overflushing the well, the sand that was placed near the wellbore will be swept away, which can cause lower conductivity near the wellbore, affecting the production. As discussed earlier, the industry standard practice in multistage hydraulic fracturing in plug and perf technique is to over flush by 10–40 BBL in horizontal wells (sometimes more depending on the operator). This practice has raised concerns about changing the near wellbore conductivity and as a result, loss in productivity. Despite this controversial practice, due to satisfactory production and economic results from various shale plays across the United States, this practice is being continued. More experimental and numerical studies must be performed to truly understand the impact of over flushing on horizontal wells with different frac fluid systems and formation properties. Besler et al. (2007) raised concern about over flushing in Bakken Shale when using cross-linked gel fluid system in the transverse fracture system. Gijtenbeek et al. (2012) concluded that over flushing in slick water frac fluid system might not be detrimental to production because of poor proppant transport. In addition, formation properties such as brittleness have a large impact on how over flushing affects production results. But it is still recommended not to overflush in horizontal wells as there could be loss in near wellbore conductivity. The impact of over-flushing has not been thoroughly studied in horizontal wells. Fig. 5.7 shows an example of a densometer used in one of the slick water frac jobs.

Frac fluid selection summary

Frac fluid selection is one of the most challenging aspects of a hydraulic frac design. A comprehensive understanding of formation properties such as Young's modulus, Poisson's ratio, and formation permeability is essential in designing a proper frac fluid system. There are advantages and disadvantages with each frac fluid system and there is not a perfect fluid system out there. Frac fluid selection alters frac geometry, the formation damage, the cleanup, and the ultimate cost of the fracturing treatment.

CHAPTER SIX

Proppant characteristics and application design

Introduction

Proppant is used to keep the fractures open after the frac job is complete. Proppant provides a high-conductivity pathway for hydrocarbons to flow from the reservoir to the well. After the frac job is completed, proppant prevents the fractures from closing due to overburden pressure. However, unpropped areas will reclose under the overburden pressure and lose their conductivity with time.

One of the most important factors in every frac job is the type of proppant used for the job. Without proppant in the formation, the formation will reclose under the overburden pressure. Pumping only water without proppant downhole might result in good initial production (IP); however, the production will decrease dramatically and the well will not be economical in the long run due to the absence of proppant to keep the fractures open. There are various types of proppant used in hydraulic fracturing; they are discussed in the following sections.

Sand

Sand is the lowest-strength proppant and is highly available and reasonably priced (it is the cheapest). Sand can typically handle closure pressure of up to 6000 psi (closure pressure is the pressure at which the fractures close). Two of the major sands used in hydraulic fracturing are known as *Ottawa* and *Brady* sands. Ottawa sand (also known as Jordan, White, and Northern) is the type of proppant used in many shale plays across the United States and it comes from the northern United States (Jordan deposits). This type of proppant is high-quality white-colored sand with monocrystalline grains. On the other hand, Brady sand, which comes from near Brady, Texas and mined from the Hickory formation outcrops, is also high-quality sand used for hydraulic fracturing. This type of sand is called "brown sand" because of its color and it is typically cheaper than Ottawa sand due to containing more impurities and having a more angular form than Ottawa sand. The quality of Brady sand is lower compared to Ottawa sand. The specific gravity of sand is typically 2.65.

Precured resin-coated sand

Resin-coated sand is considered to be an intermediate-strength proppant. Resin-coated sand is more expensive than regular sand and, therefore, economic analysis must be performed to determine the economic viability of using this type of sand. The first type of resin-coated sand is called precured resin-coated sand (PRCS). PRCS has a hard coating around the sand grains, which causes this sand to have higher conductivity as compared to uncoated sand. This type of sand is used in formations with a closure stress of between 6000 and 8000 psi. Resin-coated sand is designed to encapsulate fines, but will not bond in fractures. It is believed that this type of sand prevents the migration of crushed fines. Sand fines are created after the closure pressure is applied on the sand.

The cost of resin-coated sand could potentially be one of the primary reasons when not utilized in formations with closure pressures of more than 6000 psi. Hydraulic fracturing is not just about pumping any type of sand downhole based on the closure pressure, but it is also about cost per stage and evaluating the economic aspects of the frac job. It is very important to understand both design theory and economic evaluation of the design.

Curable resin-coated sand

Curable resin-coated sand (CRCS) has very similar properties to PRCS. One of the main applications for this type of sand is controlling the flowback. If, after the frac job and during the flowback period, a large amount of the sand that was pumped downhole travels back (i.e., flows back) to the surface, CRCS is pumped (tailed in) at the end of each frac stage to mitigate this problem. This type of sand will bind in the fractures (under closure pressure) preventing flowback of the sand to the surface after the frac job is over. In addition, this type of sand, just like PRCS, typically has a crush resistance of 6000–8000 psi. Fig. 6.1 shows curable resin-coated proppant at standard conditions while Fig. 6.2 shows the same proppant bonded under

Fig. 6.1 Curable resin-coated proppant at standard conditions.

Fig. 6.2 Curable resin-coated proppant under reservoir conditions.

reservoir conditions. Yuyi et al. (2016) experimentally tested the impact of each proppant type (sand, resin-coated, and ceramic proppant) in three deep dry Utica wells located on the same pad in order to make an economic decision on the type of sand to be used on the future pads. They concluded that based on the 2016 market conditions, about 13% and 26% uplift in EUR (from the base case) are needed to justify the incremental Capex associated with pumping resin-coated and ceramic sand, respectively. Therefore, performing such experimental testing and analysis is crucial in making an important decision for creating long-term value for the shareholders.

Intermediate-strength ceramic proppant

The next type of proppant, which is the best-quality proppant and has a higher quality than resin-coated sand, is called ceramic proppant. This type of proppant has uniform size and shape and is thermally resistant. An example of an intermediate-strength proppant is low-density fused ceramic proppant. Intermediate-strength proppant can withstand closure pressure of between 8000 and 12,000 psi. The specific gravity of intermediate-strength proppant is 2.9–3.3 (could be lower depending on the manufacturer and this variance is due to raw material sources used by different proppant manufacturers) (Economides and Martin, 2007). Ceramic proppant has a very high crush resistance. Ceramic proppant has a crush resistance that is so high that if you pour some of the sand on a flat table and beat it with a hammer as hard as you would like, the proppant will not crush and will disperse over the flat area. This demonstrates the high crush resistance of this type of proppant.

Lightweight ceramic proppant

Lightweight ceramic proppant (LWC) is not as strong as intermediate-strength proppant. This type of proppant can withstand closure pressure of 6000–10,000 psi (Economides and Martin, 2007). The specific gravity of LWC is typically 2.72 and can be as close as the specific gravity of regular sand. This type of sand provides better conductivity because of better sphericity and sieve distribution (to be discussed). Lightweight proppant also has uniform size and shape and is thermally resistant.

High-strength proppant

An example of a high-strength proppant is high-strength sintered bauxite, which is the strongest type of proppant used in the industry. It can handle a closure pressure of up to 20,000 psi and is used in deep high-pressured formations where closure pressure exceeds 10,000 psi. This type of proppant has corundum, which is one of the hardest materials known and is used in high-pressure and high-temperature environments. High-strength and intermediate-strength sintered bauxite are produced

Fig. 6.3 Ceramic proppant.

using the same manufacturing process. The main difference between the two is the raw materials used. Intermediate-strength bauxite can typically handle closure pressure of 15,000 psi while high-strength bauxite can handle closure pressure of up to 20,000 psi. Sintered bauxite typically has a specific gravity of 3.4 or greater. Fig. 6.3 shows an example of intermediate strength ceramic proppant.

Table 6.1 is the summary breakdown of the three main types of proppant that are discussed.

Table 6.1 Proppant comparisons

Regular sand	Resin-coated sand	Ceramic proppant
Cheapest	More expensive (compared to regular sand)	Most expensive
Lowest conductivity	Medium conductivity	Highest conductivity
Lowest strength	Medium strength	Highest strength
Irregular size and shape	Irregular size and shape	Uniform size and shape
Naturally occurring product	Manufactured product	Engineered and manufactured product

Proppant size

Now that the concept of proppant type is clear, the next concept that must be discussed is proppant size in unconventional shale reservoirs. There are different proppant sizes that can be used depending on the frac design and production enhancement of each proppant size. The following sizes are the most commonly used in unconventional shale reservoirs.

100 Mesh

100 mesh is very similar to baby powder since the mesh size is very small and is designed to be placed in hairline cracks of the formation. Frac jobs usually start with 100 mesh to seal off microfractures. 100 mesh also effectively decreases leak-off through any encountered cracks. 100 mesh provides a conduit for the upcoming sands by covering small microscopic cracks in the formation and erosion of perforations. Sometimes some engineers consider 100 mesh to be part of the percentage of the pad volume. This type of sand is highly recommended in naturally fractured formations. Although this type of sand is not designed for conductivity it is frequently used for sealing off microfractures, perforation erosion, and obtaining as much surface area as possible by traveling farther into the formation. 100 mesh is typically the smallest sand mesh size used during frac jobs. Many operators in different basins have completely switched to using 100% 100 mesh of the frac stage treatment due to outstanding production performance observed from pumping 100 mesh. This practice and the increase in 100 mesh demand has caused the price of 100 mesh to increase and 40/70 and other mesh sizes to decrease. Fig. 6.4 shows an example of 100 mesh sand size proppant.

40/70 Mesh

40/70 mesh is typically used after 100 mesh (if pumped). 40/70 mesh is larger in size compared to 100 mesh. Pumping this kind of sand downhole creates the required fracture length for maximum surface area and some conductivity in the fractures. Pumping 100% 100 mesh or combinations of 100 mesh and 40/70 are typically the most common sand sizes used in majority of the unconventional shale reservoirs as of the published date of this book. It is a known fact that smaller mesh sizes will have a higher crush resistance as compared to the same type of material in a larger mesh. This is because in a fixed fracture width, there are more grains in that width that are able to support the

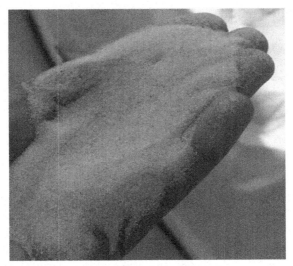

Fig. 6.4 100 mesh sand size.

stress. In other words, the stress is more evenly distributed across more grains of proppant with smaller mesh sizes. Therefore, it is crucial to take this concept into consideration when designing proppant size for any frac job.

30/50 Mesh

30/50 mesh is larger than 40/70 and as a result has greater conductivity providing larger flow paths for multiphase flow. Some companies do not run 40/70. Instead, 30/50 mesh proppant is pumped right after 100 mesh for better fracture conductivity near the wellbore especially in liquid-rich and oil windows. Others prefer to run 30/50 after 40/70 for a better transition after 100 mesh. 30/50 mesh is recommended to be tested in liquid-rich areas (high BTU). This is because of the multiphase flow effect (to be discussed). Some operators do not believe in pumping 30/50 after 100 mesh because 30/50 mesh does not travel as far as 100 mesh or 40/70 mesh into the formation because of its larger size. Stokes' law states that the distribution of proppant inside the fracture depends on its settling velocity in the fracturing fluids. In addition, some operators do not prefer using 30/50 or 20/40 mesh size proppants due to operational issues such as screening out (sanding off) at higher sand concentrations using bigger mesh sizes. Therefore, it is important to perform a risk/reward analysis to see whether the operational risk of pumping bigger sand sizes (if any) is worth the production uplift (if any) or not.

It can be determined that smaller sand particles penetrate deeper into the formation as compared to bigger sand particles. As proppant diameter increases, single-particle settling velocity increases as well. Therefore, 40/70 mesh (with smaller sand particle size) will penetrate more into the formation compared to 30/50 mesh (bigger sand particle). Some operators tail in 30/50 to achieve better conductivity in both dry and liquid-rich areas near the wellbore. Ultimately the final decision on what sand size to use must come from production data and success in each area. If production performance of the wells that only pumped 100 mesh and 40/70 is better than the production performance of the wells that used 100 mesh and 30/50 in the same geologic area, 100 mesh and 40/70 needs to be used on future wells and vice versa. In summary, it is important to have the best design based on theory and simulation (to be discussed); however, at the end of the day the sand size must be justified by existing production data to design a successful frac job. In Chapter 14 of this book, sand size can be determined based on the leak-off regime obtained from before closure analysis such as G-function or square root plots.

20/40 Mesh

20/40 mesh is typically the largest sand size used as compared to all the other sizes discussed thus far. Some operators tail in 20/40 mesh for maximizing near-wellbore conductivity. Some operators do not even run 20/40 mesh, and 100 mesh, 40/70 or 30/50 mesh is the last sand size pumped downhole. Production performance must ultimately be the deciding factor on what sand size to use in each area.

Depending on the frac job formation, design, and well spacing each frac stage requires between 200,000 and 700,000 lb of sand. If the design schedule for a stage is 400,000 lb of sand, the following are some example designs:

Example design # 1 (400,000 lb of sand/stage):
- 50,000 lb of 100 mesh
- 200,000 lb of 30/50 mesh
- 150,000 lb of 20/40 mesh

Example design #2 (400,000 lb of sand/stage):
- 120,000 lb of 100 mesh
- 230,000 lb of 40/70 mesh
- 50,000 lb of 30/50 mesh

Example design #3 (400,000 lb of sand/stage):
- 70,000 lb of 100 mesh
- 330,000 lb of 40/70 mesh

Example design #4 (500,000 lb of sand/stage):
- 500,000 lbs of 100 mesh (100% 100 mesh)

The above designs are just examples underscoring the fact that sand combinations can vary greatly depending on the design, production performance, and economics. The type(s) of sand needed for well optimization is debatable among operators, each having preferred recipes for achieving optimal production. There are different hydraulic frac software programs used to run various models to come up with the optimal sand size, type, and volume for the hydraulic frac design.

Proppant characteristics

It is important to have a basic knowledge of proppant characteristics and why some proppant types such as resin and ceramic are much more expensive as compared to regular sand. Some characteristics of proppant that are important to understand and monitor are roundness, sphericity, crush resistance, specific gravity, bulk density, acid solubility, sieve size, silt and fine particles, and clustering.

Roundness is the measure of relative sharpness of the grain corners. Improving proppant roundness results in more even stress distribution and potentially improves proppant pack porosity. *Sphericity* is the measure of how round an object is or how closely the grain approaches the shape of a sphere. The American Petroleum Institute (API)-recommended limit for sand in both roundness and sphericity is 0.6 or greater. Fig. 6.5 shows physical roundness and sphericity from Krumbein and Sloss (1963).

Crush resistance measures the fines created under a given load (exposure to stress). This can be performed in the lab by applying various stresses such as 3000, 4000, 5000 psi, etc. API recommends various percentages of generated fines for different types of sands. K-value testing is an important test that can be performed on various proppant types and sizes in order to understand the % fines generated under each specified stress. K-value is the closure stress (rounded down) under which 10% of the proppant will crush and become fines or out of the standard mesh size. To test the quality of the proppant, it is highly recommended to take a sample from one of the sand haulers onsite and send it to a renowned proppant testing company for K-value and other standard testing. Please note that this type of testing should not be performed by the sand supplier in order to maintain the integrity of the test. The standard API crush-testing procedure typically calls for a loading of $4\,lb/ft^2$ in the crush-testing apparatus. However, it is very difficult to obtain such loading

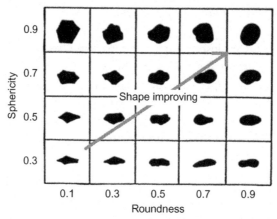

Fig. 6.5 Visual estimation of roundness and sphericity (Krumbein and Sloss, 1963). *(Modified from Saaid, I.M., Kamat, D., Muhammad, S., 2011. Characterization of Malaysia sand for possible use as proppant. Am. Int. J. Contemp. Res. 1(1), 37.)*

in a slick water frac fluid system. Therefore, it is very important to perform this crush testing at various sand loadings such as using the average fracture width near or away from the wellbore to obtain a more realistic view of crushing effect.

Specific gravity is the measurement of absolute density of individual proppant divided by the absolute density of water. The API-recommended maximum specific gravity is 2.65 for sand.

Bulk density is the volume occupied by a given mass of proppant. The API-recommended maximum for proppant is 105 lb/ft^3.

Acid solubility is the solubility of proppant in 12% HCl or 3% HF acids. Acid solubility indicates the amount of contaminants present in the proppant, in addition to relative stability of proppant in acid. The API-recommended maximum acid solubility is 2% for larger sand (30/50 mesh) and 3% for smaller sand (40/70 mesh).

Sieve analysis is a necessary test performed on the proppant throughout the frac job to ensure proper proppant size and quality control of the proppant. It indicates the size distribution of the proppant within the defined proppant size range. In this analysis, which is typically performed by the sand coordinator, 90% of the sample should fall within the designated sieve size. Not more than 0.1% should be larger than the first sieve size and not more than 1% should be smaller than the last sieve size. For example in Table 6.2, if 40/70 mesh is being tested, not more than 0.1% of the sample size test should

Table 6.2 Standard sieve openings (Ely, 2012)

US series mesh	Sieve opening (in.)	US series mesh	Sieve opening (in.)
4	0.187	25	0.0280
6	0.132	30	0.0232
8	0.0937	35	0.0197
10	0.0787	40	0.0165
12	0.0661	60	0.0098
14	0.0555	70	0.0083
16	0.0469	100	0.0059
18	0.0394	170	0.0035
20	0.0331		

Fig. 6.6 Test Sieve shaker.

be larger than 0.0165, and not more than 1% should be smaller than 0.0083. The operating company representative is responsible for verifying the properly tested sieve analysis throughout the frac treatment.

Silt and fine particles measure the amount of silt, clay, and other fine materials (impurities) present in the sample. The API recommendation for silt and fine particles is 250 FTU (formation turbidity unit) or less. Fig. 6.6 shows an example of test sieve shaker used in the laboratory to find proppant size distribution. In addition to laboratory testing, this type of test can also be easily performed in the field.

Clustering measures the degree of attachment of individual proppant grains to one another. The API-recommended maximum for clustering, which is measured by percentage weight, is 1%. One of the main reasons for this type of test is that during processing the grains were not broken apart (Ely, 2012).

Proppant particle-size distributions

The max-to-min ratio in the majority of API sieve designations is approximately 2 to 1. For example, a 20 mesh particle is roughly twice the diameter of a 40 mesh particle as can be seen in Table 6.2. A 20 mesh particle has a diameter of 0.0331 in compared to a 40 mesh particle with a diameter of 0.0165 in. (~half of 0.0331 in.). Table 6.2 shows different US sieve and their opening sizes.

Proppant transport and distribution in hydraulic fracture

During hydraulic fracturing, different proppant concentrations are pumped based on the initial frac design and to the extent that reservoir formation permits. The pumped proppants move in both horizontal and vertical directions. In the horizontal direction, proppant follows the fracture tip with the same velocity as fracturing fluid. However, in the vertical direction the proppant velocity, that is, settling velocity, is different than fluid vertical velocity due to gravitational forces and slippage between proppant particles and fluid. Proppant movement in the direction of the fracture width is commonly neglected due to scale effect (fracture width is much smaller than fracture length and height). As proppant particles settle, they fill up the fracture width and, therefore, increase the proppant concentration in the vertical cross section. There is a critical proppant concentration beyond which screening out (sanding off) occurs. Rate of proppant bank growth or screening out is a function of proppant settling velocity. Settling velocity of a single and perfectly spherical proppant particle can be obtained using Stokes' law assuming infinitely large fracture (boundary effects are neglected). Settling velocity is derived for different flow regimes based on the dimensionless Reynolds number.

If the Reynolds number is less than 2, proppant settling velocity can be obtained using Eq. (6.1).

$$\text{Proppant settling velocity Re} \leq 2.0: V_{ps} = \frac{g(\rho_p - \rho_f)d_p^2}{18\mu} \quad (6.1)$$

If the Reynolds number falls between 2 and 500, proppant velocity can be obtained using Eq. (6.2).

$$\text{Proppant settling velocity } (2 < \text{Re} < 500): V_{ps}$$
$$= \frac{20.34(\rho_p - \rho_f)^{0.71} d_p^{1.14}}{\rho_f^{0.29} \mu^{0.43}} \quad (6.2)$$

For flow regimes with high Reynolds number (i.e., >500), Eq. (6.3) will be used.

$$\text{Proppant settling velocity } (\text{Re} \geq 500): V_{ps} = 1.74\sqrt{\frac{g(\rho_p - \rho_f)d_p}{\rho_f}} \quad (6.3)$$

In Eq. (6.3), ρ_p and ρ_f stand for proppant and fracturing fluid density, μ is the fluid dynamic viscosity, d_p is proppant diameter, and V_{ps} is the uncorrected proppant settling velocity. As mentioned earlier, proppant velocity obtained using Stokes' law neglects the boundary (fracture width) effect by assuming an infinitely large fracture. It also ignores interactions between proppant particles, since it has been developed for a single particle. Gadde et al. (2004) defined a correlation to correct the proppant settling velocity for these two factors as follows:

Corrected proppant settling velocity : V'_{ps}

$$= V_{ps}\left[0.563\left(\frac{d_p}{w}\right)^2 - 1.563\left(\frac{d_p}{w}\right) + 1\right](2.37c^2 - 3.08c + 1) \quad (6.4)$$

In Eq. (6.4), V_{ps}' is the corrected proppant settling velocity and c is the proppant concentration. As proppant settling occurs, the frac fluid viscosity will change. The change in frac fluid viscosity as a function of proppant concentration can be obtained using Eq. (6.5).

Frac fluid viscosity : μ

$$= \mu_0 \left\{ 1 + \left[0.75 \left(e^{1.5n} - 1 \right) e^{-\frac{\gamma(1-n)}{1000}} \right] \frac{1.25c}{1 - 1.5c} \right\}^2 \quad (6.5)$$

In Eq. (6.5), μ_0 is uncorrected fluid viscosity to proppant concentration and n and γ are the non-Newtonian fluid constants.

Kong et al. (2015) investigated the effect of proppant settling velocity on proppant distribution and fracture conductivity in the Marcellus Shale reservoir and showed that ignoring proppant settling velocity could lead to more than 18% overestimation in dimensionless productivity index. They showed that in tighter formations and using larger proppant size the overestimation in dimensionless productivity index can be as large as 32%. A more realistic prediction of proppant distribution in hydraulic fractures can significantly help operators to design the optimum frac job. In ultra-low permeability formations such as shale with permeability less than 1 µD, there is a critical proppant size that can lead to the highest hydraulic fracturing efficiency, as shown in Fig. 6.7. In the hydraulic fracturing process, multisize proppant combinations are injected into the wellbore occasionally. Usually a smaller-size proppant is injected first, followed by a larger-size proppant. In ultra-low permeability formations such as shales, there is a critical combination of small and large proppant sizes that will result in maximum well productivity index as shown in Fig. 6.8.

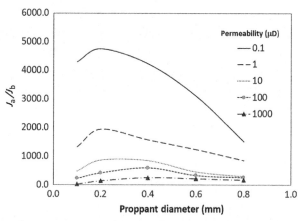

Fig. 6.7 Effect of proppant size on dimensionless productivity index for different reservoir permeability. *(Modified from Kong, B., Fathi, E., Ameri, S., 2015. Coupled 3-D numerical simulation of proppant distribution and hydraulic fracturing performance optimization in Marcellus shale reservoirs. Int. J. Coal Geol. 147–148, 35–45.)*

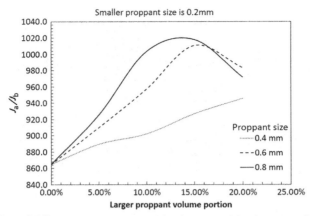

Fig. 6.8 Effect of different proppant size and volume combination on well dimensionless productivity index. *(Modified from Kong, B., Fathi, E., Ameri, S., 2015. Coupled 3-D numerical simulation of proppant distribution and hydraulic fracturing performance optimization in Marcellus shale reservoirs. Int. J. Coal Geol. 147–148, 35–45.)*

Fracture conductivity

Fracture conductivity is one of the most important concepts in hydraulic fracturing, and thus is considered in every design. Conductivity is essentially the multiplication of fracture width (ft) and proppant permeability inside the fracture (md). Proppant permeability and conductivity change under different stresses. For example, the permeability of 20/40 mesh (and ultimately conductivity) under 6000 psi of closure pressure is different compared to 10,000 psi. Conductivity is also referred to as *flowback capacity* and its unit is md-ft. Conductivity is the ability of the fractures to transmit reservoir fluid to the wellbore. As closure pressure increases, conductivity decreases. Proppant suppliers typically provide a chart for each proppant type that shows fracture conductivity on the *y*-axis vs closure pressure on the *x*-axis.

Factors that affect fracture width are proppant density, proppant loading, gel filter cake, and embedment. Also, factors that affect proppant permeability are typically proppant size, sphericity, strength, fines, and gel damage. Fig. 6.9 shows the schematic representation of a fracture and fracture width that will be used in fracture conductivity calculation.

$$\text{Fracture conductivity} = k_f \times W_f \quad (6.6)$$

where K_f is the proppant permeability, md and W_f is the fracture width, ft.

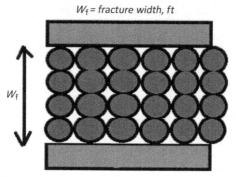

Fig. 6.9 Fracture width.

Dimensionless fracture conductivity

Dimensionless fracture conductivity is the ability of fractures to transmit reservoir fluid to the wellbore divided by the ability of the formation to transmit fluid to the fractures. Dimensionless fracture conductivity is denoted in F_{CD} and is defined as

$$\text{Dimensionless fracture conductivity}: F_{CD} = \frac{K_f \times W_f}{K \times X_f} \quad (6.7)$$

where K_f is the fracture permeability in the formation (md), W_f is the fracture width (ft), K is the formation (matrix) permeability (md), and X_f is the fracture half-length (ft).

Fig. 6.10 shows two-stage hydraulic fracturing and reservoir-stimulated volume characteristics used to calculate dimensionless fracture conductivity. Fig. 6.11 illustrates qualitative comparisons of fracture conductivity and closure pressure for different proppant types.

International Organization for Standardization conductivity test

The fracture conductivity testing that is performed for each proppant type and size is typically performed under the following conditions:
- Ohio sandstone
- 2 lb/ft² proppant loading
- Stress maintained for 50 h
- 150–200°F
- Extremely low water velocity (2 mL/min) (2% KCl water)

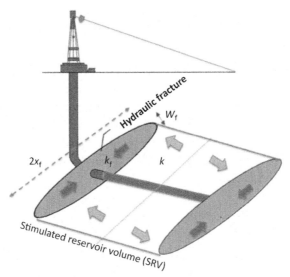

Fig. 6.10 Matrix and hydraulic fracture interactions.

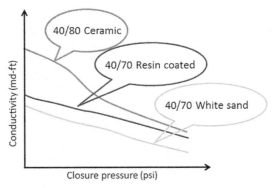

Fig. 6.11 Fracture conductivity testing.

This conductivity test accounts for proppant size, proppant strength/crush profile, some embedment, some temperature effects, and wet system. However, this conductivity test does not account for the following:
- (a) Non-Darcy flow
- (b) Multiphase flow
- (c) Reduced proppant concentration
- (d) Gel damage
- (e) Cyclic stress
- (f) Fines migration
- (g) Time degradation

Non-Darcy flow

As opposed to Darcy's law, which assumes laminar flow in the formation, non-Darcy flow is referred to as fluid flow, which deviates from Darcy flow by having a turbulent flow in the formation and especially near the wellbore. Non-Darcy flow is very common in high-rate gas wells near the wellbore. Therefore, some operators like the idea of tailing in higher-conductivity proppant at the end of each frac stage to accommodate for the non-Darcy flow effect near the wellbore.

Multiphase flow

Hydraulic fracturing usually encompasses flow of liquid (water, oil, and condensate) and gas. Fluid flow in hydraulic fracturing highly depends on the relative permeability of the formation to each of these phases. As Fig. 6.12 illustrates, by increasing the saturation of one phase, the relative permeability to that phase increases while the relative permeability to the other phase decreases. Therefore, proppant saturated with liquid is less conducive to flowing gas. This effect is not taken into account in International Organization for Standardization (ISO) conductivity testing. The importance of relative permeability comes into play in high-BTU gas reservoirs (primarily retrograde condensate reservoirs) and oil reservoirs where fluid exists as liquid at reservoir conditions. Liquid tends to accumulate in the

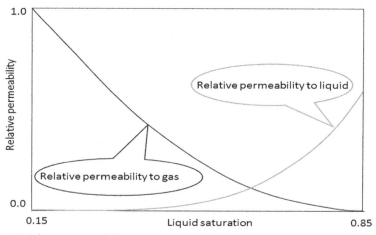

Fig. 6.12 Relative permeability curve.

fractures. This will occupy porosity that is not available for gas flow. In wet areas, higher conductivity sand is sometimes pumped near the wellbore to accommodate the relative permeability effect.

Reduced proppant concentration

An ISO conductivity test is performed using $2\,lb/ft^2$ proppant loading, which can be misleading. Proppant concentration at formation conditions is typically less than $1\,lb/ft^2$. For example, if fracture conductivity for regular sand such as 40/70 mesh under 6000 psi of closure pressure is 400 md-ft at $2\,lb/ft^2$, the conductivity at formation (which is $1\,lb/ft^2$ or less) will be much less than 400 md-ft, roughly 200 md-ft. Therefore, proppant concentration in the formation has been reduced down to 200 md-ft, as opposed to the reported 400 md-ft when the conductivity test is performed.

Gel damage

Gel damage often occurs in cross-linked jobs, where heavy viscous fluid is used during the frac job. The residual gel effect can have detrimental impact on the fracture conductivity even after using gel breaker. It is important to note that breaker loading can significantly improve the cleanup of distributed gel in the formation. In a slick water frac, gel damage is not as common unless a high concentration of linear gel was used to facilitate placing proppant into the formation. One example of gel damage is fracture face damage, which is caused by filtrate leaking into the rock. Another example of gel damage is the accumulation of residual gel. Gel residue tends to accumulate in very narrow pore throats ultimately affecting flow capacity of the fluid flow. Gel damage can also cause a reduction in effective fracture width due to filter cake buildup. Filter cake forms as frac fluid slurry leaks off into the formation. Filter cake is forced out into the fractures upon closure. Fracture width retention is extremely important to minimize gas velocities in the fracture. Decreasing gas velocity will significantly reduce pressure drop as stated in the Forchheimer equation. Gel is a non-Newtonian fluid. As opposed to Newtonian fluid where the relationship between shear stress and shear rate is linear, in non-Newtonian fluid the relationship between shear stress and shear rate is different and can be time dependent. Another form of detrimental gel damage is referred to as loss of effective fracture length due to gel plug-in tip. Since gel is a non-Newtonian fluid, it must achieve some pressure differential before being able to move. This feature of gel will cause reduction in effective fracture length by plugging the tip of the fracture. Laboratory studies suggest that proppant with better roundness, sphericity, porosity, and permeability facilitate the cleanup of gel as opposed to other types of proppant.

Cyclic stress

When producing a well, there is a pressure referred to as flowing bottom-hole pressure. Flowing bottom-hole pressure is the bottom-hole pressure inside the wellbore at flowing condition denoted as P_{wf}. This pressure is extremely important as proppant stress (stress placed on the proppant) is a function of closure pressure and flowing bottom-hole pressure. Every time flowing bottom-hole pressure changes, the proppant distribution inside the fracture rearranges and some conductivity loss can be the consequence. Fig. 6.13 shows the ideal designed proppant placement and real proppant placement in a hydraulic fracture during the frac job and after the well is in line. In an ideal case, uniform proppant distribution inside the fracture is assumed. After starting to produce from a well, hydraulic fracture width decreases due to an increase in effective stress. However, in reality, proppant distribution will not be uniform due to proppant settling velocity and fluid leak-off to the formation. As a result, there will be propped and unpropped

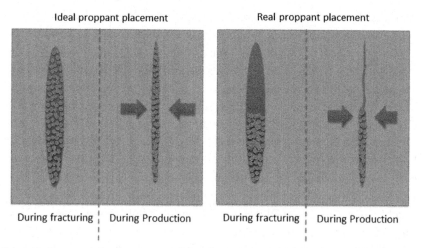

Fig. 6.13 Proppant placements in hydraulic fracturing. *(Modified from Kong, B., Fathi, E., Ameri, S., 2015. Coupled 3-D numerical simulation of proppant distribution and hydraulic fracturing performance optimization in Marcellus shale reservoirs. Int. J. Coal Geol. 147–148, 35–45.)*

regions in a hydraulic fracture. The unpropped region will be closed and fracture width in propped region will decrease due to an increase in effective stress during production. This leads to reduction in fracture conductivity.

Fines migration

Under closure pressure at downhole conditions, proppant will generate some fines (depending on the type of proppant used), which will reduce conductivity. This will highly depend on the rate of change in effective stress, which is a function of operational conditions. Unfortunately, there is a tendency to achieve high IP when the well first starts producing (initial flowback) in order to impress the investor community. Very high IP can only be achieved through very aggressive flowing bottom-hole pressure drawdown. The practice of "pulling hard" or "rip it and grip it" is common in a lot of the unconventional shale plays. This practice leads to an excessive amount of stress on proppant and as a result some proppant embedment and fines migration that will lead to loss in conductivity. Loss in conductivity is equal to loss in well productivity and ultimately revenue. Intermediate initial flow rates will lead to moderate pressure drawdown, which will result in less damage to the fracture conductivity as shown in Fig. 6.14. Eq. (6.8) shows the importance of minimizing flowing bottom-hole pressure. Minimizing flowing bottom-hole pressure can be achieved by minimizing pressure drawdown when producing the well. Belyadi et al. (2016a,b) used actual field data from eight studied wells in Utica/Point Pleasant and illustrated that

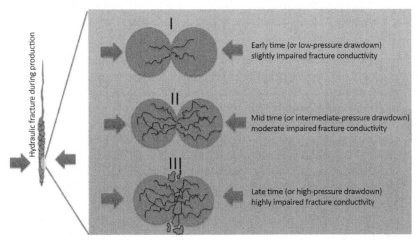

Fig. 6.14 Proppant crushing embedment.

up to 30% improvement in EUR can be achieved using a managed pressure drawdown of 15–20 psi/D casing or tubing pressure drop. In addition, they also showed that reservoir damage was most likely caused by pressure dependency of hydraulic fracture conductivity. Optimum economic rate can be determined based on a company's long-term financial metric as well as gas pricing. In essence, aggressive pressure drawdown can damage a well's performance while conservative pressure drawdown can impact the near-term economic value. Therefore, pressure drawdown schedule and optimum economic rate must be determined for each field based on a company's strategic metric and goal.

$$\text{Proppant stress} = P_{closure} - P_{wf} + P_{net} \tag{6.8}$$

where $P_{closure}$ is the closure pressure (psi), P_{wf} is the flowing bottom-hole pressure (psi), and P_{net} is the net pressure, psi.

Example

Calculate proppant stress given the two following conditions assuming net pressure is zero:
- The well is initially producing at 4500 psi flowing bottom-hole pressure and the closure pressure from DFIT is calculated to be around 6500 psi.
- The flowing bottom-hole pressure is drawn down very aggressively to about 1000 psi after the initial flowback over the course of a 2-day time period (assume closure pressure stays constant after 2 days).

Condition 1 : Proppant stress = $P_{closure} - P_{wf} = 6500 - 4500 = 2000$ psi
Condition 2 : Proppant stress = $P_{closure} - P_{wf} = 6500 - 1000 = 5500$ psi

As can be seen in this example, it turns out that proppant stress has increased from the initial 2000 psi to almost 5500 psi in just 2 days. This practice could potentially lower proppant conductivity significantly and could result in a loss in production and ultimately revenue. Therefore, it is important to study the impact of pressure drawdown on production based on the formation of interest. Highly over pressured reservoirs such as Eagleford shale as well as some parts of the deep dry Utica Shale have shown detrimental production performance impact with aggressive pressure drawdown than other formations with reservoir pressure gradient of less than 0.8 psi/ft. Ultimately, if a formation does not prove detrimental to production by aggressive pressure drawdown, it is recommended to produce the wells as fast as possible to maximize the net asset value of the field.

Time degradation

Fracture conductivity will reduce with time. The rule of thumb is that fracture conductivity will be reduced by 75% with time. This behavior, which leads to early production decline in hydraulically fractured shale reservoirs, is not well understood yet. One of the major physical phenomena that have been investigated as the possible source of early production decline is the hydraulic fracture and shale matrix permeability time-dependent characteristics, also known as creep deformation under constant loading. This has mainly been attributed to matrix and hydraulic fracture interaction with fracturing fluid.

Finite vs infinite conductivity

If F_{CD} is greater than 30, it is considered to be *infinite conductivity* and if F_{CD} is less than 30, it is considered to be *finite conductivity*. Cinco-Ley and Samaniego (1981) presented Fig. 6.15 that plots dimensionless fracture conductivity (x-axis) versus effective wellbore radius/fracture half-length (y-axis) to define the concept of dimensionless fracture conductivity.

Example
Assume the dimensionless fracture conductivity for a well in the Permian Shale is calculated to be 50. From Fig. 6.15, the effective wellbore radius can be found as follows:

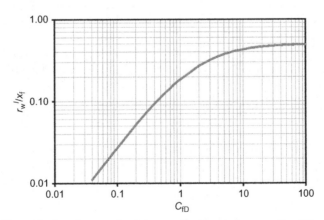

Fig. 6.15 Dimensionless frac conductivity vs effective drainage radius (Cinco-Ley and Samaniego, 1981).

$$\frac{r'_w}{X_f} = 0.5$$

Assuming a calculated fracture half-length of approximately 300 ft and substituting $X_f = 300$ ft,

$$\frac{r'_w}{300} = 0.5 \gg r'_w = 150 \text{ ft}$$

As long as the dimensionless fracture conductivity is greater than 30, the fracture is considered to be in infinite conductivity and the effective drainage radius does not change. For example, if a dimensionless fracture conductivity of 30 can be achieved using 40/70 mesh sand at downhole condition (taking into account all of the factors that the ISO conductivity test does not take into account), does pumping 30/50 mesh really matter? As long as a dimensionless fracture conductivity of 30 is reached (infinite conductivity) at downhole conditions, pumping a larger sand size is not recommended. The hardest part is figuring out the fracture conductivity after all the discussed effects are taken into account. Since the permeability in unconventional shale reservoirs is very low, it is important to note that achieving infinite conductivity is easier. However, with all the factors discussed that can reduce fracture conductivity, infinite conductivity could potentially be very difficult to obtain in some formations depending on many factors such as type of sand, type of fluid system, type of pressure drawdown, type of reservoir fluid, etc.

Example
A horizontal well with low permeability is going to be hydraulically fractured using a slick water fluid system. The matrix permeability of the reservoir is 0.0003 md (300 nd) with an estimated propped fracture half-length of 300 ft. Fracture conductivity under 6000 psi of closure pressure is estimated to be 400 md-ft from the lab ISO conductivity test (2 lb/ft^2). Calculate fracture conductivity at 1 lb/ft^2 and assume 85% reduction in conductivity due to all of the effects discussed. Calculate dimensionless fracture conductivity and specify whether the fractures are considered to be finite or infinite.

$$\text{Fracture conductivity at } 2\frac{\text{lb}}{\text{ft}^2} = K_f \times W_f = 400 \text{ md-ft}$$

$$\text{Fracture conductivity at } 1\frac{\text{lb}}{\text{ft}^2} \text{ and } 85\% \text{ reduction} = \frac{400}{2} \times (1 - 0.85)$$

$$= 30 \text{ md-ft}$$

$$\text{Dimensionless fracture conductivity} = F_{\text{CD}} = \frac{K_f \times W_f}{K \times X_f} = \frac{30}{0.0003 \times 300}$$
$$= 333$$

Since $F_{\text{CD}} > 30 \gg$ infinite fracture conductivity.

As can be seen from this example, since the formation permeability is so low even after taking into account some of the effects that can alter fracture conductivity, fractures are considered to be at infinite conductivity.

CHAPTER SEVEN

Unconventional reservoir development footprints

Introduction

Unconventional reservoir developments encompass activities such as hydraulic fracturing and wastewater deposition in underground reservoirs. These activities introduce man-made stresses that change the in situ stress condition of the underground formations leading to the cases of induced seismicity. The magnitude of induced seismicity is a function of orientation, magnitude, and relative state of the surrounding stress field. Some statistics provided by the US Geological Survey (USGS) suggest exponential growth in the cumulative number of earthquakes in the central and eastern United States since 2005, coinciding with unconventional reservoir developments in these areas. There have also been studies trying to correlate these statistics to hydraulic fracturing, withdrawal, or fluid injection by the oil and gas industry; however, there is no direct evidence and detailed studies that can prove this idea. By taking a closer look at the National Seismic Hazard Map published by the USGS in 2014, one can see that most of the earthquakes higher than a magnitude of 3 occurred in the vicinity of major fault planes that happened to also be very close to the major unconventional shale developments. Having said so, in unconventional reservoir developments, there are many cases of stress field alteration, which may impact the stability of the underground formations, faults, and any discontinuities. This might lead to hydraulic fracture and fault reactivation, or hydraulic fracturing and aquifer interaction. In ultralow-permeability shale reservoirs, which dominate most of the unconventional oil and gas resources in the United States, hydraulic fracturing treatment is absolutely essential to obtain an economic level of production. These hydraulic fracturing activities mainly introduce low-magnitude induced seismicity. These low-magnitude seismic events are used by the oil and gas industry to obtain the geometry of the hydraulic fractures. Often, these low-magnitude seismic events cannot be felt at the surface and will be limited to the treatment zone. However, in extremely rare cases due to unintended interactions between hydraulic and natural fractures, these events might have some impacts at the surface.

Environmental impacts of hydraulic fracturing are not limited to inducing seismicity. Wellbore integrity is also one of the major concerns in the oil and gas industry and is highly regulated by state legislations. Hydraulic fracturing can significantly impact the geomechanical behavior of the wells. These concerns in the oil and gas industry have become a focus of research in areas such as cement behavior and cement bond under confining pressures applied to cement during hydraulic fracturing, and hydraulic fracture pressure communications with old and abundant wells. In some cases during the hydraulic fracturing treatment, the production casing might burst due to exceeding the casing burst pressure, or flaw in the casing manufacture. This opens a great deal of discussion on casing selection and design. Other environmental impacts of unconventional resource developments can be classified in issues related to groundwater protection, wildlife impacts, community impacts, and surface disturbance. On the other hand, unconventional resource development can have an enormous positive social and political impact in terms of providing more jobs, increasing the energy security and sustainability, decreasing the pollution by providing much cleaner energy as compared to conventional developments, and in general increasing the quality of the life in the country.

Casing selection

There are four different types of casings commonly used in the industry for horizontal wells, and they are as follows:
- conductor casing,
- surface casing,
- intermediate casing, and
- production casing.

Conductor casing

The conductor casing is installed prior to the arrival of the drilling rig. This hole is usually 18″–36″ in diameter and 20′–50′ long. This casing basically keeps the top of the hole from caving in and additionally it prevents the collapse of loose soil near the surface. Moreover, it is used to help in the process of circulating the drilling fluid up from the bottom of the well. This casing needs to be either cemented or grouted in place.

Surface casing

After placing and cementing the conductor casing, the next hole size needs to be drilled before placing the surface casing. The next hole size is drilled using a drilling rig to the desired depth, which is usually anywhere between a few hundred feet and 2000 ft. This is the most crucial casing as far as the Environmental Protection Agency (EPA) is concerned since the water source is usually located in that range. As a result, to protect the water source from contamination, the EPA typically requires setting the surface casing and cementing it to at least 50′ deeper than the deepest fresh groundwater zone. In some parts of Pennsylvania, the Department of Environmental Protection (DEP) requires two surface casings to protect the coal seams as well. The main purpose of surface casing is to protect freshwater from contamination. Freshwater contamination can be caused by leaking hydrocarbon or salt water from the producing formation *if and only if* the casing or the cementing operation is not done properly. Please note that this crucial process is highly regulated by the EPA and violators are heavily fined and could be suspended from drilling and completion processes if not in compliance. In addition, the environmental agency (varies depending on the state) has to be notified 24 h before and after the cement job is started and completed to ensure a proper seal between the freshwater zone and the well. Oftentimes (depending on the state), a representative from the state will be present during the cement job to ensure the quality of the cement job and compliance with all laws and regulations. If after cementing the surface casing, cement is not received at surface, the operation cannot continue until a course of action is summarized and submitted to the state for their review and approval in order to make sure the problem is completely resolved before moving forward. Another purpose of surface casing is to make sure the drill hole is not being damaged or collapsed when drilling the next hole section of the well. If proper casing is not placed, the drilled hole could be damaged or even collapsed for many reasons that exist downhole (pressure, temperature, water invasion, etc.). Another important reason for installing the surface casing is to provide primary well control equipment to be rigged up (examples of primary well control equipment are blowout preventers). Surface casing is the first casing to provide the necessary means of installing primary well control equipment. A typical surface casing size is $13\frac{3}{8}″$.

Intermediate casing

After placing the surface casing, cementing it in place, and getting a confirmation from the environmental agency to continue operations, the next section of the hole is drilled. After drilling this section, intermediate casing is placed in the hole for many reasons. The primary reason for using intermediate casing is to minimize the hazards associated with abnormal underground pressure zones or formations that might otherwise contaminate the well, such as underground salt water deposits. This casing is often used in longer laterals so the hydrostatic pressure of the drilling fluid remains between formation and fracture pressures. Even if none of the above conditions are present, this casing is very important as insurance for any type of unexpected abnormal pressure downhole. The intermediate casing size that most of the operating companies use is 9⅝" casing.

Production casing

Finally, after placing the intermediate casing and cementing it all the way to the surface, the next section of the hole is drilled. The kickoff point (KOP) is the point at which the curve section of the wellbore is started and built. KOP is the point at which a well starts to incline using the predetermined engineering plan to get to the desired zone of interest. After building the curve, the landing point is reached. The landing point is the point at which the target formation is reached and from that depth the well can be horizontally drilled to total depth (TD). Once the TD of the well is reached, typically 5½" production casing can be run all the way to the surface and cemented in place. Depending on the formation and design, 4½", 6", or 7" production casings could also be run. A production casing, which is also called "long string," is the deepest section of the casing of a well since it goes from the desired depth all the way to the surface. This casing is basically a conduit from the surface of the well to the actual producing formation. The size of production casings depends on various considerations including lifting equipment to be used, types of completion processes required, and the possibility of deepening the well at a later date. For example, if the well is expected and designed to be deepened at a later time, the production casing should be large enough to allow passage of the drill bit later to drill the next designed section. Fig. 7.1 shows various casing string illustrations.

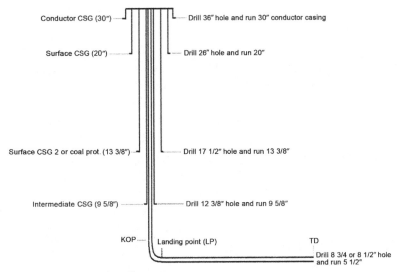

Fig. 7.1 Various casing string illustrations.

Hydraulic fracturing and aquifer interaction

This is a controversial topic amongst environmentalists concerned with the chemicals used in hydraulic fracturing and whether or not they impact drinking water sources. Hydraulic fracturing itself does not cause drinking water contamination. Hydraulic fracturing started in 1947 without a single case of drinking water contamination. The main issue that causes water contamination is bad cement jobs in the casings. After running thousands of feet of casing (steel pipe), there is always a possibility of a microannulus leakage in the casing because of a bad connection between casing joints. However, if there is not a cement bond behind the surface casing, there are oftentimes two, or in some cases three additional casings (coal seam, intermediate, and production casings) that are run and cemented to protect the surface water from any type of contamination. Cementing is a crucial part of the drilling operation. This is why the cementing operation is so highly regulated by the EPA to protect freshwater from any types of contaminates.

When hydraulic fracturing is performed, the fractures created by hydraulic horsepower (HHP) do not extend all the way to the surface. Later on in this chapter, the concept of fracture height is discussed. For example, in Marcellus Shale, the true vertical depth (TVD) of an average well is anywhere between 6500′ and 8000′ (depending on the area) and hydraulic fracturing is performed at that depth. Water sources are located between 50′ and a maximum depth of 1000′. Based on the current frac microseismic data, it is highly unlikely that the fractures created during hydraulic fracturing could grow to a length of 6000′ to upwards of 7000′, thereby contaminating the local drinking water. Fractures are naturally limited due to natural formation barriers, stresses in the rock (vertical, minimum horizontal, and maximum horizontal stresses), leak-off limits, and height growth. If we had no stresses in the earth, the fractures would easily grow to the surface when hydraulic fracturing. Operators have run microseismic in various basins and formations to identify fracture azimuth, fracture height growth, fracture length, fracture width, etc. The frac microseismic data demonstrated that the average height could be up to 1000′. Therefore, the maximum height that fractures can grow based on seismic data is still thousands of feet away from drinking water sources.

One of the main reasons for water contamination is a bad cement job, and a bad cement job nowadays would never be approved by the state's environmental agency. The industry is heavily regulated and careful regarding this matter, as it is a very sensitive subject. Having said so, introducing man-made stresses to the prestressed formation during hydraulic fracturing can cause induced seismicity. This induced seismicity might reactivate faults and discontinuities. The problem with fault reactivation or slippage is that it can extend all the way to the surface as shown in Fig. 7.2. If hydraulic fracturing causes fault reactivation or slippage, the fault will work as a flow path that can transfer frac fluid all the way to the surface. This leads to frac job failure due to huge frac fluid leak-off to the fault and can lead to severe environmental impacts. The industry knows the locations of sensitive faults, and hydraulic fracturing is not performed in the fault zones due to ineffective hydraulic fracture treatment. Therefore, frac stages thought to be in the fault zones are skipped. In addition, areas with severe geologic complexities are usually thought to produce poorly. Therefore, operators avoid those areas simply due to their economic disadvantages. Oil and gas operators have done a fantastic job protecting fresh water zones and eliminating possible hazards that might lead to environmental issues.

Hydraulic fracturing and fault reactivation

During hydraulic fracturing, the in situ stress condition of the reservoir will change. The magnitude of the change in in situ stress is directly related

Unconventional reservoir development footprints

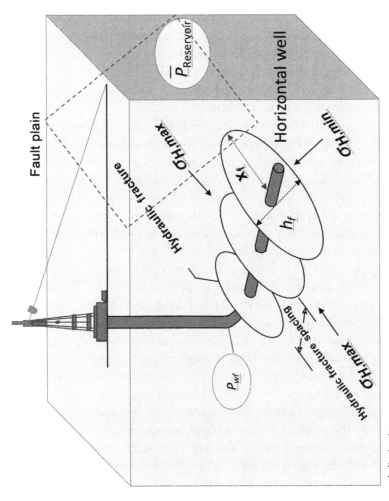

Fig. 7.2 Fracture growth limitation.

to the formation mechanical properties, induced hydraulic gradient through fracture initiation and propagation, and properties of the possible fault or discontinuities. Therefore, faults or any discontinuities in the region affected by stress change due to hydraulic fracturing might be reactivated. Fault slippage or fault motion is directly related to the coefficient of friction or friction factor. Experimental measurement of friction factor of different rock types under different stress conditions showed that the friction factor is changing in a small range between 0.6 and 1.0. Friction factor is a contact property that is measured along preexisting faults and fracture planes. As the in situ stress field is modified due to hydraulic fracturing, the friction factor along deactivated fault and fracture planes changes, which can lead to fault reactivation, instability, and rock failure.

Gao et al. (2015) performed analytical and numerical studies to investigate the stability of the identified and unidentified faults around the multistage hydraulic fracture in shale reservoirs. They showed that the stability of the fault depends on its position with respect to the hydraulic fractures. Assuming fault is in a critically stressed condition with an initial slip tendency of 0.6, there is a critical angle ($\theta = 50$ degrees) and distance r, below which there is a great possibility of fault reactivation. In other words, for a fault in a critically stressed condition with slip tendency of 0.6, if the angle between fractures and fault plain becomes less than 50 degrees ($\theta \leq 50$ degrees) and distance between the fault plain to hydraulic fracture initiation point becomes less than 2.5 times the fracture height ($r \leq 2.5$ H), there is a high potential for fault reactivation as shown in Fig. 7.3. Fig. 7.4 illustrates the impact of pressurized hydraulic fractures on stability of different regions around hydraulic fractures. High pressure on the hydraulic fracture surface leads to a decrease in slip tendency perpendicular to the hydraulic fracture plain and an increase in slip tendency parallel to the fracture plain, especially at the fracture tip. Therefore, perpendicular to the hydraulic fracture, the stability of the region increases while regions in the direction of hydraulic fracture propagation become unstable. The stable and unstable regions around multistage hydraulic fracture are shown in Fig. 7.4.

As discussed earlier, stability or failure of the fault is determined using slip tendency, which is defined as the ratio of shear to normal stress acting on fault plain as shown in Eq. (7.1):

$$\text{Slip tendency}: T_S = \frac{\tau}{\sigma_n} \geq \mu_S \tag{7.1}$$

where T_s is slip tendency, τ is shear stress, σ_n is normal stress, and μ_s is static friction factor.

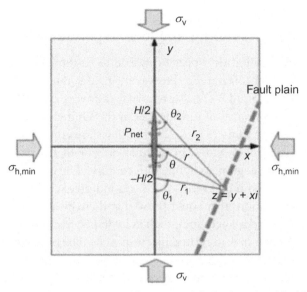

Fig. 7.3 Geometry of single hydraulic fracture and fault plain. *(Modified from Gao, Q., Cheng, Y., Fathi, E., Ameri, S., 2015. Analysis of stress-field variations expected on subsurface faults and discontinuities in the vicinity of hydraulic fracturing. SPE-168761, SPE Reserv. Eval. Eng. J.)*

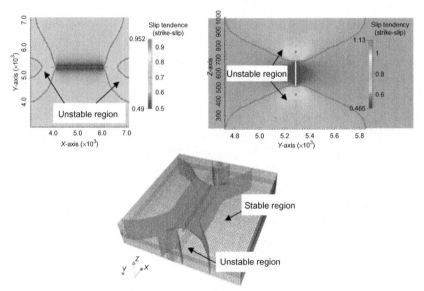

Fig. 7.4 Numerical simulation of change in slip tendency around pressurized multistage hydraulic fractures using finite element technique (Gao et al., 2015).

Hydraulic fracturing and low-magnitude earthquakes

Fault reactivation or slippage can lead to earthquake activities commonly below 1.0 in magnitude. However, in very rare cases, earthquakes with magnitude around 3 are also reported (Ellsworth, 2013). The USGS recently released the map of earthquakes in the United States with magnitude of 3 or higher from 1973 to 2014. They showed that the cumulative number of earthquakes with a magnitude of 3 or higher has significantly increased since the beginning of the 20th century. The media in the United States has quickly pointed to the oil and gas industry without any scientific and sound justifications that could show that these events are actually correlated to the oil and gas activities, such as hydraulic fracturing or disposal of contaminated water in wastewater injection wells. The oil and gas industry is heavily regulated on tracking these events using microseismic studies and there has not been any relationship between the oil and gas industry activities and large magnitude earthquakes.

Having said so, there is still a critical need for further research in this area due to the potential consequences that are associated with large-magnitude seismic events. Determining the cause and influencing factors in the occurrence of large-magnitude seismic events is essential in preventing hazards associated with these events. The main objective of these kinds of studies should be preventing harm to the public health and infrastructure by reducing or eliminating the major causes of these unintended events. The hydraulic fracturing and posthydraulic fracturing activities such as disposal of contaminated flowback water are not the only causes of induced seismicity. Hydrocarbon production from these reservoirs could also trigger seismic activities (Soltanzadeh and Hawkes, 2009). In this case, the study of complex rock and fluid interactions that influence the formation stress behavior is required. To provide an effective prediction of change in state of stress, an accurate model representation must be made using a coupled hydromechanical numerical solution. Verification of this solution shall be obtained through analytical means and progressively refined through experimental results, using real-time downhole data such as microseismic, fiber optics, and advanced imaging technologies.

CHAPTER EIGHT

Hydraulic fracturing chemical selection and design

Introduction

Chemical selection and design is another important aspect of a hydraulic fracture design. As opposed to public perception, that is, that many toxic chemicals are being pumped downhole, it is important to note that the industry has done a tremendous job developing new chemicals that are environmentally friendly and do not cause any harm to the public health and safety. Each chemical used in the hydraulic fracturing process has a very specific purpose. For instance, the slick water frac fluid system uses friction reducer (FR) to reduce the friction pressure when pumping at high rates, whereas the cross-linked frac fluid system uses linear gel and cross-linker to create the viscosity needed to place higher sand concentration proppant into the formation. Exploration and production (E&P) companies can save hundreds of thousands of dollars on chemical selection and design. An optimized chemical package including types and concentrations of each needed chemical is crucial in a successful and cost-effective frac job. Therefore, various field and laboratory tests must be performed to find the optimum design. Since chemicals are part of each frac stage cost and E&P companies are responsible for paying for the type and amount of each chemical used, it is very important to perform various field and laboratory tests to find the optimum chemical design, as will be discussed in this chapter. Chemicals used during hydraulic frac jobs are costly and running the chemical at higher concentrations than needed can add a significant amount of expenditure to each frac stage, which can add up rapidly. There are a limited number of chemicals used in hydraulic fracturing. The most commonly used chemicals in hydraulic fracturing are discussed in the following sections.

Friction reducer

FR is the most important chemical used during slick water frac jobs. FR is a type of polymer used to reduce the friction inside the pipe significantly in order to successfully pump the job under the maximum allowable surface-treating pressure. FR reduces the friction between fracturing fluids and tubular. Without FR, it is impossible to pump slick water frac jobs because of very high friction pressure inside the pipe. The high friction pressure is due to the high flow rate that is used during slick water frac jobs. The concentration of FR used varies from 0.5 to 1.0 gpt depending on the quality of FR and water. The unit for FR concentration is gpt, which stands for gallons of FR per thousand gallons of water. A measurement of 1.0 gpt means there is 1 gallon of FR in 1000 gallons of water. The quality of water has a significant impact on FR. For example, if freshwater is used for the frac job, lower concentrations of FR are needed to control the friction pressure; however, if reused water is used for the frac job, the FR needs to be run at higher concentrations to reduce the friction pressure. Another important factor that necessitates running FR at higher concentrations is the quality of FR. It is very important that operating companies discuss the type and quality of FR that a service company provides. One of the main reasons that the type and quality of FR must be monitored is the cost. All fracturing chemicals are expensive and most frac service companies typically make most of their money from chemicals. Therefore, not monitoring the quality and type of FR can cost the operator lots of money by running the FR at higher concentrations, which might not be necessary. The most commonly used FR in the industry is polyacrylamide, of which there are nonionic, cationic, and anionic types. FR comes in dry powder and liquid form with a mineral oil base. Polyacrylamide is also used for soil stabilization and children's toys. FR selection depends on:
- chemistry of source water for fracturing;
- high salinity vs fresh water = different products; and
- quality of FR (supplier or service company provider).

FR flow loop test

Economically attractive oil and gas production rates from unconventional shale reservoirs highly depend on the effectiveness of hydraulic fracturing stimulation that can provide maximum reservoir contact. This can be achieved by establishing high pumping rates to inject millions of gallons of fracturing fluid in these tight formations. However, there are technical and environmental problems associated with this technique that must be resolved. To meet the high pumping rates, the main problem is to overcome the tubular friction pressure that can reduce the hydraulic horsepower demand by 80% (Virk, 1975). This can be achieved by adding FRs to the fracturing fluid. The postfracturing flowback fluids have extremely high salinity and different concentrations of dissolved mineral and chemicals that cannot be just discharged to the environment. Treating the produced water is also extremely expensive so that most of the operators decide to recycle the produced water by directly using that for subsequent fracturing applications. The performance of FRs as salinity and total dissolved solid content of flowback fluid increases still remain as unsolved problems.

Different shale plays have different temperature and salt content and there is no one-size-fits-all formula for the amount or composition of each additive of fracturing fluids. Salt affects the functions of additives including surfactants, polymers, and gels in a complex way. The most efficient way to test its effects and get the most suitable formula is physical experiment, such as via dynamic flow loop experiment. Fig. 8.1 shows the schematic of the dynamic flow loop experiment. This setup includes 13′ of ½″ stainless steel in the direct and 13′ of ¾″ straight tubing in the return direction, 16 gallon reservoir tank, 7 HP electric motor, variable capacity pump, flow meter, overhead mixer, thermocouple, insulation, and band heater. We also think FRs can impact the proppant transport and settling, and since the effectiveness of hydraulic fracturing significantly depends on proppant displacement,

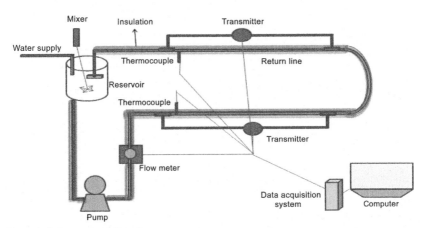

Fig. 8.1 Schematic of flow loop apparatus.

this also needs to be investigated. Numerical simulation is a powerful tool for simulation of proppant transport and displacement in hydraulic fractures as discussed earlier.

FR concentration can be determined by performing lab tests such as FR flow loop test. This test can be performed by taking a water sample that is being used during the frac job to the lab. This water sample is then run through a loop and various FR types and concentrations are tested to find the best FR type and optimum FR concentration of that particular FR. To find the best FR type, different FRs at the same concentration are tested and the one with the maximum friction reduction is selected. Next, the FR concentration of the preselected FR type is increased gradually until the addition of FR does not have a significant impact on pressure drop. This concentration is then recorded and reported to the operating company for their optimum FR concentration design. Fig. 8.2 shows an example of flow test results, where the flow and temperature are measured by flow meter and thermocouples as a function of time and are plotted against average pressure drop measured by two differential pressure transducers in the direct and return lines. During the experiment, the flow rate is kept constant and pressure drop is monitored as FR is added to the fluid system. In addition to laboratory testing and measurements, a supervised machine learning model (discussed in the ML chapter 24) can be developed from field **stage data** to analyze the impact of various parameters such as FR type, FR concentration, hole size (entry hole diameter), number of

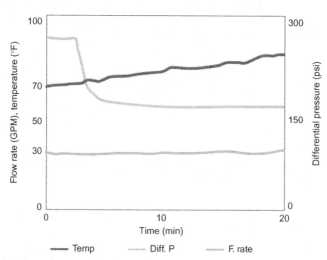

Fig. 8.2 Flow loop test results.

holes, total dissolved solids (TDS), biocide concentration, scale concentration, and geomechanical and petrophysical rock properties on total friction pressure per measured depth foot. This analysis can shed some light on the optimum concentration of FR and the breakeven FR concentration at which pumping more FR will not cause a significant reduction in friction pressure. It is also important to understand the relationship between FR, scale inhibitor, and biocide, and whether each chemical has any hindrance on one another. After training an ML model, such sensitivity can be easily performed to understand these relationships.

Pipe friction pressure

Before discussing the next chemical, it is important to understand friction pressure in slick water jobs without FR. Friction pressure inside the pipe is impacted by rate, fluid viscosity, pipe diameter, and fluid density. A smaller pipe diameter causes the friction pressure to increase. For example, if a 4½" production casing is used instead of a 5½" casing for the hydraulic fracturing treatment, friction pressure inside the pipe will increase. Fluid viscosity and density are very important parameters as well. Various concentrations of FRs are used by different operating companies in an attempt to lower the pipe and perforation friction pressure. Water quality during the frac job is an important factor to consider when designing FR concentration. If heavy flowback water is used without any treatment, more FR has to be run to reduce the friction inside the pipe.

Rate is another important parameter affecting pipe friction pressure. Rate has a proportional relationship with pipe friction pressure. This implies that by increasing the rate, pipe friction pressure will increase as well. Rate not only increases pipe friction pressure, but also it increases perforation friction pressure, which will be discussed.

Friction pressure is calculated using Eq. (8.1):

$$\text{Pipe friction pressure} = \frac{11.41 \times \text{fanning friction factor} \times \text{length of pipe} \times \text{fluid density} \times \text{flow rate}^2}{\text{inside pipe diameter}^5}$$

(8.1)

where length of pipe is measured in ft, fluid density in ppg, flow rate in bpm, and inside pipe diameter in inches.

Fanning friction factor is the most difficult parameter to find and there are various methods that can be used to come up with the fanning friction factor. There are two parameters that must be known in order to calculate fanning friction factor, including Reynolds number and relative roughness of pipe.

Reynolds number

The Reynolds number needs to be calculated in order to come up with the fanning friction factor. Slick water is considered to be a Newtonian fluid; therefore, the following equation is used to calculate the Reynolds number for Newtonian fluid:

$$N_R = \frac{1.592 \times 10^4 \times \text{rate} \times \text{fluid density}}{\text{Inside diameter of pipe} \times \text{fluid viscosity}} \quad (8.2)$$

Relative roughness of pipe

Relative roughness is the amount of surface roughness that exists inside the pipe. The relative roughness of a pipe is known as the absolute roughness of a pipe divided by the inside diameter of a pipe.

$$\text{Relative roughness} = \frac{\varepsilon}{D} \quad (8.3)$$

where ε is absolute roughness in inches and D is inside diameter of pipe in inches.

Once relative roughness and Reynolds numbers are calculated, fanning friction factor can be obtained depending on whether the flow is laminar or turbulent. There are two equations to calculate the fanning friction factor. Eq. (8.4) is for laminar flow (which means the Reynolds number is less than 2300) and is as follows:

$$\text{Fanning friction factor for laminar flow}: F = \frac{16}{N_R} \quad (8.4)$$

If the Reynolds number is more than 4000, then Eq. (8.5) is used to calculate the Darcy friction factor:

Table 8.1 Relative roughness (Binder, 1973)

Pipe material	Absolute roughness (in.)
Drawn brass	0.00006
Drawn copper	0.00006
Commercial steel	0.0018
Wrought iron	0.0018
Asphalted cast iron	0.0048
Galvanized iron	0.006
Cast iron	0.0102
Wood stave	0.0072–0.036
Concrete	0.012–0.12
Riveted steel	0.036–0.36

Darcy friction factor for turbulent flow : $f(D)$

$$= \left(\frac{1}{-1.8 \times \log_{10}\left[\left(\frac{\text{relative roughness}}{3.7}\right)^{1.11} + \left(\frac{6.9}{N_R}\right)\right]} \right)^2 \quad (8.5)$$

Please note that

Darcy friction factor $= 4 \times$ fanning friction factor ⋙ therefore:

$$\text{Fanning friction factor for turbulent flow} = \frac{f(D)}{4} \quad (8.6)$$

Once the fanning friction factor is obtained, the pipe friction pressure can be calculated. This is just one method of pipe friction calculation, which can be very tedious and time consuming. However, there are various handbooks and software available that calculate friction pressure inside the pipe considering the impact(s) of FR concentration. Please note that this calculation does *not* take the use of FR into account. This is just to demonstrate that if FR is not used during high-rate slick water jobs, it is practically impossible to pump the job. Table 8.1 shows the relative roughness of different pipe materials.

Example
Calculate pipe friction pressure if 11,000″ of 5½″, 20 lb/ft, P-110 pipe (ID = 4.778″) is run in the hole without the use of FR (assume fresh water is used for frac). Assume a relative roughness of zero inside the pipe.

Rate = 100 bpm, Freshwater density = 8.33 ppg, Freshwater viscosity =1 cp.

Step 1. The Reynolds number needs to be calculated in order to come up with the fanning friction factor:

$$N_R = \frac{1.592 \times 10^4 \times 100 \times 8.33}{4.778 \times 1} = 2,775,504$$

Step 2. Since the relative roughness inside the pipe is assumed to be zero and the Reynolds number confirms that the flow is turbulent, the Darcy friction factor for turbulent flow can be calculated:

$$f(D) = \left(\frac{1}{-1.8 \times \log 10 \left[\left(\frac{0}{3.7} \right)^{1.11} + \left(\frac{6.9}{2,775,504} \right) \right]} \right)^2 = 0.009826$$

$$\text{Fanning friction factor} = \frac{\text{darcy friction factor}}{4} = \frac{0.009826}{4} = 0.00246$$

Step 3. Now that the fanning friction factor is calculated, pipe friction pressure can be calculated using the following equation:

$$\text{Pipe friction pressure} = \frac{11.41 \times 0.00246 \times 11,000 \times 8.33 \times 100^2}{4.778^5}$$

$$= 10,314 \, \text{psi}$$

Without running FR during the frac treatment, there will be 10,314 psi of pipe friction pressure, and pumping a high-rate frac job will be virtually impossible. This example shows the importance of running FR at predetermined concentrations.

FR breaker

FR breaker is used to reduce the viscosity of FR. An example of a FR breaker is hydrogen peroxide.

Biocide

Biocide is another important chemical used in hydraulic fracturing. The primary duty of biocide is killing and controlling bacteria. Bacteria can cause instability in viscosity. The concentration of biocide typically varies between 0.1 and 0.3 gpt. Pre job water testing is performed to measure

preexisting bacteria present in the water. This test introduces an effective agent with frac source water. A change in the bottle sample directly relates to the bacteria count. Results are then used to determine the biocide concentration (gpt) required for the frac job. The most commonly used biocide product is called glutaraldehyde and this product is typically pumped as a liquid additive with hydraulic fracturing fluid. The basic types of oilfield bacteria are:
- sulfate-reducing bacteria, which is the oldest known bacteria, and which creates H_2S (hydrogen sulfide, poisonous gas) and sulfide, which can form FeS (iron II sulfide) scale;
- acid-producing bacteria, which produces corrosive acid and can adapt to aerobic or anaerobic conditions; and
- general heterotrophic bacteria, which are usually formed in aerobic conditions.

The consequences of not using biocide are:
- H_2S creation in the formation (a safety hazard for producing wells),
- microbiological influenced corrosion, and
- production restriction due to microbial growth.

Now that the different types of bacteria have been discussed, we will take a look at the types of biocide used in hydraulic fracturing and other applications:
- Oxidizing biocide causes irreversible cell damage to the bacteria. Put simply, this type of biocide burns the cell. Examples of oxidizing biocides are chlorine, bromine, ozone, and chlorine dioxide.
- Nonoxidizing biocide alters the cell wall permeability, interfering with biological processes. This type of biocide essentially gives the bacteria cell cancer, which can result in bacteria either dying or surviving. Examples of nonoxidizing biocides are aldehydes, bronopol, DPNPA, and acrolein.

Scale inhibitor

Scale inhibitor is another commonly used chemical in hydraulic fracturing. Scale inhibitor prevents iron and scale accumulation in the formation and wellbore. In addition, scale inhibitor enhances permeability by eliminating scale in the formation and casing. Scale is a white material that forms inside the pipe (casing) and restricts the flow. The concentration of scale inhibitor is usually 0.1–0.25 gpt. Scale is formed by temperature changes, pressure drops, the mixing of different waters, and agitation. An example

of a commonly used scale inhibitor is ethylene glycol (commonly used in antifreeze). The most common types of scale in the oil and gas field are as follows:
- calcium carbonate, which is the most common type of scale and, as opposed to most forms of scale, is less soluble in higher temperatures;
- barium sulfate, which forms a very hard and insoluble scale that has to be mechanically removed;
- iron sulfide, which is the most common sulfide type, and is formed as sulfate-reducing bacteria reduces sulfate; and
- sodium chloride, also known as salt, is another self-explanatory type of scale.

Linear gel

Linear gel is sometimes used with slick water frac to facilitate placing proppant into the formation. Linear gel increases the viscosity of frac fluid, adds friction reduction, and eases the proppant transport into the formation. Higher fluid viscosity increases fracture width, and proppant can be transported more easily into the formation, especially at higher sand concentrations. In addition, gelling agents such as linear gel increase fluid-loss capabilities. Typical polymer types for gelling agents are guar (G, raw guar contains 10%–13% insoluble residue), hydroxypropyl guar (HPG, 1%–3% insoluble residue), carboxymethyl hydroxypropyl guar (CMHPG, 1%–2% insoluble residue), hydroxyethyl cellulose (HEC, minimal residue), and polyacrylamide (FR, minimal residue). Typically, the less-residue gelling agents are associated with more refined gelling agents, and as a result are more expensive.

Guar is the most common linear gel that is currently being used in the industry. Guar is considered the cheapest polymer compared to the other polymers discussed above because it leaves much more insoluble residue behind. Guar is primarily grown in India and Pakistan. Guar is often harvested by hand as a secondary crop by subsistence farmers and can be used for human and cattle food. Guar seeds can be grounded into powder. Guar is typically used as slurry concentrate; however, it can also be used as dry powder that is mixed on the fly.

As previously mentioned, linear gel increases fracture width and bigger sand size and higher sand concentrations can be eased into the formation. When linear gel hits the perforations during a slick water frac, the surface-treating pressure will decrease. The decrease in surface-treating

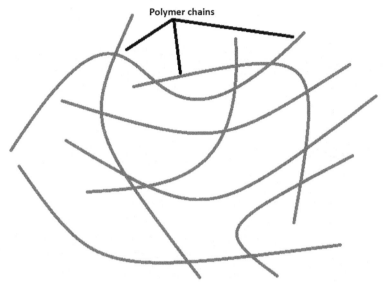

Fig. 8.3 Linear base gel (polymer chains).

pressure allows surface-treating rate to increase. For example, if a stage is being treated at 9500 psi (maximum allowable surface treating pressure for 5½″, P-110, 20 lb/ft casing) and 30 bpm, linear gel is used with slick water to overcome tortuosity, increase fracture width, decrease surface-treating pressure, and be able to get into the stage. The reason surface-treating rate can be increased when using linear gel is because linear gel increases the width of the fracture and the viscosity of the fluid, which in turn improves the proppant transport into the fractures. The concentration of linear gel used during the stage varies and is typically 5–30 lb systems depending on the severity of tortuosity. Since guar concentrate is commonly mixed at ~4 lb/gal, a 5-lb system means 5 divided by 4, which is 1.25 gpt (gallons of gel per thousand gallons of water). Fig. 8.3 illustrates the schematic of polymer chains in a linear base gel system.

Gel breaker

Gel breaker is pumped along with gelling agents so the gel will break once it has been placed into the formation. Gel breaker reduces the viscosity of the gel in the formation. Gel breaker causes the gel to break (reduces viscosity) at certain temperatures at downhole conditions. The degree of gel reduction is controlled by gel breaker type, gel concentration, breaker concentration, temperature, time, and pH. Gel breaker can be tested at the surface by heating it to the formation temperature (using a bath) to ensure proper reduction in viscosity after breaking. It is strongly recommended to do a gel breaker test to visualize the reduction in viscosity. If gel is not completely broken in the formation, it can cause serious formation damage, such as a reduction in conductivity.

Buffer

Buffer is another chemical that is used when linear gel is used. Buffer is run at predetermined concentrations based on the lab test analysis. Buffer essentially adjusts and controls the pH of the gel for maximum effectiveness. The only time it is necessary to run buffer in conjunction with gel (with slick water frac) is if the base fluid has a basic pH (pH from 8 to 14 is considered basic). If the base fluid has a basic pH, buffer must be run to bring the pH down to neutral or slightly acidic pH (6.5–7). It is the service company's quality assurance/quality control (QA/QC) responsibility to measure the pH of the fracturing fluid to determine whether buffer is needed or not. There are two types of buffer used in the industry. The first one is referred to as *acidic buffer*, which is used to speed up the gel hydration time. The second type of buffer is called a *basic buffer*, and is used with cross-linked fluid to create delayed cross-linked. The *delayed cross-linked* delays the process of cross-linked fluid to ensure less friction inside the pipe when the fluid passes through thousands of feet of pipe. Having passed through the pipe, the cross-linked fluid starts working normally in an attempt to overcome the tortuous path after the perforations and before the main body of fractures. pH measures how acidic or basic a substance is, and ranges from 0 to 14. A pH level of 7 (e.g., distilled water) is neutral. A pH of less than 7 (e.g., black coffee and orange juice) is acidic. Finally, a pH of more than 7 (e.g., bleach and baking soda) is considered to be basic. Examples of two buffers used in frac jobs are potassium carbonate and acetic acid.

Cross-linker

Cross-linker is the chemical used to create a cross-linked fluid system. When cross-linker is combined with a 20–30 lb linear gel system, cross-linked fluid is created. Cross-linker increases the viscosity of gelling agents by connecting the separate gel polymers together. Cross-linker significantly increases the viscosity of linear gel by increasing the molecular weight of the

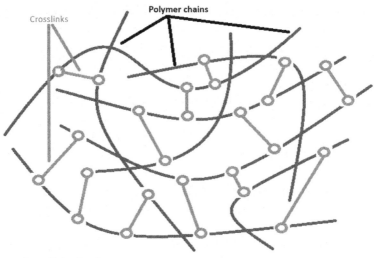

Fig. 8.4 Cross-linked gel.

base polymer by linking multiple molecules together. Cross-linker increases molecular weight without additional polymers. From an economic perspective, it is far more expensive to create heavy viscous fluid with linear gel than a cross-linked fluid system. For example, when linear gel is used to create a 150-centipoise fluid system, it is considerably more expensive than using a cross-linked fluid system to create the same viscosity. This feature of the cross-linked fluid system is considered to be the biggest advantage of using this type of fluid system. One of the disadvantages of cross-linkers is the potential increase in friction pressure. On the other hand, cross-linkers improve the fluid's ability to carry proppant and create viscosity for wider fracture geometry. Common cross-linkers are borate (high pH and moderate temperatures) and zirconate (low pH and high temperatures). Fig. 8.4 illustrates how cross-linkers link multiple molecules together.

Surfactant

Surfactants have different applications. The main application of surfactant is to reduce the surface tension of a liquid. Surface tension is the tendency of a liquid surface to resist an external force. There are various surfactants available in the industry for different usages. The most commonly used surfactants in hydraulic fracturing operations are as follows:
- Microemulsion is a type of surfactant that changes the contact angle, which results in reducing surface tension. Reducing the surface tension

results in more fluid recovery during flowback. This type of surfactant was used in the early development of Marcellus Shale to gain more fluid recovery; however, most operators stopped using surfactants as fracturing fluid in dry gas regions of the Marcellus and Barnett Shales as no benefits were observed.
- Nonemulsifiers minimize or prevent the formation and treatment fluids from emulsion. This type of surfactant is typically used in formations with oil or condensate in an attempt to separate oil or condensate from an aqueous emulsion. Some companies use nonemulsifier in liquid-rich areas of Utica Shale.
- Foamers create stable foam and allow for effective proppant transport.

Surfactants have many more applications and selection depends upon the desired goal. Examples of surfactants are methanol, isopropanol (common use: glass cleaner), and ethoxylated alcohol.

Iron control

Iron control is used to control and prevent dissolved iron in frac fluid. Iron control prevents the precipitation of some chemicals, such as carbonates and sulfates, which can plug off the formation. Examples of iron controls are ammonium chloride, ethylene, citric acid (food additive), and glycol.

CHAPTER NINE

Fracture pressure analysis and perforation design

Introduction

The next step in hydraulic fracturing stimulation is to understand the basic pressure concepts for a successful fracture design, treatment, and execution. Understanding pressure is one of the key aspects of a safe and successful frac operation. One of the most important concepts is the calculation of surface-treating pressure (STP), which is used for production casing design by completion engineers and is discussed in further detail in this chapter. Casing design is very important in new exploration areas because some operators are not able to successfully initiate hydraulic fracturing due to underestimating the expected STP and using a low-burst casing pressure size and grade. Therefore, understanding the basic pressure concepts that will be used in casing design calculation is crucial to a successful frac job. Perforation design is another important parameter in completion design. In this chapter, a special emphasis is placed on designing limited entry for optimum production enhancement in unconventional reservoirs.

Pressure (psi)

Pressure is defined as force divided by area. The unit of pressure in the oil and gas field is psi, which stands for pounds per square inch. For example, 3000 psi means 3000 lb of force over a square inch of area.

$$P = \frac{F}{A} \tag{9.1}$$

P is the pressure in (psi), F is the force in lb, and A is the area in square inches.

Hydrostatic pressure (psi)

Hydrostatic pressure is the pressure of the fluid column exerted in static condition. Hydrostatic pressure is one of the most important concepts that must be learned by heart. Hydrostatic pressure depends on the weight of fluid (ppg) and true vertical depth (TVD) of the well. In addition, 0.052 is a constant for conversion to psi. One of the most common mistakes that beginners make is using measured depth (MD) instead of TVD to calculate hydrostatic pressure in the wellbore. MD can be used for volume calculation; however, TVD has to be used for hydrostatic pressure calculation. The hydrostatic pressure can be calculated using Eq. (9.2).

$$P_h = 0.052 \times \rho \times \text{TVD} \tag{9.2}$$

P_h is the hydrostatic pressure (psi), ρ is the fluid density (ppg), and TVD is the true vertical depth (ft).

Hydrostatic pressure gradient (psi/ft)

Hydrostatic pressure gradient refers to the pressure exerted by the column of fluid per foot of TVD. For example, freshwater has a hydrostatic pressure gradient of 0.433 psi/ft, which means 0.433 psi of fluid column acts on 1 ft of TVD. Hydrostatic pressure gradient is the multiplication of 0.052 constant by fluid density (ppg). Hydrostatic pressure gradient can be calculated using Eq. (9.3).

$$P_h \text{gradient} = 0.052 \times \rho \tag{9.3}$$

P_h gradient is the hydrostatic pressure gradient (psi/ft) and ρ is the fluid density (ppg).

Example

Calculate hydrostatic pressure and hydrostatic pressure gradient for a well with the following properties:

$$\text{TVD} = 10,500 \, \text{ft}, \text{MD} = 19,500 \, \text{ft}, \rho = 8.55 \, \text{ppg}$$

Please make sure to use TVD and not MD when calculating hydrostatic pressure.

$$P_h = 0.052 \times \rho \times \text{TVD} = 0.052 \times 8.55 \times 10,500 = 4668 \, \text{psi}$$
$$P_h \text{gradient} = 0.052 \times \rho = 0.052 \times 8.55 = 0.4446 \, \text{psi/ft}$$

Instantaneous shut-in pressure (psi)

ISIP stands for instantaneous shut-in pressure, and is the pressure at which all of the pumps come offline following a hydraulic fracturing stage treatment or diagnostic fracture injection test (DFIT). ISIP can be obtained using a STP graph after each hydraulic fracture stage treatment. ISIP is extremely important to calculate for new exploration areas where hydraulic fracturing will take place in order to ultimately calculate the estimated STP. Fig. 9.1 illustrates STP, calculated bottom-hole pressure, slurry rate, blender, and formation sand concentrations. ISIP in Fig. 9.1 is the pressure as soon as all of the pumps are offline (i.e., the slurry rate goes to 0). In this figure, ISIP is approximately 4900 psi. ISIP can also be calculated using Eq. (9.4).

$$\text{ISIP} = \text{BHTP} - P_h \tag{9.4}$$

ISIP is the instantaneous shut-in pressure (psi), BHTP is the bottom-hole treating pressure (psi), and P_h is the hydrostatic pressure (psi).

At the end of a hydraulic fracturing stage when all of the frac pumps are offline, pressure drops significantly, and a water hammer effect can be seen on the pressure signal. Pressure continues to drop afterward due to fluid leak-off to the formation. The amount of pressure drop is directly related to the permeability of the formation and frac fluid viscosity. Fig. 9.2 illustrates

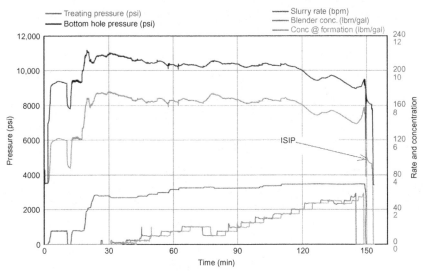

Fig. 9.1 ISIP illustration.

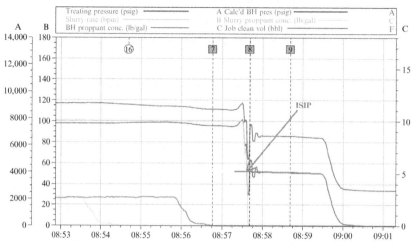

Fig. 9.2 ISIP selection.

the STP and surface-treating rate (pump injection rate) versus time. To determine the ISIP, one must draw a vertical line at the point at which surface-treating rate goes to zero and fit a straight line to the pressure fall-off after the shut-in. The intersection of the two lines yields the ISIP.

Example
Calculate ISIP with the following data:
  Bottom-hole treating pressure = 10,000 psi, TVD = 7550′, Fluid density = 8.9 ppg

$$\text{ISIP} = \text{BHTP} - P_h = 10,000 - (0.052 \times 8.9 \times 7550) = 6506 \text{ psi}$$

Fracture gradient (psi/ft)

Fracture gradient (FG), also known as frac gradient, is the pressure gradient at which the formation breaks. Frac gradient is crucial to understand in order to calculate the expected bottom-hole treating pressure (BHTP) before the start of a frac job. Eq. (9.5) can be used to calculate the frac gradient.

$$FG = \frac{\text{ISIP} + P_h}{\text{TVD}} \quad (9.5)$$

FG is the frac gradient (psi/ft), ISIP is the instantaneous shut-in pressure (psi), P_h is the hydrostatic pressure (psi), and TVD is the true vertical depth, ft.

Example

ISIP at the end of a DFIT job is obtained to be around 4500 psi. If TVD of the formation is 7500' (assuming fresh water was used during DFIT), calculate the frac gradient.

$$\text{Hydrostatic pressure} = 0.052 \times 8.33 \times 7500 = 3249 \text{ psi}$$
$$\text{Frac gradient} = \frac{4500 + 3249}{7500} = 1.033 \text{ psi/ft}$$

Bottom-hole treating pressure (psi)

BHTP is the amount of pressure required at the perforations to cause fracture extension during hydraulic fracture stimulation. BHTP is the pressure along the fracture face that keeps the fractures open. BHTP is also referred to as bottom-hole frac pressure (BHFP). Correct estimation of BHTP is essential when preparing the estimates of STP and ultimately a frac job. BHTP can be calculated using Eq. (9.6).

$$\text{BHTP} = \text{FG} \times \text{TVD}$$
$$\text{or} \qquad (9.6)$$
$$\text{BHTP} = \text{ISIP} + P_h$$

Please note that the second equation can be derived by rearranging the first equation as follows:

$$\text{BHTP} = \text{FG} \times \text{TVD} \rightarrow \frac{\text{ISIP} + P_h}{\text{TVD}} \times \text{TVD} = \text{ISIP} + P_h$$

Example

Calculate estimated BHTP if ISIP (obtained from DFIT) is 6427 psi and TVD is 8500' (assuming fresh water).

$$\text{Frac gradient} = \frac{\text{ISIP} + P_h}{\text{TVD}} = \frac{6427 + (0.052 \times 8500 \times 8.33)}{8500} = 1.189 \text{ psi/ft}$$
$$\text{BHTP} = \text{FG} \times \text{TVD} = 1.189 \times 8500 = 10,109 \text{ psi}$$

or

$$\text{BHTP} = \text{ISIP} + P_h = 6427 + (0.052 \times 8500 \times 8.33) = 10,109 \text{ psi}$$

Total friction pressure (psi)

There are various types of friction pressures that must be considered and calculated before and after treatment to derive perforation efficiency and optimum design. Friction pressures during a frac job are pipe friction pressure, perforation friction pressure, and tortuosity pressure. Total friction pressure after each frac stage can be calculated using Eq. (9.7).

$$FP_T = \text{Avg surface treating pressure} - \text{ISIP} \qquad (9.7)$$

FP_T is the total friction pressure (psi), Avg surface-treating pressure is given in psi, and ISIP is the instantaneous shut-in pressure in psi.

As the name indicates, average surface-treating pressure is the average surface-treating pressure during each hydraulic frac stage. ISIP can also be obtained after each hydraulic frac stage treatment.

Example

A frac stage was completed in Barnett Shale and the data listed below were obtained at the end of the frac stage. Calculate total friction pressure for this stage.

Average STP for the stage = 8650 psi, ISIP = 4500 psi

$FP_T = \text{Avg surface treating pressure} - \text{ISIP} = 8650 - 4500 = 4150 \, \text{psi}$

From this example, 4150 psi indicates total friction pressure which consists of pipe, perforation, and tortuosity pressures during the stage. This basically indicates that out of 8650 psi of average STP, 4150 psi is the total friction pressure. In this example, total friction pressure is about 48% of the average treating pressure. This illustrates the importance of calculating and understanding pipe, perforation, and tortuosity friction pressures. Please note that 4150 psi does include the impact of FR used during the frac job. As previously discussed, without the use of FR pumping a slick water frac stage at a high rate would be impossible.

Pipe friction pressure (psi)

Pipe friction pressure can be calculated excluding FR impacts. However, it is much more important to obtain the pipe friction pressure after FR is added to the fracturing fluid pumped in the well. This calculation depends on the type of FR provided by the service company. There are various tools that can be used to approximate pipe friction pressure depending on the type of FR used. Service companies typically perform lab tests to understand the impact of their particular FR product on pressure, and to quantify the pressure reduction caused by the FR. The pressure reduction of each friction reducer varies depending on the type and manufacturer of the product.

Perforation friction pressure (psi)

In addition to pipe friction pressure, which is one of the main considerations in hydraulic fracturing treatment design, perforation friction pressure is another important parameter in hydraulic fracturing design that needs to be calculated and considered. Perforation friction pressure can be calculated using Eq. (9.8) if optimum perforation friction pressure for a particular area is known.

$$\text{Perforation friction pressure} = \frac{0.2369 \times Q^2 \times \rho}{C_d^2 \times D_p^4 \times N^2} \quad (9.8)$$

Q is the flow rate (bpm), ρ is the fluid density (ppg), C_d is the discharge coefficient, coefficient of roundness of jet perforation, assume C_d of 0.8–0.85, D_p is the perforation diameter (in), and N is the numbers of perforations (holes).

In Eq. (9.8), as pump rate (flow rate) and fluid density increase, perforation friction pressure will also increase. On the other hand, as the number of perforations (holes) and perforation diameter increases, perforation friction pressure will decrease. Perforation diameter is also referred to as entry-hole diameter (EHD). Discharge coefficient is the measure of perforation efficiency when fluid passes through the perforations. Typically a discharge coefficient of 0.6 is assumed for new perfs and a discharge coefficient of 0.85 is assumed for eroded perfs.

Example
Calculate the perforation friction pressure with the following data:
$Q = 85 \text{ bpm}, \rho = 8.5 \text{ ppg}$, discharge coefficient of 0.8, $D_p = 0.42''$, $N = 40$

$$\text{Perforation friction pressure} = \frac{0.2369 \times Q^2 \times \rho}{C_d^2 \times D_p^4 \times N^2} = \frac{0.2369 \times 85^2 \times 8.5}{0.8^2 \times 0.42^4 \times 40^2}$$
$$= 457 \text{ psi}$$

Open perforations

Open perforations refer to the number of perforations that are actually open during a frac stage treatment. At the beginning of unconventional shale development, some companies used up to 90 perforations (holes in the casing) per treatment stage. Does this mean that all of the perfs are open during the treatment? Absolutely not. This is primarily why the industry lowered the number of perfs in an attempt to improve the number of perfs that remain open during the treatment. Each single hole can take up to 1–3 bpm depending on the formation. Designed pumping rates for slick water frac jobs are usually anywhere between 70 and 100 bpm. Therefore, completion engineers perform various calculations to derive the optimum design and perf efficiency so as to have as many holes open as possible during a frac stage treatment.

The perforation friction-pressure equation can be rearranged and the number of open holes (perfs) can be calculated using Eq. (9.9) if optimum perforation friction pressure for a particular area is known.

$$\text{Open perfs(holes)} = \sqrt{\frac{0.237 \times \rho_f \times Q^2}{\text{Perf friction pressure}\, C_d^2 \times D_p^4}} \quad (9.9)$$

Perforation efficiency

Perforation efficiency refers to the percentage of open holes either before or after a frac job. Typically the perforation efficiency during hydraulic frac jobs ranges from 30% to 80%. If 80% perforation efficiency can be obtained from a frac stage, it is considered to be an outstanding perforation design. If the designed number of perforations per stage is 45 holes, on average 50%–60% of the holes could possibly be open during the frac job. This means hydraulic frac stimulation will have only taken place through 23–27 holes out of the original 45 holes. Therefore, it is important to understand that perforation efficiency is highly dependent on the perforation design, formation type/heterogeneity, natural fracturing, and stresses around the stimulated zones. These factors could all impact the perforation efficiency that can be obtained from a well. Perforation efficiency can be calculated using Eq. (9.10).

$$\text{Perforation efficiency}(\%) = \frac{\text{Number of open holes}}{\text{Designed number of holes}} \times 100 \quad (9.10)$$

Perforation design

Another important concept in hydraulic fracturing design is the number of holes per stage. Designing the number of holes per stage in a conventional reservoir is completely different from in an unconventional shale reservoir. *Limited entry* is a term of art used in the industry and is referred to as the practice of limiting the number of perforations (holes) in a completion stage to help the development of perforation friction pressure during a frac stimulation treatment. The "choking" effect produces back pressure in the casing, which allows simultaneous entry of fracturing fluid into multiple zones of varying in situ stress states. Treatment distribution among zones can be controlled to a degree. Limited entry is known to increase perforation efficiency, and as a result, production in unconventional shale reservoirs (Cramer, 1987). Limited entry can be achieved using the following steps:

1. Determine the friction pressure of a single perforation for the limited entry design. A value of at least 200–300 psi is recommended since a value of this magnitude should be noticeable in the total STP.
2. Once the friction pressure is chosen, solve for rate per perforation to determine the rate per perf (Q/N). This new equation provides the rate per perforation needed to develop the friction pressure of a single perforation.

$$\text{Original perforation friction pressure} = P_f = \frac{0.2369 \times Q^2 \times \rho}{C_d^2 \times D_p^4 \times N^2}$$

Eq. (9.11) can be obtained by rearranging the perforation friction pressure and solving for Q/N.

$$Q/N \text{ in limited entry design} : \frac{Q}{N} = \frac{D_p^2 \times C_d \times \sqrt{\frac{P_f}{\rho}}}{0.487} \quad (9.11)$$

Please note that the value of friction pressure in a single perforation must be chosen based on the production success in an area (knowing the history of injection rates and number of perfs in different wells).

Example

Calculate the desired rate per perf (Q/N) if 260 psi is the desired perforation friction pressure per perf assuming the following data:

$D = 0.42''$, $C_d = 0.80$ (coefficient of roundness of jet perforation, 1.0 is round), $P_f = 260$ psi, $\rho = 8.33$ lb/gal

$$\frac{Q}{N} = \frac{D_p^2 \times C_d \times \sqrt{\frac{P_f}{\rho}}}{0.487} = \frac{0.42^2 \times 0.8 \times \sqrt{\frac{260}{8.33}}}{0.487} = 1.6 \, \text{bpm/perf}$$

Based on the calculated rate per perforation, the injection rate for the limited-entry fracturing treatment can be determined by taking into account the maximum allowable STP. From this example, Table 9.1 can be constructed. This table shows the number of perforations required at various injection rates if 260 psi per perforation is chosen to be the desired perforation friction pressure.

Table 9.1 Limited-entry design example

Number of perfs	Rate (bpm)
20	32
25	40
30	49
35	57
40	65
45	73
50	81

Numbers of holes (perfs) and limited entry technique

Holes in fracing are also referred to as perfs (perforations). The number of holes is important in a frac design. It was the industry's belief that more holes would result in better productivity by having more reservoir entries in unconventional shale reservoirs. However, time and actual production data have proven otherwise. Limited entry is believed to result in better perf efficiency and production. Limited entry means obtaining roughly 2 bpm (rate) or more from each hole. In the limited entry technique, EHD in the casing acts as a choke. During a frac stage, the choked flow rate through a limited number of holes produces back pressure. As a result, back pressure impacts the fracture propagation pressure. To a certain degree, this will yield a controlled treatment distribution among fractured zones. The number of holes per cluster depends on the length of the perf gun. For example, if a 1′ perf gun is used and 6 shots per foot are the designed shot density, 6 shots (holes) will be used in each cluster. If there are six clusters in one frac stage, 36 holes are used for one particular stage. The length of the perf gun varies between 1′ and 3′ depending on the operator.

Perforation diameter and penetration

A perforation diameter frequently used in shale plays can range between roughly 0.42″ and 0.58″. A 0.42″ EHD means each hole created in the casing has a diameter of 0.42″. In addition, the nominal penetration depends on the type/size, manufacturer of perforation gun, and the amount of explosives used in each gun. A common nominal penetration obtained in shale formations varies between 7″ and 45″. Deep penetration shots are believed to help bypass the near-wellbore damage (e.g., skin damage from drilling) and be closer to the virgin rock in order to establish the initial fractures.

Perforation erosion

Another important topic in the hydraulic fracturing world is perforation erosion. Does each hole that has a certain diameter (e.g., 0.42″) stay the same size after pumping thousands of pounds of proppant? The answer is no because perforations will erode and get bigger. Perforation friction pressure is dependent on erosion rate. As perforations erode, perforation friction pressure will decrease. As previously discussed, the discharge coefficient in the perforation friction-pressure equation takes into consideration whether the perfs are new or eroded when calculating perforation friction pressure.

Near-wellbore friction pressure

Near-wellbore friction pressure (NWBFP) is another term used to indicate the total pressure loss near the wellbore. NWBFP is the sum of perforation friction pressure and tortuosity. Eq. (9.12) is used to calculate NWBFP.

$$\text{Near wellbore friction pressure} = \text{perforation friction pressure} + \text{tortuosity} \tag{9.12}$$

So far, total friction pressure has been discussed as follows:

Total friction pressure = pipe friction pressure + perforation friction pressure + tortuosity

Fracture extension pressure

Fracture extension pressure is referred to as the pressure inside the fracture(s) that makes the fractures grow as pumping continues. In other words, fracture extension pressure is the pressure required to extend the existing fractures. In order to keep the fractures open while gaining length, height, and width, the fracture extension pressure must be greater than the closure pressure of the formation. Fracture extension pressure can be thought of as BHTP. These terms are used interchangeably.

$$\text{Fracture extension pressure} = \text{Frac gradient} \times \text{TVD} \qquad (9.13)$$

Closure pressure

Closure pressure is the minimum pressure required to keep the fractures open. In other words, closure pressure is the pressure at which the fracture closes without proppant in place. For example, during a hydraulic fracturing treatment, closure stress in the pay zone must exceed the BHTP in order to grow an existing fracture. This means that BHTP has to be greater than the pay zone's closure stress. Difficulty starting a stage during frac jobs could be due to not exceeding closure stress because of high closure stress in the zone of interest. As a result, it is highly recommended to land the wellbore in a zone that has higher closure pressure above and below in an attempt to keep the fractures contained. Closure pressure can be assumed to be the same as the minimum horizontal stress. Determining the closure pressure is extremely important in a hydraulic fracturing design because it helps the engineers determine the type of sand needed for the job. Closure pressure can be determined from a DFIT or step-rate test.

A step rate test is performed before the frac job and is used to determine the fracture extension pressure (P_{EXT}). Fracture extension pressure is normally slightly higher than fracture closure pressure. The first method in determining the closure pressure is a step-up test, which is part of a step-rate test. Therefore, this test is useful in figuring out the upper boundary of closure pressure.

Procedure

1. Water is typically pumped using the lowest possible rate a pump can handle (usually 0.5–1 bpm). Once the desired rate has been reached, wait for pressure stabilization and then record the exact pressure and rate.
2. After getting the exact pressure and rate, step up the rate to 1.5, 2, 3, 5, and 10 bpm and record the stabilized pressure at each rate.

If done correctly, this test is very simple to perform. Please note that this test needs to be done before starting treatment for accurate results. Conducting this test after a frac stage treatment will yield inaccurate results that cannot be used for determining the closure pressure. Fig. 9.3 illustrates how to determine the fracture extension pressure, which can be used to estimate an upper boundary value for the closure pressure. This test usually takes 15 min to conduct.

The second method that can be used to determine closure pressure is an injection falloff test. In this test, the fluid is injected at a constant rate and the well is then shut in. The pressure will naturally fall below the closure pressure, and eventually the fractures close. The permeability of the formation will determine the time to closure. The lower the permeability, the longer it takes to reach closure pressure. The time to reach closure is a function of

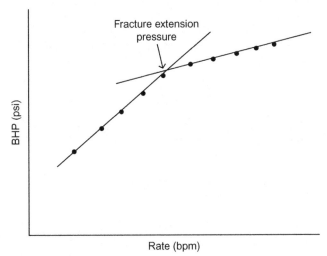

Fig. 9.3 Fracture extension pressure.

pump time and estimated permeability. Since this type of test requires longer shut-in time for unconventional shale reservoirs due to low permeability, the practical application is limited to higher perm reservoirs. The time to reach closure (Barree, 2013) can be approximated using Eq. (9.14).

$$\text{Time to reach closure} = \frac{0.3 \times \text{pump time}}{\text{estimated perm}} \quad (9.14)$$

Pump time is given in minutes and estimated permeability is in md.

Example
Calculate the time to reach closure if 5 min of pump time was conducted in 0.003 and 0.03 md rock: Time to closure @ 0.003 md = $\frac{0.3 \times \text{pump time}}{\text{estimated perm}} = \frac{0.3 \times 5}{0.003} = 500$ min = 8.33 h

$$\text{Time to closure @ 0.03 md} = \frac{0.3 \times \text{pump time}}{\text{estimated perm}} = \frac{0.3 \times 5}{0.03} = 50 \text{ min}$$
$$= 0.833 \text{ h}$$

Shut-in bottom-hole pressure (y-axis) is plotted versus square root of time (x-axis) to determine the fracture closure pressure. The fracture closure pressure is the point at which the flow deviates from a straight line.

Example
Estimate the fracture closure pressure using Table 9.2:

BHP (y-axis) versus square root of time (x-axis) needs to be plotted. The closure pressure is illustrated in Fig. 9.4 where the plotted line deviates from the linear line. In this example, closure pressure is approximately 3600 psi.

Another commonly used method to calculate closure pressure is from the DFIT analysis, which will be covered in detail later in the book.

Table 9.2 Bottom-hole pressure (BHP) vs Sqrt(time) example

Shut-in time min	BHP psi	Sqrt(time) min
0	4300	0.00
1	4073	1.00
4	3840	2.00
6	3740	2.45
8	3605	2.83
10	3470	3.16
12	3350	3.46

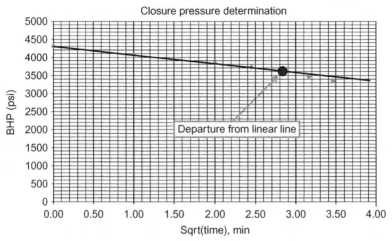

Fig. 9.4 Closure pressure determination from injection fall-off test.

Net pressure

Net pressure is one of the most important pressures to consider in hydraulic fracturing. Net pressure is the energy required for propagating fractures and creating width during the frac job and refers to the excess pressure over the frac pressure required to extend the fractures. Net pressure is essentially the difference between the fracturing fluid pressure and the closure pressure and is the driving mechanism behind fracture growth. The more pressure inside a fracture, the more potential there is for growth. The term net pressure is only used when the fracture is open. If the fracture is closed, net pressure is equal to 0. Net pressure depends on various parameters such as Young's modulus, fracture height, fluid viscosity, fluid rate, total fracture length, and tip pressure. Net pressure is also referred to as process zone stress and can be calculated using Eq. (9.15) or Eq. (9.16).

$$P_{net} = BHTP - P_c$$
Net pressure, Eq. (9.1)   or   (9.15)
$$P_{net} = BHISIP - P_c$$

P_{net} is the net pressure (psi), BHTP is the bottom-hole treating pressure (psi), P_c is the closure pressure which is approximately minimum horizontal stress (psi), and BH ISIP is the bottom-hole ISIP (psi), BH ISIP = ISIP + P_h.

$$P_{net} = \frac{E^{3/4}}{h}(\mu \times Q \times L)^{1/4} + P_{tip} \tag{9.16}$$

E is the Young's modulus, psi; h is the fracture height, ft; Q is the rate, bpm; L is the total fracture length, ft; and P_{tip} is the Fracture tip pressure, psi.

As can be seen from Eq. (9.16), Young's modulus is raised to the power of 3/4 while the fluid rate, viscosity, and total fracture length are only raised to the power of 1/4. This shows that Young's modulus has more impact on net pressure compared to viscosity, rate, and length. As a result, the Young's modulus measurement of a formation is a key parameter in fracture propagation. Fracture-tip pressure is a quantity that is not easy to find, however, different numerical simulations depending on a wide range of assumptions (e.g., fracture tip with or without fluid lag) will provide estimates of the fracture tip pressure, Bao et al. (2016). In hydraulic fracturing, a dynamic gap zone between fracture tip and fracturing fluid following the tip exists which can impact the fracture tip pressure.

When net pressure (y-axis) versus time (x-axis) is plotted on the log-log plot during a live frac stage treatment, a net pressure chart can be constructed. A net pressure chart is also referred to as a Nolty chart, and is used during the hydraulic frac treatment to follow various pressure trends throughout the stage. Net pressure charts are used to estimate various fracture propagation behavior at different points in time. As previously discussed, since net pressure is the driving mechanism behind the fracture growth, it can be used to predict the fracture dimension. Company representatives in the field rely heavily on the Nolty chart during the treatment since it is very accurate in conventional reservoirs. In unconventional reservoirs, this chart is still a useful tool to determine the fracture propagation, but it is not as accurate as it is in the conventional reservoirs. Fig. 9.5 shows the

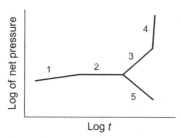

Fig. 9.5 Net pressure interpretation.

concept of net pressure during the treatment, which can be used to make critical decisions.
- If the pressure response in the Nolty chart is similar to trend #1, it is an indication of contained height and unrestricted length extension during the treatment (slightly positive slope).
- If the slope of the net pressure line is zero (trend #2), it represents contained height and possibly opening up more fractures with fluid loss. It indicates a less-efficient length extension.
- Trend #3 pressure response during treatment is bad news because the formation is giving up and there is a high possibility of a tip screening-out (sanding-off) if sand is not cut on time.
- Trend #4 is basically a full screen-out and the pump room needs to be ready to come offline as soon as the pressure starts rising dramatically to avoid exceeding the pressure limitations on the casing and equipment.
- Trend #5 illustrates uncontrolled fracture height growth.

Net pressure typically ranges between 100 and 1400 psi. In some instances, net pressure could be higher. If net pressure is much higher than 1400 psi, this could be due to near-wellbore restriction or large tip plasticity.

Example
Estimate net pressure if closure pressure is obtained from a step-rate test to be 6500 psi and ISIP is 4700 psi. The well has a TVD of 6800' and used an 8.8-ppg frac fluid to pump the job.

$$\text{BH ISIP} = \text{ISIP} + P_h = 4700 + (0.052 \times 8.8 \times 6800) = 7811 \text{ psi}$$
$$P_{net} = \text{BH ISIP} - P_c = 7811 - 6500 = 1312 \text{ psi}$$

Surface-treating pressure (psi)

STP, also known as wellhead treating pressure (WHTP) is the pressure at the surface during a hydraulic fracturing treatment. STP during a hydraulic fracturing treatment is the real-time pressure obtained from the surface pressure transducer on the main line. A transducer uses pulsation to get the real-time pressure during a hydraulic fracture treatment. STP can be estimated using Eq. (9.17).

$$\text{STP} = \text{BHTP} + P_f - P_h + P_{net} \qquad (9.17)$$

BHTP is the bottom-hole treating pressure (psi), P_f is the total friction pressure (psi), P_h is the hydrostatic pressure psi; and P_{net} is the net pressure (psi).

It is important to estimate the STP in order to have enough hydraulic horsepower (HHP) on site during the frac job. The HHP needed for the job is a function of surface-treating rate and STP. Once STP is estimated and the designed rate is known, HHP can be calculated using Eq. (9.18).

$$HHP = \frac{WHTP \times R}{40.8} \qquad (9.18)$$

WHTP is the wellhead-treating pressure, psi; and R is the surface-treating rate, bpm.

By rearranging the STP equation, BHTP can be solved as shown in Eq. (9.19).

$$BHTP = STP - P_f + P_h - P_{net} \qquad (9.19)$$

Example

You are the completion engineer responsible for determining the anticipated STP for a hydraulic frac treatment in a low-permeability field with the following data. Assuming a designed rate of 80 bpm, how much HHP is needed for the job? If each pump has 2250 HHP, how many pumps will be needed for the job?

ISIP = 7500 psi (from DFIT test), TVD = 10,500′, water density = 8.6 ppg, pipe friction pressure = 4221 psi (calculated assuming 1 gpt FR, 80 bpm, 20,000′ pipe MD, and 4.778″ casing ID), D_p = 0.42″, N = 36 perfs, C_d = 0.8, ΔP_{net} = 0 psi (Assume net pressure is 0)

Step 1. Calculate hydrostatic pressure:

$$P_h = 0.052 \times \rho \times TVD = 0.052 \times 8.6 \times 10,500 = 4696 \, psi$$

Step 2. Calculate frac gradient:

$$FG = \frac{ISIP + hydrostatic\ pressure}{TVD} = \frac{7500 + 4696}{10,500} = 1.16 \, psi/ft$$

Step 3. Calculate BHTP:

$$BHTP = Frac\ gradient \times TVD = 1.16 \times 10,500 = 12,180 \, psi$$

Step 4. Calculate perforation friction pressure:

$$Perf\ friction = \frac{0.2369 \times Q^2 \times \rho}{C_d^2 \times D_p^4 \times N^2} = \frac{0.2369 \times 80^2 \times 8.6}{0.8^2 \times 0.42^4 \times 36^2} = 505 \, psi$$

Step 5. Calculate STP:

$$STP = BHTP + P_f - P_h + P_{net} = 12,180 + (4221 + 505) - 4696 + 0$$
$$= 12,210 \, psi$$

Step 6. Calculate HHP needed for the job:

$$\text{HHP} = \frac{\text{WHTP} \times R}{40.8} = \frac{12,210 \times 80}{40.8} = 23,941 \text{ psi}$$

Step 7. Calculate total number of pumps if each pump is 2250 HHP:

$$\text{Total \# of pumps} = \frac{23,941}{2250} = 10.6$$

Typically a 20% safety factor is added to the calculated number to make sure enough HHP is available in the event that some of the pumps malfunction during the job. Therefore, 13 pumps will be needed for this job. Please note that perforation friction pressure calculated above assumes that all of the perforations are open and taking fluid during the treatment. It is recommended to take some precaution and assume that only a percentage of the total designed perforations will be taking fluid (e.g., 60%) and as a result estimate the new perforation friction pressure.

To give some perspective on HHP used during a hydraulic frac job by assuming 16 pumps for the job with 2250 HHP for each pump, this is equivalent to about 72 Corvettes.

Production casing design

Once the estimated STP is obtained, the next crucial step is to design the production casing. Please note that production casing must be designed using both estimated STP and estimated breakdown pressure, whichever one is expected to be higher. Wild Well Control Technical Data Book (WWCTDB) (https://wildwell.com/wp-content/uploads/technical-data-book.pdf) can be used to find casing burst pressure for various weights, sizes, and grades. Let's assume that the anticipated STP is expected to be 9500 psi (and higher than breakdown pressure) and you are tasked to find a 5 ½″ production casing to withstand such pressure during the frac treatment. Looking at the "Casing Strength" table in the WWCTDB, a P-110 and 20 #/ft casing can withstand 12,640 psi casing burst pressure. It is also crucial to take 80% of the casing burst pressure into account as a **safety factor** which would yield:

$$80\% \text{ of casing burst pressure} = 80\% \times 12,640 = 10,112 \text{ psi}$$

As demonstrated, 5 1/5″, 20 #/ft, and P-110 casing string can withstand 10,112 psi which is sufficient to handle the anticipated 9500 psi. What if the anticipated STP during frac treatment was 11,300 psi? As shown in the table, P-110 and 23 #/ft has a casing burst pressure of 14,520 psi. Repeating the same analysis would yield:

$$80\% \text{ of casing burst pressure} = 80\% \times 14,520 = 11,616 \text{ psi}$$

Therefore, P-110 and a higher casing weight of 23 #/ft must be used in this example to withstand the anticipated STP of 11,300 psi and incorporate the 80% safety factor. There are other casing grades that are excluded from WWCTDB but can be easily obtained via coordination with drilling department and various vendors. The most **cost-effective** casing that can withstand the anticipated STP or breakdown pressure (whichever one is higher) incorporating the discussed safety factor must be used for a successful production casing design.

As lateral length increases, total pipe friction pressure also increases. Lower treating rate during slick water frac jobs than desired could result. For example, if achieving 100 bpm was feasible when the total MD was 15,000′, extending the total MD to 30,000′ (due to longer lateral lengths) would lead to a restricted treating rate of less than 100 bpm if the same production casing grade, weight, and size is used. Therefore, it is crucial to redesign the production casing as well as wellhead and surface equipment to avoid sacrificing treating rate. Sacrificing treating rate during slick water frac jobs is **only justified** when the field-testing results do not show any detrimental impact on production outcome. Some of the ways to mitigate sacrificing treating rate are as follows:

- Redesigning casing **weight** and **grade** to withstand higher casing burst pressure by keeping the same casing size. Wellhead and surface equipment used during frac jobs must also be redesigned.
- Increasing **casing size** to reduce the pipe friction pressure. Wellhead must also be redesigned.
- Increasing **FR concentration** if and only if increasing the FR concentration has a direct impact on lowering total friction pressure (and as a result lowering STP). In many circumstances, it can be noted that increasing the FR concentration (e.g., by 0.2 gpt) can significantly lower total friction pressure.

These are some common solutions when dealing with longer lateral length wells and designing the right production casing without sacrificing frac treatment design.

Rate step-down test workflow:

Another approach to understand perforation efficiency such as the impact of the number of holes (perforations), perforation phasing, EHD is to perform a rate step-down analysis (RSD). Massaras and Dragomir (2007) proposed a new rate RSD which will be used to illustrate a practical example of such proposal via a step by step guide and example. The procedure is as follows:

(1) In a slick water frac, get up to the designed rate (e.g., 100 bpm). Hold the rate for 10–20 s or until the pressure is stabilized. Afterwards, record the rate and pressure at each step depending on the number of steps recorded. For a three-point rate step-down test, three stabilized rates and pressures will be needed, for a four-point rate step-down test, four stabilized rates and pressures will be needed, and for a five-point rate step-down test, five stabilized rates and pressures will be needed. In addition, record the ISIP when the rate goes to zero. For example, for a four-point rate step-down test, record the stabilized pressure at 100 bpm, 80 bpm, 60 bpm, 40 bpm, and ISIP when the pump rate goes to zero. Please make sure that all chemicals (that are normally used) such as FR, biocide, scale inhibitor, etc. are being used when performing the step-down test.

(2) This test can be conducted on multiple stages throughout the lateral length of a well. Some operators perform such test on every stage which can become time consuming. Therefore, to save time and money, pick number of stages along the lateral length of the well and that should provide enough information about the perforation design.

(3) Always validate the test using production data or any other possible analysis. For example, if perforation design A had an average perforation efficiency of 90% while perforation design B had an average perforation efficiency of 60% (while all other completions parameters remained identical), does the production performance of perforation design A also show an improvement in EUR/ft (or any other production metric that your respective corporation uses) as compared to perforation design B? This provides confirmation and validation of the validity of the step-down test.

(4) Calculate total friction pressure based on MD of the stage, pump rate, etc. Total friction pressure will vary depending on the type of FR, water quality, as well as many other factors. Operators must work with the service providers to derive an accepted total friction pressure to be used for various purposes. A bottom-hole gauge during frac stage treatment can also be used to validate such conversion.

(5) After calculating total friction pressure for each step rate, calculate BHTP as follows:

$$\text{BHTP} = \text{STP} + P_h - P_f$$

(6) Calculate actual near-wellbore pressure differential as follows ($\Delta P_{\text{near-wellbore}}$):

$$\Delta P_{\text{near-wellbore}} = \text{BHTP at each step} - \text{BH ISIP}$$

(7) Calculate coefficient of perforation friction as shown below. In this equation, N_p is the number of perforations. Number of perforations will be iterated (altered) until a solution is achieved using the excel solver.

$$K_{pf} = 0.2369 \frac{\rho}{N_p^2 D^4 C_d^2}$$

(8) Assume a coefficient of near wellbore friction (K_{nw}). This will be one of the parameters that will be iterated (altered) in later stages when using the excel solver.

(9) Calculate the calculated near-wellbore pressure differential as follows ($\Delta P_{\text{calculated\_near-wellbore}}$):

$$\Delta P_{\text{calculated\_near-wellbore}} = \left(K_{pf} \times \text{Rate}^2\right) + \left(K_{nw} \times \text{Rate}^{\frac{1}{2}}\right)$$

(10) Calculate the absolute value of the difference between actual near-wellbore pressure differential and calculated near-wellbore pressure differential:

$$\left| \Delta P_{\text{near-wellbore}} - \Delta P_{\text{calculated\_near-wellbore}} \right|$$

(11) Calculate perforation friction pressure as follows:

$$\text{Perf friction pressure} = K_{pf} \times \text{Rate}^2$$

(12) Calculate tortuosity friction pressure as follows:

$$\text{Tortuosity friction pressure} = K_{nw} \times \text{Rate}^{\frac{1}{2}}$$

Fracture pressure analysis and perforation design

(13) Calculate parameter X which will be used as a cell to be minimized when using excel solver:

$$X = \left(\text{sum product} |\Delta P_{\text{near-wellbore}} - \Delta P_{\text{calculated near-wellbore}}|\right)^{1/2}$$

(14) Use the excel solver to perform the following action:

Objective function:
- Minimize X

Constraints:
- Number of open perforations that is being calculated must be less or equal to the total number of designed perforations
- K_{nw} must be greater or equal to 0

By changing:
- Number of open perforations
- Coefficient of near wellbore friction (K_{nw})
- Discharge coefficient (if the stage was performed at the end of the stage)

Use GRG nonlinear (generalized reduced gradient) method to find a solution using the excel solver.

Rate step down test example

A 4-point rate step-down test was performed at the beginning of a frac stage and summary of the test result is listed in the following table.

TVD = 9000′
Stage MD = 14,000′
Fluid density = 8.6 ppg
Number of perforations = 60 holes
Hole diameter = 0.42″
Discharge coefficient $C = 0.8$

Rate (BPM)	Surface pressure (psi)	Total friction pressure (psi)
86	8687	1999
81	8473	1824
63	7684	1268
43	6960	780
0	6013	0

(1) Calculate BHTP for each step rate as follows:

$$\text{BHTP@step rate 1} = \text{STP} + P_h - P_f$$
$$= 8687 + (0.052 \times 9000 \times 8.6) - 1999 = 10{,}713 \, \text{psi}$$
$$\text{BHTP@step rate 2} = 8473 + (0.052 \times 9000 \times 8.6) - 1824 = 10{,}673 \, \text{psi}$$
$$\text{BHTP@step rate 3} = 7684 + (0.052 \times 9000 \times 8.6) - 1268 = 10{,}441 \, \text{psi}$$
$$\text{BHTP@step rate 4} = 6960 + (0.052 \times 9000 \times 8.6) - 780 = 10{,}204 \, \text{psi}$$
$$\text{BH ISIP} = 6013 + (0.052 \times 9000 \times 8.6) - 0 = 10{,}038 \, \text{psi}$$

(2) Calculate actual near-wellbore pressure differential as follows $(\Delta P_{\text{near-wellbore}})$:

$$\Delta P_{\text{near-wellbore}} \text{ @step rate 1} = \text{BHTP at each step} - \text{BH ISIP}$$
$$= 10713 - 10038 = 675 \, \text{psi}$$
$$\Delta P_{\text{near-wellbore}} \text{ @step rate 2} = 10673 - 10038 = 636 \, \text{psi}$$
$$\Delta P_{\text{near-wellbore}} \text{ @step rate 3} = 10441 - 10038 = 403 \, \text{psi}$$
$$\Delta P_{\text{near-wellbore}} \text{ @step rate 4} = 10204 - 10038 = 166 \, \text{psi}$$
$$\Delta P_{\text{near-wellbore}} \text{ @ISIP} = 10038 - 10038 = 0 \, \text{psi}$$

(3) Calculate the coefficient of perforation friction as follows:

$$K_{pf} = 0.2369 \frac{\rho}{N_p^2 D^4 C_d^2} = 0.2369 \frac{8.6}{60^2 \times 0.42^4 \times 0.8^2} = 0.02841$$

(4) Assume a coefficient of near wellbore friction of 10.

(5) Calculate the calculated near-wellbore pressure differential as follows $(\Delta P_{\text{calculated\_near-wellbore}})$:

$$\Delta P_{\text{calculated\_near-wellbore}} \text{ @step rate 1} = \left(K_{pf} \times \text{Rate}^2\right) + \left(K_{nw} \times \text{Rate}^{\frac{1}{2}}\right)$$
$$= \left(0.02841 \times 86^2\right) + \left(10 \times 86^{\frac{1}{2}}\right)$$
$$= 301 \, \text{psi}$$

$$\Delta P_{\text{calculated\_near-wellbore}} \text{ @step rate 2} = \left(0.02841 \times 81^2\right) + \left(10 \times 81^{\frac{1}{2}}\right)$$
$$= 274 \, \text{psi}$$

$$\Delta P_{\text{calculated\_near-wellbore}} \text{ @step rate 3} = \left(0.02841 \times 63^2\right) + \left(10 \times 63^{\frac{1}{2}}\right)$$
$$= 190 \, \text{psi}$$

$$\Delta P_{\text{calculated\_near-wellbore}} @ \text{step rate } 4 = \left(0.02841 \times 43^2\right) + \left(10 \times 43^{\frac{1}{2}}\right)$$
$$= 117\,\text{psi}$$
$$\Delta P_{\text{calculated\_near-wellbore}} @ \text{ISIP} = \left(0.02841 \times 0^2\right) + \left(10 \times 0^{\frac{1}{2}}\right) = 0\,\text{psi}$$

(6) Calculate the absolute value of the difference between actual near-wellbore pressure differential and calculated near-wellbore pressure differential:

$$\text{Step rate } 1 = \left|\Delta P_{\text{near-wellbore}} - \Delta P_{\text{calculated near-wellbore}}\right| = 675 - 301 = 374\,\text{psi}$$
$$\text{Step rate } 2 = |636 - 274| = 361\,\text{psi}$$
$$\text{Step rate } 3 = |403 - 190| = 213\,\text{psi}$$
$$\text{Step rate } 4 = |166 - 117| = 50\,\text{psi}$$
$$\text{Step rate ISIP} = |0 - 0| = 0\,\text{psi}$$

(7) Calculate perforation friction pressure as follows:

$$\text{Perf friction pressure} @ \text{step rate } 1 = K_{pf} \times \text{Rate}^2 = 0.02841 \times 86^2 = 208\,\text{psi}$$
$$\text{Perf friction pressure} @ \text{step rate } 2 = 0.02841 \times 81^2 = 184\,\text{psi}$$
$$\text{Perf friction pressure} @ \text{step rate } 3 = 0.02841 \times 63^2 = 111\,\text{psi}$$
$$\text{Perf friction pressure} @ \text{step rate } 4 = 0.02841 \times 43^2 = 52\,\text{psi}$$
$$\text{Perf friction pressure} @ \text{ISIP} = 0.02841 \times 0^2 = 0\,\text{psi}$$

(8) Calculate tortuosity friction pressure as follows:

$$\text{Tortuosity friction pressure} @ \text{step rate } 1 = K_{nw} \times \text{Rate}^{\frac{1}{2}} = 10 \times 86^{\frac{1}{2}} = 92\,\text{psi}$$
$$\text{Tortuosity friction pressure} @ \text{step rate } 2 = 10 \times 81^{\frac{1}{2}} = 90\,\text{psi}$$
$$\text{Tortuosity friction pressure} @ \text{step rate } 3 = 10 \times 63^{\frac{1}{2}} = 79\,\text{psi}$$
$$\text{Tortuosity friction pressure} @ \text{step rate } 4 = 10 \times 43^{\frac{1}{2}} = 65\,\text{psi}$$
$$\text{Tortuosity friction pressure} @ \text{ISIP} = 10 \times 0^{\frac{1}{2}} = 0\,\text{psi}$$

(9) Calculate parameter X which will be used as a cell to be minimized when using excel solver:

$$X = (\text{sum product}|\Delta P_{\text{near-wellbore}} - \Delta P_{\text{calculated near-wellbore}}|)^{1/2}$$

$$X = ((374 \times 374) + (361 \times 361) + (213 \times 213) + (50 \times 50))^{\frac{1}{2}} = 564\,\text{psi}$$

(10) Use the excel solver to do the following action:

Objective:

The objective function is to minimize X which is essentially to minimize the calculated near wellbore pressure differential and actual near wellbore pressure differential.

Constraint:

Make sure to include the following two constraints:
- Number of open perforations that is being solved must be less than total designed number of perforations
- Coefficient of near-wellbore friction (k_{nw}) must be equal or greater than 0

By changing:
- Coefficient of near-wellbore friction (K_{nw})
- Number of open perforations

Use the GRG nonlinear solver to solve for the optimum solution. After running the solver for this example, the following parameters will be obtained:
- Number of open perforations = 33
- Coefficient of near-wellbore friction (K_{nw}) = 2.87

Perforation efficiency can then be calculated as follows:

$$\text{Perforation efficiency} = \frac{\text{Number of open perforations}}{\text{Total number of designed perforations}} = \frac{33}{60} = 55\%$$

The coefficient of near-wellbore friction and number of perforations are altered using the excel solver until the model converges and finds a solution which will result in population of Table 9.3. In addition, Fig. 9.6 illustrates the difference between actual and near wellbore pressure differential after finding a solution using the excel solver. As shown, the same process can be repeated for other rate step down tests to validate the result. It is always a good practice to plot "perforation efficiency" vs "stage number" to make

Table 9.3 Summary of the result after using excel solver Rate step-down analysis

Rate (bpm)	Surface pressure (psi)	ΔP_nearwellbore (psi)	ΔP_nearwellbore_cal (psi)	abs(ΔP_nearwellbore − ΔP_nearwellbore_cal)	ΔP_perf (psi)	ΔP_tort (psi)
86	8687	675	696	21	670	27
81	8473	636	620	16	594	26
63	7684	403	381	22	358	23
43	6960	166	185	18	166	19
0	6013	0	0	0	0	0

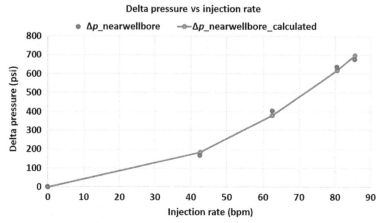

Fig. 9.6 Comparison between actual and calculated near wellbore pressure differential **after** running the excel solver.

sure the average perforation efficiency is in line across all stages and remove the outliers due to poor testing conditions. This is a powerful technique to compare perforation efficiency of various perforation designs across different wells.

CHAPTER TEN

Fracture treatment design

Introduction

Now that the concept of various pressures has been discussed, the next topic that will be discussed is fracture treatment design. In this chapter, various frac schedule concepts and calculations will be presented to design a frac treatment schedule that can be used in the field to pump a frac job. This chapter will primarily focus on designing a slick water and foam fracture treatment schedule with example problems that can be followed and applied. The workflow presented in this chapter can be used and applied to generate various fracture treatment schedules for testing various completions designs.

Absolute volume factor (AVF, gal/lbs)

Absolute volume factor (AVF) refers to the absolute volume that a solid occupies in water. For example, pouring 1 lb of Ottawa sand (2.65 specific gravity) into 1 gallon of water will displace 0.0453 gallons of water. The AVF depends on the density of the frac fluid and the specific gravity of the proppant used, and is calculated using Eq. (10.1).

$$\text{Absolute volume factor} = \frac{1}{\text{Absolute density}} = \frac{1}{\rho_f \times SG} \quad (10.1)$$

where AVF (absolute volume factor) is measured in gal/lb, ρ_f is the fluid density, ppg, and SG is the specific gravity of proppant.

As can be seen from the AVF equation, as specific gravity and fluid density increase, AVF decreases.

Example
Calculate AVF of Ottawa sand with SG of 2.65 considering freshwater density of 8.33 ppg.

$$\text{Absolute volume factor} = \frac{1}{8.33 \times 2.65} = 0.0453 \, \text{gal/lb}$$

Note: 0.0453 gal/lb is commonly used for hydraulic fracturing design schedules when regular sand is utilized.

Calculate AVF of sintered bauxite with a specific gravity of 3.4 (assuming freshwater).

$$\text{Absolute volume factor} = \frac{1}{8.33 \times 3.4} = 0.0353 \, \text{gal/lb}.$$

Dirty (slurry) vs clean frac fluid

In hydraulic fracturing operations, two terms are frequently used. The first one is referred to as *clean volume*, which means only water and chemicals make up the volume. The second commonly used term is *dirty (slurry) volume*, which means combinations of water, sand, and chemicals make up the volume. In addition, *clean rate* refers to the rate of the clean side (water and chemicals) and the *dirty rate* refers to the rate of the dirty side (water, sand, and chemicals). The slurry rate is typically read from a flow meter located on the blender and the clean rate is normally calculated using Eq. (10.2).

$$\text{Clean rate} = \frac{\text{Slurry rate}}{1 + (\text{sand concentration} \times \text{AVF})} \quad (10.2)$$

where slurry rate and clean rate are measured in bpm, sand concentration in ppg, and AVF in gal/lb.

Example
Calculate clean rate during a 3-ppg sand concentration if the slurry rate is 94 bpm (from flow meter) and Ottawa sand (SG=2.65) is being used.

$$\text{AVF for ottawa sand} = \frac{1}{2.65 \times 8.33} = 0.0453 \, \text{gal/lb}$$

$$\text{Clean rate} = \frac{\text{Slurry rate}}{1 + (\text{sand concentration} \times \text{AVF})} = \frac{94}{1 + (3 \times 0.0453)} = 83 \, \text{bpm}$$

The clean rate is *always less* than the slurry rate because if *only* water is being pumped downhole, the rate will be less as compared to the mix of water and sand.

Slurry (dirty) density (ppg)

Slurry density is the density of water and sand that are being pumped downhole. Slurry density has a direct impact on the hydrostatic pressure inside the casing during a frac job. As slurry density increases, hydrostatic pressure increases as well. If only water is being pumped downhole, the hydrostatic pressure of the column of water can be calculated using the hydrostatic pressure equation. However, when various sand concentrations are added to the water at various stages of hydraulic fracturing, slurry density must be considered in the hydrostatic pressure calculation.

$$\text{Slurry density} = \frac{\text{Base fluid density} + \text{sand conc.}}{1 + (\text{sand conc.} \times \text{AVF})} \qquad (10.3)$$

where base fluid density are measured in ppg, sand concentration in ppg, and AVF in gal/lb.

Assuming every parameter in the surface-treating pressure equation stays constant during a hydraulic fracturing stage, as slurry density (sand concentrations) and hydrostatic pressure increase, the surface-treating pressure must decrease. This is because surface-treating pressure is inversely related to hydrostatic pressure. During the flush stage, after all of the designed sand volume is put away and only fluid is being pumped, surface-treating pressure is usually increased (pressure increase depends on the sand concentration). This is due to hydrostatic pressure decreasing (when sand is no longer being pumped), and as a result surface-treating pressure increasing.

Example
Calculate slurry density and hydrostatic pressure of 2.5 ppg Ottawa sand mixed with freshwater at a true vertical depth (TVD) of 7450′. How much hydrostatic pressure will be attained if sand is cut and the well is flushed with only freshwater?

$$\text{Slurry density} = \frac{\text{Base fluid density} + \text{sand conc.}}{1 + (\text{sand conc.} \times \text{AVF})} = \frac{8.33 + 2.5}{1 + (2.5 \times 0.0453)}$$
$$= 9.73 \, \text{ppg}$$

Hydrostatic pressure of the slurry fluid can be calculated:

$$P_{h, \text{slurry fluid}} = 0.052 \times 7450 \times 9.73 = 3769 \, \text{psi}$$

Hydrostatic pressure of *only* fresh water can also be calculated:

$$P_{h,\text{fresh water}} = 0.052 \times 7450 \times 8.33 = 3227\,\text{psi}$$

Therefore:

Surface treating pressure increase $= 3769 - 3227$
$= 542\,\text{psi surface pressure increase}$

This example shows the importance of sand concentration in relation to surface-treating pressure monitoring during the frac job. As soon as the extra hydrostatic pressure created by the various sand concentrations is cut, the surface-treating pressure will increase.

Stage fluid clean volume (BBLs)

Clean volume refers to the volume of water and chemicals. Stage fluid clean volume is the amount of clean volume for every proppant stage concentration. For example, after finishing the acidization and pad stages, the proppant stage is started. In slick water frac, proppant stage concentration starts with low proppant concentration of 0.1–0.25 ppg. Each proppant stage concentration can have varying clean volume. For example, 500 BBLs of frac fluid can be the designed clean volume for a 0.25 ppg proppant stage. After staging up to a 0.5 ppg proppant stage, clean volume can now be 450 BBLs depending on the job design. The amount of clean volume for each proppant stage concentration is determined from the contact surface area that would like to be created and is typically obtained using a hydraulic frac software or from an optimized schedule designed using production data. The amount of water to pump is also a function of water availability and ease of transportation, well spacing (interlateral spacing), formation properties, and distance from adjacent producing wells. Sometimes in an attempt to mitigate fracture communication between new wells and producing wells in an area, the amounts of sand and water are both reduced to avoid fracture interference (also called frac hit). For example, if a horizontal well has been producing for the last 4 years and a pad consisting of six horizontal wells will be hydraulically fractured right next to the producing well located 750' apart, it is vital to adjust the schedule accordingly to mitigate fracture communication with the depleted nearby well. If the sand schedule is not properly designed and altered to facilitate this concern, this could have a detrimental production consequence to the depleted producing well. The concept of frac hit and mitigation is discussed in Chapter Twenty-One.

Stage fluid slurry (dirty) volume (BBLs)

As previously mentioned, slurry volume refers to the volume of water, proppant, and chemicals. Stage fluid slurry volume at different proppant concentrations can be calculated and is provided to the field personnel as part of the frac schedule. Slurry volume is always more than the clean volume since sand is considered part of the volume. Stage fluid slurry volume can be calculated using Eq. (10.4).

Slurry volume : Dirty volume
$$= \text{Clean volume} + (\text{sand conc.} \times \text{clean volume} \times \text{AVF})$$
(10.4)

where clean volume is measured in BBLs, sand concentration is in ppg, and AVF is in gal/lb.

Example
Calculate the dirty volume needed at 2 ppg Ottawa sand if 250 BBLs of clean volume is used.

Dirty volume = Clean volume + (sand conc. × clean volume × AVF)
$$= 250 + (2 \times 250 \times 0.0453) = 273 \, \text{BBLs}$$

As can be seen in the above calculation, the dirty volume is 23 BBLs more than the clean volume. This is because the 2 ppg Ottawa sand is used in that stage of treatment. As sand concentration increases throughout each stage, dirty volume increases too.

Stage proppant (lbs)

The next step in creating a frac schedule is to calculate stage proppant at different concentrations. Stage proppant is basically the amount of proppant that needs to be calculated at various proppant stage concentrations. For example, at 1 ppg proppant concentration, stage proppant might be 20,000 lbs of sand depending on the clean volume. Stage proppant is a function of proppant concentration and stage fluid clean volume. Stage proppant can be calculated using Eq. (10.5).

Stage proppant = 42 × proppant conc. × stage fluid clean volumes (10.5) where stage proppant is measured in lbs, proppant concentration is in ppg, and stage fluid clean volume is in BBLs.

Example
Calculate stage proppant at 2 ppg proppant concentration if 340 BBLs of clean volume is designed for this particular proppant stage.

$$\text{Stage proppant} = 42 \times \text{proppant conc.} \times \text{stage fluid clean volume}$$
$$= 42 \times 2 \times 340 = 28,560 \text{ lbs}$$

Sand per foot (lb/ft)

Sand per foot is the amount of sand per foot that can be calculated on both stage and well levels (assuming geometric design).

$$\text{Sand per foot} = \frac{\text{total sand per stage}}{\text{stage length}} = \frac{\text{total sand per well}}{\text{perforated lateral length}} \quad (10.6)$$

Water per foot

Water per foot is the amount of water per foot that can be calculated on both stage and well levels (assuming geometric design).

$$\text{Water per foot} = \frac{\text{total water per stage}}{\text{stage length}} = \frac{\text{total water per well}}{\text{perforated lateral length}} \quad (10.7)$$

Example
In a particular region of the Barnett Shale, 800 lbs/ft of sand and 40 BBL/ft of water has been determined as the optimum sand and water per foot based on the actual production data in 400' frac stage spacing. Calculate total sand and water per stage.

$$\text{Total sand per stage} = 800 \times 400 = 320,000 \text{ lbs of sand per stage}$$
$$\text{Total water per stage} = 40 \times 400 = 16,000 \text{ BBLs of water per stage}$$

Sand-to-water ratio (SWR, lb/gal)

Sand-to-water ratio (SWR) is another important metric in frac design. Total sand divided by total water per stage will yield the SWR. A lower SWR means a higher percentage of water in relation to sand. A higher SWR indicates more aggressive stages by pumping higher amounts of sand in relation to water. Typically, the SWR in slick water fracs ranges from 0.7 to 1.7. A SWR in more viscous fluid type systems such as cross-linked jobs could be much higher.

$$\text{Sand-to-water ratio} = \text{SWR} = \frac{\text{total sand}}{\text{total water}} \qquad (10.8)$$

where total sand is in lbs and total water is in gal.

Example
Calculate SWR for a well with 400,000 lbs of sand per stage and 8500 BBLs of water.

$$\text{SWR} = \frac{\text{total sand}}{\text{total water}} = \frac{400{,}000}{8500 \times 42} = 1.12 \,\text{lb/gal}$$

Slick water frac schedule

Completions engineers design hydraulic frac jobs. Different hydraulic fracturing software that can be used to design an optimum job is available in the industry. The purpose of this section is not to dig deep into the derivation of equations and calculations, but to understand the basic concepts of a frac schedule provided to the field personnel for execution. The idea behind optimum fracture design is to spend the least amount of money and get the most out of the reservoir by stimulating and contacting as much reservoir rock as possible. The best and most comprehensive design is obtained by investigating completed wells and comparing this data to the production performance of the wells. Computer modeling can be run to solve for optimum design. However, the production performance of the well should dictate the completions design that is chosen for the well.

Every horizontal well is divided into many stages. The number of stages for each well depends on the lateral length. Normally, as lateral length increases, the number of stages increases as well. For example, a well with a 4000' lateral length could have 20 frac stages (depending on the design) but a well with an 8000' lateral length could have 40 frac stages. Therefore, hydraulic fracturing occurs throughout multiple stages to stimulate and contact as much reservoir rock volume as possible. During slick water frac jobs, every stage can use anywhere from 150,000 to 800,000 lbs of proppant. The amount of proppant pumped downhole is massive. For example, an 8000' horizontal well that has 40 stages and uses 400,000 lbs of proppant per stage will need 16 million lbs of sand. In addition to proppant, water will be needed. The average amount of water per stage depends on many design factors such as stage length, amount of sand, treatment difficulty, etc. Typically, a stage can use anywhere between 4000 and 14,000 BBLs of water in slick water jobs. In cross-linked jobs, less water is required since high viscous fluid carries the slurry fluid into the formation. For example, an 8000' horizontal well that has 40 stages and uses 8000 BBLs of water per stage will need 320,000 BBLs (13.44 million gallons) of water. These examples are discussed to give perspective on the total amount of sand and water that must be used to stimulate these low-permeability reservoirs during slick water frac jobs.

Example

Fill in Table 10.1 using the following assumptions, and calculate the SWR, sand/ft, water/ft, pad %, % 100 mesh, and % 40/70 mesh.
Proppant type = Ottawa sand, SG = 2.65, Stage length = 350'
Stage fluid slurry volume sample calculation @ 0.25 ppg and 600 BBLs:

$$\text{Dirty volume} = \text{clean volume} + (\text{sand conc.} \times \text{clean volume} \times \text{AVF})$$
$$= 600 + (0.25 \times 600 \times 0.0453) = 607 \text{ barrels}$$

% of total clean volume sample calculation @ 0.25 ppg and 600 BBLs:

$$\% \text{ of total clean volume} = \frac{\text{stage fluid clean volume}}{\text{total clean volume}} \times 100 = \frac{600}{8236} \times 100$$
$$= 7.3\%$$

Stage proppant sample calculation at 0.25 ppg and 600 BBLs:

$$\text{Stage proppant} = 42 \times \text{proppant conc.} \times \text{stage fluid clean volume}$$
$$= 42 \times 0.25 \times 600 = 6300 \text{ lbs}$$

Fracture treatment design   157

Table 10.1 Slick water schedule example
85 bpm, 396,554 lbs

Stage name	Pump rate bpm	Fluid name	Stage fluid clean vol. BBLs	Stage fluid slurry BBLs	% of total clean vol. %	Prop conc. ppg	Stage proppant lbs	% of total prop. %	Cumulative prop. lbs	Stage time min
Pump ball	15	Slick water	300			0				
5% HCl acid	85	Acid	60			0				
Pad	85	Slick water	410			0				
100 mesh	85	Slick water	600			0.25				
100 mesh	85	Slick water	550			0.5				
100 mesh	85	Slick water	375			0.75				
100 mesh	85	Slick water	550			1				
100 mesh	85	Slick water	450			1.25				
100 mesh	85	Slick water	500			1.5				
40/70 mesh	85	Slick water	450			0.5				
40/70 mesh	85	Slick water	365			0.75				
40/70 mesh	85	Slick water	365			1				
40/70 mesh	85	Slick water	455			1.25				
40/70 mesh	85	Slick water	350			1.5				
40/70 mesh	85	Slick water	379			1.75				
40/70 mesh	85	Slick water	389			2				
40/70 mesh	85	Slick water	380			2.25				
40/70 mesh	85	Slick water	360			2.5				
40/70 mesh	85	Slick water	299			2.75				
40/70 mesh	85	Slick water	299			3				
Flush	85	Slick water	350			0				
		Total clean volume	8236	BBLs						

% of total proppant sample calculation at 0.25 ppg and 600 BBLs:

$$\% \text{ of total proppant} = \frac{\text{stage proppant}}{\text{total proppant}} \times 100 = \frac{6300}{396,554} \times 100 = 1.6\%$$

Stage time sample calculation at 0.25 ppg and 600 BBLs:

$$\text{Stage time} = \frac{\text{stage fluid slurry volume}}{\text{pump rate}} = \frac{607}{85} = 7.14 \text{ min}$$

Sand per foot calculation:

$$\text{Sand per foot} = \frac{\text{total sand per stage}}{\text{stage length}} = \frac{396,554}{350} = 1133 \text{ lb/ft}$$

Water per foot calculation:

$$\text{Water per foot} = \frac{\text{total water per stage}}{\text{stage length}} = \frac{8236}{350} = 24 \text{ barrels/ft}$$

SWR calculation:

$$\text{SWR} = \frac{\text{total sand}}{\text{total water}} = \frac{396,554}{8236 \times 42} = 1.15 \text{ lb/gal}$$

Pad % calculation:

$$\text{Pad\%} = \frac{\text{pad volume}}{\text{total slurry volume excluding acid and ball}} \times 100 = \frac{410}{7876} \times 100 = 4.94\%$$

Please note that some completions engineers do include acid volume as part of the pad volume calculation but this example excludes acid volume as a percentage of the pad volume. Table 10.2 shows the completed slick water schedule for this problem. This slick water example format is very similar to the provided treatment schedule for job execution in the field. As previously discussed, the designed sand and water volumes are heavily dependent on the success in each area as well the optimum economic sand schedule that yields the highest net present value (NPV). For instance, if pumping higher sand and water loadings in a particular field would yield better well performance results and justifies spending additional capital expenditure on higher sand and water loadings, more sand and water loadings will be used for that particular area. In essence, the incremental gain obtained from production results must justify spending the additional capital on higher sand and water loadings in order to economically justify pumping such schedules. This discussion will be heavily discussed in Chapter Twenty-One.

Fracture treatment design

Table 10.2 Completed slick water schedule answer 85 bpm, 396,554 lbs

Stage name	Pump rate bpm	Fluid name	Stage fluid clean vol. BBLs	Stage fluid slurry vol. BBLs	% of total clean vol. %	Prop conc. ppg	Stage proppant lbs	% of total prop. %	Cumulative prop. lbs	Stage time min
Pump ball	15	Slick water	300	300	3.64	0	0		0	20.0
5% HCl acid	85	Acid	60	60	0.73	0	0		0	0.7
Pad	85	Slick water	410	410	4.98	0	0		0	4.8
100 mesh	85	Slick water	600	607	7.29	0.25	6300	1.6	6300	7.1
100 mesh	85	Slick water	550	562	6.68	0.5	11,550	2.9	17,850	6.6
100 mesh	85	Slick water	375	388	4.55	0.75	11,813	3.0	29,663	4.6
100 mesh	85	Slick water	550	575	6.68	1	23,100	5.8	52,763	6.8
100 mesh	85	Slick water	450	475	5.46	1.25	23,625	6.0	76,388	5.6
100 mesh	85	Slick water	500	534	6.07	1.5	31,500	7.9	107,888	6.3
40/70 mesh	85	Slick water	450	460	5.46	0.5	9450	2.4	117,338	5.41
40/70 mesh	85	Slick water	365	377	4.43	0.75	11,498	2.9	128,835	4.44
40/70 mesh	85	Slick water	365	381	4.43	1	15,330	3.9	144,165	4.49
40/70 mesh	85	Slick water	455	481	5.52	1.25	23,888	6.0	168,053	5.66
40/70 mesh	85	Slick water	350	374	4.25	1.5	22,050	5.6	190,103	4.40
40/70 mesh	85	Slick water	379	409	4.60	1.75	27,857	7.0	217,959	4.81
40/70 mesh	85	Slick water	389	424	4.72	2	32,676	8.2	250,635	4.99
40/70 mesh	85	Slick water	380	419	4.61	2.25	35,910	9.1	286,545	4.93
40/70 mesh	85	Slick water	360	401	4.37	2.5	37,800	9.5	324,345	4.71
40/70 mesh	85	Slick water	299	336	3.63	2.75	34,535	8.7	358,880	3.95
40/70 mesh	85	Slick water	299	340	3.63	3	37,674	9.5	396,554	3.99
Flush	85	Slick water	350	350	4.25	0	0	0.0	0	4.12
Total clean volume			8236	BBLs						

Continued

Table 10.2 Completed slick water schedule answer—cont'd 85 bpm, 396,554 lbs

Stage name	Pump rate bpm	Fluid name	Stage fluid clean vol. BBLs	Stage fluid slurry vol. BBLs	% of total clean vol. %	Prop conc. ppg	Stage proppant lbs	% of total prop. %	Cumulative prop. lbs	Stage time min
		Sand/water ratio	1.15	lb/gal						
		Pad percentage	4.94	%			Total stage time (min)	118.4		
		100 mesh (lbs)	107,888	OR	27%					
							Stage length (ft)	350		
		40/70 mesh (lbs)	288,666	OR	73%		Water/ft	24	BBL/ft	
		Total (lbs)	396,554				Sand/ft	1133	lb/ft	

Foam frac schedule and calculations

Nitrogen gas measurements are reported in standard cubic feet (SCF). When pressure is exerted on nitrogen gas, the volume of nitrogen gas will decrease. In contrast, when heat is applied to nitrogen gas, the volume of nitrogen gas will increase. Nitrogen that is brought on location during foam frac jobs is in liquid form. When nitrogen is pumped downhole, it will be exposed to both pressure and temperature at downhole conditions. Since temperature and pressure affect this gas in opposite ways, it is very important to use bottom-hole conditions to calculate nitrogen volume. Volume factor tables or charts can be used to calculate how many SCF of nitrogen gas is equal to one barrel of liquid. To obtain the volume factor of nitrogen at downhole conditions, bottom-hole treating pressure (BHTP) and bottom-hole static temperature (BHST) must be available. The volume factor of nitrogen can be approximated using Eq. (10.9).

$$\text{Standard cubic feet of nitrogen per barrel of liquid}: \frac{\text{SCF}}{\text{BBL}}$$
$$= \left(\frac{Z \text{ factor at standard condition} \times \text{standard temperature} \times \text{BHTP}}{Z \times (\text{BHST} + 460) \times \text{atmospheric pressure}} \right)$$
$$\times 5.615$$

(10.9)

where SCF/BBL is the standard cubic feet of nitrogen per barrel of volume, Z factor at standard conditions = 1, standard temperature = 520°R (60 + 460), BHTP is bottom-hole treating pressure in psi, Z is the compressibility factor at downhole condition, BHST is bottom-hole static temperature in °F, and atmospheric pressure = 14.7 psia.

In addition, there are other charts and plots available in various handbooks that can be used to obtain the volume factor of nitrogen at different pressures and temperatures.

Foam volume

Foam volume can be calculated when clean fluid volume (volume of water) is available. Foam volume is calculated using Eq. (10.10).

$$\text{Foam volume} = \frac{\text{liquid volume}}{1 - \text{FQ}} \quad (10.10)$$

where foam volume is measured in BBL, liquid volume in BBL, FQ is the foam quality in %.

Example
A foam frac requires a total of 600 barrels of foam volume. Calculate the total clean volume (volume of water) required assuming a 70% foam quality.

$$\text{Foam volume} = \frac{\text{liquid volume}}{(1 - FQ)} \rightarrow 600 = \frac{\text{liquid volume}}{1 - 70\%}$$
$$= 180 \text{ BBLs of liquid volume}$$

Nitrogen volume

Before being able to calculate nitrogen volume, the nitrogen volume factor at downhole pressure and temperature must be known. Once the nitrogen volume factor is calculated, the nitrogen volume for the job can be calculated using Eq. (10.11).

$$\text{Nitrogen volume} = \text{clean foam volume} \times VF \times FQ \quad (10.11)$$

where nitrogen volume, BBLs, clean foam volume, BBLs, VF is the nitrogen volume factor, SCF/BBL, and FQ, %.

Example
Calculate nitrogen volume assuming 625 barrels of clean foam volume using the following parameters:
BHTP = 2500 psi, BHST = 125°F, FQ = 70%
Step 1: First, nitrogen volume factor at 125°F and 2500 psi must be obtained using Eq. (10.9). From this equation, nitrogen volume factor is 810 SCF/BBL.

Step 2: Calculate nitrogen volume assuming 70% foam quality:

$$\text{Nitrogen volume} = \text{clean foam volume} \times VF \times FQ = 625 \times 810 \times 70\%$$
$$= 354,375 \text{ SCF}$$

Blender sand concentration

During foam frac jobs, the sand concentration at the blender must be much higher than the sand concentration at downhole conditions. This is because the slurry fluid carrying the sand from the blender will be diluted with nitrogen. Therefore, blender sand concentrations need to be calculated using Eq. (10.12).

$$\text{Blender sand concentration} = \frac{\text{BH sand concentration}}{1 - \text{FQ}} \quad (10.12)$$

where blender sand concentration is in ppg, BH sand concentration is in ppg, and FQ is in %.

Example
The bottom-hole sand concentrations for a foam frac job are designed at 0.5, 1, 1.5, and 2 ppg. Calculate blender sand concentration at these concentrations assuming 75% foam quality.

$$\text{Blender sand concentration} @ 0.5 = \frac{0.5}{1 - 75\%} = 2\,\text{ppg}$$

$$\text{Blender sand concentration} @ 1 = \frac{1}{1 - 75\%} = 4\,\text{ppg}$$

$$\text{Blender sand concentration} @ 1.5 = \frac{1.5}{1 - 75\%} = 6\,\text{ppg}$$

$$\text{Blender sand concentration} @ 2 = \frac{2}{1 - 75\%} = 8\,\text{ppg}$$

Slurry factor (SF)

Slurry factor (SF) is one of the most important calculations that must be performed for designing a foam frac job. Since the proppant concentration at bottom hole and blender is different, SF at surface (blender) and bottom hole must both be calculated. Since adding sand to fluid on foam jobs decreases foam quality, the SF calculation becomes very important. Increasing sand concentration will decrease clean rate when designing a foam job schedule.

$$\text{Slurry factor}\,(SF) = 1 + (\text{sand concentration} \times AVF) \qquad (10.13)$$

where sand concentration is in ppg and AVF is absolute volume factor, gal/lb.

Example

Calculate SF at bottom hole and surface (blender) assuming regular sand with SG of 2.65 at various BH sand concentrations of 1 and 2 ppg, assuming 70% foam quality.

$$AVF = \frac{1}{2.65 \times 8.33} = 0.0453\,\text{gal/lb}$$

$$\text{Slurry factor}\,@\,1\,\text{ppg BH conc.} = 1 + (\text{sand conc.} \times AVF)$$
$$= 1 + (1 \times 0.0453) = 1.0453$$
$$\text{Slurry factor}\,@\,2\,\text{ppg BH conc.} = 1 + (2 \times 0.0453) = 1.0906$$

The 1- and 2-ppg bottom-hole sand concentrations are calculated to be 3.33 and 6.67 ppg sand concentrations at the blender, assuming 70% foam quality.

$$\text{Slurry factor}\,@\,3.33\,\text{ppg blender conc.} = 1 + (3.33 \times 0.0453) = 1.151$$
$$\text{Slurry factor}\,@\,6.67\,\text{ppg blender conc.} = 1 + (6.67 \times 0.0453) = 1.302$$

Clean rate (no proppant)

Clean rate (assuming no proppant) during pad has to be calculated when designing a foam frac job. Clean rate during pad and no proppant can be calculated using Eq. (10.14).

$$\text{Clean rate}\,(\text{no proppant}) = \text{Foam rate} \times (1 - FQ) \qquad (10.14)$$

where clean rate assuming no proppant is measured in bpm, foam rate is also known as downhole rate is in bpm, and FQ is in %.

Example

The foam rate for a foam job is designed at 30 bpm. Calculate the clean rate during pad assuming 75% foam quality.

$$\text{Clean rate}\,(\text{no proppant}) = \text{foam rate} \times (1 - FQ) = 30 \times (1 - 75\%)$$
$$= 7.5\,\text{bpm}$$

Clean rate (with proppant)

Once clean rate with no proppant is calculated, clean rate with proppant at different bottom-hole concentrations must also be calculated when designing a sand schedule for a foam frac job. Clean rate (with proppant) is calculated using Eq. (10.15).

$$\text{Clean rate (with proppant)} : \text{Clean rate}_{\text{proppant}} = \frac{\text{Clean rate}_{\text{no proppant(pad)}}}{SF_{BH}} \quad (10.15)$$

where clean rate$_{\text{proppant}}$ is the clean rate with proppant measured in bpm, clean rate$_{\text{no proppant}}$ is the clean rate during pad in bpm, and SF_{BH} is the slurry factor at bottom-hole sand concentration.

Example
The bottom-hole sand concentration during a foam frac job is at 2 ppg. The design foam rate was 25 bpm. Assuming a 68% foam quality and regular sand with 2.65 SG, calculate clean rate during 2 ppg bottom-hole sand concentration.

$$\text{Clean rate(no proppant)} = \text{Foam rate} \times (1 - FQ) = 25 \times (1 - 68\%)$$
$$= 8 \, \text{bpm}$$

$$\text{Slurry factor} = 1 + (\text{sand concentration} \times AVF) = 1 + (2 \times 0.0453)$$
$$= 1.0906$$

$$\text{Clean rate}_{\text{proppant}} = \frac{\text{Clean rate}_{\text{no proppant(pad)}}}{SF_{BH}} = \frac{8}{1.0906} = 7.34 \, \text{bpm}$$

Slurry rate (with proppant)

The next important calculation when designing a foam frac job is the slurry rate with proppant calculation. Slurry rate with proppant can be calculated using Eq. (10.16).

$$\text{Slurry rate (with proppant)} : \text{Slurry rate}_{\text{proppant}} = \left(\frac{\text{Clean rate}_{\text{no proppant(pad)}}}{SF_{BH}}\right) \times SF_{\text{blender}} \quad (10.16)$$

where slurry rate$_{proppant}$ is the slurry rate with proppant measured in bpm, clean rate$_{no\ proppant}$ is the clean rate during pad in bpm, SF_{BH} is the slurry factor at bottom-hole sand concentration, and $SF_{blender}$ is the slurry factor at blender sand concentration.

Example
Calculate slurry rate with proppant assuming a 72% foam quality and clean rate (no proppant) of 6 bpm during 1.5 ppg bottom-hole sand concentration. Assume regular sand with SG of 2.65.

$$SF_{BH} = 1 + (\text{sand concentration} \times AVF) = 1 + (1.5 \times 0.0453) = 1.068$$

$$\text{Blender sand concentration} = \frac{\text{BH sand concentration}}{1 - FQ} = \frac{1.5}{1 - 72\%}$$
$$= 5.36\,\text{ppg}$$

$$SF_{blender} = 1 + (5.36 \times 0.0453) = 1.242$$

$$\text{Slurry rate}_{proppant} = \left(\frac{\text{Clean rate}_{pad}}{SF_{BH}}\right) \times SF_{blender} = \left(\frac{6}{1.068}\right) \times 1.242$$
$$= 6.98\,\text{bpm}$$

Nitrogen rate (with and without proppant)

The next step in designing a foam frac job is to calculate nitrogen rate with and without proppant. Nitrogen rate without proppant can be calculated using Eq. (10.17).

$$\text{Nitrogen rate}(\text{no proppant}) = \text{dirty foam rate} \times VF \times FQ \quad (10.17)$$

where nitrogen rate assuming no proppant is measured in SCF/min, dirty foam rate is the designed foam rate in bpm, VF is the nitrogen volume factor in SCF/BBL, and FQ is in %.

In addition, nitrogen rate with proppant is calculated using Eq. (10.18).

$$\text{Nitrogen rate}(\text{with proppant}) = (\text{dirty foam rate} - \text{slurry rate}) \times VF \quad (10.18)$$

where foam rate is measured in bpm, slurry rate is in bpm, and VF is the nitrogen volume factor in SCF/BBL.

Fracture treatment design

Example
A foam frac job is scheduled to have a foam rate of 32 bpm with a slurry rate of 8.2 bpm. Nitrogen volume factor is calculated to be 1001 SCF/BBL and foam quality scheduled for the job is 70%. Calculate the nitrogen rate with and without proppant for this particular stage.

$$\text{Nitrogen rate(pad)} = \text{Dirty foam rate} \times \text{VF} \times \text{FQ} = 32 \times 1001 \times 70\%$$
$$= 22{,}422 \, \text{SCF/min}$$

$$\text{Nitrogen rate(with proppant)} = (\text{Dirty foam rate} - \text{slurry rate}) \times \text{VF}$$
$$= (32 - 8.2) \times 1001 = 23{,}824 \, \text{SCF/min}$$

Example
You are a completions engineer responsible for designing a foam frac schedule for a coalbed methane (CBM) well. Assuming the following properties and schedule, calculate the rest of the foam frac schedule.
ISIP = 2150 psi, hydrostatic pressure = 1350 psi, BHST = 100°F, FQ = 70%, SG = 2.65 (regular sand), dirty foam rate (bottom hole) = 30 bpm

Stage name	BH proppant conc.	Dirty foam volume
	ppg	BBLs
Acid	0.00	6.0
Pad	0.00	40.0
20/40	1.00	30.0
20/40	1.50	30.0
20/40	2.00	30.0
20/40	2.50	30.0
20/40	3.00	30.0
Flush	0.0	45.0

Step 1: Calculate nitrogen volume factor based on BHTP and BHST:

$$\text{BHTP} = P_h + \text{ISIP} = 1350 + 2150 = 3500 \, \text{psi}$$

Nitrogen volume factor at 3500 psi and 100°F is approximately **1139 SCF/BBL** using Eq. (10.9).

Step 2: BH proppant concentration for each proppant stage is provided. Calculate blender proppant concentration for each proppant concentration using the equation below.

$$\text{Blender sand concentration} = \frac{\text{BH sand concentration}}{1 - \text{FQ}}$$

Stage name	BH proppant conc. ppg	Blender proppant conc. ppg
Acid	0.00	0.00
Pad	0.00	0.00
20/40	1.00	$1/(1-70\%) = 3.33$
20/40	1.50	$1.5/(1-70\%) = 5$
20/40	2.00	$2/(1-70\%) = 6.67$
20/40	2.50	$2.5(1-70\%) = 8.33$
20/40	3.00	$3/(1-70\%) = 10$
Flush	0.0	0.00

Step 3: Calculate bottom-hole SF for each proppant stage:

$$\text{AVF} = \frac{1}{2.65 \times 8.33} = 0.0453 \, \text{gal/lb}$$

Slurry factor $= 1 + (\text{sand concentration} \times \text{AVF})$

Stage name	BH proppant conc. ppg	BH slurry factor
Acid	0.00	$1 + (0 \times 0.0453) = 1$
Pad	0.00	$1 + (0 \times 0.0453) = 1$
20/40	1.00	$1 + (1 \times 0.0453) = 1.05$
20/40	1.50	$1 + (1.5 \times 0.0453) = 1.07$
20/40	2.00	$1 + (2 \times 0.0453) = 1.09$
20/40	2.50	$1 + (2.5 \times 0.0453) = 1.11$
20/40	3.00	$1 + (3 \times 0.0453) = 1.14$
Flush	0.0	$1 + (0 \times 0.0453) = 1$

Step 4: Calculate clean foam volume simply by taking dirty foam volume (provided) and dividing it by the calculated bottom-hole SF.

Stage name	Dirty foam volume BBLs	BH slurry factor	Clean foam volume BBLs
Acid	6.0	$1 + (0 \times 0.0453) = 1$	$6/1 = 6$
Pad	40.0	$1 + (0 \times 0.0453) = 1$	$40/1 = 40$
20/40	30.0	$1 + (1 \times 0.0453) = 1.05$	$30/1.05 = 28.7$
20/40	30.0	$1 + (1.5 \times 0.0453) = 1.07$	$30/1.07 = 28.09$
20/40	30.0	$1 + (2 \times 0.0453) = 1.09$	$30/1.09 = 27.51$
20/40	30.0	$1 + (2.5 \times 0.0453) = 1.11$	$30/1.11 = 26.95$
20/40	30.0	$1 + (3 \times 0.0453) = 1.14$	$30/1.14 = 26.41$
Flush	45.0	$1 + (0 \times 0.0453) = 1$	$45/1 = 45$

Fracture treatment design

Step 5: Calculate clean fluid volume using the following equation:

$$\text{Clean fluid volume} = \text{Clean foam volume} \times (1 - FQ)$$

Stage name	Clean foam volume BBLs	Clean fluid volume BBLs
Acid	6/1 = 6	6 × (1 − 0%) = 6
Pad	40/1 = 40	40 × (1 − 70%) = 12
20/40	30/1.05 = 28.7	28.7 × (1 − 70%) = 8.61
20/40	30/1.07 = 28.09	28.09 × (1 − 70%) = 8.43
20/40	30/1.09 = 27.51	27.51 × (1 − 70%) = 8.25
20/40	30/1.11 = 26.95	26.95 × (1 − 70%) = 8.08
20/40	30/1.14 = 26.41	26.41 × (1 − 70%) = 7.92
Flush	45/1 = 45	45 × (1 − 70%) = 13.5

Step 6: Calculate surface SF for each proppant stage:

Stage name	Blender proppant conc. ppg	Surface slurry factor
Acid	0.00	1 + (0 × 0.0453) = 1
Pad	0.00	1 + (0 × 0.0453) = 1
20/40	1/(1 − 70%) = 3.33	1 + (3.33 × 0.0453) = 1.15
20/40	1.5(1 − 70%) = 5	1 + (5 × 0.0453) = 1.23
20/40	2/(1 − 70%) = 6.67	1 + (6.67 × 0.0453) = 1.30
20/40	2.5/(1 − 70%) = 8.33	1 + (8.33 × 0.0453) = 1.38
20/40	3/(1 − 70%) = 10	1 + (10 × 0.0453) = 1.45
Flush	0.00	1 + (0.0453) = 1

Step 7: Calculate dirty fluid volume for each proppant stage as shown below:

Dirty fluid volume = clean fluid volume × surface(blender)slurry factor

Step 8: Calculate amount of sand for each blender sand concentration as shown below:

$$\text{Amount of sand} = \text{clean fluid volume in gallons} \times \text{blender proppant concentration}$$

Step 9: Calculate nitrogen volume for each stage as shown below:

$$\text{Nitrogen volume} = \text{clean foam volume} \times VF \times FQ$$

Stage name	Clean fluid volume		Surface slurry factor	Dirty fluid volume
BBLs	BBLs			BBLs
Acid	$6 \times (1 - 0\%) = 6$		$1 + (0 \times 0.0453) = 1$	$6 \times 1 = 6$
Pad	$40 \times (1 - 70\%) = 12$		$1 + (0 \times 0.0453) = 1$	$12 \times 1 = 12$
20/40	$28.7 \times (1 - 70\%) = 8.61$		$1 + (3.33 \times 0.0453) = 1.15$	$8.61 \times 1.15 = 9.91$
20/40	$28.09 \times (1 - 70\%) = 8.43$		$1 + (5 \times 0.0453) = 1.23$	$8.43 \times 1.23 = 10.34$
20/40	$27.51 \times (1 - 70\%) = 8.25$		$1 + (6.67 \times 0.0453) = 1.30$	$8.25 \times 1.30 = 10.74$
20/40	$26.95 \times (1 - 70\%) = 8.08$		$1 + (8.33 \times 0.0453) = 1.38$	$8.08 \times 1.38 = 11.14$
20/40	$26.41 \times (1 - 70\%) = 7.92$		$1 + (10 \times 0.0453) = 1.45$	$7.92 \times 1.45 = 11.51$
Flush	$45 \times (1 - 70\%) = 13.5$		$1 + (0 \times 0.0453) = 1$	$13.5 \times 1 = 13.5$

Stage name	Blender proppant conc.	Clean fluid volume	Sand (lbs)		
	ppg	BBLs	Stage		CUM
Acid	0.00	$6 \times (1 - 0\%) = 6$	$6 \times 42 \times 0 = 0$		0
Pad	0.00	$40 \times (1 - 70\%) = 12$	$12 \times 42 \times 0 = 0$		0
20/40	$1/(1 - 70\%) = 3.33$	$28.7 \times (1 - 70\%) = 8.61$	$8.61 \times 42 \times 3.33 = 1205$		1205
20/40	$1.5/(1 - 70\%) = 5$	$28.09 \times (1 - 70\%) = 8.43$	$8.43 \times 42 \times 5 = 1770$		$1205 + 1770 = 2975$
20/40	$2/(1 - 70\%) = 6.67$	$27.51 \times (1 - 70\%) = 8.25$	$8.25 \times 42 \times 6.67 = 2311$		$2975 + 2311 = 5286$
20/40	$2.5/(1 - 70\%) = 8.33$	$26.95 \times (1 - 70\%) = 8.08$	$8.08 \times 42 \times 8.33 = 2830$		$5286 + 2830 = 8115$
20/40	$3/(1 - 70\%) = 10$	$26.41 \times (1 - 70\%) = 7.92$	$7.92 \times 42 \times 10 = 3328$		$8115 + 3328 = 11,443$
Flush	0.00	$45 \times (1 - 70\%) = 13.5$	$13.5 \times 42 \times 0 = 0$		11,443

Fracture treatment design

Stage name	FQ %	Clean foam volume BBLs	Nitrogen (SCF) Stage	CUM
Acid	0	6/1 = 6	6 × 0 × 1139 = 0	0
Pad	70	40/1 = 40	40 × 70% × 1139 = 31,892	31,892
20/40	70	30/1.05 = 28.7	28.7 × 70% × 1139 = 22,882	31,892 + 22,882 = 54,774
20/40	70	30/1.07 = 28.09	28.09 × 70% × 1139 = 22,397	54,774 + 22,397 = 77,172
20/40	70	30/1.09 = 27.51	27.51 × 70% × 1139 = 21,932	77,172 + 21,932 = 99,104
20/40	70	30/1.11 = 26.95	26.95 × 70% × 1139 = 21,486	99,104 + 21,486 = 120,589
20/40	70	30/1.14 = 26.41	26.41 × 70% × 1139 = 21,057	120,589 + 21,057 = 141,647
Flush	70	45/1 = 45	45 × 70% × 1139 = 35,879	141,647 + 35,879 = 177,525

Step 10: Dirty foam rate (designed rate) is given in the problem to be 30 bpm. Clean foam rate can be calculated as follows:

$$\text{Clean foam rate} = \frac{\text{dirty foam rate}}{\text{BH slurry factor}}$$

Stage name	BH slurry factor	Rate (bpm)	
Dirty foam rate			Clean foam rate
Acid	$1 + (0 \times 0.0453) = 1$	30	$30/1 = 30$
Pad	$1 + (0 \times 0.0453) = 1$	30	$30/1 = 30$
20/40	$1 + (1 \times 0.0453) = 1.05$	30	$30/1.05 = 28.70$
20/40	$1 + (1.5 \times 0.0453) = 1.07$	30	$30/1.07 = 28.09$
20/40	$1 + (2 \times 0.0453) = 1.09$	30	$30/1.09 = 27.51$
20/40	$1 + (2.5 \times 0.0453) = 1.11$	30	$30/1.11 = 26.95$
20/40	$1 + (3 \times 0.0453) = 1.14$	30	$30/1.14 = 26.41$
Flush	$1 + (0 \times 0.0453) = 1$	30	$30/1 = 30$

Step 11: Clean fluid rate for each proppant stage can be calculated as follows:

$$\text{Clean fluid rate} = \text{clean foam rate} \times (1 - FQ)$$

Stage name	FQ	Rate (bpm)		
	%	Dirty foam rate	Clean foam rate	Clean fluid rate
Acid	0	30	$30/1 = 30$	$30 \times (1 - 0\%) = 30$
Pad	70	30	$30/1 = 30$	$30 \times (1 - 70\%) = 9$
20/40	70	30	$30/1.05 = 28.70$	$28.70 \times (1 - 70\%) = 8.61$
20/40	70	30	$30/1.07 = 28.09$	$28.09 \times (1 - 70\%) = 8.43$
20/40	70	30	$30/1.09 = 27.51$	$27.51 \times (1 - 70\%) = 8.25$
20/40	70	30	$30/1.11 = 26.95$	$26.95 \times (1 - 70\%) = 8.08$
20/40	70	30	$30/1.14 = 26.41$	$26.41 \times (1 - 70\%) = 7.92$
Flush	70	30	$30/1 = 30$	$30 \times (1 - 70\%) = 9$

Step 12: Dirty fluid rate for each proppant stage can be calculated as follows:

$$\text{Dirty fluid rate} = \text{clean fluid rate} \times \text{blender(surface)SF}$$

Step 13: Calculate nitrogen rate for each stage using either of the equations listed below:

$$\text{Nitrogen rate} = \text{clean foam rate} \times FQ \times VF \text{ or Nitrogen rate}$$
$$= (\text{dirty foam rate} - \text{slurry rate}) \times VF$$

Fracture treatment design

Stage name	Surface slurry factor	Rate (bpm)		
		Clean fluid rate	Dirty fluid rate	Dirty foam rate
Acid	$1 + (0 \times 0.0453) = 1$	30	$30 \times (1 - 0\%) = 30$	$30 \times 1 = 30$
Pad	$1 + (0 \times 0.0453) = 1$	30	$30 \times (1 - 70\%) = 9$	$9 \times 1 = 9$
20/40	$1 + (3.33 \times 0.0453) = 1.15$	30	$28.70 \times (1 - 70\%) = 8.61$	$8.61 \times 1.15 = 9.91$
20/40	$1 + (5 \times 0.0453) = 1.23$	30	$28.09 \times (1 - 70\%) = 8.43$	$8.43 \times 1.23 = 10.34$
20/40	$1 + (6.67 \times 0.0453) = 1.30$	30	$27.51 \times (1 - 70\%) = 8.25$	$8.25 \times 1.30 = 10.74$
20/40	$1 + (8.33 \times 0.0453) = 1.38$	30	$26.95 \times (1 - 70\%) = 8.08$	$8.08 \times 1.38 = 11.14$
20/40	$1 + (10 \times 0.0453) = 1.45$	30	$26.41 \times (1 - 70\%) = 7.92$	$7.92 \times 1.45 = 11.51$
Flush	$1 + (0 \times 0.0453) = 1$	30	$30 \times (1 - 70\%) = 9$	$9 \times 1 = 9$

Stage name	FQ %	Rate Dirty foam rate	Clean foam rate (bpm)	Nitrogen rate (SCF/min)
Acid	0	30	30/1 = 30	0
Pad	70	30	30/1 = 30	30 × 70% × 1139 = 23,919
20/40	70	30	30/1.05 = 28.70	28.7 × 70% × 1139 = 22,882
20/40	70	30	30/1.07 = 28.09	28.09 × 70% × 1139 = 22,397
20/40	70	30	30/1.09 = 27.51	27.51 × 70% × 1139 = 21,932
20/40	70	30	30/1.11 = 26.95	26.95 × 70% × 1139 = 21,486
20/40	70	30	30/1.14 = 26.41	26.41 × 70% × 1139 = 21,057
Flush	70	30	30/1 = 30	30 × 70% × 1139 = 23,919

Step 14: The last step is to calculate pump time for each stage as follows:

$$\text{Pump time} = \frac{\text{dirty foam volume}}{\text{dirty foam rate}}$$

Stage name	Dirty foam volume BBLs	Rate (bpm) Dirty foam rate	Time (min) Pump time	Total
Acid	6.0	30	6/30 = 0.2	0.2
Pad	40.0	30	40/30 = 1.33	0.2 + 1.33 = 1.53
20/40	30.0	30	30/30 = 1	1.53 + 1 = 2.53
20/40	30.0	30	30/30 = 1	2.53 + 1 = 3.53
20/40	30.0	30	30/30 = 1	3.53 + 1 = 4.53
20/40	30.0	30	30/30 = 1	4.53 + 1 = 5.53
20/40	30.0	30	30/30 = 1	5.53 + 1 = 6.53
Flush	45.0	30	45/30 = 1.5	6.53 + 1.5 = 8.03

The foam schedule for this example is summarized in Table 10.3.

Table 10.3 Foam design schedule example

Stage name	Proppant concentration (ppg)			Foam volumes (BBLs)		
	BH proppant conc.	FQ (%)	Blender prop conc.	Dirty foam volume	BH SF	Clean foam volume
Acid	0.00	0	0.00	6.0	1.00	6.00
Pad	0.00	70	0.00	40.0	1.00	40.00
20/40	1.00	70	3.33	30.0	1.05	28.70
20/40	1.50	70	5.00	30.0	1.07	28.09
20/40	2.00	70	6.67	30.0	1.09	27.51
20/40	2.50	70	8.33	30.0	1.11	26.95
20/40	3.00	70	10.00	30.0	1.14	26.41
Flush	0.0	70	0.00	45.0	1.00	45.00

Fluid volume (BBLs)			Sand (lbs)		Nitrogen (SCF)	
Clean fluid volume	Surface SF	Dirty fluid volume	Stage	CUM	Stage	CUM
6.00	1.00	6.00	0	0	0	0
12.00	1.00	12.00	0	0	31,892	31,892
8.61	1.15	9.91	1205	1205	22,882	54,774
8.43	1.23	10.34	1770	2975	22,397	77,172
8.25	1.30	10.74	2311	5286	21,932	99,104
8.08	1.38	11.14	2830	8115	21,486	120,589
7.92	1.45	11.51	3328	11,443	21,057	141,647
13.50	1.00	13.50	0	11,443	35,879	177,525

Rate					Time	
Dirty foam rate	Clean foam rate	Clean fluid rate	Dirty fluid rate	Nitrogen rate	Pump time	Total
30.0	30.00	30.00	30.00	0	0.20	0.20
30.0	30.00	9.00	9.00	23,919	1.33	1.53
30.0	28.70	8.61	9.91	22,882	1.00	2.53
30.0	28.09	8.43	10.34	22,397	1.00	3.53
30.0	27.51	8.25	10.74	21,932	1.00	4.53
30.0	26.95	8.08	11.14	21,486	1.00	5.53
30.0	26.41	7.92	11.51	21,057	1.00	6.53
30.0	30.00	9.00	9.00	23,919	1.50	8.03

CHAPTER ELEVEN

Horizontal well multistage completion techniques

Introduction

Multistage hydraulic fracturing, along with drilling longer horizontal wells, has tremendously helped the industry to make unconventional shale resources, which used to be uneconomical, economically profitable. This requires a vast acknowledgment to the industry that has been effortlessly testing various concepts in the unconventional shale reservoirs in order to make the process safer, cost effective, and environmentally friendlier. This is just the beginning and there are so many more new advancements in technology and science that the industry will see for the years to come in these resources.

There are two commonly used completion (frac) methods in the industry. The first one is referred to as "conventional plug and perf," which is the most used completion method. The second type is called "sliding sleeve," which is less frequently used and can be seen more often in shale oil plays such as the Bakken shale (although lots of operators have stepped away from using the sliding sleeve technique in the Bakken). The choice of which type of frac technique to use depends on the operator's success along with the economics of each particular technique and technology. If sliding sleeve works better in certain areas from economical, operational, and production perspectives, sliding sleeve should be used. However, if plug and perf causes an increase in production by a big proportion without any operational issues and economical concerns, plug and perf must be used. This is driven by each company's success and philosophy on each technique. One important lesson that the industry has learned since the late 1990s developing the unconventional shale reservoirs is that data should be the biggest driven and deciding factor of every engineering and operational decision. There are substantial amounts of complexity and heterogeneity in the unconventional shale reservoirs that would dictate using data to drive the business forward instead of relying on opinions and theories that may or may not function.

The advantage of using big data has been noticeable among various operators. This is where the use of artificial intelligence and machine learning (AI&ML) to extract hidden pattern from years of unconventional development comes into play. Chapter 24 will walk through a few case studies in which (AI&ML) is applied to create shareholder value.

Conventional plug and perf

Conventional plug and perf is the most commonly used method in unconventional shale plays. Composite bridge (frac) plugs are used for isolation between frac stages. Plug and perf is a completion system that uses perforation guns with a composite bridge plug using wireline. Once at the desired measured depth, a composite bridge plug is set, and each perforation gun is pulled up to the designed depths until all of the perforation guns are fired. Each perforation gun represents a cluster. After firing all of the perforation guns, wireline is pulled out of the hole (POOH). Conventional plug and perf can be done using cemented or uncemented casings. This method involves multiple perforation clusters per stage. This method is also known to be a slow and repetitive perforation and stimulation process. Conventional plug and perf is slow because after every frac stage, wireline must stab onto the well. The plug and perforation guns are sent downhole to set the composite bridge plug, shoot the guns (clusters), and finally pull out of the hole. The wireline process for stage isolation and perforation can take anywhere between 2 and 4 h depending on measured depth, crew efficiency, wireline speed, etc. For example, if a well has 40 stages, this process must be performed 40 times. If each wireline run is assumed to be about 3 h, 120 h are spent on frac stage isolation and perforation. This is 5 days' worth of frac stage isolation and perforation on a well with 40 stages. The industry uses a zipper frac technique where while one well is being fraced, another well is being perforated in an attempt to improve the operational efficiency when using the conventional plug-and-perf technique on multiwell pads. The conventional plug-and-perf technique has been very successful in the industry from a production perspective. Otherwise, given the slow progress of the conventional plug-and-perf method, the industry would have moved away from this technique. The industry has also developed other efficient techniques such as the dissolvable ball and plug, which will dissolve at downhole conditions by reducing drill-out time. Some operators have also used dissolvable plugs on a deeper portion of the well (toward the toe of a well) in an attempt to eliminate using a snubbing unit on longer lateral length wells.

This allows the use of coiled tubing, well known for having MD limitations, for drill out in lieu of using a snubbing unit.

Composite bridge (frac) plug

Composite bridge (frac) plugs are used for isolation between frac stages in the conventional plug-and-perf method. The main reason composite bridge plugs are used is because these types of plugs can be easily and rapidly drilled out after the frac job is over. After setting the plug, a ball is dropped and pumped downhole until seated inside of the composite bridge plug. The ball is typically pumped downhole at 10–15 bpm and as soon as the ball is seated inside of the plug, there is a spike in the surface-treating pressure. The spike in surface-treating pressure confirms the ball sitting in the plug. Once the ball is seated, the previous stage has now been isolated and the treatment for the new stage can commence. Fig. 11.1 shows the composite bridge plug and associated components including burn charge, frac plug setting tool, and pump down ring. Fig. 11.2 displays the perforation guns used in the conventional plug-and-perf technique. Fig. 11.3 shows the inside view of the same perforation guns presented in Fig. 11.2. Fig. 11.4 shows

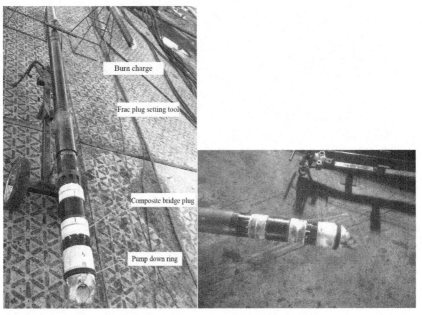

Fig. 11.1 Composite bridge plug.

Fig. 11.2 Perforation guns.

Fig. 11.3 Inside view of perforation guns.

the frac ball seated in the composite bridge plug. Fig. 11.5 shows the schematic of the one-stage hydraulic fracturing (plug to plug) with four clusters.

Stack fracing

Stack fracing involves fracing one stage, and then waiting for the wireline to perforate the next stage on the same well before being able to frac again. In this type of frac, one well is completed at a time. Stack fracing is very common in exploration areas where only one well is located on a pad. Therefore, frac crews pump a stage and wait for the wireline to set the plug and perforate

Fig. 11.4 Frac ball inside of a composite bridge plug for frac stage isolation.

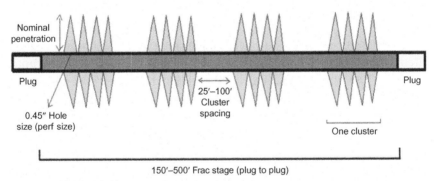

Fig. 11.5 Plug-and-cluster spacing example.

the next stage by performing routine maintenance on their equipment. Once the wireline is done setting the plug, perforating, and pulling out of the hole, the frac crew will proceed to pumping the next stage. This continues until all of the stages are completed on the same well. The main disadvantage of stack fracing is the wireline run waiting time between stages. As previously mentioned, stack fracing is only used if there is only one well on the pad. Therefore, zipper fracing is recommended in pads with multiple wells.

Zipper fracing

Zipper fracing refers to fracing a stage on one well while perforating and setting the plug on another well. Zipper fracing can be performed on multiple wells at one time. One of the main advantages of zipper fracing is saving time and money by continuously fracing and perforating. Zipper fracing is very common in the majority of the shale plays.

Simultaneous frac

Simultaneous frac is not as commonly used as zipper or stack frac. In this type of frac, two wells are simultaneously fracked at the same time. This requires a great deal of both coordination and equipment onsite. In addition, the pad has to be large enough to fit all of the frac equipment for this enormous job. The advantage of simultaneous frac stems from turning wells in line faster which will result in producing a higher net present value for the field.

Sliding sleeve

Sliding sleeve is also known as fracturing sleeve. Sliding sleeve is an alternative to plug and perf, and is used to stimulate multistage horizontal wells through holes/ports. This method is operated by a ball and baffle. When the ball lands on the baffle, the inner sleeve is opened and activated. This provides the flow path for the fracturing fluid. This type of frac typically has one opening (cluster) per stage. Multiport technology is also available to mimic plug and perf with multiple clusters. The biggest advantage of sliding sleeve is timing. Since there is no need to send composite plug and perforation guns downhole, it saves a tremendous amount of time, which is equal to saving money. The system can either be cemented or uncemented.

Sliding sleeve advantages

An advantage that sliding sleeve is known for is reduction in stimulation cycle times. Given no wireline will be needed for frac stage isolation and perforation, stage after stage can be completed as long as sand and water logistics keep up. Sliding sleeve also reduces water usage and over displacement (flush stage). Sliding sleeve is also known for maximizing near-wellbore conductivity and can be used with dissolvable frac balls.

> ### Sliding sleeve disadvantages

Some of the biggest disadvantages of sliding sleeve are mechanical issues. Anything mechanical can fail, and mitigation processes can be very costly. Another disadvantage is the limited number of stages in cemented applications. In today's conventional plug-and-perf methodology, clusters (perforation guns) are sometimes placed 20′ apart to maximize the contact area. However, depending on the service provider, the number of stages can be limited. Since the sleeves are run downhole with the casing, each joint of casing is usually 40′–45′ in length. Therefore, the sleeves cannot be placed closer than 40′ unless special casing is ordered, which can be expensive. Hole conditioning is another crucial step before running sliding sleeve with casing in the hole. Finally, the industry has limited experience with sliding sleeve as compared to the tried and true conventional plug-and-perf method. Sliding sleeve can be divided into different types. The most common ones are described in the following sections.

Toe sleeve/valve

Toe sleeve is a pressure-operated valve that creates flow path without any intervention of the wireline.

Single-entry-point frac sleeve

The single-entry-point frac sleeve system is operated by a ball and baffle. Frac balls are dropped in sequence of smallest to largest in order to activate the sleeve. Ball trailer or pneumatic ball launcher can be used to launch the balls from the surface.

Multientry-point frac sleeve

As opposed to single-entry point, multientry-point frac sleeve allows multiple entry points in a single stage without the use of plug and perf. The idea is to mimic the plug-and-perf design by using sliding sleeve technology. One ball can open more than one sleeve. This technique is very similar to the conventional plug and perf and every entry point is similarly referred to as a "cluster."

Hybrid design

Hybrid design uses a combination of frac sleeves and plug and perf. The first half of the well (toe section) uses sliding sleeve and the second half (heel section) uses plug and perf. The Bakken shale is an ideal example of the hybrid design. Since the lateral lengths of Bakken wells are typically in excess of 8000′ and coiled tubing is limited by depth that can be reached, some operators use the hybrid design to facilitate this process.

Frac stage spacing (plug-to-plug spacing)

Frac stage refers to the space from plug to plug in a vertical or horizontal well. In many formations across the United States the horizontal lateral length of a well is divided into many stages to optimize production. This is why hydraulic fracturing is often referred to as multistage hydraulic fracturing, that is, each well has many stages depending on the horizontal length of the well, design, and economic calculations. Therefore, the next interesting subject in hydraulic fracturing is the number of stages necessary to maximize production in horizontal wells. When hydraulic fracturing started, some companies tried to perform a single-stage frac job with no success, thus causing the need for multistage hydraulic fracturing in various formations across the United States.

Shorter stage length

In conventional plug-and-perf technique, the industry standard for plug-to-plug spacing is anywhere between 150′ and 300′. Operating companies have used various frac spacing designs (e.g., 150′, 200′, 300′, etc.) to come up with the optimum production that yields the best economic outcome based on actual production data. Some companies believe in shorter frac stage spacing such as 150′–200′. This type of frac is referred to as shorter stage length (SSL) due to shorter plug-to-plug spacing. For example, if the lateral length of a well is 6000′ and 200′ plug-to-plug spacing is chosen, hydraulic fracturing will take place in 30 stages. This means the process of setting plug, perforating, and fracing needs to be performed 30 times on one well. One of the main factors associated with frac spacing is economics. Every frac stage is very costly and depends on various factors such as service provider, amount of sand, water, chemicals pumped, market conditions, etc. For example, if a stage is pumped using 250,000 lbs of sand with the associated water and chemicals, it will be less expensive compared to a stage that uses 500,000 lbs of sand with the associated water and chemicals. Stage spacing in different areas and formations is ultimately dictated by production success and economic analysis.

The concept of SSL was heavily applied and tested since 2013 because shorter spacing between frac stages often yields higher initial productions (IPs) with steeper decline and in some areas shallower decline. In some areas, steeper decline is not noted and decline percentage can sustain itself. For example, the IP from a standard spacing of 300′ is about 6 MMSCF/day with an initial annual secant decline of 62%; however, the IP from a well with 150′ spacing (SSL) may yield an IP of 8 MMSCF/day with a similar or steeper decline of more than 62%. Sometimes the decline percentage could be shallower than historically observed using SSL depending on the effectiveness of the completions design. The incremental production volume that is initially gained is basically the time value of money and as a result would sometimes be more economically beneficial in certain areas as long as the decline percentage can sustain itself. The time value of money is a concept that means that money available today is worth more than the same amount of money in the future due to its potential earning capacity. SSL does not work everywhere, and the probability of success will depend on the formation properties. In some areas, SSL works very well and a production uplift of 10%–40% can be seen. However, in other areas, no production uplift can be seen from SSL especially in complex geologic areas. The question then is whether the incremental gain in production offsets the additional capex (capital expenditure) required for any particular design. If so, and the required funding is available, then a more optimized design must be used.

To determine the most economic option, economic analysis must be performed using both methods in each area. The biggest challenge in SSL design is the incremental capital (capex) that must be spent. The additional IP and estimated ultimate recovery (EUR) gained from SSL must be enough to offset the incremental capital spent on the well. The decision to use SSL or standard stage spacing is truly an economic decision and, therefore, economic analysis must be the deciding factor and not the IP or EUR of the well. SSL is truly area dependent.

Why does SSL not work everywhere?

1. Not enough gas originally in place to be recovered
2. Complex geology may add complications

3. Heavily naturally fractured regions
4. Higher permeability of the rock
5. Lower pore pressure and pumping too much water can be detrimental in some areas
6. Lower pore pressure means less gas in place (GIP), particularly in dry gas areas
7. Hydraulic fracture stage interaction and competing fractures.

Cluster spacing

There are various clusters in each frac stage. A cluster is referred to as a perforating gun. If there are five clusters in one frac stage, there are five perforation guns in that stage that are usually evenly distributed (geometric design). The industry average for the number of guns (e.g., clusters) in the unconventional reservoirs is anywhere between 3 and 20 clusters that are equally spaced in each stage. For example, if six clusters are used for a 300' frac stage, the cluster spacing is 50'. The industry average for cluster spacing is 20'–60'. Every operator has its own theory regarding the number of clusters and holes in a frac stage. The major deciding factor in choosing the number of clusters is formation permeability, GIP, and perforation efficiency. A general rule of thumb is that if the formation permeability is higher than usual, fewer clusters will be needed. In contrast, if the formation permeability is lower, more clusters will be necessary. The goal is to achieve the maximum surface area between clusters. In addition, if GIP in a particular area is not significant, fewer clusters and stages will be needed to release the hydrocarbon.

Some operating companies believe that the spacing between clusters needs to be minimized to gain the most surface area out of each zone. On the other hand, others believe that a lower number of clusters are necessary to achieve longer fracture networks by forcing the hydraulic fracturing energy to go to a limited number of clusters. Every operator justifies its theory with its production results. Having shorter cluster-to-cluster spacing has shown to maximize the production outcome in certain areas. Hydraulic fracturing operation in shorter cluster-to-cluster spacing has sometimes shown to be more difficult due to competing fractures or communication between clusters.

Refrac overview

Refrac refers to a second fracture stimulation on a well with existing production data and is another important topic that the industry has been experimenting with in various shale plays since 2011. Discussions on refrac are very common in a low commodity-pricing environment where plenty of time is available for analyzing the previous frac jobs with poor completion designs. In addition, instead of investing more funds into drilling and completing a new well, refrac could potentially offer better economics in areas with excellent reservoir quality and pressure. Refrac has caused a substantial production increase in many of the shale plays including but not limited to Marcellus, Haynesville, Eagle Ford, Barnett, and Bakken. The primary reasons behind refrac are as follows:

1. To implement new or enhanced completion design: Many wells were completed using old completion designs such as $400'-500'$ stage spacing, low sand/ft, high number of perforations per stage, high or low number of clusters per stage, etc. The combination of the above designs has caused a large percentage of unstimulated (virgin) rock that has not been touched yet. Refracing and implementing new completion designs such as reduced stage spacing, reduced cluster spacing, limited entry design, more sand/ft, etc., could potentially improve the production performance in some areas.

2. Contact more surface area by adding diversion, perforation, and reorientation: One of the most common refrac methods is using diversion (special bimodal degradable particulate), which is offered by various service companies. The basic concept behind using diversion is to pack off the currently open perforations to effectively stimulate the unstimulated perforations, allowing breakdown into new areas of the reservoir. In addition to diversion, adding new perforations and reorientation could aid in increasing the contact area.

3. Bypass skin damage caused by scale, fines migration, and iron/salt deposition: Skin damage can be caused by scale accumulation in the pipe and formation, salt and iron deposition, or simply by fines migration. The improper use of a chemical package when hydraulically fracturing a well could cause a detrimental impact to the long-term productivity of a well.

4. Wells that did not use managed pressure drawdown (especially in over-pressured reservoirs) and caused proppant crushing, embedment, and conductivity reduction: Unmanaged pressure drawdown has shown detrimental impact to the productivity of the wells in many different shale formations especially in overpressured reservoirs. Aggressive pressure drawdown can cause proppant crushing, proppant embedment, fines migration, cyclic stress, and pressure-dependent permeability effects. Pressure-dependent permeability is very important to consider in

overpressured reservoirs. This is due to the pore volume reduction since the natural compaction process is incomplete. Therefore, the available flow area is reduced and permeability decreases with pressure. Refrac has shown to be successful on some of the wells that did not originally follow a managed pressure drawdown in overpressured formations such as the Haynesville and Eagle Ford Shales.

5. Increase conductivity and restore conductivity loss: One of the most unknown segments of a hydraulic fracture design and production evaluation is how the conductivity loss with time affects the production performance and the economics of the wells. Almost all exploration and production (E&P) companies are interested in the first 5–10 years of producing life of a well because economically speaking that is when 80%+ of the value is returned to the shareholders. Therefore, if conductivity loss in the fracture or near the wellbore during this time period does not severely impact the production, it is not a subject that is often discussed. However, if conductivity loss occurs sooner rather than later within the sensitive economic timeframe, it is very important to understand both the mechanism behind this loss and ways to mitigate this issue on future completion designs. Applying refrac on wells that are believed to have encountered some kind of near-wellbore or fracture conductivity loss due to various factors such as unmanaged drawdown, scale accumulation, non-Darcy effect, proppant crushing and embedment, fines migration, liquid trapping, liquid loading, fracture face skin, convergence skin, etc., has shown to restore conductivity loss and production enhancement in many refracs tested to date.

6. Change in fluid system could be successful refrac candidates: Another important reason behind a successful refrac could be the implementation of a different type of frac fluid system that is more compatible with the formation of interest. For example, if a well was originally hydraulically fractured using a cross-linked fluid system without successful production results, other frac fluid systems such as slick water could cause significant increase in production by performing refrac. It is important to note that if the area is not a rich area from both reserve and geologic perspectives, performing refrac is not recommended. Rodvelt et al. (2015) analyzed seven Marcellus wells that were refractured in Greene County, Pennsylvania, and noted 65%–123% increase in reserve from refracing these Marcellus wells located in a geologically superior area with high reservoir pressure, poor original completion design properties, and excellent reservoir properties by using diversion material for the refrac.

When evaluating wells for refrac, keep the following guidelines in mind:
- Select wells with high remaining reserves and excellent geologic areas.
- Focus on the wells with old completion designs such as wells with larger stage spacing and minimal proppant mass.
- Stay away from wells with mechanical integrity issues, as this can get very costly.
- Select a great refrac candidate first in an attempt to add value to the entire prospect in the event it is successful and can be repeated on all of the remaining wells in the area. Some E&P companies also assign a present value on their refrac candidates and potentials when divesting assets.
- Poor wells often make bad refrac candidates, unless there is solid evidence that the original frac design, materials, or implementation was a failure.
- Stay away from low-pressure or depleted reservoirs where frac fluid recovery will be very challenging.
- Stay away from wells with excellent original design and implementation and focus on the ones with poor designs first, as they are many wells with poor designs that must be fixed first.
- Stay away from poor reservoir quality wells as refracing might not economically generate any additional value due to the poor area.
- Always run economic analysis using the existing refrac wells as analogous wells.

CHAPTER TWELVE

Completions and flowback design evaluation in relation to production

Introduction

After obtaining sufficient production data for analysis, one of the most important aspects of completions optimization is evaluating the completions design. Typically, 6 months to 1 year of data (depending on data quality) is needed to evaluate each completions test in unconventional shale reservoirs. There are various tools that can be used to evaluate the productivity of a well in unconventional shale reservoirs. Calculating the estimated ultimate recovery (EUR) using various types of decline curve analysis (DCA) or using rate transient analysis (RTA) is widely used to determine the flow capacity and strength of a well in conjunction with one another to tie back to completions design. One of the most important plots used to determine the flow capacity of a well is referred to as a superposition plot, from which flow capacity or $A\sqrt{K}$ of each well is determined. The flow capacity of a well can be determined by plotting pseudo $\Delta P/q$ on the y-axis vs material balance square root of time (CUM/q) on the x-axis to determine the slope of the linear portion of the plot, which is inversely proportional to $A\sqrt{K}$. $A\sqrt{K}$ is one of the most important parameters that can be used to determine the strength of a well in unconventional shale reservoirs based on their completions design, reservoir quality, pressure drawdown management, and other variables. $A\sqrt{K}$ is basically the contacted surface area multiplied by the effective permeability of the contacted rock. $A\sqrt{K}$ in unconventional reservoirs is the equivalent of kh in conventional reservoirs. $A\sqrt{K}$ is obtained from square root or superposition plots and it is a function of initial reservoir pressure, flowing bottom-hole pressure (with time), production rate (with time), porosity, gas viscosity, total compressibility, and reservoir temperature. The industry has found out that there is a direct correlation between $A\sqrt{K}$ and EUR from all of the analyses that have been performed in the past. As opposed to DCA, which assumes constant flowing bottomhole pressure, drainage area, permeability, skin, and existence of boundary dominated flow, $A\sqrt{K}$ analysis takes pressure and rate with time as well as other reservoir properties into account for accurate determination of a well's strength. $A\sqrt{K}$ is also used to rank the best to worst performers in each field and make important completions design decisions for the company. There are many important parameters that must be taken into consideration when evaluating the production aspect of a completions test. They are described in the following sections.

Landing zone

The landing zone of a well is extremely important to evaluate in an attempt to find the optimum target zone for each field. In theory, the best landing zone would have high resistivity, low water saturation, low formation density, high total organic contents (TOCs), low clay content, high effective porosity, high Young's Modulus, and low Poisson's ratio (brittle from a fracability perspective). It is very challenging to find a formation that has all of the aforementioned properties. Therefore, in new exploration areas, different landing zones (keeping all of the other variables constant) must be tested to understand the production performance of each landing zone. In addition, it is important to understand in situ stress around each landing zone from various logging suites such as sonic log. Target zones are typically $5'-15'$. The ideal landing zone should have excellent barriers above and below the target zone to stay in the rich target rock for as long as possible. Frac barriers are very important in staying within the target area for the maximum production optimization. Susquehanna County, Pennsylvania is known for having excellent frac barriers below and above the Marcellus, with outstanding production performance as a result. The landing zone is heavily area dependent and numerous logs and testing must be performed to understand the best landing zone in each area. The best optimizing technique is to pick two or three different zones in the rich rock and test each landing zone to determine the target zone that yields the best production performance. It is well known that picking the proper landing zone has a substantial impact on the production performance and economic viability of a well in unconventional shale reservoirs. Therefore, special emphasis should be placed on choosing the proper target zone for optimum production enhancement.

Stage spacing

Stage spacing is another important completions design parameter that must be evaluated and understood from production analysis and evaluation. The idea is to contact as much surface area as possible within the clusters, while minimizing fracture interference (competing fractures). Stage spacing, just like any other parameter in unconventional shale reservoirs, is area dependent. For instance, if 150' stage spacing is optimum in one area, it does not necessarily mean that it will be optimum in other areas. Therefore, stage spacing needs to be obtained depending on the formation properties. Tighter stage spacing requires more capital expenditure; therefore, it is important to determine the percentage uplift in production from tighter stage spacing to justify the additional capital. In a commodity-pricing environment where completions capital is cheaper than usual, no significant uplift is required to justify tighter stage spacing. However, in an environment where completions capital is expensive, higher percentage uplift in production is needed to economically justify tighter stage spacing. Therefore, economic analysis must be performed at various points in time based on different oil/condensate/NGL/gas pricing and capital expenditure associated with each design. For example, if 150' stage spacing is the optimum design based on $2.5 MM completions capital expenditure given a well with 7000' lateral length, this stage spacing might not be optimum when compared to a completions capital of $5 MM because a higher percentage uplift in the production curve will be needed to justify the higher capital, depending on the commodity pricing. Therefore, production performance and economic analysis in each area must be the sole deciding factor behind choosing the optimum stage spacing.

Cluster spacing

Cluster spacing is another important factor that needs to be considered when evaluating the production capacity of a well. In a conventional plug-and-perf technique, the distance between clusters should be optimized in a manner that yields the highest production result. There are typically 3–20 clusters per stage depending on stage spacing, formation properties, and company philosophy. The number of clusters must be tested in each area to understand the impact of cluster spacing on production performance.

Number of perforations, entry-hole diameter, and perforation phasing

From a perforation design perspective, the number of perforations, entry-hole diameter (EHD), and perforation phasing are three of the most important perforation design parameters that should be tied back to production results, if any testing is being performed. A rule of thumb that is used in the industry is that the EHD must be at least 6 times the maximum sand grain size used during the job to prevent sand grains from bridging and screening out during the fracture treatment. It is also known that the perforations should be within 30 degrees of the maximum principle stress orientation to reduce near-wellbore tortuosity and potential treatment issues throughout the frac job. Therefore, a perforating gun with 60-degrees phasing will orient the perforations to be within 30 degrees of the maximum principle stress direction, especially since the exact perforation orientation is rarely known. Some companies use other phasing angles such as 0-, 90-, 120-, and 180-degrees phasing. As previously discussed, the number of perforations is typically designed based on the limited entry technique for optimum production enhancement in unconventional reservoirs. However, various testing on the number of perforations, EHD, and perforation phasing must be done to understand the outcome of each test. It is very important to test one variable at a time to understand the sole impact of that parameter on production. Step down tests (discussed in Chapter Nine) must also be included to reveal the perforation efficiency associated with each design. Fig. 12.1 shows different perforation phasing.

Sand and water per foot

The amount of sand and water per foot are other important design parameters that must be tested to understand their impact on production,

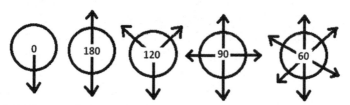

Fig. 12.1 Perforation phasing.

and economic viability of determining the optimum sand-and-water-per-foot design. The sand-and-water-per-foot design, just like all of the other parameters that have been discussed, are also area dependent. In areas with higher pore pressure and exceptional formation properties, higher sand and water per foot could potentially help the production performance of a well. On the other hand, in areas with lower pore pressure and poorer formation properties, pumping high volumes of sand and water might not be the ideal production enhancement solution. Therefore, various sand-and-water-per-foot designs must be tested to determine the optimum economic design for each area. For instance, 1000 #/ft, 1500 #/ft, 2000 #/ft, 2500 #/ft, 3000 #/ft, etc. (sand per foot) designs must be tested in new areas with limited data. One exercise that can be performed before pumping higher sand and water loadings is to determine the percentage uplift in production needed to justify the additional capital that will be spent on higher sand and water loadings. Once production data are available, percentage uplift for each design can be easily determined, and the optimum economical sand-and-water-per-foot design can be found for each area.

Another important factor in a downturn market with minimal frac activity is the water disposal cost. In a continuous frac environment, flowback water along with stored water can be continuously used on future frac jobs. This eliminates the cost spent on water disposal. Depending on water infrastructure and trucking fees, the water can be continuously used instead of disposing of the water. However, in a downturn market where plenty of water is available and the water storage capacity is full, water will have to be disposed of. Water disposal might exceed the trucking fees. Therefore, many companies will continue to frac in a downturn market just to avoid paying a significant amount of money for water disposal, which can become costly depending on the area.

Proppant size and type

Proppant size and type are other important factors when analyzing production results. The decision to choose between different proppant schedules such as pumping 100 mesh and 40/70 vs pumping 100 mesh, 40/70, and tailing in with 30/50 or 20/40 mesh should solely depend on the production performance (including economic analysis) of each proppant design. Sometimes, the availability and price of each proppant size can have a direct impact on the sand size selection. Some fields have shown that pumping a large percentage of 100 mesh yields the best production results.

This could be due to the area being heavily naturally fractured and 100 mesh proppant traveling farther into the formation due to its size as compared to 40/70, 30/50, or 20/40. In theory, it might not make sense to pump a large percentage of 100 mesh due to higher closure pressure in a particular area, but if pumping a large percentage of 100 mesh is justified based on production data while saving money, more 100 mesh must be pumped to acknowledge the production results and more studies need to be done to understand the physical mechanism that results in excessive production.

Proppant type is another important factor that must be determined. The proppant type to be used in formations that have closure pressure of less than 6000–7000 psi is straightforward as regular sand is typically used in those formations. However, the discussion on whether ceramic or resin-coated sand is needed in higher closure pressure areas will depend on the closure pressure of the area that can be obtained from diagnostic fracture injection test (DFIT). In theory, if closure pressure is more than 8000 psi, ceramic proppant is recommended to avoid proppant crushing and embedment. However, the economics of pumping ceramic comes into play because ceramic is considered to be very expensive and pumping a large percentage of ceramic proppant might not be economically feasible. In addition, the amount of ceramic needed in overpressured formations with high closure pressure is determined from testing various percentages of ceramic proppant. Proppant-type testing will need to be performed on new exploration areas from day one in an attempt to understand the impact of each proppant type in relation to production. Many operators in different basins have tested various proppant types such as regular sand, resin-coated, and ceramic proppants to understand the impact of each proppant in higher closure pressure formations. Each field is unique due to the heterogeneity of the unconventional plays, in the sense that ceramic proppant might be absolutely necessary in some fields but it might not be economically justified in others, even with higher than 8000 psi closure pressure.

Bounded vs unbounded (inner vs outer)

Another very important aspect of production evaluation in relation to completions design is whether a well is surrounded by other wells on both sides (bounded) or a well is not bounded by other wells on one or two sides. In Fig. 12.2, the B well is surrounded by A and C wells. Therefore, the B well is considered to be a bounded (inner) well. The A and C wells are both considered to be unbounded from one side, and finally the D well is

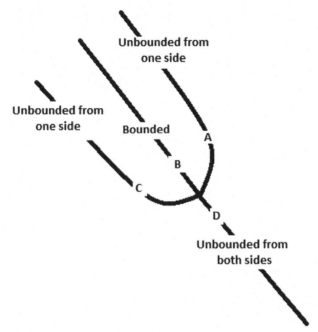

Fig. 12.2 Bounded vs unbounded example.

unbounded from both sides. The D well is also referred to as a "standalone" well. Production data in some fields has shown that unbounded wells from either one side or two sides have better production performance depending on the sequence in which the wells were hydraulically fractured (frac order). Belyadi et al. (2015) analyzed more than 100 Marcellus wells in the same geologic area and showed that unbounded wells are the best performers in the field. Frac order could also have an impact on fracture propagation by creating a pressure barrier or stress shadow effect around the wellbore and helping fracture propagation outward into the unbounded virgin rock. By assuming that all four wells in Fig. 12.2 have identical completions design, production performance of the wells could potentially be different if the B and D wells are zipper fractured first, followed by zipper fracturing A and C. By creating a stress shadow effect around the B well, the fractures will propagate outward when the A and C wells are hydraulically fractured. This could hypothetically cause better fracture half-length and more contacted surface area on the two unbounded wells. On the other hand, zipper fracturing A and C wells first followed by zipper fracturing B and D wells will create more complexity around the B well. The D well is a standalone well and should not have a direct impact on the production performance of the A, B, and C wells. In an attempt to take advantage of unbounded wells, some companies pump more sand and water per foot on the unbounded wells to contact as much surface area as possible especially if the surrounding area around the unbounded wells will not be hydraulically fractured anytime soon. Therefore, it is very important to take all of the aspects of the completions design including frac order, bounded, and unbounded wells into account when testing various completions designs and tying them back to production performance. It is also extremely important to avoid testing a test design on an unbounded or standalone well and comparing them to other bounded wells. Otherwise, the result of such testing will be very biased and null.

Up dip vs down dip

Another important parameter that should be taken into consideration when analyzing production data is whether a well was drilled up dip or down dip, especially in liquid-rich fields with undulations. A well with an inclination of more than 90 degrees is called up dip and a well with an inclination of less than 90 degrees is called down dip (Fig. 12.3). Some fields have seen better production results from up-dip wells while others have witnessed better production results from down-dip wells. In some fields, it is very challenging to see the impact of up dip vs down dip wells due to many other completions design changes. Fig. 12.3 illustrates the difference between up-dip and down-dip inclinations.

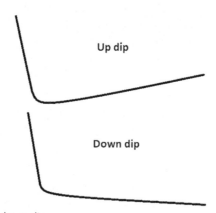

Fig. 12.3 Up dip vs down dip.

Well spacing

Well spacing, or inter-lateral spacing, is also extremely important when analyzing and comparing the production results from various wells in the same field. Well spacing can range anywhere from 300' to 1500' depending on the rock characteristics (especially permeability), fracture half-length, fracture conductivity, gas pricing, capital expenditure, operating costs (OPEX), and many other parameters. It is crucial to select the optimum well spacing for each area based on production and economic evaluation of each area. Gas pricing has a tremendous impact on well spacing. Higher gas pricing indicates tighter well spacing, while lower gas pricing indicates wider well spacing. Higher capital expenditure dictates wider well spacing but lower capital expenditure indicates tighter well spacing. Various analytical and numerical models can be run to find the optimum economical well spacing for each area. From a production perspective, it is imperative to make sure that well spacing between all of the experimental wells is taken into account when wells with various completions designs are being tested. From a completions design perspective, well spacing should have a direct impact on completions design, that is, on the amount of sand and water/ft, stage spacing, cluster spacing, etc. Exploration and production (E&P) companies design well spacing for each area based on actual field testing in addition to reservoir modeling techniques such as RTA, analytical, or numerical simulators. Belyadi et al. (2016a,b) performed a well-spacing analysis sensitivity in which history matching was done on a dry Utica well in an exploration area and fracture and reservoir simulators along with economic analysis were used to find the optimum well spacing for the area. They concluded that well spacing is heavily influenced by fracture half-length, conductivity, effective permeability, gas pricing, capital expenditure (Capex), and operating costs (OPEX). They also concluded that well spacing is heavily area dependent and a spacing that might be optimum today may not be optimum in the future. In addition to well spacing, lateral length is another important parameter to consider when analyzing production data. The industry has been moving more toward drilling longer lateral wells (8000+ ft) in order to save Capex/ft. As lateral length increases, the economics of the wells typically improve unless severe production impairment is observed from drilling longer lateral wells. Economic analysis should be performed to understand the amount of production that can be lost and still yield better economic results by drilling longer lateral wells. For example, even if drilling a 12,000′ lateral length well causes a 5% EUR/ft reduction as compared to a 7000′ well due to completions efficiency (and other factors) in longer lateral wells, it might still be more economical to lose insignificant amount of reserve but create a higher value for the shareholders as significant amount of capital can be saved by drilling longer lateral wells. Sometimes a company's acreage position does not allow drilling long lateral wells in some units. Production results in some fields have shown no loss in production by increasing lateral lengths while other places have shown a percentage reduction in EUR/ft as lateral length increases. This must be evaluated from area to area to make the best possible decision for the company. Some public acquisitions were intended to increase the potential lateral length from each well and produce tremendous value for the shareholders. The impact of increased lateral length across a field is significant. A straightforward economic analysis can reveal the significance of increasing lateral length across the field due to cost savings associated with longer laterals. Therefore, if companies can take advantage of small or large acquisitions at the right price to increase their lateral length across multiples pads or a field, significant value can be generated for the shareholders. For more discussions on wells spacing and completions design, please refer to Chapter Twenty-One.

Water quality

Water quality used during frac jobs is another debatable topic that many companies are trying to understand. Some of the very important parameters that are used to compare against production performance are total dissolved solids (TDS), water conductivity, and the amount of chlorides that can be measured by taking a water sample prior to every stage. The importance of water analysis becomes more complex when 100% produced water with high TDS ($>120,000$ ppm) is used for the job. High-TDS water could cause some issues with friction reducer (FR) selection for the job, and various FRs must be tested in the lab and field to determine the best FR type and concentration with high-TDS produced water. Sometimes the best FR selection from the laboratory analysis might not perform the same when tested in the field. Therefore, a contingency plan should be available with FR selection, especially during high-rate slick water frac jobs. This is why the artificial intelligence and machine learning techniques (AI&ML) can reveal very important info about the impact of TDS on FR type and concentration.

Other fracture treatment metrics that are recommended to be reviewed are:
- average treating pressure and rate trends
- breakdown pressure and instantaneous shut-in pressure (ISIP) vs % proppant placed
- breakdown pressure vs average treating pressure
- number of clusters and perf diameter vs % proppant placed
- fluid types and volumes vs % proppant placed

Flowback design

Flowback design is vital and can be just as important as completions design particularly in overpressured formations. In essence, the way a well is flowed back and produced is just as crucial as the way a well is hydraulically fractured. The notion that unconventional shale wells can be produced just like conventional wells is not correct due to the possibility of losing the integrity of the pumped proppant because of proppant crushing, proppant embedment, geomechanical effect (overpressured reservoirs), fines migration, cyclic stress, near-wellbore conductivity loss, and non-Darcy effects. Therefore, proper care must be taken to prevent proppant damage in the formation, and maintaining the integrity of the well's performance for decades to come. After hydraulic fracturing is finished, the drill-out phase takes place. Drill-out is defined as a poststimulation phase where coil tubing or stick tubing is utilized to clean the wellbore before flowback and production.

New technologies have recently been developed in which dissolvable plugs are used between stages in a conventional plug-and-perf technique. The use of dissolvable plugs eliminates the need for the drill-out phase and flowback through third-party equipment can take place. Sometimes a cleanup run is performed even after using dissolvable plugs to make sure that there is no debris in the hole. After the drill-out or cleanup period (if any), flowback takes place. Flowback is defined as postdrill-out phase (if any) where the well is flowed through the third-party equipment and cleaned up before turning the well over to permanent production equipment. The flowback procedure is typically provided by production or completions engineers with feedback from reservoir engineers. This procedure needs to be provided in a manner that proppant integrity is not sacrificed in any shape or form by following a pressure drawdown, depending on the reservoir characteristics and pore pressure. One rule of thumb that can be used during flowback is to stay within the critical drawdown pressure limit. Critical drawdown pressure is defined in Eq. (12.1).

$$\text{Critical drawdown pressure} = \text{closure pressure} - \text{reservoir pressure} \qquad (12.1)$$

The most important part during the life of a well is when a well goes from a full column of fluid to a full column of gas during the flowback period. Therefore, special care must take place to avoid any proppant damage during this period by not exceeding critical drawdown pressure during flowback. The difference between closure and reservoir pressure is defined as critical drawdown pressure because when this pressure is exceeded, stress will start to be applied on the proppant. It is essential to avoid placing stress on the proppant during the flowback period where lots of events, including cleanup/proppant flowback and conversion of the full column of water to gas, occur. As the well cleans up and the gas cut starts occurring, casing pressure builds up until the peak casing pressure is reached. The critical drawdown pressure count begins as soon as the peak casing pressure along with stabilized water and tubing pressure (if there is any tubing in the well) are reached. After deducting the critical drawdown pressure from the peak casing pressure, the pressure drawdown must be in a manner that is controlled with time thereafter. Uncontrolled, sharp pressure drawdown has shown detrimental impact on production in various shale plays such as Utica, Haynesville, and Eagle Ford shale plays. There is a balance between how heavily a well is curtailed back and sacrificing the long-term productivity of the well by pulling hard on the well. Therefore, an economic analysis must be run to understand the impact of managed pressure drawdown on the net asset value and obtain the optimum economic rate at which a well should be produced (Belyadi et al., 2016a,b). An analysis that can be performed to make an educated decision for the company is to determine the amount of uplift in EUR needed to justify producing the well at a lower rate. This analysis is heavily influenced by gas pricing. In a low commodity-pricing environment, depending on each company's goal and strategy, it could potentially be economically justifiable to heavily curtail the wells when there is a potential future upside in pricing. However, in a high commodity-pricing environment, it is important to understand and run various economic sensitivities to obtain the optimum economic rate at which a well must be produced. For instance, if the price of gas is $6/MMBTU and there is only a 5% reduction in EUR by producing the well at a higher incremental rate of 5 MMSCF/day, the company might be economically better off to cause 5% damage to the long-term productivity of the well (negligible) and make more gas up front to take advantage of the time value of money. However, if production data shows a 30% reduction in EUR by producing the well at a higher incremental rate of 5 MMSCF/day, economic analysis must be performed to thoroughly understand this impact and determine the optimum economic rate that will create value for the shareholders.

Flowback equipment

Flowback equipment is used during flowback and is provided by a third-party flowback company. To save costs, some E&P companies have started flowing back their wells through third-party equipment such as sand traps, choke manifolds, and gas buster tanks when on fluid production and when the well is initially flowed back until the gas cut is reached. Once the gas cut is reached, the well can be diverted to permanent production equipment. Taking a slowback approach, wells must be managed to minimize sand production and stay within the limits of gas production units (GPUs) for water production. During a regular flowback job using a third-party flowback company, the following flowback equipment is typically used.

Choke manifold

A choke manifold is used to control the flow of a well by providing back pressure. There are two types of chokes used in flowback operations. The first type of choke, which is more commonly used, is called an *adjustable choke*. The adjustable choke has two parts, called the *valve* and *seat*. The valve and seat on an adjustable choke wash out quickly after flowing back lots of sand; however, they can be replaced very easily by diverting the well flow to a different direction and replacing the valve or seat. In addition, adjustable chokes are operated using a wheel and changing the choke size is very easy to perform. The second type of choke is called a *bean choke*. Bean chokes come in various sizes and to achieve the required size, the insert inside the choke must be replaced. The bean choke consists of a replaceable insert (also referred to as the bean) made from steel. The inserts are manufactured with various hole diameters and are available in different sizes. Recently, automated chokes have also become common in the industry to improve efficiency and to remotely operate the wells. Fig. 12.4 shows a choke manifold used during the flowback operation.

Fig. 12.4 Choke manifold.

Sand trap

The sand trap is typically located right after the choke manifold on a multiwell pad and is used to prevent erosive materials such as proppant from entering the equipment, in an attempt to prevent washout and damage to the equipment. Flowback fluid (water + gas + sand + oil) coming out of the choke manifold typically goes to a sand trap. Since proppant has a heavier density, it will fall down to the bottom of the vessel. Sand is then removed through the outlet located at the bottom of the vessel. The blowdown line located at the bottom of the vessel is used to pump the sand every so often to the flowback tanks depending on the amount of sand that is flowed back. Sand traps typically have pressure ratings of 2.8–10 K psi.

In single-well applications, the sand trap is placed upstream of the choke manifold (before the choke manifold). In this scenario, the sand trap should be able to handle 20% above the maximum anticipated wellhead pressure. On the other hand, the sand trap is located downstream (after the choke manifold) on multiwell pad applications and lower pressure rating will be required since fluid coming out of the choke manifold has lower pressure. It is better to run the sand trap upstream (before) the choke manifold in any application. This is simply because the differential pressure between wellhead and choke manifold is less, therefore, resulting in slower velocity. However, on multiwell pads, having a sand trap before the manifold would require a sand trap per well, which would be very costly; this is why the sand trap is run after the choke manifold. In addition, sometimes due to the pressure limitations on the sand trap, the sand trap is run after the choke manifold. Another reason for placing the sand trap before the choke manifold is to prevent washing out the choke manifold. Producing substantial sand can wash out the seat and stems (used to control the choke) inside the choke manifold. Seats and stems wash out very quickly after encountering a large amount of sand and to prevent paying a lot of funds for damaged and washed

Fig. 12.5 Sand trap.

out equipment, a sand trap can be placed upstream of the choke manifold during drill-out and flow operations. Any sand trap used during the operation must have a mechanical pressure-relief system referred to as a pop-off. A pop-off releases the pressure if the pressure exceeds the limitation. In addition, sand traps must be inspected yearly and should also have a bypass system in the event of failure. Fig. 12.5 shows a sand trap used during flow back operation.

High-stage separator

High-stage separators are divided into three main types. Vertical, horizontal, and spherical separators are well-known separators throughout various operations. Horizontal separators are more commonly used in a variety of operations. Separators can be two, three, or four phases. A two-phase separator separates fluid from the wells into gas and total liquids. Since water has a heavier density than gas, it leaves the vessel at the bottom while gas leaves the separator at the top. A three-phase separator separates fluid into gas, water, and oil. The first partition in a three-phase separator is used for water removal. One of the primary differences between two-phase and three-phase separators is that additional weir is used in a three-phase separator to control the oil/water interface. Finally, a four-phase separator (not commonly used) has the ability to separate sand, water, oil, and gas. First of all, the sand falls down into the first partition as soon as it enters the inlet diffuser (because of heavier weight). Oil and water are directed to the second partition (middle portion of the separator). Since water has a higher density, it remains in the second partition and is dumped through a dump line to the flowback tank. Oil on the other hand has a lighter density as compared to water and reaches the third partition, where it is dumped to a low-stage separator that is located right after the high-stage separator. For instance, in Marcellus Shale operations, some areas are known to only produce dry gas and therefore a two-phase separator is used. Sometimes a three-phase separator is used in dry gas windows and two partitions are used to remove water in the event water production is expected to be high. However, if an area is known to produce water, condensate, and gas, it would not be feasible to separate the liquid (condensate and water) from each other in a two-phase separator. In this particular scenario, a three-phase separator would be necessary to efficiently separate water, condensate, and gas. The majority of separation occurs at the inlet diffuser. Adequate settling partitions allow turbulence to fall down and liquid to fall out. Liquid capacity of a separator depends on the retention time of the fluid in the vessel. For a good separation to occur, sufficient time to obtain equilibrium condition between liquid and gas must be met. It is important to note that even after going through the sand trap, there might be some residual sand that was not caught in the sand trap and found its way to the separator. This will be caught in the first partition of the four-phase separator and is basically an added safety against erosion and washout on the vessel.

Every separator has pressure, rate (velocity), and volume limitations depending on the company's manufacturer. The most commonly used units are 720, 1440, and 2000 psi units in different operations. Please note that separator operating pressure must not exceed 75% of its maximum operating pressure for an extended period of time. For instance, if 1440 psi separator is used, it is important not to exceed 1080 psi for an extended period of time as a safety factor. The most important parameters when it comes to units are the rate and the volume that each separator can handle. Different separators can handle various volumes of liquid and gas. The rule of thumb for having sufficient number of separators onsite is that all of the separators must be able to handle 40% over the maximum anticipated production rate (liquid and gas).

This safety factor can be increased in multiwell pads and exploration wells. For example, if an eight-well pad is anticipated to produce 64 MMSCF/day, separators should be able to handle at least 90 MMSCF/day or preferably more due to being a multiwell pad. Separators typically have an electronic gas-flow measurement with a backup mechanical meter (Barton meter) in the event the electronic one fails. Another type of safety equipment that must be installed on separators is a check valve, which is rigged up (R/U) right after the exit gas line on the separator. Completions, facilities, or production engineers in the office are typically responsible for designing the necessary flowback equipment for drill-out, production tubing, and flowback operations. A separator has various regulators and regulators are employed to reduce pressure to the areas of the tank. The most common areas that need regulations are as follows:

- *Liquid level controller (LLC)* is a pneumatic controller to evacuate fluid from the separator. When the level of water or oil gets to a predefined limit recommended by the manufacturer and set by the operator, it will automatically dump the water to the flowback tank and it will dump the oil or condensate to the low-stage separator. **Dump valve** is a working valve actuated by a LLC to evacuate fluid. There are two dump valves located on a three-phase separator (water and oil). Dump valve acts like a "toilet flush" when it dumps. The water and oil dumps must have manual bypass in addition to pneumatic.
- The *scrubber pot* removes traces of liquid droplets from gas stream. When gas leaves the vessel at the top, it passes through mist extractor (removes mist from gas stream) followed by scrubber pot, which removes the liquid droplets in the gas stream. The scrubber pot is essentially a water knockout unit used for conditioning fuel gas supply. The supply gas must be dry for separator control or failure can be the consequence if it is not. In the wet gas scenarios, a dryer is run after the scrubber pot to avoid this issue. This regulator prevents the gauge located on the scrubber pot from any moisture as well.
- *Back pressure regulator (BPR)* is located on the separator to hold pressure inside the separator as needed. For instance, if sales line pressure (which goes to the compressor station) is 700 psi, there needs to be a pressure higher than 700 psi on the separator back pressure to send the gas to the sales line. All units must have a pneumatic and manual BPR. Other mechanical controls of a separator are as follows:
- *Mechanical pop-off* is used to prevent overpressuring the separator and rupture. A mechanical pop-off is set slightly above the operating pressure. Pop-offs on the separator are inspected annually or if ruptured. If

mechanical pop-off gets activated, it will send the gas to the flare to prevent overpressuring the separator.
- *Meter* is used to measure the flow. Flow measurements must be electronic with mechanical backup. There are typically a minimum of two flow meters in place to measure the flow.

The fluid rate, also called liquid capacity, of a separator is related to the retention time and liquid settling volume. The amount of liquid retention time needed in a vessel governs liquid capacity of a separator. Factors affecting separation are operating pressure, operating temperature, and flow stream composition. Eq. (12.2) is often used to calculate liquid capacity of a separator for separator design during flowback operation. Fig. 12.6 shows three horizontal separators located on a multiwell pad during a flowback operation. Fig. 12.7 illustrates a four-stage horizontal separator from an inside view.

Fig. 12.6 Horizontal separators.

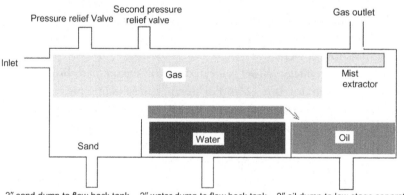

Fig. 12.7 Four-phase horizontal separator from inside.

Figs. 12.8–12.10 depict the LLC, BPR, and mechanical pop-off on a horizontal separator, respectively.

$$\text{Liquid capacity}: \quad W = \frac{1440 \times V}{t} \qquad (12.2)$$

where W is the fluid rate (liquid capacity), BBL/day, V is the liquid settling volume, BBL, and t is the retention time, min.

Fig. 12.8 Liquid level controller (LLC).

Fig. 12.9 Back pressure regulator (BPR).

Fig. 12.10 Mechanical pop-off.

Low-stage separator

Anytime there is a possibility of producing crude oil, condensate, or wet gas, a low-stage separator is needed. The line coming out of the oil partition of the high-stage separator (horizontal separator) will go into a low-stage separator to give oil more retention time before going to the oil tanks.

Flare stack

A flare stack is used to burn off flammable gas in certain events. One example is not having the necessary pipeline infrastructures. For instance, in Utica Shale operation (Ohio), many operating companies are still exploring the shale play and some companies do not have the pipeline infrastructure to commercially produce and sell the gas in new exploration areas. As a result, any commercial gas produced from any well is flared. Another example is not having the proper equipment to handle the volume of gas that is being produced. Sometimes the sales line cannot handle the produced volume of gas for whatever reasons and some of the gas is sent to flare. Ideally, any company would like to sell every bit of the gas produced; however, sometimes combinations of various reasons lead to sending the gas to the flare stack. The flare stack is essentially a safety precaution that needs to be used during various shale operations. In the Bakken Shale, gas is occasionally flared since the infrastructure for gas is not there. Bakken Shale is primarily a shale oil formation and this is the primary reason that noncommercial amount of gas produced with oil is flared. Typically there is a line that comes out of the high-stage separator and goes into the flare stack. In addition, there is another line that comes out of the low-stage separator and goes to the flare stack. Some requirements for flare stacks (Fig. 12.11) are as follows:

- Diameter of the flare stack needs to be at least $6''$.
- Height of flare stack has to be at least $40'$ for safety reasons depending on the amount of gas that is expected to be flared.
- Flare must be equipped with a check valve and it must also have an auto-ignition system.

Fig. 12.11 Flare.

Fig. 12.12 Oil tanks (upright tanks).

Oil tanks (upright tanks)

Oil tanks are only employed when there is a possibility of producing commercial oil or condensate. The entire purpose of having oil tanks on location is to store the commercial oil or condensate that is being produced. The number of oil tanks depends on the anticipated production volumes. The regular-size oil tanks typically have a capacity of 250 or 450 barrels (Fig. 12.12). There are trucks that are lined up throughout the day to transfer the oil or condensate from the oil tanks. Oil tanks must be closed top for safety reasons. Vapor produced from oil or condensate is heavier than air and can be fatally ignited if open tanks are used. The vapor produced has to be sent to a vapor destructive unit (VDU), which essentially flares any vapor produced from the liquid during flowback. The VDU must be rigged up and used on any location that has a possibility of producing liquid. The VDU is essentially a type of flare.

Flowback equipment spacing guidelines

1. All ignition sources *must* be 100′ from the flowback tanks. In addition, flowback tanks must be 100′ from the wellhead. The ignition sources should also be upwind of tanks if possible.
2. Choke manifold should be at least 50′ from the wellhead.
3. Low- and high-stage separators, along with sand traps, should be placed 75′ from the wellhead and 100′ from flowback tanks.
4. Flare stack should be 100′ from wellhead and flowback tanks.
5. Grounding is very important during flowback as it drains away any unwanted buildup of static electricity.

6. Bonding is also used to connect all metallic equipment to prevent buildup of static electricity.

Tubing analysis

One of the last steps of completions in unconventional shale reservoirs is running production tubing. Production tubing is used to efficiently remove water from the well until critical rate is reached, at which point the well starts liquid loading and some type of artificial lift will be needed. A well starts liquid loading when the production velocity is unable to carry the liquids from the bottom hole to the surface. Not efficiently and properly removing the liquids from the bottom hole will cause liquid buildup and as a result will lower production, finally killing the well.

The most commonly used production tubing sizes are 2 7/8″ and 2 3/8″. Some operators do not run production tubing on their well from day 1 in order to produce as much volume as possible through the casing (typically 5½″), especially in very prolific areas and longer lateral length wells. Running production tubing in a well will limit the amount of rate that can be produced through the tubing depending on various factors such as reservoir pressure, wellhead flowing pressure, water rate, turbulent factor (n value), etc. Therefore, nodal and economic analyses are performed to decide whether production tubing is needed and its size, or simply produce the well through casing until critical velocity of the casing is reached and tubing will have to be run to efficiently unload the water through the tubing. Fig. 12.13 illustrates an inflow performance curve (IPR) vs different tubing performance curves (TPCs) (various pipe sizes) including 5½″ casing, 2 7/8″, and 2 3/8″ assuming a reservoir pressure of 5200 psi, n value of 0.5 (fully turbulent), gas rate of 13 MMSCF/day, wellhead flowing pressure of 3500 psi, and water-to-gas ratio (WGR) of 40 BBL/MMSCF. The intersection of the IPR curve with various TPCs indicates the operating point of a particular well at an instantaneous point in time. This analysis is performed at various operating conditions for both IPR and TPCs to determine tubing sizing, drawdown, and compression of a particular well. Other analyses such as erosional and unloading velocity calculations across the lateral are also performed to prevent exceeding the erosional velocity and efficiently unloading fluid from the horizontal section of the wellbore.

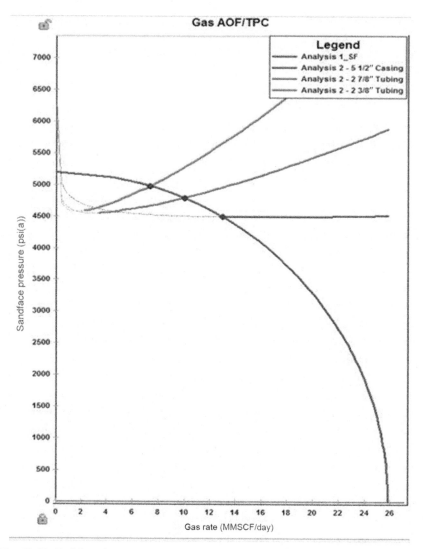

Fig. 12.13 Nodal analysis.

CHAPTER THIRTEEN

Rock mechanical properties and in situ stresses

Introduction

In general, rock mechanics is a branch of geomechanics where the main focus is on rock deformation and possible failure of rock due to the applied manmade or natural forces. This has been a topic of studies in different earth sciences and engineering programs. In the oil and gas industry and particularly in the field of hydraulic fracturing, the rock and fluid interactions have become a major topic of studies in which fracture initiation, propagation, and geometry due to the applied hydraulic force are investigated. This requires an advanced understanding of formation in situ stress conditions and stress behavior around the fracture as it generates and propagates to the formation. Stress, strain, and deformation are essential parameters required for characterization of mechanical properties of the rock. In this section of the book, various concepts of rock mechanics and interactions between induced and in situ stresses, especially during hydraulic fracturing, will be discussed.

Young's modulus (psi)

Young's modulus is a measurement of stress over strain. Simply put, Young's modulus is the slope of a line on stress vs strain plot. When hydraulic fracturing occurs, Young's modulus can be referred to as the amount of pressure needed to deform the rock. Young's modulus measures a rock's hardness, and the higher the Young's modulus, the stiffer the rock. For a successful hydraulic frac job to occur, higher Young's modulus is required. A higher Young's modulus indicates that the rock is brittle and will help to keep the fractures open for better production after the frac job. Examples of materials with high Young's modulus would be glass, diamond, granite, etc. These materials tend to be very hard but are prone to brittleness.

On the other hand, examples of materials with low Young's modulus would be rubber and wax, which are very flexible and resistant to brittleness. The Young's modulus in various unconventional shale plays varies, and the brittleness of the rock will determine the type of frac fluid system to be chosen for the job. Young's modulus can be measured by using a sonic log or core data. Core data yields static Young's modulus and sonic log represents dynamic Young's modulus in Eq. (13.1).

$$\text{Static Young's modulus from core analysis}: E = \text{Young's modulus} = \frac{\sigma}{\varepsilon_{xx}} \quad (13.1)$$

where σ is the stress, psi and ε_{xx} is the strain.

Another method for calculating Young's modulus is using a sonic log. The equation listed below can be used to calculate dynamic Young's modulus from a sonic log. Dynamic Young's modulus must then be converted to static Young's modulus.

- Formation modulus calculation

$$\text{Formation modulus}: G = 1.34 \times 10^{10} \times \frac{\rho_b}{\Delta t_s^2} \quad (13.2)$$

where G is the formation modulus, psi, ρ_b is the bulk density, g/cc, Δt_s is the shear wave travel time, μs/ft.

- Dynamic Young's modulus calculation:

$$\text{Dynamic Young's modulus from log analysis}: E = 2G(1+\nu) \quad (13.3)$$

where E is the dynamic Young's modulus, psi, G is the formation modulus, psi, and ν is the Poisson's ratio.

- Larry Britt came up with a correlation to convert dynamic Young's modulus from log to static Young's modulus as shown in Eq. (13.4) (King, 2010).

$$\text{Static Young's modulus conversion}: E_{\text{static}} = 0.835 \times E_{\text{dynamic}} - 0.424 \quad (13.4)$$

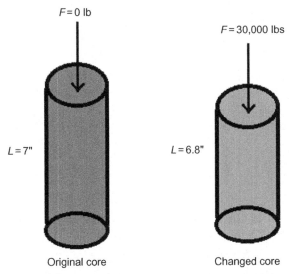

Fig. 13.1 Young's modulus example.

Example

A core sample was taken and sent to the lab. After applying 30,000 lbs of force to the core cross-sectional area of 0.3 in.^{2}, the length of the core decreased from 7″ to 6.8″ as shown in Fig. 13.1. Calculate Young's modulus from this core test.

$$\text{Stress} = \sigma = \frac{F}{A} = \frac{30{,}000}{0.3} = 100{,}000 \, \text{psi}$$

$$\text{Strain} = \varepsilon_{xx} = \frac{\Delta L}{L} = \frac{7 - 6.8}{7} = 0.02857$$

$$E = \text{Young's modulus} = \frac{\sigma}{\varepsilon_{xx}} = \frac{100{,}000}{0.02857} = 3.5 \, \text{MMpsi}.$$

Poisson's ratio (ν)

Poisson's ratio measures the deformation in the material in a direction perpendicular to the direction of the applied force. Essentially Poisson's ratio is one measure of a rock's strength that is another critical rock property related to closure stress. Poisson's ratio is dimensionless and ranges between 0.1 and 0.45. Low Poisson's ratio, such as 0.1–0.25, means rocks fracture easier whereas high Poisson's ratio, such as 0.35–0.45, indicates the rocks are harder to fracture. Please note that Poisson's ratio changes from layer to layer. The best formations to hydraulically fracture have the lowest Poisson's ratios. Poisson's ratio can be measured from a core sample. A core sample is taken to the lab and a compressive force is applied. Afterward, the height and diameter changes are measured (strain in x- and y-directions) and Eq. (13.5) is used to calculate Poisson's ratio.

$$\text{Poisson's ratio, core analysis}: \nu = -\frac{\varepsilon_y}{\varepsilon_x} = \frac{\text{Radial strain}}{\text{Axial strain}} \qquad (13.5)$$

ε_x = Strain in the x-direction, which means how much material is deformed when a stress is applied. Compressive strength is positive.

ε_y = How much material has been deformed after the stress application and is negative because of being tensile strain.

Poisson's ratio can also be calculated by running a sonic log in the depth of interest. The sonic log provides the shear and compression wavelength travel time, which are used in the calculation of Poisson's ratio using Eqs. (13.6), (13.7).

$$\text{Poisson's ratio, log analysis}: \nu = \frac{0.5R_\nu^2 - 1}{R_\nu^2 - 1} \qquad (13.6)$$

where R_ν is:

$$R_\nu \text{ calculation}: R_\nu = \frac{\Delta t_s}{\Delta t_c} \qquad (13.7)$$

where Δt_s is the shear wave travel time, μs/ft and Δt_c is compression wave travel time, μs/ft. Due to large amount of log data, machine learning is a viable approach in predicting shear and compression wave travel times. This development, in turn, would eliminate the need to run expensive sonic logs but still yield the ability to accurately predict shear and compression wave travel times that can then be used to calculate Young's modulus, Poisson's ratio, and minimum horizontal stress. Having access to thousands of rows of log data creates a perfect opportunity to train a supervised ML model to accurately predict geomechanical properties. The rock properties that can be included in the model as input variables are gamma ray, deep resistivity, neutron porosity, photoelectric effect, and bulk density. The outputs of the model would be shear and compression wave travel times. Machine learning is essentially used to predict a multioutput shear and compression wave travel times which then can be used to estimate Young's modulus, Poisson's ratio, and minimum horizontal stress.

Example

A core sample is taken from a Marcellus Shale formation. The sample height is 10″ and the diameter is 3″. After applying a compressive force of 150,000 lbs, the height decreases by 0.15″ and the diameter increases by 0.007″. Calculate the Poisson's ratio of the sample.

Strain in the x- and y-directions need to be calculated:

$$\varepsilon_x = \frac{\Delta L}{L} = \frac{0.15}{10} = 0.015$$

$$\varepsilon_y = \frac{\Delta D}{D} = \frac{0.007}{3} = 0.0023.$$

Finally, Poisson's ratio can be calculated.

$$\text{Poisson's ratio} = \nu = -\frac{\varepsilon_y}{\varepsilon_x} = \frac{\text{Radial strain}}{\text{Axial strain}} = \frac{0.0023}{0.015} = 0.16.$$

Example

Calculate Poisson's ratio and Young's modulus with the following data obtained from the sonic log:

Bulk density = 2.6 g/cc, Δt_s =115 μs/ft, Δt_c = 67 μs/ft

$$R_\nu = \frac{\Delta t_s}{\Delta t_c} = \frac{115}{67} = 1.72$$

$$\text{Poisson's ratio} = \nu = \frac{0.5 R_\nu^2 - 1}{R_\nu^2 - 1} = \frac{(0.5 \times 1.72^2) - 1}{1.72^2 - 1} = 0.24$$

$$\text{Formation modulus} = G = 1.34 \times 10^{10} \times \frac{\rho_b}{\Delta t_s^2} = 1.34 \times 10^{10} \times \frac{2.6}{115^2}$$
$$= 2.63 \times 10^6 \text{ psi}$$

$$\text{Dynamic Young's modulus} = E = 2G(1 + \nu)$$
$$= (2 \times 2.63 \times 10^6)(1 + 0.24) = 6.5 \times 10^6 \text{ psi}$$

Fracture toughness (psi/$\sqrt{\text{in}}$)

Fracture toughness modulus is another indicator of a rock's strength in the presence of a preexisting flaw. For example, glass has a high strength, but the presence of a small fracture reduces the strength. Therefore, glass has low fracture toughness. Fracture toughness is an important consideration in hydraulic fracture design. Fracture toughness is an essential parameter in very low fluid viscosity (water) and very low modulus formations. A low fracture toughness value indicates that materials are undergoing brittle fractures, while high values of fracture toughness are a signal of ductility. Fracture toughness ranges from 1000 to 3500 psi/$\sqrt{\text{in}}$. Fracture toughness is measured in the laboratory and is denoted by K_{IC}. Formations with low Poisson's ratio, low fracture toughness, and high Young's modulus are typically the best candidates for slick water hydraulic frac. Fig. 13.2 shows the schematic of a sample under vertical stress and therefore changes in length and width of the sample that can be used to calculate the Poisson's ratio as described in Eq. (13.5).

Brittleness and fracability ratios

Brittleness and fracability ratios are very important to compute and understand hydraulic fracture design. Calculating Young's modulus and Poisson's ratio separately does not give a clear understanding of the brittleness and fracability of the rock. Therefore, various equations have been developed to combine both parameters into one single variable. The

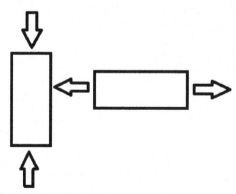

Fig. 13.2 Poisson's ratio illustration.

simplest way to find the brittleness of the rock is by taking the ratio of Young's modulus over Poisson's ratio (PR); the higher E/PR, the higher the brittleness. As previously mentioned, various equations were developed for both fracability and brittleness ratios and Eqs. (13.8), (13.11) are examples of brittleness and fracability ratios that were developed primarily for Barnett Shale. The following brittleness ratio was generated after Rickman et al. (2008):

$$\text{Brittleness ratio} = \frac{\left(\left(\frac{E_{static}-1}{7}\right) \times 100\right) + \left(\left(\frac{\nu - 0.4}{-0.25}\right) \times 100\right)}{2} \qquad (13.8)$$

where E_{static} is the static Young's modulus and ν is Poisson's ratio.

Fracability ratio was generated after Goodway et al. (2010) using Eq. (13.11) that is a function of incompressibility constant λ, Eq. (13.9), and rigidity constant μ, Eq. (13.10):

$$\text{Incompressibility constant}: \lambda = \frac{E_{static} \times \nu}{(1+\nu)(1-2\nu)} \qquad (13.9)$$

$$\text{Rigidity constant}: \mu = \frac{E_{static}}{2(1+\nu)} \qquad (13.10)$$

$$\text{Fracability ratio} = \frac{\lambda}{\mu} = \frac{\text{Incompressibility constant}}{\text{Rigidity constant}} \qquad (13.11)$$

where λ is used to relate rock's resistance to fracture dilation and μ describes rock's resistance to shear failure.

For the rock to be brittle and fracable in Barnett shale, the brittleness ratio has to be more than 50 and fracability ratio must be less than 1. These two equations were developed based on the Barnett Shale reservoir, which has the best E and Poisson's ratio as compared to other shale plays across North America. If the sonic log is available, obtain dynamic E and Poisson's ratio from the sonic log. Afterward, convert dynamic E to static E. Finally, use brittleness and fracability ratio equations to calculate these two parameters in each section (usually 6″) to determine the optimum placement for your lateral well from a completions perspective.

Example
Given the 20 samples below, calculate brittleness, fracability, E/PR ratios, and determine which 10 consecutive zones are the best zones for hydraulic fracturing from a rock brittleness and fracability perspective.

Table 13.1 Brittleness and fracability ratios example

	Provided			Results				
Sample	Static modulus	PR	Brittleness	λ	μ	Fracability	YM/PR
1	4.8	0.33	41.1	3.50	1.80	1.94	14.5
2	5.3	0.35	40.7	4.58	1.96	2.33	15.1
3	4.5	0.27	51.0	2.08	1.77	1.17	16.7
4	3.5	0.22	53.9	1.13	1.43	0.79	15.9
5	3.3	0.25	46.4	1.32	1.32	1.00	13.2
6	5	0.3	48.6	2.88	1.92	1.50	16.7
7	4.5	0.27	51.0	2.08	1.77	1.17	16.7
8	4.1	0.23	56.1	1.42	1.67	0.85	17.8
9	4.3	0.26	51.6	1.85	1.71	1.08	16.5
10	4	0.19	63.4	1.03	1.68	0.61	21.1
11	3.5	0.33	31.9	2.55	1.32	1.94	10.6
12	3.5	0.32	33.9	2.36	1.33	1.78	10.9
13	3.2	0.39	17.7	4.08	1.15	3.55	8.2
14	4.1	0.29	44.1	2.19	1.59	1.38	14.1
15	4.1	0.33	36.1	2.99	1.54	1.94	12.4
16	4.1	0.34	34.1	3.25	1.53	2.13	12.1
17	3.3	0.37	22.4	3.43	1.20	2.85	8.9
18	4.1	0.36	30.1	3.88	1.51	2.57	11.4
19	3.8	0.28	44.0	1.89	1.48	1.27	13.6
20	3.9	0.32	36.7	2.63	1.48	1.78	12.2

PR, Poisson's ratio; YM, Young's modulus.

The best zone from a hydraulic fracturing perspective is the zone with the highest E and the lowest PR. This basically means the highest brittleness, the lowest fracability, and the highest E/PR ratios. In Table 13.1, the first 10 consecutive samples must be targeted when drilling the well. This does not take into account other formation properties. In reality, brittleness and fracability along with other formation properties must be taken into account before deciding the landing zone of a well. But for the sake of this example, all the other parameters have been excluded.

Vertical, minimum horizontal, and maximum horizontal stresses

The next exciting concept in hydraulic fracture design is the various principal stresses that exist within the rock. There are three principal stresses that exist within a rock.

Vertical stress

Vertical stress, also referred to as overburden stress, is the sum of all the pressures applied by all of the different rock layers. Every formation contains fluid and rock and each one must be accounted. Porosity correlation can be simply used to define the amount of space that is occupied by fluid vs the amount of space that is occupied by a rock. The average density of the rock can simply be calculated using Eq. (13.12).

Average formation density: $\rho_{avg} = \rho_{rock}(1 - \phi) + \rho_{fluid}\phi$ (13.12)

where ρ_{avg} is the average formation density, ppg, ρ_{rock} is the rock density, ppg, ρ_{fluid} is the fluid density, ppg, and ϕ is the porosity, fraction.

Now that the average formation density is known, the magnitude of the vertical stress in an isotropic, homogeneous, and linearly elastic formation can be calculated using Eq. (13.13).

Vertical stress $= \sigma_V = 0.05195 \times \rho_{avg} \times H$ (13.13)

where ρ_{avg} is the average formation density, ppg and H is the height of layer or TVD in ft, 0.05195 is the conversion from ppg to psi/ft.

Eq. (13.13) can be rewritten as follows if average formation density is reported in lb/ft^3:

$$\text{Vertical stress} = \frac{\rho \times \text{TVD}}{144}$$

Example

The following data is provided from core analysis. Using this data, calculate vertical (overburden stress) at the zone of interest (8000′):

0′–4000′ → Zone 1 = 9% porosity, rock density = 21.5 ppg, fluid density = 8.35 ppg

4000′–6000′ → Zone 2 = 12% porosity, rock density = 23.6 ppg, fluid density = 8.6 ppg

6000–8000 → Zone 3 = 7.5% porosity, rock density = 22.4 ppg, fluid density = 8.4 ppg

Layer 1: $\rho_{avg} = \rho_{rock}(1 - \phi) + \rho_{fluid}\phi = 21.5(1 - 9\%) + (8.35 \times 9\%) = 20.32$ ppg

Layer 2: $\rho_{avg} = \rho_{rock}(1 - \phi) + \rho_{fluid}\phi = 23.6(1 - 12\%) + (8.6 \times 12\%) = 21.8$ ppg

Layer 3: $\rho_{avg} = \rho_{rock}(1 - \phi) + \rho_{fluid}\phi = 22.4(1 - 7.5\%) + (8.4 \times 7.5\%) = 21.35$ ppg.

Now calculate the incremental vertical stress (overburden stress) for each layer:

Layer 1: $\sigma_v = 0.05195 \times \rho_{avg} \times H = 0.05195 \times 20.32 \times 4000 = 4222.5$ psi

Layer 2: $\sigma_v = 0.05195 \times \rho_{avg} \times H = 0.05195 \times 21.8 \times (6000 - 4000) = 2265.0$ psi

Layer 3: $\sigma_v = 0.05195 \times \rho_{avg} \times H = 0.05195 \times 21.35 \times (8000 - 6000) = 2218.3$ psi.

Total vertical stress at 8000′:

Total vertical stress = 4222.5 + 2265.0 + 2218.3 = 8706 psi.

Total vertical stress gradient at 8000′:

$$\text{Vertical stress gradient} = \frac{8706}{8000} = 1.09 \, \text{psi/ft}.$$

In the real world, it is very challenging to obtain rock density and fluid density at various depths. Therefore, a density-logging tool can measure the density of the formation every half-foot. A density-logging tool is typically not run all the way to the surface and is only run a few thousand feet around the zone of interest. The vertical stress gradient is typically between 1 and 1.1 psi/ft depending on the depth and porosity. In a given formation, the higher the porosity and the shallower the depth, the lower the vertical stress. In contrast, the lower the porosity and the deeper the depth, the higher the vertical stress.

Minimum horizontal stress

Minimum horizontal stress is approximated as fracture closure pressure. Units of stress and pressure are both psi. This is not a coincidence because stress and pressure are fundamentally related. The primary difference is that pressure acts in all directions equally, while stress only acts in the direction of the force. Since effective horizontal stress is a direct result of the effective vertical stress, Poisson's ratio determines the amount of stress that can be transmitted horizontally. Minimum horizontal stress or fracture closure pressure can be obtained from either Poroelastic theory, diagnostic fracture injection test (DFIT) or approximated using Eq. (13.14) (if rock properties are available and fracture is vertical):

Minimum horizontal stress : $\sigma_{h,min}$

$$\cong \frac{\nu}{1-\nu} \times (\sigma_v - \alpha P_p) + \alpha P_p + \sigma_{Tectonic} \tag{13.14}$$

where ν is the Poisson's ratio, σ_v is the vertical stress, psi, α is the Biot's constant and dimensionless value, P_p is the pore pressure, psi, and σ_{Tectonic} is the tectonic stress, psi.

As can be seen in Eq. (13.14), Poisson's ratio, vertical stress, Biot's constant, and pore pressure primarily affect approximated minimum horizontal stress. The tectonic stress is important in tectonically active areas and can be obtained from the difference between measured stress obtained using DFIT and calculated using the theory equation.

Biot's constant (poroelastic constant)

Biot's constant, also known as poroelastic constant, measures how effectively the fluid transmits pore pressure into rock grains. Biot's constant ranges between 0 and 1. In an ideal case where porosity does not change as pore pressure and confining pressure change, Biot's constant can be calculated using Eq. (13.15).

$$\text{Biot's constant}: \alpha = 1 - \frac{C_{\text{matrix}}}{C_{\text{bulk}}} \tag{13.15}$$

where C_{matrix} is the compressibility of the matrix and C_{bulk} is the compressibility of the matrix and pore space.

When porosity is high, rock formation (bulk) is very compressible compared to the matrix of the rock. This will cause the ratio of $C_{\text{matrix}}/C_{\text{bulk}}$ to approach zero resulting in Biot's constant of 1. In contrast, when porosity is low, $C_{\text{matrix}}/C_{\text{bulk}}$ approaches 1 resulting in Biot's constant of 0.

Biot's constant estimation

If a value of porosity is known and there is no information on geomechanical properties of the rock such as bulk modulus and Poisson's ratio, a rough estimate of Biot's constant can be found using Eq. (13.16).

$$\text{Biot's constant estimation}: \alpha = 0.64 + 0.854 \times \phi \tag{13.16}$$

where ϕ is the porosity fraction.

Example

A formation has a Poisson's ratio of 0.25, overburden pressure of 9000 psi, pore pressure gradient of 0.67 psi/ft, true vertical depth (TVD) of 8500′, and porosity of 8.5%. Assuming tectonic stress of 400 psi, calculate closure pressure. $\alpha = 0.64 + 0.854 \times \phi = 0.64 + (0.854 \times 8.5\%) = 0.713$

Pore pressure = pore pressure gradient × TVD = 0.67 × 8500 = 5695 psi

$$\text{Closure pressure} = \sigma_{h,\min} \cong \frac{\nu}{1-\nu} \times (\sigma_\nu - \alpha P_p) + \alpha P_p + \sigma_{\text{Tectonic}}$$

$$= \frac{0.25}{1 - 0.25} \times (9000 - 0.713 \times 5695) + 0.713 \times 5695 + 400$$

$$= 6107 \text{ psi}$$

Maximum horizontal stress

Maximum horizontal stress is more challenging to calculate. Maximum horizontal stress can be determined from the relationship presented by Haimson and Fairhurst (1967). They showed the relationship between the magnitude of near-wellbore stress and the magnitude of horizontal stress through breakdown pressure.

For penetrating fluid (slick water), Eq. (13.17) can be used to calculate maximum horizontal stress.

Breakdown pressure for penetrating fluid :

$$P_b = \frac{3 \times (\sigma_{\min} - P_R) - (\sigma_{\max} - P_R) + T}{\left(2 - \alpha\left(\frac{1-2\nu}{1-\nu}\right)\right)} + P_R \quad (13.17)$$

where P_b is the breakdown pressure, psi, $\sigma_{\min}$ is the min horizontal stress, psi, α is the Biot's constant, P_R is the reservoir pressure, psi, ν is the Poisson's ratio, and T is the tensile stress.

For nonpenetrating fluid (gelled fluid), the following equation Eq. (13.18) can be used to calculate maximum horizontal stress.

Breakdown pressure for nonpenetrating fluid :

$$P_b = 3 \times (\sigma_{\min} - P_R) - (\sigma_{\max} - P_R) + P_R + T \quad (13.18)$$

where P_b is the breakdown pressure, psi, $\sigma_{\min}$ is the min horizontal stress, psi, P_R is the reservoir pressure, psi, and T is the tensile stress, psi.

Example

Calculate vertical and approximated minimum horizontal stresses given the following data and Table 13.2.

Average overburden density = 160 lbs/ft^3, Biot's constant = assume 1, Tectonic stress = 200 psi

Table 13.2 True vertical depth (TVD), Poisson's ratio, and pore pressure gradient

Formation	TVD (ft)	Poisson's ratio	Pore pressure gradient (psi/ft)
Overlaying shale	7350	0.28	0.64
Sandstone	7400	0.22	0.64
Underlying shale	7450	0.28	0.65

Vertical stress for each layer must be calculated first, as follows:

$$\text{Overlaying shale vertical stress(psi)} = \frac{\rho \times \text{TVD}}{144} = \frac{160 \times 7350}{144} = 8167\,\text{psi}$$

$$\text{Sandstone vertical stress(psi)} = \frac{160 \times 7400}{144} = 8222\,\text{psi}$$

$$\text{Underlying shale vertical stress(psi)} = \frac{160 \times 7450}{144} = 8278\,\text{psi}.$$

As can be seen from the calculated overburden pressures (above), as the TVD of the rock layer increases the overburden pressure (vertical stress) increases as well.

The minimum horizontal stress of each rock layer is calculated, as follows:

Overlying shale layer:

$$\text{Pore pressure} = 0.64 \times 7350 = 4704\,\text{psi}$$

$$\sigma_{h,\min} \cong \frac{0.28}{1-0.28} \times (8167 - 4704) + 4704 + 200 = 6250\,\text{psi}.$$

Sandstone layer:

$$\text{Pore pressure} = 0.64 \times 7400 = 4736\,\text{psi}$$

$$\sigma_{h,\min} \cong \frac{0.22}{1-0.22} \times (8222 - 4736) + 4736 + 200 = 5919\,\text{psi}.$$

Underlying shale layer:

$$\text{Pore pressure} = 0.65 \times 7450 = 4843\,\text{psi}$$

$$\sigma_{h,\min} \cong \frac{0.28}{1-0.28} \times (8278 - 4843) + 4843 + 200 = 6379\,\text{psi}.$$

Example

Calculate approximated minimum and maximum horizontal stresses of a formation with the following properties:

$v = 0.24$, Vertical stress gradient = 1.1 psi/ft, TVD = 11,500′, Pore pressure gradient = 0.65 psi/ft, Tensile stress = 250 psi, Breakdown pressure = 10,500 psi, Biot's constant of 1, Assume slick water fluid

Overburden stress = vertical stress gradient × TVD = 1.1 × 11,500
= 12,650 psi

Pore pressure = pore pressure gradient × TVD = 0.65 × 11,500 = 7475 psi

$$\sigma_{h,\min} \cong \frac{0.24}{1-0.24} \times (12650 - 7475) + 7475 + 250 = 9359\,\text{psi}$$

$$P_b = \frac{3 \times (\sigma_{\min} - P_R) - (\sigma_{\max} - P_R) + T}{\left(2 - \alpha\left(\frac{1-2\nu}{1-\nu}\right)\right)} + P_R$$

$$= 10{,}500 = \frac{3 \times (9359 - 7475) - (\sigma_{\max} - 7475) + 250}{2 - \left(\frac{1-2(0.24)}{1-0.24}\right)} + 7475$$

$$\sigma_{\max} = 9398\,\text{psi}.$$

Various stress states

There are three different types of geologic environments that can be determined from min, max, and vertical stress magnitudes. These three fault environments are as follows:

1. Normal fault environment:

$$S_V \geq S_{H,\max} \geq S_{h,\min}$$

2. Strike-slip (shear) environment:

$$S_{H,\max} \geq S_V \geq S_{h,\min}$$

3. Reverse (thrust) fault environment:

$$S_{H,\max} \geq S_{h,\min} \geq S_V$$

Fracture orientation

Fracture is always created and propagated (grows) perpendicular to the least principal stress (minimum horizontal stress). Fracture orientation is influenced by various factors such as overburden pressure, pore pressure, tectonic forces, Poisson's ratio, Young's modulus, fracture toughness, and rock compressibility. It is extremely important to understand the principal stresses acting on the rock in the formation of interest for a successful frac job. Engineers, petrophysicists, geologists, and geoscientists are in charge of understanding and calculating the principal stresses. There are two types of fractures that can be achieved via hydraulic fracturing. The first is referred to as a *longitudinal fracture*, which is essentially one big fracture, and the second is called *transverse fractures*, which are a combination of long, narrow fractures.

Transverse fractures

In almost all of the unconventional shale plays across the country, the goal is to create transverse fractures due to the stress directions, magnitudes, production, and economic feasibility. To create transverse fractures, the well needs to be drilled (placed) parallel to the minimum horizontal stress or perpendicular to the maximum horizontal stress. This means the fractures will propagate (grow) perpendicular to the minimum horizontal stress. Stress directions can be typically obtained from a fracture microseismic, formation microimager (FMI) log, or in the worst-case scenario a world stress map, which is free and widely available. The world stress map is a very useful tool that engineers and geologists use to understand various in situ stresses. A world stress map is used to understand the direction of maximum horizontal stress in a particular region of interest. Therefore, after finding the region of interest where fracing needs to take place, the wells must be drilled perpendicular to the stress direction on the world stress map to create transverse fractures (perpendicular to maximum horizontal stress). For example, when looking at development maps in the Marcellus and Utica/Point Pleasant shale plays, almost all of the wells are drilled NW-SE because from the world stress map, the maximum horizontal stress is facing NE-SW in these areas. Therefore, to create transverse fractures, the wells must be drilled perpendicular to the NE-SW direction.

Longitudinal fractures

To create a longitudinal fracture, the well needs to be drilled parallel to the maximum horizontal stress or perpendicular to the minimum horizontal stress. This means the fractures will propagate parallel to the minimum horizontal stress and perpendicular to the maximum horizontal stress, which is exactly the opposite of transverse fractures. Longitudinal fractures are

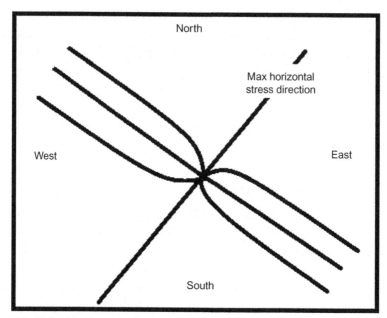

Fig. 13.3 Wells drilled perpendicular to max horizontal stress.

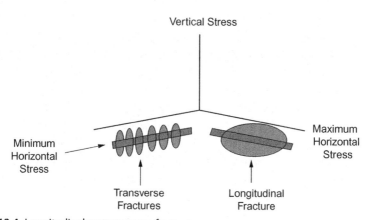

Fig. 13.4 Longitudinal vs transverse fractures.

typically created at shallower depths. Fractures created in Bakken, Eagle Ford, Marcellus, Utica, and Barnett Shales, along with in many other shale plays, are confirmed to be transverse fractures from frac microseismic data (Figs. 13.3 and 13.4).

Example
A horizontal well is going to be drilled and hydraulically fractured. The vertical overburden stress gradient is calculated to be 1 psi/ft. One of the principal horizontal stresses has a gradient of 0.7 psi/ft in the direction of N45°E while the other one has a gradient of 0.85 psi/ft in the direction of N45°W. If the goal is to achieve transverse fractures, sketch the direction of the horizontal well and transverse fractures.

The minimum horizontal stress gradient in the direction of N45°E is 0.7 and 0.85 psi/ft is the maximum horizontal stress in the direction of N45°W. To create transverse fractures, the well must be drilled perpendicular to the maximum horizontal stress direction, which means the well must be drilled perpendicular to N45°W as shown in Fig. 13.5. Hydraulic fractures (transverse fractures) on the other hand will grow perpendicular to the minimum horizontal stress.

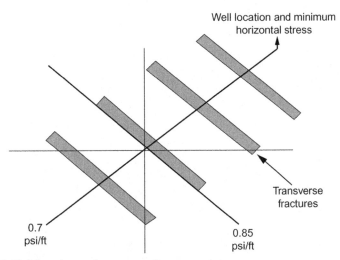

Fig. 13.5 Well location and transverse fractures.

CHAPTER FOURTEEN

Diagnostic fracture injection test

Introduction

Diagnostic fracture injection test (DFIT) has become very popular in unconventional shale reservoirs. DFIT is the most commonly used technique in unconventional shale reservoirs to determine various completions and reservoir properties for optimum fracture design. The idea is to create a small fracture by pumping 10–100 BBLs of water at 2–10 bpm and monitor pressure falloff for a specific period of time. DFIT is typically performed a few weeks before the start of a frac job depending on formation permeability. The time of shut-in after pumping the DFIT will be dependent on the formation permeability and the pump time, which in turn translates into the time it takes to reach pseudoradial flow. After pumping the DFIT test, enough monitoring time should be allowed to reach pseudoradial flow to determine various reservoir properties. Some of the completions properties that can be obtained from DFIT are instantaneous shut-in pressure (ISIP), fracture gradient, net extension pressure, fluid leak-off mechanism, time to closure, closure pressure (minimum horizontal stress), approximation of maximum horizontal stress, anisotropy, fluid efficiency, effective permeability, transmissibility, and pore pressure. It is strongly recommended not to use any volume in excess of 50 BBLs in nanodarcy permeability formations as it might delay the time it takes to reach pseudoradial flow. If permeability is higher, more fluid as high as 100 BBLs can be pumped and still reach pseudoradial flow just in time. The main purpose is to contact the whole net pay to get accurate completions and reservoir properties. The following guideline is an estimation of the shut-in time (post-DFIT shut-in) needed to reach pseudoradial flow and accurately calculate reservoir properties.

$$
\begin{array}{ll}
1\,\text{day} & \text{if } k > 0.1\,\text{md} \\
1\,\text{week} & \text{if } k > 0.01\,\text{md} \\
2\,\text{weeks} & \text{if } k > 0.001\,\text{md} \\
1\,\text{month} & \text{if } k > 0.0001\,\text{md}
\end{array}
$$

Most wells in unconventional reservoirs are shut-in anywhere between 2 to 6 weeks in order to reach pseudoradial flow.

Typical DFIT procedure

The following procedure is typically used when performing a DFIT:
- DFIT can be performed through perforations (toe stage) or toe initiation tools.
- If DFIT is performed through perforations, run in the hole (RIH) with TCP (tubing conveyed perforations) guns and perforate the toe stage using 6–10 shots.
- If DFIT is performed through the toe initiation tool, no perforation will be needed.
- Fresh or potassium chloride (KCl) water can be used depending on the percentage of clay in the formation. KCl water must be used if the formation is prone to swelling.
- Install the surface self-powered intelligent data retriever (SPIDR) gauge (or any other types of high-resolution gauges) to get accurate pressure measurement (1 psi resolution gauge). If enough money is available in the budget, a bottom-hole pressure gauge is recommended instead for more accurate pressure recording.
- Load the hole with fresh or KCl water.
- Once the hole (casing) is filled, continue pumping at the designed rate until formation breakdown occurs.
- After formation breakdown, continue pumping at 2–10 bpm until the desired DFIT volume is achieved (should not exceed 100 BBLs depending on the permeability).
- It is very important to continuously pump at a constant rate after breakdown because DFIT calculation assumes a continuous rate.

DFIT data recording and reporting

Record the following data from DFIT:
- type and specific gravity of fluid that was pumped
- pump rate (bpm) during breakdown and while pumping the designed volume
- ISIP
- formation breakdown pressure
- start and end time

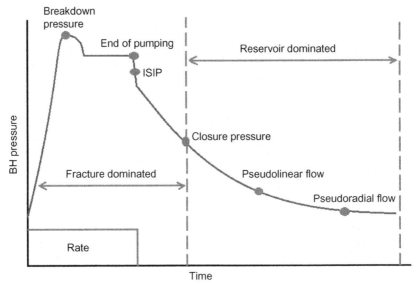

Fig. 14.1 Typical fracture injection test.

- total pump time
- volume pumped after the breakdown
- any unexpected events, that is, shutdowns and how the test was restarted, casing and/or surface equipment leaks, pressure spikes while pumping, initial DFIT gauge pressure, and time reading, etc.

Fig. 14.1 illustrates a typical fracture injection test that is divided into two sections. The first section is fracture dominated and the second section is reservoir dominated. From the fracture-dominated section, completions properties can be determined, and from the reservoir-dominated section, essential reservoir properties can be obtained.

Before-closure analysis

The first analysis in DFIT is called before-closure analysis (BCA), which means the analysis is performed right until closure occurs. The three main plots used for BCA are:
- Square root plot: plot BHP (y-axis) vs square root of time (x-axis)
- Log–log plot: plot log (BH ISIP-BHP) vs log of time
- BHP vs G-function time

Please note that bottom-hole pressure is typically calculated from surface pressure when a surface pressure gauge is used during DFIT. Therefore,

make sure the correct fluid density is used for bottom-hole pressure calculation as it will have a significant impact on the outcome of the DFIT analyses. Time is defined as the time since ISIP. The main idea is to use different types of diagnostic plots to make sure consistent results are obtained from three different BCA plots. All plots must be used in conjunction with one another for better estimation of properties and majority of DFIT interpretation commercial software packages have that capability. Derivatives in DFIT analysis are used as an aid in the straight-line segment of the decline curve.
- First derivative:
  - yields the slope of the curve,
  - a constant slope yields a straight line,
  - yields local minima and maxima.
- Second derivative:
  - yields the curvature of the decline curve.

Square root plot

A square root plot is commonly used to determine the closure pressure. When the square root of time (x-axis) vs the bottom-hole pressure (y-axis) is plotted, the linear portion of the plot will lie along a straight line going through the origin. The point at which deviation from the straight line occurs on the superposition plot (second derivative) is referred to as closure pressure. Every square root plot will have three main curves: pressure curve, first derivative, and second derivative (also referred to as superposition). Deviation from the straight line on the pressure curve is used to define minimum closure pressure. In addition, deviation from the smart line going through the origin on the second derivative curve is referred to as fracture closure. In Fig. 14.2, the *blue curve* (dark gray curve in print version) is the pressure curve, the *green curve* (light gray curve in print version) is the first derivative curve, and the *red curve* (gray curve in print version) is the second derivative (superposition curve).

To identify fracture closure, a linear extrapolated line from the origin is drawn on the second derivative curve (*black line*). Fracture closure can be approximated when the second derivative curve deviates from the linear line. After identifying fracture closure on the second derivative curve, draw a vertical line from the fracture closure point until the pressure curve is intersected as shown in *red* (gray in print version). After intersecting the pressure curve, closure pressure can be read on the y-axis.

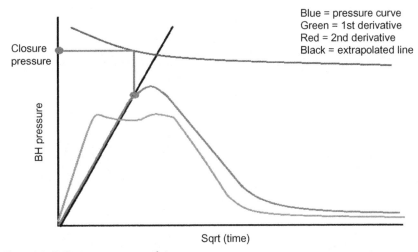

Fig. 14.2 BHP vs square root of time.

In Fig. 14.3 the deviation from the linear extrapolated line going through the origin on the second derivative is referred to as closure pressure, which is around 6845 psi in this example. In addition, closure time is also around 463 min.

Log-log plot (log (BH ISIP-BHP) Vs log (time))

A log-log plot is derived from a square root plot. This plot should be sufficient to identify closure and various flow regimes before and after closure. Various flow regimes on the second derivative of the log-log plot can be determined:

Before-closure analysis:

Half-slope line (1/2 slope) = Corresponds to linear flow regime

Quarter-slope line (1/4 slope) = Corresponds to bilinear flow regime

After-closure analysis:

Negative half-slope line (−1/2) = Corresponds to linear flow

Negative three-fourth (−3/4) = Corresponds to bilinear flow

Negative unit slope (−1) = Corresponds to pseudoradial flow

The log-log plot shows a positive ½ slope on the second derivative curve before closure. In some rare instances, it shows a positive ¼ slope on the second derivative before closure. Closure occurs by the change in slope from positive to negative on the second derivative curve. Pseudolinear flow is indicated when the second derivative curve shows a negative ½ slope in conjunction with a negative 1.5 slope on the first derivative curve. Pseudoradial

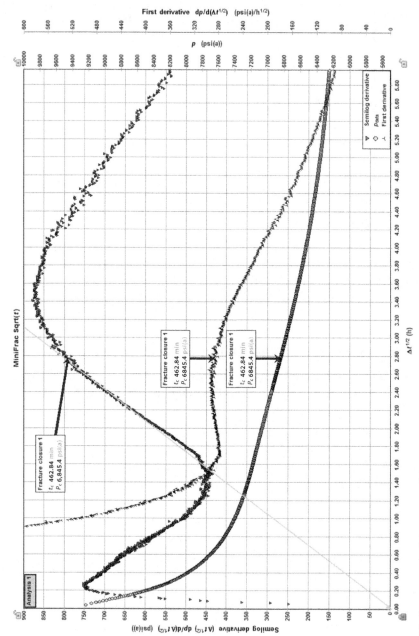

Fig. 14.3 Square root of time example.

flow is indicated when the second derivative curve displays a negative unit slope in conjunction with a negative 2 slope on the first derivative curve (Barree et al., 2007).

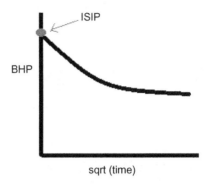

$$y = mx + b$$

$$\text{BHP} = m\left(\sqrt{\text{time}}\right) + \text{ISIP} \rightarrow \text{BHP} - \text{ISIP} = m\left(\sqrt{\text{time}}\right)$$

$$\Delta P = m \times \text{time}^{1/2} \rightarrow \log\Delta P = \log\left(m \times \text{time}^{1/2}\right)$$

$$\log(\Delta P) = \frac{1}{2}\log(\text{time}) + \log(m)$$

In the log–log plot example shown in Fig. 14.4, the *blue curve* (dark gray curve in print version) represents delta pressure, the *green curve* (light gray curve in print version) represents the first derivative, and the *red curve* (gray curve in print version) represents the second derivative. As can be seen on the second derivative, the slope of the curve changes from being positive to negative. The slope of the open fracture line on the second derivative is ½. Any derivation from this ½ slope line means the fracture would have changed or in this case closed. This represents closure occurrence and that point can be picked as the fracture closure pressure. Negative 1 slope (unit slope) on the second derivative is also an indication of pseudoradial flow. When pseudoradial flow is reached, more confidence is obtained when calculating various reservoirs properties, especially pore pressure.

Fig. 14.5 is another log–log plot example where the deviation from positive ½ slope line to negative on the second derivative can be used to determine closure pressure. Closure pressure is identified to be 16,627 psi.

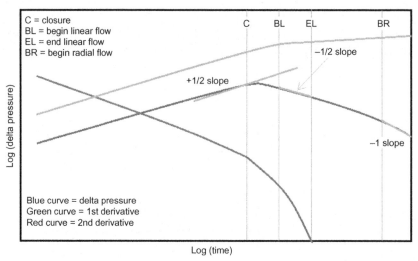

Fig. 14.4 Log-log plot.

In addition, the beginning and the end of linear flow (after closure) can be determined from negative ½ slope line. This well appears to have reached pseudoradial flow but more time will be needed to monitor the pressure falloff to have confidence with the scattered data during pseudoradial flow, which can be identified by the −1 slope line as shown on the plot. Note that in the log-log plot, derivative slopes before closure are always positive, but after closure slopes are negative. Unit slope indicates storage before closure but −1 slope indicates pseudoradial flow after closure (Barree et al., 2007).

Fig. 14.6 is another log-log plot example where closure is reached by observing the deviation from the positive ½ slope to negative on the second derivative (9250 psi = closure pressure), but due to other operational and data recording issues, pseudoradial flow cannot be observed from this example. The results from this DFIT must not be used to perform any types of after-closure analyses. Data recording and monitoring is the key to a successful DFIT interpretation.

G-function analysis

G-function is a variable related to time. G-function (x-axis) vs BHP (y-axis) can be plotted to determine various fracture and formation properties such as fracture closure, fluid efficiency, effective permeability, and leak-off mechanism. G-function assumes constant fracture height, constant pump rate, and

Diagnostic fracture injection test 241

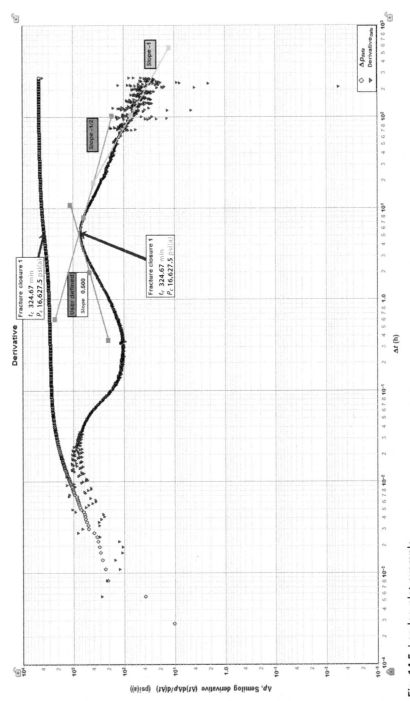

Fig. 14.5 Log-log plot example.

242 Hydraulic fracturing in unconventional reservoirs

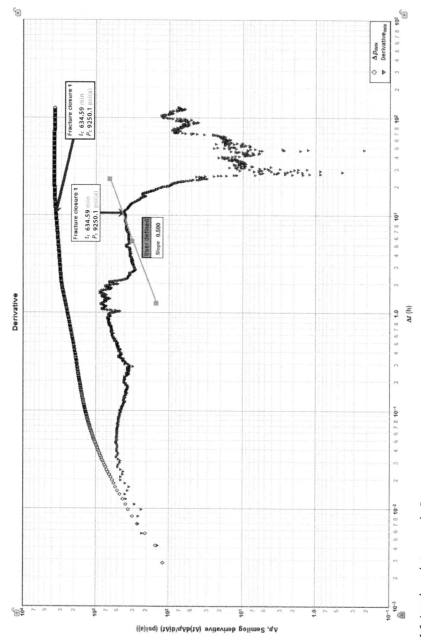

Fig. 14.6 Log-log plot example 2.

stoppage of fracture propagation when pumping stops. Eq. (14.1) can be used to approximate G-function time:

G-function time :
$$G(\Delta t_D) = \frac{4}{\pi}[g(\Delta t_D) - g_0]$$
$$g(\Delta t_D) = \frac{4}{3}(1 + \Delta t_D)^{1.5} - \Delta t_D^{1.5}; \beta = 1.0$$
$$g(\Delta t_D) = (1 + \Delta t_D)\sin^{-1}(1 + \Delta t_D)^{-0.5} + \Delta t_D^{0.5}; \beta = 0.5$$
$$\Delta t_D = \frac{t - t_p}{t_p}$$

(14.1)

where t is the shut-in time, minutes and t_p is the total pump time, minutes.

A β value of 1.0 refers to tight formations with low fluid leak-off, while a β value of 0.5 refers to high-permeability formations with high leak-off.

It is important to note that the G-function at shut-in (ISIP) is zero. For example, if total pump time is 5 min ($t_p = 5$ min), t at ISIP will be equal to 5 as well. Therefore, G-function at ISIP is equal to zero. G-function time starts at ISIP. The following steps can be used to find closure pressure on the G-function time:

1. Look for local maximum on the first derivative.
2. Look for deviation from the straight line on the pressure curve.
3. Look for deviation from the straight line going through the origin on the second derivative curve.
4. Closure occurs where the second derivative curve deviates from the straight line.

Fluid leak-off regimes on G-function plot

There are four unique leak-off regimes that can be noted from the G-function plot. The first type of leak-off on a G-function plot is referred to as "normal leak-off," which refers to the occurrence of leak-off through a homogeneous rock matrix (not the typical signature for unconventional reservoirs). Other leak-off types on a G-function plot are pressure-dependent leak-off (PDL), height recession leak-off (transverse storage), and fracture tip extension, which are discussed in detail below.

Pressure-dependent leak-off

Pressure-dependent leak-off (PDL) typically occurs in hard naturally fractured rocks. PDL can indicate complexity during hydraulic fracturing since the rock is naturally fractured. There is a pressure that, when exceeded, will control the opening in natural fractures. Once this pressure is exceeded, the surface area to leak-off will also increase. Since this pressure is driving the leak-off process, it is referred to as PDL. PDL is the most common type of leak-off regime in unconventional shale reservoirs due to the existing natural fractures.

PDL can be easily identified on the G-function plot. The easiest way to identify PDL is when the second derivative curve shows a concave down feature above the extrapolated line going through the origin as shown in Fig. 14.7. In the G-function plot below, *blue* (dark gray in print version) represents the pressure curve, *green* (light gray in print version) represents the first derivative, and *red* (gray in print version) represents the second derivative. When a straight-line (*black*) going through the origin is plotted on the second derivative, there is a concave down feature (hump) above the extrapolated line. This represents the existence of natural fractures, which will result in a complex fracture system and indicates that fluid leaks off faster than expected for a normal bi-wing fracture. As soon as the second derivative curve gets back on the normal linear trend on the extrapolated line, that point is referred to as PDL pressure. This pressure can also be used as an approximation to maximum horizontal stress if and only if natural fractures are believed to be perpendicular to the created hydraulic fractures. Once this point is identified, a vertical line can be drawn from that point to where the pressure curve is intersected. After the intersection of the vertical line with

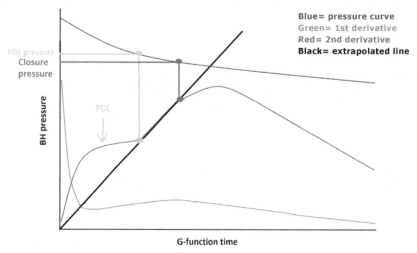

Fig. 14.7 Pressure-dependent leak-off.

the pressure curve is identified, PDL pressure or maximum horizontal stress can be read on the y-axis as shown in Fig. 14.7.

The next phenomenon that happens on the G-function plot is closure. Fracture closure occurs when the second derivative deviates from the straight line going through the origin. Once that point is identified on the G-function plot, draw a vertical line until the pressure curve is intersected. After the intersection of the pressure curve with the vertical line (as shown in *red* (gray in print version)) is identified, closure pressure can be read on the y-axis. Closure pressure is regarded as the minimum horizontal stress. To summarize this section, PDL pressure represents maximum horizontal stress (if natural fractures are perpendicular to hydraulic fractures) and closure pressure represents minimum horizontal stress. Anisotropy can be calculated using Eq. (14.2).

$$\text{Anisotropy} = \text{maximum horizontal stress} - \text{minimum horizontal stress} \qquad (14.2)$$

When the difference between the maximum and minimum horizontal stresses is small (e.g., 200 psi), the created fractures are expected to be complex. In contrast, when the difference is large, the created fractures are expected to be bi-wing. PDL describes the fluid leak-off into fissures that open at a higher pressure than the fracture closure pressure. As a general rule of thumb, when the difference between PDL and closure pressures is <5% of the closure, it is an indication that fracture complexity will be created.

Fig. 14.8 shows an example of a G-function plot with a PDL signature. This signature is illustrated by the concave downward feature above the extrapolated line going through the origin on the second derivative curve. Closure pressure can be determined when the second derivative deviates from the linear extrapolated line going through the origin. The closure pressure is around 6845 psi (G-function closure time = 29.4).

Dealing with PDL PDL is very common in naturally fractured formations, especially in unconventional shale reservoirs. Consider pumping a smaller sand size and concentration (100 mesh and 40/70) and longer frac stages. Pumping smaller sand size, such as 100 mesh, bridges natural fissures and reduces the chance of screening out. This will increase the fluid efficiency by preventing high fluid leak-off through natural fractures.

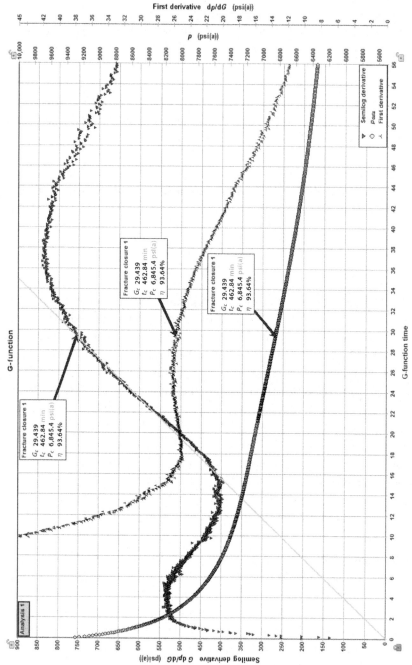

Fig. 14.8 G-function plot with PDL signature example.

Height recession leak-off

The second leak-off regime that can be noted from the G-function plot is height recession. Height recession occurs when the fracture height is decreasing during closure because of contact with an impermeable zone. When these conditions are met, some strange behavior can happen within the fracture. As the fracture closes, the upper and lower zones close more quickly because of the higher stress. However, because permeability is so low in the zones above and below, the fluid is pushed into the main section of the fracture instead of leaking off into the formation. This reduces the apparent leak-off rate in the fracture and, therefore, reduces the leak-off rate from the wellbore into the fracture. In essence, fluid leaks off slower than expected for a normal bi-wing fracture. The resultant fracture may be very narrow and tall (Fig. 14.9).

Identifying height recession leak-off The second derivative curve on the G-function plot has a concave upward shape below the trend line going through the origin. This concave upward feature below the extrapolated line going through the origin is an indication of height recession.

Dealing with height recession leak-off If height recession happens, consider reducing rate and proppant amount. Also, lowering sand concentration to bridge off impermeable zones can tremendously help when dealing with height recession.

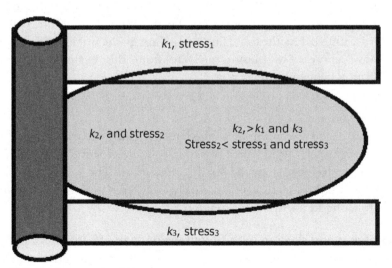

Fig. 14.9 Height recession behavior.

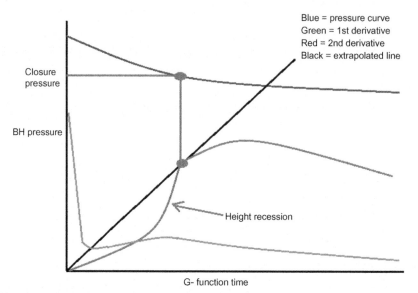

Fig. 14.10 Height recession leak-off.

In Fig. 14.10, *blue* (dark gray in print version) is the pressure curve, *green* (light gray in print version) is the first derivative, and *red* (gray in print version) is the second derivative. The concave upward feature on the second derivative below the extrapolated line going through the origin represents height recession.

Fig. 14.11 shows another example of a G-function plot with a height recession signature. This feature can be determined from the concave upward feature below the trend line going through the origin on the second derivative curve. The closure pressure from this example is around 16,743 psi (G-function time = 18.4).

Tip extension
The fracture-tip extension is a phenomenon that occurs in very low-permeability reservoirs in which a fracture continues to propagate even after the injection has been stopped with the well shut-in. The energy that is typically released through leak-off is actually transferred to the tip of the fractures resulting in fracture-tip extension.

Identifying and dealing with tip extension leak-off When fracture-tip extension exists, the extended straight line on the second derivative curve does not pass through the origin. In addition, no sustained negative slope on the second derivative curve can be seen. Since no negative slope on the

Diagnostic fracture injection test

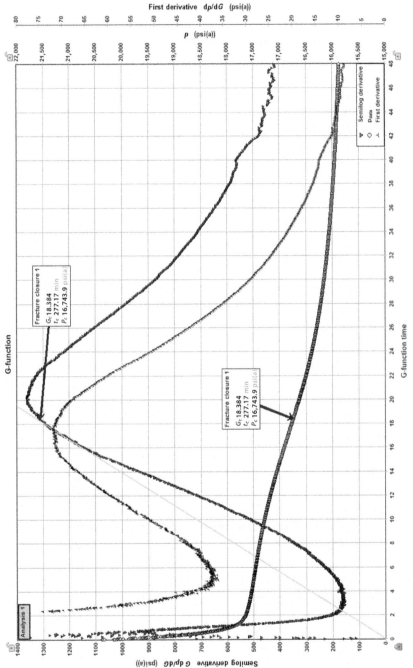

Fig. 14.11 G-function plot with height recession signature example.

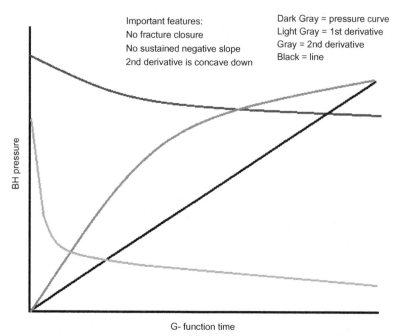

Fig. 14.12 Tip extension leak-off.

second derivative can be realized, fracture closure behavior is not seen when encountering tip extension. The best way to deal with tip extension behavior leak-off is to increase pad volume to assist in creating longer frac lengths.

Fig. 14.12 illustrates a tip extension leak-off behavior. *Blue* (dark gray curve in print version) is the pressure curve, *green* (light gray curve in print version) is the first derivative, and *red* (gray in print version) is the second derivative. As can be seen on the second derivative curve, no sustained negative slope or fracture closure can be seen from this behavior.

Effective permeability estimation from G-function plot G-function plot is a powerful tool that can be used to obtain various fracture and formation properties. In addition to calculating closure pressure and identifying various leak-off regimes, effective permeability can be calculated from the G-function plot using Eq. (14.3) (Barree et al., 2007).

Effective permeability using the G-function plot:

$$k = \frac{0.0086\mu\sqrt{0.01P_Z}}{\phi C_t \left(\dfrac{G_c Er_p}{0.038}\right)^{1.96}} \tag{14.3}$$

where k is the effective permeability to reservoir fluid, md, μ is the viscosity of injected fluid used during DFIT, cp, P_z is the net extension pressure or process zone stress, BHISIP $- P_c$ (psi), ϕ is the porosity, fraction, C_t is the total compressibility, 1/psi, G_c is the closure G-function time, E is the Young's modulus, MMpsi, and r_p is the storage correction factor.

For normal and PDL, assume r_p of 1 and for height recession and tip extension, assume $r_p < 1$.

In addition to the effective permeability calculation, fluid efficiency can also be calculated using the G-function time at fracture closure using Eq. (14.4).

$$\text{Fluid efficiency using G} - \text{function time} : \text{Fluid efficiency} = \frac{G_c}{2 + G_c} \tag{14.4}$$

where G_c is the G-function time at fracture closure.

Example

Estimate the effective permeability and fluid efficiency from the following parameters obtained from a G-function plot:

Injected fluid viscosity $(\mu) = 1$ cp, BHISIP $= 7748$ psi, $P_c = 6338$ psi, $C_t = 0.0000234$ 1/psi, $G_c = 29.012$, $E = 3.5$ MMpsi, $r_p = 1$ (since PDL exists), Porosity $= 10\%$

$$P_z = \text{BHISIP} - P_c = 7748 - 6338 = 1410 \text{ psi}$$

$$k = \frac{0.0086\mu\sqrt{0.01 P_z}}{\phi C_t \left(\frac{G_c E r_p}{0.038}\right)^{1.96}} = \frac{0.0086 \times 1 \times \sqrt{0.01 \times 1410}}{0.1 \times 0.0000234 \times \left(\frac{29.012 \times 3.5 \times 1}{0.038}\right)^{1.96}}$$

$$= 0.00265 \text{ md}$$

$$\text{Fluid efficiency} = \frac{29.012}{29.012 + 2} = 0.9355 \text{ or } 93.55\%$$

A high fluid efficiency of ~94% and a permeability of 0.00265 md indicate high fluid efficiency and low fluid leak-off.

What would the permeability and fluid efficiency have been if the G-function time at fracture closure was 0.554 instead of 29.012?

Replacing the G-function time of 0.554 with 29.012 in the same equation, and keeping all of the other parameters the same, the effective permeability can be solved as follows:

$$k = \frac{0.0086 \times 1 \times \sqrt{0.01 \times 1410}}{0.1 \times 0.0000234 \left(\frac{0.554 \times 3.5 \times 1}{0.038}\right)^{1.96}} = 6.2 \text{ md}$$

$$\text{Fluid efficiency} = \frac{0.554}{0.554 + 2} = 0.2169 \text{ or } 21.69\%$$

If G-function time was 0.554, the formation would have a lower fluid efficiency and higher effective permeability, which is an indication of high fluid leak-off.

After-closure analysis

After-closure analysis (ACA) refers to various methods used to determine reservoir properties after the closure has occurred. The first step in ACA is determining various flow regimes from DFIT analysis. This can be performed using a log-log plot and determining pseudolinear and pseudoradial flows from the log-log plot. Once pseudolinear and pseudoradial flow regimes are identified on the log-log plot, pore pressure, transmissibility, and permeability can then be estimated using various techniques that will be discussed. On certain occasions, if sufficient time is not allowed after DFIT, only pseudolinear flow can be reached and pseudoradial flow will not be reached. In those instances, reservoir pore pressure can be estimated from pseudolinear flow, but this pressure (obtained from pseudolinear flow) is optimistic. Pseudoradial flow occurs after pseudolinear flow. Once this flow regime is reached, pore pressure can be determined using a Horner plot or other available pressure transient methods.

Horner plot (one method of ACA)

Horner analysis uses the log of Horner time on the x-axis vs bottom-hole pressure on the y-axis to calculate pore pressure and reservoir permeability. Note that the y-axis is plotted on the Cartesian axis and logarithmic scale is applied to the x-axis. Horner time is defined in Eq. (14.5).

$$\text{Horner time} = \frac{t_p + \Delta t}{\Delta t} \qquad (14.5)$$

where t_p is the fracture propagation time, minutes and Δt *is the* elapsed shut-in time, minutes.

As shut-in time increases, Horner time decreases. As the shut-in time approaches infinity, Horner time approaches 1. A straight-line extrapolation to the y-intercept (at Horner time of approximately 1) yields reservoir pressure (pore pressure). One of the biggest limitations with a Horner plot is that pseudoradial flow must be reached or Horner analysis is not recommended to be used. Once pseudoradial flow is identified, the slope of the straight extrapolated line is referred to as m_H. The point at which the extrapolated line reaches the y-intercept (as shown below) is pore pressure. The slope of the Horner plot (m_H) can be used to estimate reservoir transmissibility (kh/μ) and subsequently reservoir permeability using Eq. (14.6).

$$\text{Reservoir transmissibility}: \frac{kh}{\mu} = \frac{162.6(1440)q}{m_H} \qquad (14.6)$$

where kh/μ is the reservoir transmissibility, md.ft/cp, k is the reservoir permeability, md, h is the net pay height, ft, μ is the far-field fluid viscosity (*not* injected fluid viscosity), cp, m_H is the slope of the Horner plot, psi, and q is the average injected fluid rate, bpm.

By assuming a far-field fluid viscosity and net pay height in Eq. (14.6), reservoir effective permeability can be calculated. Fig. 14.13 illustrates a Horner analysis.

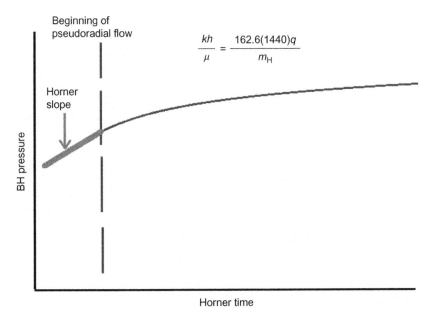

Fig. 14.13 Horner analysis.

Example

Calculate the reservoir transmissibility and reservoir permeability given the following Horner plot and data. Also, estimate the reservoir pressure and reservoir pressure gradient assuming a true vertical depth (TVD) of 8250′.

Avg pump rate = 7 bpm, $h = 100'$, far-field fluid viscosity $(\mu) = 0.0452$ cp, $m_H = 568{,}564$

From Fig. 14.14, Horner slope is 568,564 (mH), therefore,

$$\text{Transmissibility} = \frac{kh}{\mu} = \frac{162.6(1440)q}{m_H} = \frac{162.6 \times 1440 \times 7}{568{,}564} = 2.88 \text{ md.ft/cp}$$

$$\text{Effective permeability} = k = \frac{2.88 \times 0.0452}{100} = 0.0013 \text{ md}$$

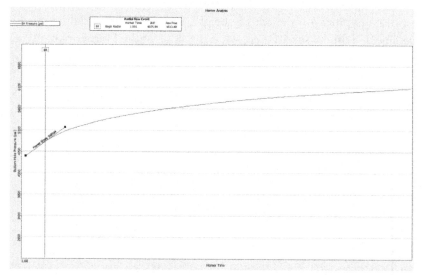

Fig. 14.14 Horner analysis example.

The extrapolation of the pseudoradial flow line (straight line) to the y-intercept yields reservoir pressure of roughly 4600 psi in this example. Therefore, the reservoir pressure gradient is 4600 psi divided by 8250′ which yields 0.56 psi/ft.

Linear flow-time function vs bottom-hole pressure (another method of ACA)

In addition to Horner analysis, reservoir pressure can be determined from the linear flow-time function (x-axis) vs BHP (y-axis). Linear flow-time function is described in Eq. (14.7).

Linear flow-time function: $F_L(t, t_c) = \dfrac{2}{\pi} \sin^{-1} \sqrt{\dfrac{t_c}{t}}$ for $t \geq t_c$ (14.7)

where t_c is the time to closure, minutes and t is the total pump time, minutes.

A straight-line extrapolation from the linear flow yields an estimated pore pressure from the linear flow-time function plot. In other words, once after-closure pseudolinear flow is observed during shut-in, the intercept of the extrapolated straight line through the pseudolinear flow data provides an estimate of the pore pressure. Reservoir pore pressure extrapolation is valid and no direct information of transmissibility can be obtained from this analysis. If pseudoradial flow is not obtained from DFIT analysis, this plot can be used to estimate the reservoir pressure (Fig. 14.15).

Radial flow-time function vs BHP (another method of ACA)

Radial flow-time function can also be used to calculate reservoir pressure along with transmissibility when true pseudoradial flow is identified. Radial flow-time function is defined in Eq. (14.8).

$$F_R(t, t_c) = \dfrac{1}{4} \ln\left(1 + \dfrac{X t_c}{t - t_c}\right),$$
$$X = \dfrac{16}{\pi^2} \cong 1.6$$
(14.8)

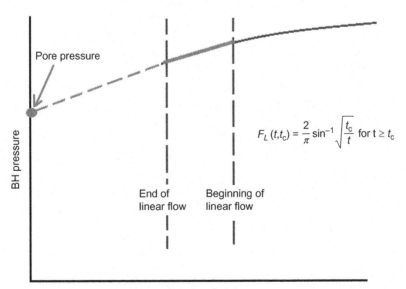

Fig. 14.15 Linear flow-time function plot (ACA).

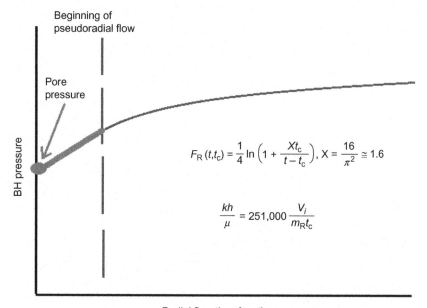

Fig. 14.16 Radial flow-time function plot.

where t_c is the time to closure, minutes and t is the total pump time, minutes. In addition to reservoir pressure, when the pseudoradial flow period is properly identified, far-field transmissibility can also be calculated by knowing the slope of the extrapolated line, time to fracture closure, and total volume injected during the test. Transmissibility using a radial flow-time function plot can be obtained using Eq. (14.9) (Fig. 14.16).

$$\text{Transmissibility using the radial flow-time function}: \frac{kh}{\mu} = 251{,}000 \frac{V_i}{m_R t_c} \qquad (14.9)$$

where V_i is the injected fluid during the test, BBLs, m_R is the derived slope, 1/psi, t_c is the time to closure, minutes, h is the net pay, ft, and μ is the far-field fluid viscosity, cp.

CHAPTER FIFTEEN

Numerical simulation of hydraulic fracturing propagation

Introduction

Hydraulic fracturing has been accepted as a technique with a variety of applications. These applications include measurement of in situ stress (Hayashi and Haimson, 1991), underground storage of hazardous materials (Levasseur et al., 2010), heat production from geothermal reservoirs (Legarth et al., 2005), and barrier walls to prevent containment from transportation (Murdoch, 2002). Currently, one of the most important applications of hydraulic fracturing is to improve the recovery of unconventional hydrocarbon reservoirs. Hydraulic fracturing is a coupled process including (1) the deformation of the solid medium, where the fracture width is dependent on the fluid pressure in a global manner and it has the property of nonlocality; (2) the fluid flow within the fracture, which is a nonlinear function of fluid pressure and fracture width. These two fundamental properties produce notorious difficulty when investigating hydraulic fracturing.

The conventional methods for the numerical simulation of hydraulic fracturing are the boundary element and finite element methods. The discontinuous displacement (DD) method, which is a variant of the boundary element method, has been greatly used for this purpose. However, it is found to be difficult in the event of complex structures. Compared to the boundary element method, the finite element method has greater flexibility but requires extensive computational power. Recently, advanced techniques such as condensation technique and parallel computing approaches are used to overcome the limitation of well-developed numerical schemes of fracture propagation models (Bao et al., 2014, 2015, 2016).

While significant effort has been put toward the simulation of fracture propagation and fluid flow during injection (Mobbs and Hammond, 2001; Yamamoto et al., 1999; Phani et al., 2004), fracture geometry after flowback has been merely studied. Fracture geometry after flowback is a function of proppant distribution and closure stress, which is significantly different from fracture geometry after injection stops. Proppant transport and distribution in hydraulic fracture is a nonlinear function of injection rate, proppant size, density, and frac fluid properties, that is, viscosity and density. Therefore, for hydraulic fracturing optimization, a fully coupled numerical simulation coupling governing equations describing fracture opening, fluid flow and leak-off, and solid transport is required. This numerical simulation should also handle different proppant size and density injection with different pumping schedules in order to increase the efficiency of hydraulic fracture stimulation and enhance oil and gas recovery.

Stratigraphic and geological structure modeling

Field stratigraphic and geological structure modeling is necessary to obtain a robust and consistent geometry and petrophysical properties of the formations under study. Three-dimensional (3D) geological models can also be used to provide knowledge of distributions of rock mechanical properties and in situ stress, including maximum horizontal stress, minimum horizontal stress, vertical stress, Young's modulus, Poisson's ratio, and tensile strength. Detailed knowledge of distribution of petrophysical properties is critical to locate the initiation of hydraulic fractures and to evaluate the evolution of fracture-geometry configuration. A number of different commercial software packages are available that provide macromodels for imaging purposes or detailed petrophysical models for reservoir simulation purposes (Aziz and Settari, 1979). They can also be used for stratigraphic studies where the presence of meanders, channels, faults, and discontinuities is important (Mallet, 2002). Conventionally there are two different approaches that have been taken for building geological models and populating the data, namely, statistical approaches including different forms of kriging (Xu et al., 1992) and deterministic geometries obtained from seismic or ground-penetrating radar (GPR) data. Stochastic or deterministic interpolations need to be compared with control horizon geometries obtained from measurements of petrophysical properties using drilling data and well logs, such as spontaneous potential, gamma ray, resistivity, density, and sonic velocity logs. Sufficient well logs and drilling data are necessary to obtain well-defined control horizons providing the required level of accuracy in 3D geological modeling. However, these studies are relatively expensive and provide limited data if abundant wells are not available.

Conventional methods of data analysis obtained from well logs and seismic are knowledge-driven and ignore underlying physical relationships between correlated petrophysical parameters. Recently, advanced techniques such as decision trees (DTs), support vector machines (SVMs), data mining, and artificial neural networks (ANNs) have been used for the identification of sedimentary facies and lithology variation using limited log and core data. However, the applications of these techniques also need special attention since they might not consider important geological phenomena embedded in geological formations. Therefore, a combination of advanced mathematical and knowledge-driven techniques is required to obtain sound geological models.

Development of hydraulic fracturing simulators

Multistage hydraulic fracturing of horizontal wells enabled the oil and gas industry to economically enhance production from unconventional resources, especially organic-rich shale reservoirs. The process involved the creation of multiple fractures in any single stage of hydraulic fracturing by injecting significant amounts of fracturing fluid and proppant at high pressures. This process is expected to generate highly conductive flow paths for oil and gas to flow from the reservoir to the production well. After reaching the predesigned fracture length, the injection will stop and fracturing fluid will be produced during the flowback process. However, injected proppants will remain in the fracture to prevent fracture closure due to overburden pressure. There have been extensive studies on hydraulic fracturing optimization to generate maximum oil and gas production from unconventional resources. These studies mainly focus on the impact of different reservoir and operational parameters on the efficiency of hydraulic fracturing, multiple hydraulic fracturing interactions, and hydraulic and natural fracture interactions (Ozkan et al., 2009; Olson and Dahi 2009; Cheng, 2012). They showed that the change in local stresses due to earlier hydraulic fracturing stages or preexisting natural fractures can significantly impact the dimensions and orientation of subsequent fractures.

Numerical schemes used to model the hydraulic fracture propagation and optimization are mostly based on the theory of linear elastic fracture mechanics (LEFM), which was developed in the 1920s and introduced fundamental equations governing the process of hydraulic fracturing. The major assumption of LEFM is an isotropic and linear elastic formation. This neglects deformation at the fracture tip or it assumes the deformation at the fracture tip is negligible as compared to the fracture dimensions. This assumption is not valid for fracture tip behavior in soft formations with significant plastic deformation. In this case, the crack-tip plasticity (CTP) method might be more applicable. As the plastic properties of the formation increase, hydraulic fracturing and fracture propagation in the formation becomes difficult to perform since most of the energy that otherwise would be used for fracture propagation will be absorbed by the formation. Even though the CTP technique is more promising for modeling fracture-tip behavior, it has not been used due to its complexity. Modifications have been applied to the theory of LEFM to capture some nonlinear fracture-tip behaviors.

In modeling fracture propagation using LEFM, the stress field near the fracture tip will be calculated and compared with fracture toughness, which is the formation property that needs to be obtained through experimental studies. When the stress field exceeds fracture toughness, the fracture propagates inside the material. In addition to fracture toughness, one also needs to have a good understanding of variables such as normal and shear stresses, strain, Young's modulus, Poisson's ratio, tensile strength, and yield strength to understand the basics of the theory of elasticity.

Stress "σ" is defined as force or load per area and can be shown using Eq. (15.1):

$$\text{Stress}: \sigma = \frac{F}{A} \qquad (15.1)$$

where F is the force applied to the cross-sectional area A. The dimension of stress is the same as pressure and can be measured in pascals (newtons per square meter) in SI units or psi (pound per square inch) in field units. The component of stress applied perpendicular to the surface area is called normal stress, usually shown by σ, and the component of stress parallel to the surface area is called shear stress, or τ. 3D space stress encompasses nine components that can be shown using a 3×3 matrix as follows:

$$\begin{bmatrix} \sigma_{xx} & \tau_{xy} & \tau_{xz} \\ \tau_{yx} & \sigma_{yy} & \tau_{yz} \\ \tau_{zx} & \tau_{zy} & \sigma_{zz} \end{bmatrix}$$

By studying the behavior of shear stresses acting on an infinitesimally small volume, one can show that $\tau_{xy} = \tau_{yx}$, $\tau_{xz} = \tau_{zx}$, and $\tau_{yz} = \tau_{zy}$, which is the basis of the theory of shear stress reciprocity. Based on the theory of shear stress reciprocity, changing the shear stress indicates only changes in the direction of shear stress and does not change the magnitude of the shear stress. A coordinate system in which stresses are calculated can be transformed to any coordinate system given that the shear stress components of the total stress becomes zero and only the diagonal component of stress remains through coordinate transformation. In this case, the normal stresses in the x-, y-, and z-directions are called principal stresses, where σ_1 is the maximum and σ_3 is the minimum principal stress, shown as follows:

$$\begin{bmatrix} \sigma_1 & 0 & 0 \\ 0 & \sigma_2 & 0 \\ 0 & 0 & \sigma_3 \end{bmatrix}$$

Principal stresses can be obtained from components of the general stress matrix. In a two-dimensional (2D) system, the maximum "σ_1" and minimum "σ_2" principal stresses can be obtained as follows:

$$\sigma_1 = \left(\frac{\sigma_x + \sigma_y}{2}\right) + \sqrt{\left(\frac{\sigma_x - \sigma_y}{2}\right)^2 + \tau_{xy}^2}$$

$$\sigma_2 = \left(\frac{\sigma_x + \sigma_y}{2}\right) - \sqrt{\left(\frac{\sigma_x - \sigma_y}{2}\right)^2 + \tau_{xy}^2}$$

Strain "ϵ" is used to quantify the deformation of solid material, which is defined as a relative change in displacement in the x-, y-, and z-directions as follows:

$$\epsilon = \frac{dL}{L}$$

In this equation, dL is defined as a change in displacement and L is the initial length. The stress and strain relationship is defined using constitutive equations such as Hooke's law, which assumes a linear relationship between applied load and displacement in the range of the elastic behavior of the material. To quantify this relationship, Young's modulus "E" is used, which is a ratio of stress to strain and is an indication of formation stiffness.

$$E = \frac{\sigma}{\epsilon}$$

Tensile strength is defined as the maximum stress formation can bear before it breaks. This is the point of stress at which the formation is permanently damaged. Yield strength is the formation property and is defined as stress at which the formation starts to deform plastically. Even though LEFM have been used extensively in hydraulic fracture simulation, they suffer from a high computational cost and decreased accuracy when predicting the fracture-tip behavior. LEFM especially cannot predict the formation failure ahead of the fracture tip. This is due to the fact that LEFM only considers the local stress criteria at the fracture tip (i.e., where the fracture propagates when stress intensity factor KI overcomes the fracture toughness K_{IC}).

On the other hand, cohesive zone models (CZMs) are more suitable to model fracture-tip behavior. CZM extends the fracture tip area to a "cohesive zone" ahead of the fracture tip within which the fracture propagation processes occur gradually. Cohesive zone modeling is based on the determination of two important parameters: cohesive strength and separation energy. This introduces both strength and energy criteria for fracture propagation, and enables CZM to predict formation failure ahead of fracture tip. These parameters can be measured experimentally or obtained using numerical simulations developed for interface behavior predictions. Different methods other than LEFM and CZM are also used for hydraulic fracturing simulation such as crack-tip open displacement (CTOD), but these are not as common as the first two techniques. Gao et al. (2015) applied the DD technique using a boundary element model to investigate the changes that occur in multiple hydraulic fracture pressures on local stress changes and any geological discontinuities such as faults. However, their model assumed a predefined/fixed fracture length and pressure at the fracture surface and neglected the poroelastic effect of the formation. Morrill and Miskimins (2012) applied the finite element technique to optimize the fracture spacing, neglecting the fracture interactions.

Different numerical simulators have been developed in spite of LEFM or CZM to simulate fracture propagation, fracture geometry, and magnitude and the direction of stress change around hydraulic fractures. These are either 2D, pseudo-three-dimensional (pseudo-3D), or 3D hydraulic fracturing models depending on the complexity of the problem and the amount of information available. These models are useful when studying the general behavior and physics of the simplified hydraulic fracturing process. The following is a brief discussion of the different models available that enable the development of more accurate hydraulic fracturing simulators.

2D hydraulic fracturing models

Hydraulic fracture geometry is a complex function of initial reservoir stress conditions (global and local), reservoir rock properties such as heterogeneous and anisotropic rock mechanical properties (Young's modulus and Poisson's ratio), permeability, porosity, natural fracture system, and operational conditions such as injection rate, volume, and pressure. To model this complicated process, specific assumptions have been made to simplify the problem while capturing the major characteristics of hydraulic fracture geometry. For this, scientists had first assumed the hydraulic fracturing process would occur in a homogeneous and isotropic formation that would lead to a symmetric, bi-wing fracture from the point or line source of the injecting fluid. There are three common fracture modeling methods introduced based on these assumptions: (1) the Khristianovic-Geertsma de Klerk (KGD) model, (2) the Perkins and Kern (PKN) model, and (3) the radial fracture geometry or penny-shaped model (Abe et al., 1976).

The KGD model assumes a 2D plane-strain model in a horizontal plane with a constant fracture height that is larger than the fracture length. In the KGD model, an elliptical horizontal cross-section and rectangular vertical cross-section are assumed where the fracture width is independent of the fracture height and is constant in the vertical direction. The rock stiffness is also only considered in the horizontal plane. Fig. 15.1 shows the schematic representation of fracture geometry in the KGD model.

The PKN model assumes a constant fracture height independent of fracture length. In the PKN model a 2D plane-strain model is assumed in the vertical plane where the fracture has an elliptical cross-section both in the horizontal and vertical directions. Unlike the KGD model, the PKN model assumes a fracture height much smaller than the fracture length. The PKN model also assumes the hydraulic fracturing energy applied by the fluid injection would only be consumed by an energy loss from fluid flow

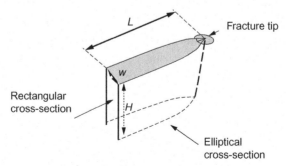

Fig. 15.1 A schematic diagram of Khristianovic-Geertsma de Klerk (KGD) fracture geometry.

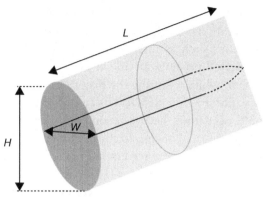

Fig. 15.2 A schematic diagram of Perkins and Kern (PKN) fracture geometry.

(viscosity-dominated regime) and ignores fracture toughness. Fig. 15.2 shows the schematic of fracture geometry in a PKN model.

The PKN and KGD models assume fluid flow in a fracture as a one-dimensional (1D) problem in the direction of the fracture propagation or fracture length governed by the lubrication theory and Poiseuille's law. They assume that the fracture is confined and there is no change in horizontal stress, reservoir pressure, and temperature.

The third model used to simulate hydraulic fracture propagation in a 2D plane is called the penny-shaped or radial fracture model. This model has found application in shallow formations where overburden stress became equal to minimum horizontal stress. In this case, the symmetric geometry was assumed to be at the point of line-injection source. In this model, the injection rate and fluid pressure within the fracture are assumed to be constant. Fig. 15.3 shows the schematic representation of fracture geometry assuming the penny-shaped model.

In all hydraulic fracturing models, the hydraulic fracture propagation is a function of injection of fracturing fluid "Q_0" from the injection point or injection line representing the well perforations. As a result, a bi-wing

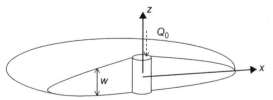

Fig. 15.3 A schematic diagram of radial fracture geometry.

symmetric fracture is assumed to propagate in the formation perpendicular to minimum principal stress "σ_0" of the formation. The fracture width generated is therefore a function of effective stress, which is the difference between the pore pressure and minimum principal stress "σ_0" ($P_e = P_f - \sigma_0$). Effective pressure is a good indicator of fracture width and is a likely indicator of well performance after hydraulic fracturing. The higher this effective pressure measured during the hydraulic fracturing, the better well productivity is expected.

Fluid flow in hydraulic fractures

Fluid flow in hydraulic fracture is governed by 1D- or 2D Poiseuille's law and lubrication theory. In fluid dynamics, lubrication theory is used for cases where fluid flows through a media where 1D is significantly smaller than another is considered. In hydraulic fracturing, this translates into having a fracture width much smaller than the fracture height and length. Given a 2D model, 1D fluid flow along the fracture length is assumed, and can be shown as follows:

$$q = -\frac{w^3}{12\mu}\nabla p_f$$

In this equation, q is the flow rate, μ is the fluid viscosity, w is the fracture width, and (∇p_f) is the gradient of fracture pressure defined in a direction of fracture length. Assuming incompressible fracturing fluid and fluid leak-off governed by Carter's leak-off model, Eq. (15.2) describes the conservation of fluid in the fracture.

$$\partial w / \partial t + \nabla \cdot q + \xi = 0$$

$$\text{Mass conservation in fracture:} \quad \frac{\partial w}{\partial t} + \nabla \cdot q + \xi = 0 \qquad (15.2)$$

It is very common in the oil and gas industry to attribute the fluid leak-off to the surrounding formations using Carter's model as described by the following equation:

$$\text{Carter's leak-off model:} \quad \xi(x, t) = \frac{2C}{\sqrt{t - t_0(x)}} \qquad (15.3)$$

where C is the leak-off coefficient, t is the time, and t_0 is the fracture-tip arrival time. The boundary conditions for the fluid flow equation can be obtained by assuming a constant flow rate of $Q_0/2$ at the injection point of the symmetric bi-wing fracture and a zero flow rate at the fracture tip (assuming no fluid lag or zero flow rate at waterfront having fluid lag). Fluid lag refers to a zone between the fluid front and the fracture tip. Depending on formation permeability and mechanical properties, fluid lag may not be present, which is due to a fracture tip and fluid front moving with the same velocity.

Solid elastic response

The solid elastic response of the medium is governed by three equations: equilibrium condition, constitutive law, and geometry, which can be defined using Eqs. (15.4)–(15.6), respectively. The equilibrium condition is defined as follows:

$$\text{Equilibrium condition}: \nabla \cdot \sigma + g = 0 \tag{15.4}$$

The constitutive law of linear elasticity is governed by

$$\text{Constitutive law}: \sigma(x) = k : \varepsilon(x) \tag{15.5}$$

The geometry, which is a function of solid displacement, is expressed as

$$\text{Displacement}: \varepsilon = \left[\nabla D + (\nabla D)'\right]/2 \tag{15.6}$$

By definition, σ is the stress tensor, g is the gravitational acceleration, κ is the elastic stiffness, ε is the strain tensor, and D is the displacement. Superscript "'" here denotes the transpose of the matrix. The stress boundary conditions also need to be defined based on specific upper, lower, and fracture surface conditions.

Pseudo-3D hydraulic fracturing models

Even though using 2D models is useful to understand the fundamentals of hydraulic fracturing, they cannot be used for practical purposes. Therefore, pseudo-3D models are developed with the assumption of a constant fracture height as described in the PKN model. Two different models are introduced to consider variations in fracture height, namely, the equilibrium and dynamic height pseudomodels. At equilibrium height, pseudo-3D models with a uniform pressure distribution in vertical cross sections are assumed. In these models, toughness criteria for fracture propagation (i.e., the fracture propagates when stress intensity factor K_I overcomes fracture

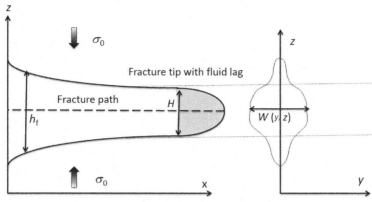

Fig. 15.4 Schematic diagram of pseudo-3D fracture model.

toughness K_{IC}), are also considered. Given a dynamic height, pseudo-3D models with 2D fluid flows (parallel and perpendicular to fracture path) are assumed (note in 2D models, fluid flow is in one dimension along with the fracture path), and fracture height calculation followed KGD model solutions (Dontsov and Peirce, 2015). Pseudo-3D models are more practical than 2D models since they consider fracture height variation as a function of location in the direction of fracture propagation and time. However, they are still restricted to certain geometries and follow the plane-strain conditions in each cross-section perpendicular to the fracture path. They also have different accuracies in different hydraulic fracturing regimes. For example, these models are inaccurate in toughness-dominated hydraulic fracturing regimes due to local elasticity assumption. These models are also inaccurate in viscosity-dominated hydraulic fracturing regimes due to viscous losses perpendicular to fracture path. Fig. 15.4 shows the schematic representation of pseudo-3D hydraulic fracture.

Different hydraulic fracturing regimes are defined based on energy dissipation through the process of hydraulic fracturing due to fracture toughness or viscous flow of fluid in the fracture. A parameter called K_m independent of time is used to distinguish between these two energy dissipation regimes. High K_m value represents a toughness-dominated regime ($K_m > 4$) and low K_m refers to a viscosity-dominated regime, that is, $K_m < 1.0$. A K_m value between 1 and 4 is referred to as an intermediate case. Eq. (15.7) shows the definition of K_m.

$$K_m = \frac{4K_{IC}\sqrt{2/\pi}}{(E/1-\nu^2)}\left[\frac{(E/1-\nu^2)}{12\mu Q_0}\right]^{1/4} \quad (15.7)$$

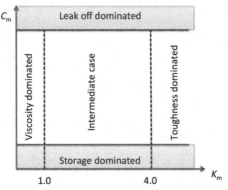

Fig. 15.5 Different hydraulic fracturing regimes.

In Eq. (15.7), K_{IC} is the fracture toughness and rock property, E is the Young's modulus, ν is the Poisson's ratio, Q_0 is the injection rate, and μ stands for the dynamic viscosity of the fluid. Hydraulic fracturing regimes divided into leak-off-dominated or storage-dominated processes are quantified using the parameter "C_m." Given nonzero fluid leak-off to the formation, C_m value will vary between zero and infinity. A higher C_m value denotes higher fluid leak-off in the formation and therefore lower fracture efficiency. C_m is a function of time and is defined in Eq. (15.8).

$$C_m = 2C\left[\frac{(E/1-\nu^2)t}{12\mu Q_0^3}\right]^{1/6} \quad (15.8)$$

Fig. 15.5 shows the schematic representation of different hydraulic fracturing regimes (Bunger et al., 2005).

3D hydraulic fracturing models

Extending the pseudo-3D models to full 3D models, in addition to the high computational cost associated with full 3D models, would be difficult considering the nonlocal dependency of fracture geometry to fluid pressure and confining stresses. The exact model should be able to fully combine multiphase fluid flow in the fracture, leak-off to the formation, and rock deformation. This can be achieved by solving sets of coupled partial differential equations governing this multiphysics process simultaneously. Finding the solution for fully coupled problems is an extremely difficult task, and therefore simplified models have been introduced where one-way coupling or weak coupling was previously used. In one way coupling, partial differential equations governing the fluid flow in the fracture are solved at each time step and pressure distributions are obtained. Pressure distributions will then be used as an initial condition for differential equations governing the rock deformation and fracture propagation. In this technique, the fracture pressure distributions at each time step are assumed to be independent of rock deformations; therefore, they will not be updated due to a change in fracture geometry. There is also an intermediate technique between fully coupled and one-way coupling called weak coupling. In this technique, similar to one-way coupling, fracture pressure distribution at each time step is calculated independent of change in the fracture geometry at that time step. However, after certain time intervals, the fracture pressure distribution is updated based on fracture geometry.

Fracture-tip behavior and dynamics of fluid lag at the fracture tip create additional difficulties in hydraulic fracturing simulation due to the dynamic boundary conditions imposed at the fluid front following the fracture tip. In the literature, there has been a great effort to take these effects into consideration by different research groups (Garagash 2006, 2007; Adachi and Detournay, 2008; Shen, 2014; Dontsov and Peirce, 2015). The conventional method implemented in these studies is the DD method, which is a modified version of the boundary element method and can be applied to simulate models with arbitrary fracture geometry. In this technique, displacement along the fracture propagation path will be discretized into a series of elements where displacement is assumed to be constant for each element. Having an analytical solution based on Green's function that describes the displacement and stress tensor relationship for a single element, the total displacement will be calculated by the summation of all displacements in each element. The advantage of this method is that the key equations for the coupled process are built upon the fracture surface rather than on the whole model. This significantly reduces the computational cost of numerical simulation. The disadvantage of the technique is to find its nonlocal kernel function when the model has a complex structure (Siebrits and Peirce, 2002).

Recently, different finite element methods have been used for hydraulic fracturing simulation. These techniques have more flexibility when compared to the DD method, in that they do not require explicit calculation of the kernel function. In these techniques, two coupled nonlinear finite element equations are defined. One is describing the elastic response of the elastic medium, and the other describes the fluid and solid transport within the fracture. The first system will be used to find the relationship between the net pressure in the fracture and fracture width, and the second system to simulate the fluid and solid transport in the fracture. The investigation will be accomplished by solving the coupled equations from the two systems using the Newton-Raphson iteration algorithm.

Hydraulic and natural fracture interactions

Hydraulic fracturing in naturally fractured formations is completely different from hydraulic fracturing in homogeneous and isotropic formations, which is assumed in most of the numerical simulators. This is due to the interactions between hydraulic and natural fractures. In the presence of natural fractures, and depending on their density and major direction with respect to local minimum and maximum in situ stresses, hydraulic fracture might cross the natural fracture, locally merge with natural fractures, and break out in a short distance, or completely follow natural fracture directions. Different experimental studies on hydraulic fracture and natural fracture interactions showed that a hydraulic fracture tends to cross the natural fractures, if they approach the natural fractures at a high angle (close to perpendicular) and also where there is a significant difference between fracture pressure and natural fracture stresses. If a hydraulic fracture reaches the natural fractures with a low angle and similar stress conditions, the natural fractures will open up and the hydraulic fracture merges with natural fractures (Lamont and Jessen, 1963; Daneshy, 1974; Blanton, 1982). In the oil and gas industry, microseismic data and core characterization and imaging have been used to map the hydraulic fractures and investigate the natural fracture and hydraulic fracture interactions. These studies in shale gas reservoirs show that it is not uncommon to have complex fractures generated instead of the expected conventional symmetric and bi-wing hydraulic fractures. However, microseismic studies are expensive and not available for every frac job. There are also significant concerns regarding the upscaling of the experimental studies from laboratory scale to actual field applications. Therefore, different numerical and analytical techniques have been used to investigate these effects. Potluri et al. (2005) studied the effect of natural fractures on hydraulic fracture propagation using Warpinski and Teufel's criteria (1987) and concluded that the hydraulic fracture will pass the natural fractures, if the normal stress on the natural fractures is higher than the rock fracture toughness. They also defined different criteria for a hydraulic fracture when it merges with a natural fracture and extends from the natural fracture tip and when it merges and breaks out after a short distance based on the angle of the hydraulic and natural fracture, fracture toughness, and hydraulic and natural fracture pressures. Recent studies also investigated the major parameters impacting the hydraulic fracturing behavior in the presence of natural fractures using numerical simulations based on the extended finite element method (XFEM). Dahi and Olson (2011) investigated the interactions between hydraulic fractures and cemented and uncemented natural fractures using XFEM. They showed that the anisotropy in a stress field can significantly enhance the hydraulic fracture and natural fracture interactions and suggested that further detailed studies are required to quantify these impacts.

Hydraulic fracture stage merging and stress shadow effects

Developing unconventional resources such as organic-rich shale reservoirs using horizontal well technology and multistage hydraulic fracturing introduced a whole new area of research in both academia and industry to optimize these activities. One of the major concerns in designing and optimizing the multistage hydraulic fracturing jobs is the merging hydraulic fracturing stages. This has not been seen using commercial hydraulic fracturing numerical simulators due to oversimplified assumptions made during their developments. Therefore, detailed studies on quantifying the magnitude of induced stresses and reorientation of the stress fields during multistage hydraulic fracturing jobs are required.

Recently, different numerical and analytical studies have been published concerning the magnitude of stress change and stress reorientation in single pressurized fractures or multiple fractures. Cheng (2012) studied the change in fracture geometry due to the change in stress field around three pressurized fractures. In her model, she assumed fixed fracture lengths and used the DD method to quantify the change in the fracture width as a function of the change in the stress field around the pressurized fractures. Soliman et al. (2004) used an analytical technique to calculate the magnitude of the stress change around multistage hydraulic fractures. In general, the magnitude of the stress change around a propagating fracture is a function of fracture dimensions, target, and over- and underlying formation characteristics such as fracture length, width, height, the formation's Poisson's ratio, relative magnitude of the target formation's Young's modulus, and under- and overlaying formations, magnitude, and direction of in situ stresses.

Fisher et al. (2004) introduced the stress shadowing effect in multistage hydraulic fracturing where the local maximum and minimum horizontal stresses are changed due to hydraulic fracture propagation. These changes in local state of stress will highly impact the subsequent hydraulic fracture paths and will result in hydraulic fracture stage merging or deviation depending on the magnitude of the change. If the pressure applied during hydraulic fracturing at the fracture surface falls between local minimum and maximum horizontal stresses, it will not be expected to have a huge change in local stresses and subsequent fracture paths. However, if the pressure exceeds the maximum horizontal stress, a phenomenon called principal stress reversal will occur, leading to a significant change in subsequent hydraulic fracture paths. Taghichian (2013) studied the stress shadowing around single and multiple pressurized fractures in confined and unconfined environments. He demonstrated that there is a nonlinear and direct relationship between fracture pressure, formation's Poisson's ratio, and stress shadow size. Increasing the fracture pressure leads to an increase in the stress shadow zone but a decrease in gradient. Waters et al. (2009) showed that the shadow effect around a single hydraulic fracture leads to locally increased compressive stresses perpendicular to the fracture propagation plane. This leads to reorientation of local maximum stresses and thus unintended change in the direction of the subsequent fractures, if they happen to fall in the stress shadow zones. Therefore, optimized hydraulic fracture spacing is required to increase the efficiency of hydraulic fracturing stimulation. The effect of stress shadowing is investigated not only in single-well multistage hydraulic fracturing but also in multihorizontal well fracturing. Recently, new publications have been focused on multihorizontal well stimulations in which simultaneous hydraulic fracturing of parallel horizontal wells is studied. Mutalik and Gibson (2008) showed that this technique could increase the efficiency of the stimulation between 21% and 100%. Rafiee et al. (2012) applied a similar concept to zipper frac to increase the efficiency of stimulation by using a staggered pattern. Other studies in this line tried to precondition the stress field using outer hydraulic fractures to prevent deviation in the middle stages, such as the work published by Roussel and Sharma (2011).

CHAPTER SIXTEEN

Operations and execution

Introduction

A frac job in general is a massive operation because it takes a lot of manpower and equipment to accomplish the job. Slick water frac requires high pump rates. As a result, lots of high-pressure pumps are required for each pad. The design and fracture modeling of a hydraulic frac job for optimum production enhancement is important; however, being able to operate and execute the design treatment schedule is more important. Therefore, exploration and production (E&P) companies typically develop the best operations and execution practices for each field in order to minimize nonproductive time (NPT), reduce unnecessary capital expenditures, eliminate safety accidents and compliance issues, and increase operational efficiency in an attempt to optimize the economics of the project. A frac job is logistically a large operation that will need a lot of discipline and coordination. There are lots of opportunities for improvement throughout the completions operation. When historically going back in time and reviewing the first 10 wells drilled and completed in each shale play, it can be easily seen that with time, efficiency and savings were obtained. Some of the first wells in each field cost two to four times more than the current capital expenditure in the same field for the same completions design. This can be tied back to the learning curve associated with comprehending the formation and coming up with ideas to minimize encountering any issues related to the operations of the hydraulic frac job. Drilling and completing one expensive well in a new exploration area typically does not scare a true E&P company because they have enough experience, expertise, and knowledge to know that with time, efficiency and savings will be obtained.

Water sources

There are a few sources for frac water. The most common water sources are as follows:

- freshwater from rivers, lakes, etc.,
- municipal water,
- reused flowback and produced water (100% produced water),
- treated water from a treatment facility, and
- mixing fresh and reused water from flowback and producing wells.

Water storage

As discussed earlier, water is one of the most important aspects of hydraulic fracturing. Water storage is used to make up for lack of supply, or to hold water for reuse for upcoming frac jobs. There are multiple ways to store water used during the frac job and the most common methods are as follows:

1. **Centralized impoundment (in-ground pits)** can be built on a side of the pad and water can be stored. The capacity of in-ground pits can vary from pit to pit. With stricter environmental rules and regulations, some states do not give permits for building centralized impoundments. In addition, in-ground pits can be very expensive to build, monitor, and reclaim. More regulations have been applied to in-ground pits since 2012 restricting their use due to possible detectable leakage concerns into the ground. Centralized impoundments come in various sizes and can typically hold 5 + million gallons of water ($\sim$ 120,000 BBLs). Freshwater centralized impoundment is typically single-lined but the impaired water impoundment is usually double-lined with leak detection.

2. **Aboveground storage tanks (ASTs)** have become more common since 2011 because of ease of building and monitoring associated with this type of water storage. Instead of going through the hassle of in-ground pits that must be reclaimed once done, ASTs can be built in 2–3 days and function exactly the same as in-ground pits. One of the biggest disadvantages of AST is the high cost as compared to in-ground pits but many E&P companies have been pushed to get away from using in-grounds pits in some states. Some states that do not have stricter environmental rules still use in-ground pits as they are considerably cheaper. Water storage regulations associated with in-ground pits highly depend on each state. ASTs can be rigged down in a few days depending on the size of the tank. Since regulations are becoming harder and a lot of operators are getting away from using centralized impoundments, AST has become more common in the past few years. The primary reason from an environmental perspective for using an AST is the ease and ability to detect any leaks much easier than in-ground pits. Typical sizes for ASTs

are 10, 20, 40, and 60K BBLs. In Pennsylvania, an OG71 permit is needed to use an AST for impaired water. Temporarily storage of freshwater in an AST requires a single liner while impaired water requires a double liner and secondary containment.

3. **Centralized tanks (tank batteries)** are essentially combinations of many frac tanks that are connected to each other through a manifold. Tank batteries can be anywhere from 5 to 60 or more frac tanks (depending on location) that are connected through a manifold. The capacity of each frac tank is typically 500 BBLs and enough frac tanks must be located onsite for continuous operation. For example, if each stage requires 8000 BBLs of water, 16 frac tanks (assuming a 500-BBL frac tank) are required for hydraulically fracturing only one stage. In slick water frac using conventional plug and perf, typically three to eight stages are done per day (depending on the amount of sand designed per stage). Therefore, depending on the pump rate (bpm) coming into the frac tanks, the number of frac tanks for the job can be determined. Since water is being continuously pumped from a surrounding pit, AST via buried or aboveground temporary or permanent water lines, there are only five to six frac tanks on frac location. Occasionally, where permanent or temporary water infrastructures are not in place, water must be trucked to the location, which can get expensive; there are also environmental or location impacts associated with trucking water to the location. Figs. 16.1–16.3 show examples of an in-ground pit, AST, and frac tank batteries.

Fig. 16.1 In-ground pit.

Fig. 16.2 Aboveground storage tank (AST).

Fig. 16.3 Tank batteries.

Water delivery

Water is delivered in two ways:
1. Pipeline. In developed fields with lots of frac activities, there are either buried or aboveground water pipelines that are used for water delivery and transfer. Some commonly used water pipelines are 8″–16″ HDPE, PE4710. This pipeline comes in various ratings including DR7 (315 psi), DR9 (250 psi), and DR11 (200 psi).

2. Trucking. In undeveloped fields or areas where water infrastructure does not exist, water is trucked to location. For instance, if a well in a new exploration area with no water infrastructure is being completed and eight stages are expected to be done per day using 9000 BBLs of water per stage, 720 trucks per day will be needed to deliver water on location (assuming truck capacity of 100 BBLs). This can get very costly and have some local impacts.

Pipeline and pump system design for water delivery uses Bernoulli's principle. Bernoulli's principle for incompressible flow is shown in Eq. (16.1): h_L or head loss is the energy losses in the system from pipe, friction (material), bends, pipeline size changes, etc. Due to pipeline distance, pipeline friction is the main energy loss.

$$\text{Bernoulli's principle for incompressible flow}: \frac{V_1^2}{2g} + Z_1 + \frac{P_1}{\rho g} - h_L = \frac{V_2^2}{2g} + Z_2 + \frac{P_2}{\rho g} \quad (16.1)$$

In Eq. (16.1), V is the fluid velocity at a point, Z is the elevation of the point from baseline, P is the pressure at the point of interest, and ρ is the density of the fluid. The methods discussed in this section are commonly used to store reused water or freshwater. In the winter, it is very important to have sufficient brine water onsite to prevent the frac iron from freezing. In addition, sufficient heaters have to be used for certain equipment to avoid the possibility of freezing, and as a result, this slows down the operation.

Hydration unit ("hydro")

The hydration unit, also referred to as "hydro" in the field, is a big tank used to provide sufficient time for hydration of linear gel. If gel is used in some stages to overcome tortuosity along with other benefits associated with using gel, the hydration unit provides the gel enough time to hydrate. Without a hydration unit, pumping gel is not possible. The importance of the hydration unit becomes more evident in the winter months when gel will need a longer time to hydrate due to the cold weather. If gel is not part of the job design, a hydration unit will not be needed. Some operators do not believe in pumping gel in slick water frac jobs and do not use hydration units in their equipment rig up. A hydration unit is located right after the frac tanks where water is stored. There are typically five to seven frac tanks that are connected through a manifold right before the hydration unit. The hydration unit has a suction side that sucks the water from the frac tanks and a discharge side that releases the water to the next equipment (blender, will be discussed). Some hydration units have a discharge pump but some do not. It is not a necessity to have a discharge pump on the hydro since the suction pump of the following equipment located after the hydro (blender) is used to suck the water out of the hydro. There are injection ports located on the suction side of the hydro in the event gelling agents need to be started. In addition, chemicals are stored in the special containers called "totes." There are small liquid additive (LA) pumps (such as stator or positive displacement pumps) located on the chemical totes to pump the chemicals to the hydro through the injection ports. Therefore, when gel is started, linear gel and buffer are pumped via the LA pumps through the injection ports to the suction side of the hydro. A hydration unit's capacity is typically between 170 and 220 BBLs depending on the type and size of the hydro. It takes 170–220 BBLs of fluid (to give sufficient time for gel hydration) from the time gel is started until it leaves the hydro. Therefore, it takes some time from the time gel is started until gel reaches the perforations.

Example
Calculate the volume it takes for the gel to hit the perforations assuming the following parameters:

Bottom perf measured depth (MD) = 15,500′ with a casing capacity of 0.0222 BBL/ft, Hydration unit capacity = 180 BBLs, Surface line volume capacity = 50 BBLs

$$\text{Casing capacity} = 0.0222 \times 15,500 = 344 \text{ BBLs} \rightarrow 344 + 50 + 180 = 574 \text{ BBLs}$$

Therefore, it takes 574 BBLs for the gel to reach the perforations as soon as it starts.

This example shows the importance of starting gel at the right time since gel will not yield an instantaneous relief. If surface-treating pressure increases rapidly during a slick water frac job, starting gel to increase fracture width and reducing the pressure will not help because it takes some time for the gel to reach the perforations. Typically, it is recommended to cut sand and start over again to prevent a costly screen-out. This example shows the importance of starting gel early enough to see the impact.

Blender

The blender is the heart of the frac operation. The blender is used to mix water, proppant, and some chemicals in the blender tub before sending the slurry fluid downhole. The blender is typically located right after the hydro in a frac setup. There is a tub on every blender and there is an agitator at the bottom of the tub. The tub agitator consists of two sets of blades on a shaft. The main function of the agitator is to keep the proppant suspended in the fluid without carrying air. If the agitator speed is too low, there is a high possibility of proppant building up and settling at the bottom of the tub and suddenly getting picked up as a sand slug. If the agitator speed is too high, it can entrain air in the fluid, causing the booster pump to pick up air, and as a result, causing a decrease in boost pressure.

The heart of a blender is the centrifugal pump. The main reason centrifugal pumps are utilized is because they are very tolerant to abrasive frac fluids which will result in an increase in the life of the pump. As previously discussed, millions of pounds of proppant are pumped when hydraulically fracturing a well and it is very important to use quality pumps, such as centrifugal pumps, on a blender. A centrifugal pump consists of one or more impellers equipped with vanes. The impeller is located on a rotating shaft and fluid enters the pump at the center of the impeller. Fig. 16.4 illustrates a centrifugal pump from inside.

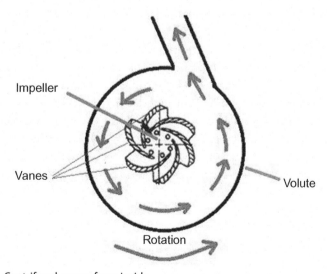

Fig. 16.4 Centrifugal pump from inside.

There are two centrifugal pumps located on the blender. The first one is called a suction pump, which sucks the water from the hydro and sends it to the blender tub. The second pump is called a discharge pump (also referred to as a "boost pump"), which sends the mixed slurry from the tub to high-pressure pumps. In other words, a blender has two sides:

1. **Suction side (clean side)**. A suction pump is located on this side of the blender. A centrifugal suction pump sucks the frac water from the hydro and sends it to the tub. This side is also referred to as the "clean side" because proppant has not yet been mixed up on this side and only frac water and chemicals enter the blender.
2. **Discharge side (dirty side)**. Combinations of water, chemicals, and proppant exit this side and this is why it is referred to as the "dirty side." The boost pump (or discharge pump) is actually the means of providing the rate (boost) for all of the high-pressure pumps located after the blender.

Sand master (sand mover, sand king, or sand castle)

Since millions of pounds of proppant are required for each pad (depending on the number of frac stages), proppant has to be stored onsite using sand masters. Sand masters have different bins used for placing various types of proppant and mesh sizes (note: some sand masters do not have any bins). Proppant is delivered on location using sand cans or sand trucks. Depending on the state regulations and guidelines, each sand truck on average can hold 40,000–50,000 lbs of proppant because of the weight limitations. For example, if a frac design of a pad consists of 100 stages and each stage is designed to use 400,000 lbs of proppant, 800 sand trucks (assuming each sand truck can haul 50,000 lbs) must travel to the pad to blow the proppant into the sand masters throughout the job. Placing the proppant into the sand master is called blowing sand. There are new technologies that do not require blowing proppant anymore; instead, proppant can be placed into sand masters through gravity. During a large-sized hydraulic frac job, sand trucks are entering and exiting the pad all day long to provide the proppant needed for the job. The sand master is also referred to as the sand mover, sand king, or sand castle. There are other sand systems such as "sand storm system," "sand box system," "arrows up system," and "klun system," but the purpose of all of these systems is the same regardless of the names and shapes of the movers. They are all used to store proppant on location. The "sand storm system," shown in Fig. 16.5, contains silos, and each silo can typically

Fig. 16.5 Sand storm system.

Fig. 16.6 Sand box system.

hold more than 400K of proppant. They are equipped with electronic weight counters. Their biggest disadvantage is that they are bulky and unsuitable for smaller locations. Another type of sand system used during hydraulic frac jobs is referred to as "sand box system." In this set up, four sand boxes are placed over the T-belt (discussed next) as shown in Fig. 16.6. The advantage of the "sand box system" is that they are efficient and easy to store on location. The main disadvantage is that the sand bins constantly shift and must be tracked (capacity is typically 40K per bin).

Fig. 16.7 Arrows up system.

Another type of sand system is referred to as the "arrows up system" in which three sand bins sit over the hopper (will be discussed) as shown in Fig. 16.7. The advantage of this system is that, just like the sand boxes, they are easy and efficient to store on location, and they do contain electronic weight counters for the bins. The main disadvantage of this system, just like the "sand box system," is constantly shifting sand bins. Finally, another type of sand system is called the "klun system" which is essentially tall silos as shown in Fig. 16.8. Each silo can accommodate ~300K or more with an electronic weight counter.

T-belt

Proppant falls out of the sand bins onto the T-belt. Once on the T-belt, proppant is carried through the T-belt and falls into the blender hopper. Once at the blender hopper, proppant gets picked up via the blender's sand screws and is dumped into the blender tub. The sand screws are another important part of the blender: they are the means of picking up proppant and dumping the proppant into the blender tub. Therefore, sand screws are essential in conveying proppant to the blender tub. Fig. 16.9 illustrates the point at which proppant gets dumped on the T-belt and carried via the T-belt until it reaches the blender hopper. Fig. 16.10 illustrates the point at which proppant is dumped into the blender hopper, the point where proppant is picked up through the sand screws from the blender hopper, and finally the point at which proppant is dumped into the blender tub where proppant, water, and chemicals will be mixed. Fig. 16.11 demonstrates a blender hopper full of proppant along with sand screws.

Fig. 16.8 Klun system.

Sand screws

As previously mentioned in a hydraulic frac job, proppant concentration is gradually increased each time proppant hits the perforations (more aggressive design schedules do not wait for the proppant to hit the perforations and proppant is staged up faster) depending on the design schedule and pressure response throughout the stage. There are typically two to three sand screws on a blender depending on the blender manufacturer and type. In slick water frac, two screws are typically used and the third one is a backup. The third screw is normally used with very high proppant concentration frac jobs such as cross-linked jobs. Every blender has a maximum rpm per screw that can be obtained from the blender manufacturer. The reason more than one rpm is

Fig. 16.9 Sand masters and T-belt.

Fig. 16.10 Hopper and blender screws.

needed for the job is to be able to pump higher sand concentrations at higher rates. The most commonly used sand screws have 12″ and 14″ diameters. Normally, the maximum output for a 12″ screw is approximately 100 sacks per min (one sack of proppant is equal to 100 lbs), and for a 14″ screw is 130 sacks per min with a maximum rpm of 350–360 (different depending on the

Fig. 16.11 Sand screws with proppant in the hopper.

blender manufacturer and type). Note that maximum rpm of 350–360 is only for one screw and since typically two screws are used in slick water frac jobs, up to 700 rpm can be obtained to fulfill the client's needs and design schedule. Different types of proppant yield different pounds per revolution (PPR). PPR decreases at higher sand concentrations and as the screws wear out. For example, if proppant delivery of Ottawa sand is about 36 PPR with a brand new screw, as sand concentration increases, PPR decreases. In addition, if lower sand concentration is run (e.g., 0.25 ppg) and PPR is run at 29 (if typical is 36), there could be a high possibility that the screws are worn out.

$$\text{Round per minute (rpm) calculation}: \text{rpm} = \frac{Q \times SC \times 42}{PPR} \quad (16.2)$$

where rpm is the round per minute, Q is the slurry rate, bpm, SC is the sand concentration, ppg, and PPR is the pounds per revolution.

Eq. (16.2) is constantly used in the field to calculate the amount of rpm needed on the screws to achieve the designed sand concentration. For example, after pumping the designed volume of pad, and once the sand stage is ready to start, the person responsible for adjusting the sand screws on the blender is notified by radio to bring his/her rpm to a certain value to achieve the required proppant concentration requested by the operating company's designed proppant schedule. This is referred to as running the blender on "manual." On the other hand, the majority of service companies run their blender screws on "auto" for simplicity. Running the blender on auto means entering the proppant concentration needed on the blender, which will automatically calculate the required rpm. This is the preferred method since every time the slurry rate is changed throughout the stage for any reason, the auto system calculates the new rpm needed and adjusts the screws. For example, if a pump is dropped for any reason (e.g., mechanical issues), a new rpm is automatically calculated. If a manual system was being used, the new rpm would need to be manually calculated and changed on the blender, which might take some time. It is strongly recommended that any service company knows how to run the blender in both auto and manual. This way, if there are any issues throughout the stage while running the sand system in auto, a manual system can be substituted and the frac stage can continue instead of coming offline while the problem is being fixed.

Example
Calculate the rpm needed at 100 bpm slurry rate if 0.25 ppg proppant concentration is to be achieved with 36 PPR.

$$\text{rpm} = \frac{Q \times SC \times 42}{PPR} = \frac{100 \times 0.25 \times 42}{36} = 29 \, \text{rpm}$$

The rpm needed to achieve 0.25 ppg proppant concentration is 29 rpm. At this stage, typically the company representative waits until proppant hits the perforations to see the reaction of the formation. If everything looks promising on the surface-treating pressure chart, proppant concentration is increased by increasing rpm. Let's assume the next designed proppant concentration is 0.5 ppg and the rate had to be dropped to 94 bpm. Calculate the new rpm needed to achieve this concentration.

$$\text{rpm} = \frac{Q \times SC \times 42}{PPR} = \frac{94 \times 0.5 \times 42}{36} = 55 \, \text{rpm}$$

As can be seen in this example, as proppant concentration increases, rpm increases as well. Please note that if, throughout the stage, rate is increased or decreased, rpm needs to be adjusted as well if the system is not set up for auto. Rate is directly proportional to rpm and as rate increases or decreases, rpm increases or decreases as well. There is a person called a treater who is responsible for calculating the new rpm every time rate, proppant concentration, or PPR is altered if and only if the manual rpm system is used. If the auto system is used, the only parameter that needs to be entered is proppant concentration and everything else will be automatically calculated.

PPR is typically adjusted throughout the job as well to stay at the required proppant concentration and volume. Throughout the job, there is a person on the sand master who takes proppant straps (the amount of proppant that is left in the bin or sand master). As previously discussed, newer sand master systems can actually measure the amount of proppant pumped out of each bin or sand master via a scale or electronic weight counter. There are two ways to measure proppant. The first one is the calculated amount located on the frac monitors inside the frac van from the blender screws. The second one is through the gentleman/lady on the sand master who measures the amount of proppant that is left in the bin (or in the newer system via a scale). For example, let's assume that after taking a bin strap, the person responsible for keeping track of the proppant announces 30,000 lbs of proppant has been pumped. On the other hand, the monitor located in the frac van shows the total proppant pumped is 40,000 lbs (calculated from sand screws). This means less proppant has been pumped compared to the proppant volume that must have been pumped. This condition is referred to as sand light. In this situation, PPR needs to be decreased in order to increase the rpm and catch up with the required proppant.

In contrast, if the monitor shows 20,000 lbs instead, proppant is referred to as sand heavy because more proppant (30,000 lbs) was actually pumped. In this case, PPR is increased in order to decrease rpm and slow down the actual amount of proppant. The difference in proppant can be easily caught up if and only if the difference is 5000–15,000 lbs at lower proppant concentrations. If the difference is drastic, such as 30,000 lbs, it is strongly recommended not to catch up. If proppant is running 30,000 lbs light and we have 50,000 lbs of proppant left to go in a frac stage, it means more proppant needs to be pumped in order to compensate for the lack of proppant concentration accuracy throughout the stage. This can be devastating because proppant is now run at very high concentrations in an attempt to catch up, which can cost the operating company a screen-out. Therefore, it is very important to remind the person in charge that if proppant is light or heavy by a drastic amount, do not try to catch up and run it as it is all the way to the end of the frac stage. The most important part about blender screws is the difference between rpm and PPR. As can be seen from Eq. (16.2), rpm and PPR are inversely proportional. When PPR decreases, rpm increases and more proppant will be pumped. PPR is typically decreased by a small amount when proppant is running light. When PPR increases, rpm decreases and proppant will be pumped slower. PPR is typically increased by a small amount when proppant is running heavy. Fig. 16.12 is another example of sand screws and a blender tub where sand, water, and some of the chemicals get mixed.

Fig. 16.12 Blender tub and blender screws.

Chemical injection ports

There are multiple chemical injection ports located on the blender. Some chemicals such as friction reducer (FR), biocide, scale inhibitor, etc. can be directly pumped from LA pumps to the suction side or tub of the blender. Other chemicals such as cross-linker have to be pumped by the LA pump to the injection ports located on the discharge side of the blender. This is because the tub agitator is not able to mix such a viscous fluid in the tub. In addition, surfactant has to be injected in the discharge side as well because special types of surfactants can foam up and block the view in the blender tub. Therefore, depending on the chemical and frac setup, some chemicals are pumped into the suction side while others are pumped into the discharge side. Fig. 16.13 shows chemical totes where various chemicals used during the hydraulic frac job are stored on location. As can be seen from Fig. 16.13, frac chemical totes are located on containment, which prevents any type of spill from reaching the ground. Any type of spill on the containment can be easily cleaned up with no environmental damage, however, the majority of companies take many precautions to avoid any type of spill regardless of its amount on the containment. Many companies even tie incentives back to any type of spill in an attempt to make sure that all of the employees place 100% of their effort in having zero environmental incidents and staying in compliance with all of the environmental regulations and laws.

Fig. 16.13 Chemical totes.

The containments are inspected constantly throughout the frac job to make sure that there are no holes in the containment. Any holes in the containment are reported and fixed immediately.

Densometer ("denso")

The densometer, also referred to as the "denso," measures the density of the frac fluid going downhole. There is always one densometer on any blender located on the discharge side of the blender to read the proppant concentration of the fluid coming out of the blender tub. This is the most accurate way of measuring the proppant concentration that is being pumped downhole. Some service companies try to hide this value from the client (operating company) and instead show a corrected proppant concentration value since this value fluctuates throughout the stage due to the level of proppant in the hopper dropping or gaining. Therefore, to make the service look better, service companies usually hide this value and instead show a corrected value on the screen inside the frac van. There is one more densometer located at the end of the main line. This densometer is usually used to make sure all of the proppant has cleared the surface lines. Once all of the proppant is cleared from the surface lines, the flush stage starts. In the flush stage, all of the proppant must be placed into the formation to make sure the wireline can go downhole to set the plug and perforate the next stage in a conventional plug-and-perf setup.

Missile

The missile is located right after the blender. Water, proppant, and chemicals are sent to the missile with the boost rate obtained from the boost pump on the blender. The missile is a big manifold that allows multiple hoses and frac irons to be connected. This reduces the amount of frac lines and hoses used during the frac job. The missile has two sides. The low-pressure side is where frac fluid is transferred from the missile to frac pumps. The high-pressure side is where the slurry comes out of the frac pumps and goes back to the missile. The low-pressure side of the missile has a low pressure of normally 60–120 psi. The discharge hoses can be used to transfer the water from the blender to the missile and the missile to the frac pumps. The high-pressure side of the missile is approximately the frac pressure obtained from the surface-treating pressure chart. The main reason this side is called the high-pressure side is to denote the difference (Fig. 16.14).

Below are the simplified steps of transferring slurry frac fluid from the boost pump (located on the blender) to the wellhead:

1. The boost pump located on the blender provides the rate needed to transfer the slurry to the missile and all the pumps on location.
2. Slurry fluid (water + proppant + chemical) enters the missile and subsequently the frac pumps via discharge hoses.

Fig. 16.14 High-pressured iron and low-pressured hose on the missile.

3. High-pressure frac pumps are used to shoot the slurry fluid back to the high-pressure side of the missile via frac irons.
4. The missile is a big manifold that takes several frac lines (depending on the size of the missile) and turns them into two to six lines depending on the rate needed for the job.

Frac manifold (isolation manifold)

Frac manifold is sometimes used in zipper frac operation to isolate one well from another. While a frac job is performed on one well, the other well can be safely perforated by having two barriers. The frac manifold isolates the frac well from the wireline well with two barriers. The first barrier is a hydraulic valve located on the frac manifold and is operated via the accumulator. The second barrier is a manual valve operated manually. Two-leg or three-leg frac manifolds are the most commonly used manifolds in the industry. Each leg has hydraulic and manual valves as shown in Fig. 16.15. A two-leg frac manifold is used to zipper frac two wells while a three-leg frac manifold is used to zipper frac three wells at a time. A frac manifold is not a necessity in frac operations since zipper frac can be done without it. However, the use of a frac manifold eliminates rigging up and rigging down frac irons between stages. As a result, using a frac manifold saves time and in the oil and gas industry, time is money. Therefore, the decision on whether to use a frac manifold or not is an economic decision based on economic analysis. Fig. 16.16 illustrates an overview of the frac equipment from a frac site.

Fig. 16.15 Three-leg frac manifold.

Fig. 16.16 An overview of frac site.

Frac van (control room)

The frac van is where the company representative ("company man") along with the frac supervisor oversees the entire operation via various charts. There are various charts that are monitored during hydraulic frac jobs and crucial decisions are taken based on those charts. The most important charts monitored during the frac job are as follows:

1. **Surface-treating pressure chart.** This chart typically has surface-treating pressure, bottom-hole pressure, slurry rate, blender concentration (sand concentration at the blender), and bottom-hole proppant concentration (calculated sand concentration at formation). Surface-treating pressure is read directly from the transducer on the main line. Bottom-hole pressure is a calculated pressure using the bottom-hole surface-treating pressure equation. Some companies display the actual blender concentration from the densometer on the blender. Slurry rate shown on the chart is obtained from the flow meter located on the blender. Fig. 16.17 shows an example of a typical slick water stage along with treating pressure, bottom-hole pressure, slurry rate, and sand concentration.

2. **Net bottom-hole pressure (NBHP) chart.** NBHP or Nolty chart is another main diagram illustrated in the frac van. In conventional reservoirs, the Nolty chart is used to make critical decisions based on pressure

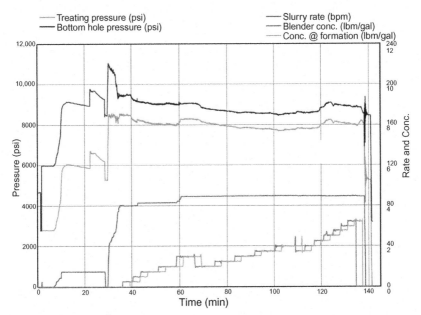

Fig. 16.17 Pressure chart.

trends. Even though this chart was mainly developed for conventional reservoirs, it is still widely used and followed during unconventional plays, such as various shale plays across the United States, as part of tradition.

3. **Chemical (chem.) chart.** This chart illustrates each particular chemical and the amount of chemical used throughout the stage. It is very important to monitor all of the chemicals that are being pumped downhole throughout the stage to make sure the right type and amount of chemical concentrations are being used. In addition, as soon as the personnel in charge of the chemical starts having issues at any point during the job, it can be easily seen on the chemical chart and immediate actions must be taken to correct the problem. For example, FR is one of the most important chemicals that must be run throughout the stage during slick water frac to reduce pipe friction. If at any point throughout the stage FR is not pumped downhole due to equipment malfunctioning or any other reason, it is very important to cut sand and flush the well right away since pumping the job without FR is impossible. Also a few buckets of FR must be placed right by the blender tub to survive the flush stage in the event FR is lost. If FR is not being pumped during the flush because there are no FR buckets by the blender tub, rate must be significantly

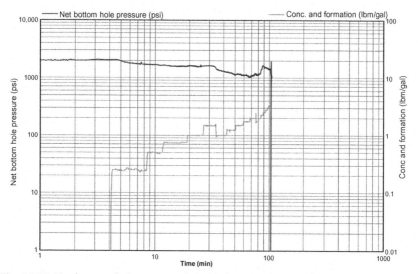

Fig. 16.18 Net bottom-hole pressure (NBHP) chart.

dropped to stay below the maximum allowable surface-treating pressure. Significantly dropping the rate will eventually cause the formations to give up and might result in a costly screen-out. Figs. 16.18 and 16.19 illustrate NBHP behavior and chemical chart in a slick water frac stage, respectively.

Overpressuring safety devices

Frac operations can be very complicated and unpredictable. In frac operations, precautionary actions must be taken to prevent overpressuring the iron, casing, and equipment during the frac job. As a result, the following two minimum precautionary actions are taken by service companies to prevent any unpleasant consequences such as parting iron, bursting the casing, blowing up the wellhead, and other well-control issues during the frac operation:

1. **Pump trips.** Pump trips can be easily placed on all of the pumps in the event of an emergency and to prevent overpressuring the iron, wellhead, and casing. Pump trips are determined by the operator and vary from operating company to company. For example, if maximum allowable surface-treating pressure during frac stage treatments is set to be 9500 psi, pump trips are staggered between 9500 and 9900 psi depending on the operator's preferences and guidelines. Pump trips need to be

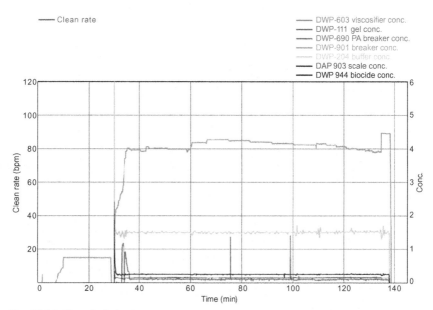

Fig. 16.19 Chemical chart.

staggered in a pressure range. If the trips on all of the pumps are set at 9400 psi, all of the frac pumps will trip at the same time in the event the pressure exceeds 9400 psi for a short period of time. The main reason for not wanting all of the pumps to trip at the same time is because a lot of the time the pressure is still under control and can be controlled by dropping one pump at a time instead of giving up and bringing all of the pumps offline. Fig. 16.20 shows a typical frac equipment setup plan. Fig. 16.21 illustrates a frac stage that was screened out due to a 1000 psi pressure spike. All of the pumps tripped out even with the pump trips being staggered. In this situation, pump trips avoided overpressuring the iron, casing, and equipment by catching the pressure spike and bringing all of the pumps offline. In these occasions where the formation completely gives up, the pump operator does not have sufficient time to take any action no matter how fast the reaction occurs. The pressure spike shown in Fig. 16.21 happened in 1 s, and this shows the importance of having mechanical and automatic pressure control equipment during a frac job to ensure the safety of the frac operation.

2. **Pressure-relief valves (PRVs).** PRVs, also known as pop-offs, are another safety precautionary action taken to prevent overpressuring the iron, casing, and equipment in the event pump trips fail or as a

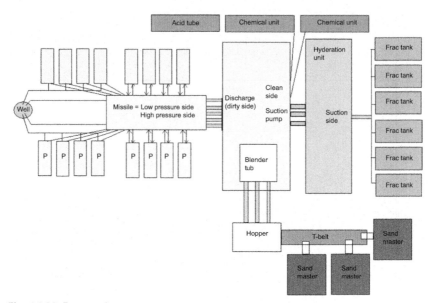

Fig. 16.20 Frac equipment setup.

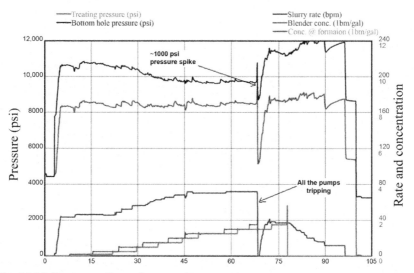

Fig. 16.21 Pressure spike and pump trips.

secondary safety preventative. PRVs can be easily set to any specific pressure and will go off as soon as that particular set pressure is reached. For example, if PRV is set at 9900 psi during the frac job, PRVs will go off and release the pressure as soon as 9900 psi is reached. The first type of PRV used during a hydraulic frac job is referred to as mechanical

Fig. 16.22 Mechanical pop-off.

pop-off. Mechanical pop-offs have been known to malfunction. Therefore, there are new patent technologies such as frac release valve (FRV) through Safoco that have been introduced that guarantee the activation of pop-offs at a certain pressure, and have been replacing conventional mechanical pop-offs. It is crucial to stay out of the pop-off area (danger zone) when pops are going off due to the release of pressure into the atmosphere. It is recommended to rig up the frac iron from the pop-offs to the flowback tank to prevent releasing any kind of pressure to the atmosphere as a safety precaution. Fig. 16.22 shows a mechanical pop-off.

Pressure transducer

Pressure transducers are used to measure the surface-treating pressure during frac jobs. Each frac pump has a pressure transducer on the discharge side. In addition, real-time surface-treating pressure is obtained via the transducer located on the main line. Pressure transducers are usually covered with a plastic cover to prevent them from getting wet during summer or winter. Some pressure transducers are known to yield inaccurate readings when wet. Figs. 16.23 and 16.24 are two examples of pressure transducers.

Check valves and manual valves

Flapper or dart check valves are very commonly used in frac operations. Every frac pump used during the frac job needs to have a check valve and a manual valve on the discharge side. Check and manual valves located

Fig. 16.23 Pressure transducer.

Fig. 16.24 Pressure transducer on a pump.

on the discharge side of the pump provide isolation between the pump and the rest of the equipment and iron. For example, if a frac iron located before the check valve on a frac pump parts (comes apart) during the live frac stage treatment, fluid will take the path of least resistance. Without check and manual valves, pumps cannot be isolated and all of the pumps have to come offline to fix the problem. When all of the pumps have to come offline to fix the problem without being able to flush the well, a costly screen-out could be the consequence. Check valves located on the discharge side of the pump isolate the pumps from the rest of the equipment so the frac operation can continue in the event of any leakage on the iron. In some occasions, flapper or dart-type check valves fail due to pumping millions of pounds of abrasive fluid. In those particular events, the manual valve located after the check valve can be closed to continue the operation without having to come offline. Note that some companies do not have a manual valve (wheel valve) after the check valve. Having a manual valve right after the check valve on each pump is highly recommended in the event the first check valve fails. Check valves fail and leak all the time. Therefore, it is very important to have a manual valve right after the check valve on each pump.

Water coordination

A frac operation is an enormous operation that needs lots of organization and coordination from both service and operating companies. As previously mentioned, each frac stage uses lots of water, proppant, and chemicals. To give some perspective on the amount of water used during slick water frac jobs, it is important to note that every two stages that the industry pumps on average is equivalent to one Olympic-sized swimming pool, which has a capacity of approximately 15,724 BBLs. One of the most challenging parts of the frac operation is obtaining sufficient water needed for the frac job. Since millions of gallons of water will be pumped per stage, each operating company usually has a water group to make sure a water transfer plan is scheduled, known, and determined before the actual operation begins to minimize any downtime. Water is essentially pumped into the pit, AST, or tank battery via polyvinyl chloride (PVC) or poly lines. From that point, water is transferred to onsite working tanks via PVC or poly lines. The rate at which water is transferred into the in-ground pit, AST, or tank battery depends on the amount of water used per stage and the number of stages to be completed per day. For example, if six frac stages are estimated to be pumped per day and each frac stage uses approximately 8000 BBLs of water, 48,000 BBLs of water must be transferred and pumped to maintain the pit level at all times. This means water needs to be pumped at approximately 33.3 bpm to keep up with the frac job and avoid any NPT. Therefore, water transfer and coordination are not as easy as they sound and require 24-h supervision during the frac job to make sure a sufficient amount of water is available to be pumped every day.

Sand coordination

Every frac stage can use anywhere between 100,000 and 700,000 lbs of proppant and it is very important to have an excellent sand coordinator in charge to make sure an adequate amount of proppant is delivered on time for a continuous frac operation to take place. Proppant is stored in sand masters and each sand master has a limited capacity. The capacity will be based on the type of sand system used for the job. As a result, sand trucks are constantly delivering proppant during the frac job. It is the sand coordinator's responsibility to make sure proper size and type of proppant are placed into each sand master. No frac stage must be started until enough proppant is present on location to pump a stage. For example, if the designed proppant stage is 300,000 lbs and there are only 230,000 lbs on location and two sand trucks are estimated to be on location at any moment, it is not a good practice to start a stage without having the total amount of designed proppant available. There is always the possibility of sand trucks breaking down or delaying for various reasons. Therefore, until the total amount of proppant is present in the sand masters, no stage must be started.

Chemical coordination

Having a sufficient amount of chemicals per stage is another important aspect of the hydraulic frac operation. Eq. (16.3) is used to find the amount of each chemical used during each stage. Based on the chemical usage per stage, more chemicals can be ordered and coordinated throughout the day.

$$\text{Chemical needed per stage} = 0.042 \times \text{chemical concentration} \times \text{clean volume} \quad (16.3)$$

where chemical needed/stage is in gallons, chemical concentration is chemical concentration that will be pumped during the stage, gpt, and clean volume is total estimated clean volume per stage in barrels.

Example

How many gallons of FR are needed if 8000 BBLs of water are estimated to be pumped in a stage at 1.5 gpt FR concentration?

$$\text{FR (gallons)} = 0.042 \times 1.5 \times 8000 = 504 \text{ gallons of FR}$$

The person in charge of chemical coordination needs to have a sufficient amount of chemicals onsite for at least twice the volume of the calculated number. In addition, the number of hours of downtime can be completely prevented by performing a simple calculation. Time is money in the oil and gas industry and whenever there is downtime due to lack of water, sand, or chemical coordination, money is lost. NPT caused by the service company has to be reported and recorded for the end-of-the-year evaluation and continuous improvement.

Stage treatment

As previously mentioned, a frac operation is big, stressful, exciting, and live. One of the main aspects of hydraulic fracturing is that typically each stage treats differently. This makes hydraulic fracturing an interesting operation. Formal education and understanding the theories definitely help as far as visualizing and estimating the formation treatment. However, the most important aspect of the hydraulic fracturing job is experience. This is the main reason the majority of companies across the United States hire very experienced people to be in charge of the operation. Some companies hire two people to be in charge of a slick water frac operation due to the extent of the job.

Tips for flowback after screening out

An important tip known in the field is to avoid a screen-out for the first few stages because there is not enough energy downhole to have a successful flowback after screening out. Typically, when a well screens out on the first few stages, the possibility of a successful flowback is very low. The energy downhole needed to clear the wellbore from proppant when flowing back is not available during very early screen-outs. Without this energy, it is not possible to successfully flow a well back. Not being able to flow a well back is very costly and time consuming. It usually takes at least a day to rig up (R/U) coiled tubing, perform a clean out run, and rig down (R/D) coiled tubing. In some instances, coiled tubing cannot reach all the way to the bottom depth where the screen-out occurred (from torque and drag model analysis). Therefore, a snubbing unit has to be used to perform a clean-out run. Thus, due to time and expense, special care must be taken to avoid the possibility of screening out on the first few stages or stages (depths) where reaching coiled tubing is not possible in long lateral wells. The industry has been moving toward drilling longer lateral wells (lateral lengths in excess of 8000 ft) since drilling longer lateral wells is significantly better from an economic perspective, as long as insignificant or no detrimental damage is observed in production results.

When screened out, it is very important to flow the well back within minutes to prevent the proppant settling in the heel. For a safe and efficient flowback operation to take place, it is important to have a safety and operational meeting with the flowback crew on a daily basis. This way the crew knows their responsibility and the company representative's expectations when screening out. It is not recommended to have a meeting after screening out because flowback needs to start within minutes of a screen-out for a successful flowback operation.

The idea behind flowing a well back after screening out is to have a balance of enough flowback rate while not pulling more proppant from the formation and previous zones into the wellbore. Flowing a well back in different formations varies. However, it is recommended to flow back at 8–10 bpm when the flowback is taking place through a 5½″ casing and 5–7 bpm through a 4½″ casing. Flowing the well hard is the best way to clean the wellbore from sand but as previously mentioned getting sand from the formation and previous stages must be minimized and avoided. When slick water is used, a minimum of two hole volumes (plug depth) needs to be flowed back before attempting an injection test. In cross-linked gel frac, 1½ hole volume is probably sufficient before attempting to perform an injection test.

Flowback tanks and lines must be rigged up in a manner that is ready to accept high fluid rates and large sand volumes without having to shut-in during flowback for any adjustments. Flow back after screening out is like the cementing operation. Once the operation starts, it is to be continued without any stoppage unless there is an absolute emergency. The reason being is that having to shut-in during flowback truly jeopardizes the chances of success. A sufficient number of flowback tanks must be available to flow the well back 2–4 hole volumes at a high rate. If only one gas-buster tank is available, there needs to be a transfer line (such as poly line) accessible to pump out the flowback fluid to an existing pit. Essentially, it is very important to have enough room available to flow a well back without having to shut-in. If more than one flowback tank is used, equalizing hoses must be high enough on the tank to avoid getting plugged up with high volumes of sands.

After flowing the well back for a minimum of two hole volumes, returns must be monitored to make sure sand is not being recovered anymore. For accurate volume measurements, periodic straps must be taken from the tanks to make sure proper flowback rate is obtained.

Postscreen-out injection test

An injection test after screening out requires patience and experience. The key to success in injection testing is taking sufficient time and fluid to slowly increase rate. Increasing the rate rapidly as pressure drops dramatically has proven to be unsuccessful in many different shale plays. Increasing rate quickly causes too much sand to be picked up at a time and as a result sending sand slug at lower rate and plugging off the perforations.

Below is the recommended postscreen-out injection test procedure:

1. Roll over a pump truck at the lowest possible rate (usually 1.5–2 bpm depending on the pump).
2. Once pressure is stabilized, increase the rate to 3–4 bpm by pumping 5 lbs linear gel and 1–2 gpt FR. After about 100–150 BBLs, increase the gel to a 10-lb system for two full wellbore volumes. Having the 10-lb gel will help land the ball softly (if not retrieved during flowback) and prevent a dramatic pressure spike, which can cause all of the pumps to trip out.
3. Afterward, walk up the rate 1 bpm at a time for ½ to 1 full wellbore volume depending on the pressure reaction. For example, it is important not to exceed 8000 psi (if max pressure is 9500 psi) to have plenty of room for pressure to increase and roll over. The name of the game is patience.
4. Do not increase rate more than 2 bpm at a time even if pressure decreases dramatically. As previously mentioned, grabbing too much rate at a time causes too much sand to be picked up at one time and may cause failure.
5. After reaching about 10–12 bpm, cut gel and keep the FR running. Hold rate until all of the pumped gel clears the perforations. Continue working the rate up very slowly as pressure allows until 30–35 bpm is reached; 30–35 bpm is the rule of thumb and the desired rate among operating companies in conventional plug-and-perf operations to ensure clean wellbores and pumping wireline down for the next stage.
6. Once 30–35 bpm is reached, pump 100–200 BBL gel sweep until the sweep clears the perforations.

A typical post-screen-out injection test should take at least 5 h and can be as long as 10–15 h depending on the MD of the screened-out stage.

Frac wellhead
Tubing head (B section)

The tubing head is one of the main components of a wellhead, which is placed after the drilling process is over and before frac operations start. Tubing head is used to land the production tubing, and the back-side pressure (casing pressure) can be monitored throughout the life of the well via the wing valves located on the tubing head. The tubing head also provides the means of attaching the Christmas tree to the wellhead. The casing head is referred to as the "A section," the tubing head is called the "B section," and the Christmas tree is referred to as the "C section." Fig. 16.25 shows a tubing head with production tubing hung inside the tubing hanger.

Lower master valve (last resort valve)

The lower master valve is used to control the flow of fluid from the wellbore and is located directly above the tubing head. This valve is the last valve that needs to be operated as the last chance in the event that all of the primary well-control barriers fail. It is very crucial to perform bolt check and other visual inspections on this and all of the other valves frequently during the frac job. During the frac job operation, a lot of proppant is being pumped downhole at a high pressure and rate, which can cause the bolts or threads to gradually come loose. To prevent this issue, regular

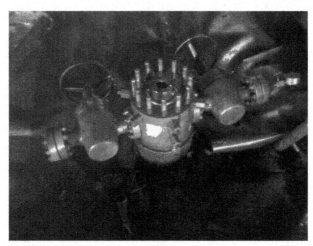

Fig. 16.25 Tubing head with production tubing hung inside the tubing hanger.

visual inspections on all of the valves are recommended and a must. In addition, every valve has a certain number of turns to open or close provided by the manufacturer. The most important valve in the frac operation is the lower master valve. If this valve washes out (by pumping abrasive fluid) during the frac job, there is a major well-control issue. If this valve starts leaking, wireline needs to be stabbed on the wellhead as soon as possible to set a kill plug inside the casing to control the flow of fluid before it gets worse. If the valve washout is uncontrollable, the site needs to be evacuated as soon as possible and a well-control company should be called to control the live well.

Hydraulic valve

The hydraulic valve is another important valve during frac jobs because in the event of an emergency during the frac operation, this valve can be hydraulically closed from the accumulator, which has to be located at least 100' from the wellhead for safety purposes. Some operators do not use a hydraulic valve to save cost. The use of a hydraulic valve is recommended in the event of an emergency situation where a frac line close to the wellhead comes apart (parts). Please note that the hydraulic valve is only designed to close on open wells with no tubing or wireline in the hole. The hydraulic valve will easily shear wireline in the hole even though it is not designed for this type of application. Therefore, it is important to label this valve accurately to prevent cutting wireline by accident during a plug-and-perf operation. Fig. 16.26 shows the hydraulic valve.

Fig. 16.26 Hydraulic valve.

Flow cross

The flow cross is located above the hydraulic valve and in the event of a screen-out, the well can be flowed back through either side of the flow cross. One of the main applications of a flow cross is to flow a well back after screening out during frac jobs. Flow cross can also be used to pump down wireline in zipper frac operations. In a zipper frac operation, since one well is being fracked while the other one is being perforated, one side of the flow cross could be used to pump the wireline down to the desired depth. Any line that comes off of the flow cross must have an ESD, which stands for emergency shutdown valve. The ESD needs to be function tested or cycled before the operation begins. It is absolutely necessary to rig up the ESD in the event of an emergency or parting iron at surface. In those emergency situations, ESD is automatically or hydraulically closed without approaching the wellhead. There are two types of ESD valves used in the industry. The first one is referred to as pneumatic ESD, which will automatically close once a substantial pressure drop has occurred during the flowback. The second type of ESD, which is not as commonly used as a pneumatic one, is referred to as hydraulic ESD. Hydraulic ESD will not automatically shut itself down in the event of an emergency and the small accumulator located away from the well has to be used. The ESD accumulators are recommended to be placed 100' from the wellhead. Fig. 16.27 shows a flow cross used as part of the frac wellhead during frac jobs. Fig. 16.28 shows pneumatic vs hydraulic ESD.

Manual valve (upper master valve, frac valve, swab valve, top valve)

The manual valve is also referred to as the upper master valve, frac valve, swab valve, or top valve, and is located above the flow cross. This valve is typically the main valve used for opening and closing the well during a frac job. For example, after each frac stage is completed, the manual valve is closed and the pressure above the manual valve is bled off. If this valve fails

Fig. 16.27 Flow cross, 2″ and 4″ sides.

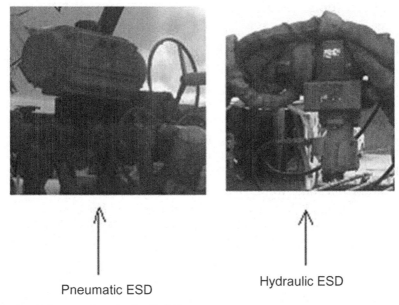

Fig. 16.28 Pneumatic vs hydraulic ESD.

Fig. 16.29 Manual valve.

or leaks, a hydraulic valve or manual valve can be used to close the well and pressure can be bled off above the hydraulic valve while the manual valve can get greased, fixed, or replaced. Finally, if the hydraulic valve fails as well, the last resort (lower master valve) will be used to close the well (Fig. 16.29).

Frac head (goat head)

The frac head, also referred to as the goat head, is located above the manual valve and typically 2–6 frac irons (referred to as candy canes) are connected to the goat head for the frac job. The goat head is the head of the frac job and is the means of injecting water, sand, and chemicals at high pressure and rate into the well. During slick water frac jobs, high rate is required to pump each stage; therefore, to get to the desired rate, two to six $4''$ or $3''$ lines are rigged up to the goat head to obtain the desired rate. The main application of frac head (goat head) in frac and perforation operation is being able to R/U 2–6 lines to achieve the designed rate. The rule of thumb for obtaining the maximum rate through each line is $OD^2 \times 2$. For instance, one $4''$ line yields a maximum rate of 32 bpm; therefore, four $4''$ lines yield a maximum rate of 128 bpm. In cross-linked fluid system jobs, not much rate is required. Therefore, there is no need to rig up four $4''$ lines. Essentially, the rig-up for each frac type and formation is different. If the designed rate for a cross-linked job is 50 bpm, only two $4''$ lines are required to perform the job. Fig. 16.30 is a four-way entry frac wellhead used during frac jobs. Fig. 16.31 illustrates a typical wellhead configuration during the frac job.

Fig. 16.30 Four-way entry frac head (goat head).

Fig. 16.31 Typical frac wellhead.

CHAPTER SEVENTEEN

Decline curve analysis

Introduction

Economic analysis is one of the most important aspects of any sales, acquisition, drilling, and completions design in any of the shale plays across North America. In fact, it is so important that the decision to complete a well using new technology will solely depend on the economics of the well. The first step in performing any type of economic analysis is to forecast the expected production volumes with time. Forecasting the production volumes and well behavior with time can be quite challenging especially in new exploration areas or areas with limited production data. In this chapter, primary methods for determining production volumes with time are discussed.

Decline curve analysis

Decline curve analysis (DCA) is used to predict the future production of oil and gas, and it has been widely used since 1945. Arnold and Anderson (1908) presented the first mathematical model of DCA. Cutler (1924) also used the log-log paper to obtain a straight line for hyperbolic decline, so the curve shifted horizontally. Larkey (1925) proposed the least-squares method to extrapolate the decline curves. Pirson (1935) proposed the loss ratio method and concluded that the production decline curve rate/time has a constant loss ratio. Furthermore, Arps (1944) categorized the decline curve using the loss ratio method, and he then defined the rate/time and rate/cumulative production. He defined three types of decline curve models: exponential, harmonic, and hyperbolic decline curves. The hyperbolic decline curve can be considered as a general model, and exponential and harmonic decline curves can be derived from it. The decline curve consists of three parameters [q_i, D_i, and b] that could be found from production data.

Furthermore, the following differential equation was used to define the three decline curve models:

$$\text{Decline curve differential equation}: d = -\frac{1}{q}\frac{dp}{dt} = Kq^b \qquad (17.1)$$

b is the hyperbolic decline exponent and K is the proportionality constant.

In Eq. (17.1) d is called decline factor that is a slope of the natural log of production rate versus time. The decline curve equations assume that production decline is proportional to reservoir pressure decline. In addition, conventional DCA assumes constant flowing bottom-hole pressure, drainage area, permeability, skin, and existence of boundary-dominated flow. Most of these assumptions are not valid in unconventional shale reservoirs. The reason DCA is still widely used is because it is an easy and quick tool to estimate production decline (rate) with time on producing and nonproducing wells. In today's business model, DCA drives the business by providing near- and long-term production forecasts and booking economic reserves. In fact, various forms of DCA are taught in short courses for reserve booking and estimation. Other tools such as rate transient analysis (RTA) and numerical simulation can also be used to forecast the future behavior of wells. Machine learning approaches can also be used to predict two years of production volume over time using various supervised algorithms such as a complex nonlinear multilayered artificial neural network.

Anatomy of decline curve analysis

There are a few crucial parameters used in DCA that are as follows.

Instantaneous production

Instantaneous production (IP) is measured in MSCF/day or BBL/day. IP is often mistaken for 24-h initial production. However, IP in DCA refers to the IP rate at a point in time that the well has been able to reach.

Nominal decline (D_i)

Nominal decline is the instantaneous slope of the decline.

Effective decline (D_e)

Effective decline is the percentage change in flow rate over a time interval. Effective decline is usually calculated 1 year from time zero. For example, if

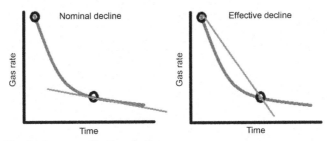

Fig. 17.1 Nominal versus effective decline.

the IP of a well is 15 MMscf/day and after 1 year, the flow rate is 4 MMscf/day, the effective decline is 73.3%. Effective decline is the percentage reduction in production volumes over a 1-year period. The lower the nominal decline, the less it varies from an effective decline. Effective decline is defined in Eq. (17.2). Fig 17.1 illustrates the difference between nominal and effective decline.

$$\text{Effective decline}(D_e) = \frac{q_i - q}{q_i} \quad (17.2)$$

Hyperbolic exponent (*b*)

The *b* value is referred to as the hyperbolic exponent and reduces effective decline over time. Hyperbolic exponent is the rate of change of the decline rate with respect to time. In other words, hyperbolic exponent is the second derivative of the production rate with respect to time. Fig. 17.2 illustrates that as *b* value increases, the rate of deceleration of effective decline increases as well. In addition, as effective decline decreases, *b* value will have less of an impact.

Shape of the decline curve

The most important parameter in DCA that drives the shape of the decline curve is the *b* value. Table 17.1 shows the approximate range of *b* values for various reservoir drive mechanisms. As can be seen, unconventional shale and tight gas reservoirs typically have a *b* value in excess of 1 due to the long transient period caused by low permeability. In conventional reservoirs that have hyperbolic decline, *b* is typically between 0 and 1 depending on the reservoir drive mechanism.

Before discussing different types of decline curves, it is very important to understand different well-flow behaviors in multistage horizontal fracs.

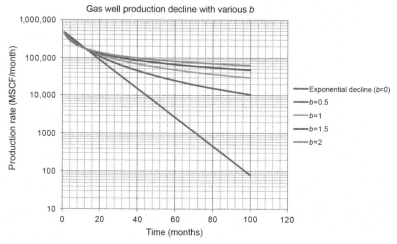

Fig. 17.2 Gas well production decline with various b.

Table 17.1 Reservoir drive mechanism vs b values

b value	Reservoir drive mechanism
0	Single-phase liquid expansion (oil above bubble point)
0.1–0.4	Solution gas drive
0.4–0.5	Single-phase gas expansion
0.5	Effective edge water drive
0.5–1	Layered reservoirs
>1	Transient (tight gas, shales)

Unsteady (transient flow) state period

Transient flow is observed in unconventional shale reservoirs with low permeability (<0.1 md) and it is the time period when reservoir boundaries have no effect on pressure behavior. The reservoir acts as if it is infinite in size. Wellbore storage effect does take place during this period. In general, transient flow is defined as the pressure pulse traveling through the reservoir without any interference by the reservoir boundaries.

Late transient period

This is the period of time that separates the transient state from the steady or pseudosteady state. It is when the well drainage radius has reached some parts of the reservoir boundaries.

Pseudosteady state (boundary-dominated flow)

Pseudosteady state occurs when there is boundary-dominated flow and the transient period ends. The boundary-dominated flow is a flow regime that starts when the drainage radius of the well reaches the reservoir boundaries. Boundary-dominated flow is a late-time flow behavior when the reservoir is in a state of pseudoequilibrium. One of the aspects that makes the unconventional shale production analysis quite challenging is that the flow stays in transient mode for a very long period of time. As a result, determination of fracture geometry is difficult from modern production analysis such as RTA.

Primary types of decline curves

There are three primary types of decline curves and they are as follows.

Exponential decline

When production rate (y-axis) versus time (x-axis) is plotted on a semi-log plot, the plot will be a straight line or exponential. In exponential decline, b is equal to 0. Exponential decline is also known as "constant-rate" decline. Exponential decline has two terms. The first term is the initial production (IP) rate and the second one is the decline rate. Decline rate in exponential decline refers to the rate of change of production with time, which stays constant.

Hyperbolic decline

When production rate (y-axis) versus time (x-axis) is plotted on a semi-log plot, the plot will be a curved line. Hyperbolic decline has three terms. The first term is referred to as the IP rate, the second term is initial decline rate @ IP rate, and finally the third term is hyperbolic exponent or b value. In hyperbolic decline as opposed to exponential decline (where the decline rate stays constant with time), the decline rate decreases as a function of the hyperbolic exponent with time. This is because the data shows a hyperbolic behavior on a semi-log plot. Hyperbolic decline rate varies and is typically 40%–80% depending on many factors such as reservoir pressure, reservoir characteristics, completions characteristics, pressure drawdown strategy, etc. The decline rate depends on the way a well is produced. The way in which a well is produced is just as important as the way in which a well is completed. The harder that a well is produced (higher pressure drawdown), the sharper the decline rate. The slower that a well is produced (minimizing pressure drawdown), the shallower the decline rate. For example, two wells side by side with the exact same completions design and formation properties can have varying decline rates depending on the way in which each well was produced. The well with higher pressure drawdown will have a higher decline rate (e.g., 85%) and the well with lower pressure drawdown will have a lower decline rate (e.g., 55%). For comparison reasons between wells, it is very important to produce all wells in the same manner operationally. Figs. 17.3 and 17.4 are examples of exponential and hyperbolic declines.

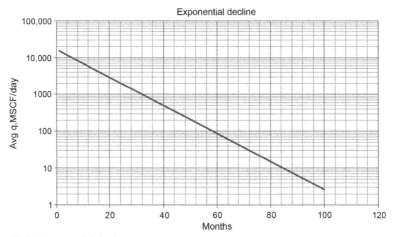

Fig. 17.3 Exponential decline.

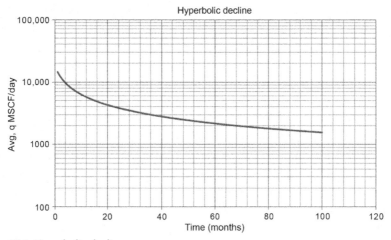

Fig. 17.4 Hyperbolic decline.

Harmonic decline

Harmonic decline occurs when the b value is equal to 1 and the decline rate of change is constant.

Modified hyperbolic decline curve (hybrid decline)

With the development of unconventional shale reservoirs, choosing only hyperbolic decline could cause an overestimation of estimated ultimate recovery (EUR). This is because hyperbolic decline without limit tends to overestimate cumulative production during the life of a well. In an attempt to account for this, modified hyperbolic decline is typically used in unconventional shale reservoirs and reserve booking. Reserve engineers will typically transition a decline curve to an exponential decline to compensate for this overestimation. The transition to an exponential decline in later stages of production is called the terminal decline. Terminal decline is the rate at which the hyperbolic decline switches from hyperbolic to exponential decline. For example, if the initial D_e (annual effective decline) for a Haynesville Shale well is 65%, once D_e reaches around 4%–11%, the hyperbolic decline switches over to exponential decline. Determination of terminal decline for reservoirs that have not produced long enough is very challenging. Companies typically assume the terminal decline to be anywhere between 4% and 11%. The higher the terminal decline, the faster the transition from hyperbolic to exponential and as a result, lower EUR. In areas with limited production data, higher terminal decline is assumed to be conservative. Fig. 17.5 shows the difference between a hyperbolic

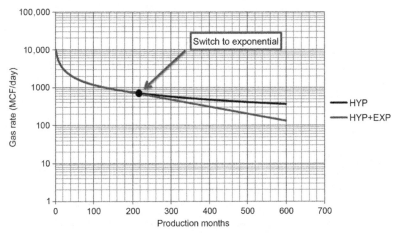

Fig. 17.5 Hyperbolic versus modified hyperbolic decline.

decline and a modified hyperbolic decline in which terminal decline is assumed to be 5%. As can be seen, as soon as annual effective decline (D_e) reaches 5%, hyperbolic decline switches to exponential for the remaining life of the well (50 years in this example).

Other DCA techniques

There are other types of DCA techniques that were developed recently. Some of those techniques are as follows:

1. **Power law exponential decline model (PLE):** PLE decline was developed by Ilk et al. (2008) by modifying Arps' exponential decline (Seshadri and Mattar, 2010). This methodology was developed specifically for tight gas wells to model the decline in a transient period of production data. The PLE decline model is defined in Eq. (17.3) (McNeil et al., 2009):

$$\text{Power law exponential decline model}: q = q_i e^{\left[-D_\infty t - \frac{D_1}{n} t^n\right]} \quad (17.3)$$

Eq. (17.3) can be reduced to power law loss ratio as defined in the following:

$$q = q_i e^{[-D_\infty t - D_i t^n]}$$

In Eq. (17.3) D_1 is the decline constant at specific time such as 1 day, D_∞ is the decline constant at infinite time, D_i is the initial decline rate % per year, and n is the time exponent. The PLE method does not consider the b value as a constant value but as a declining function in contrast to the Arps method. Moreover, by using the PLE model it is easier to match the production data in transient and boundary-dominated regions without overestimating the reserve (McNeil et al., 2009).

2. **Stretched exponential:** A newer variation of the Arps model adding a bounding component to provide a limit on EUR was developed by Valkó (2009). Stretched exponential rate time relationship is defined as follows:

$$\text{Stretched exponential decline}: q = q_i \exp\left[-\left(\frac{t}{\tau}\right)^n\right] \quad (17.4)$$

τ is a characteristic time for stretched exponential and n is the time exponent. This technique is similar to the PLE model, however, it ignores the behavior at late times. Stretched exponential has an advantage over PLE by providing the cumulative time relation as follows:

Cumulative time relationship: $Q = \dfrac{q_i \tau}{n} \left\{ \Gamma\left[\dfrac{1}{n}\right] - \Gamma\left[\dfrac{1}{n}, \left(\dfrac{t}{\tau}\right)^n\right] \right\}$ (17.5)

3. Duong decline

Duong (2011) developed the rate decline analysis for fractured shale reservoirs. In this model, the long-term linear flow was taken into consideration. This model is defined based on Eq. (17.6).

$$\text{Duong decline model}: q(t) = q_i t(a, m) + q_\infty \quad (17.6)$$

Parameters a and m are determined by using Eq. (17.7):

Determination of parameters "a" and "m" in

$$\text{Duong decline model}: \dfrac{q}{G_p} = a t^{-m} \quad (17.7)$$

q is the flow rate, volume/time, a is the intercept in log-log plot of $\dfrac{q}{G_p}$ vs t, and G_p is the cumulative gas production.

Furthermore, a plot of q versus $t\,(a, m)$ should provide a straight line with a slope of q_1 and intercept of q_∞:

$$t(a, m) = t^{-m} \exp\left[\dfrac{a}{1-m}\left(t^{1-m} - 1\right)\right]$$

Note that q_∞ can be positive, zero, or negative depending on the operating conditions. A cumulative gas production can be determined using q_∞ is equal to zero as:

$$G_p = \dfrac{q_1 t(a, m)}{a t^{-m}}$$

Duong examined different types of wells such as tight, dry, and wet gas to prove the accuracy of his model. He also found that most of the shale models have a values ranging from 0 to 3 and m values ranging from 0.9 to 1.3. His model yields reasonable estimation of cumulative production compared to the power law and Arps models.

Arps decline curve equations for estimating future volumes

As previously discussed, nominal decline is simply the conversion of the effective decline.

Exponential decline equations

Nominal decline as a function of effective decline is written in Eq. (17.8).

$$\text{Monthly nominal exponential}: D = -\ln\left[(1-D_e)^{\frac{1}{12}}\right] \quad (17.8)$$

D is the monthly nominal exponential (1/time), and D_e is the annual effective decline (1/time).

Exponential decline rate equation can also be written in Eq. (17.9).

$$\text{Exponential decline rate}: q_{\text{exponential}} = \text{IP} \times e^{-D \times t} \quad (17.9)$$

$q_{\text{exponential}}$ is the exponential decline rate (MSCF/day), IP is the initial production (instantaneous rate) (MSCF/day), D is the monthly nominal exponential (1/time), and t is the time in months.

Exponential example: Calculate the production rate of an exponential decline at the end of two years if IP is 800 MSCF/day and annual exponential effective decline (D) is 6%.

D = Monthly nominal exponential

$$= -\ln\left[(1-D)^{\frac{1}{12}}\right] = -\ln(1-6\%)^{\frac{1}{12}} = 0.515\%$$

$$\text{Exponential } q @ \text{the end of 2 years} = \text{IP} \times e^{-D \times t} = 800 \times e^{-0.515\% \times 24}$$

$$= 707 \text{MSCF/day}$$

Hyperbolic decline equations

Initial decline rate can be defined in three ways for hyperbolic decline. Nominal, tangent effective, and secant effective decline equations can be used in defining initial decline rate. Secant effective decline is the preferred methodology used in unconventional shale reservoirs. Fig. 17.6 illustrates the difference between secant and tangent effective decline rates.

Nominal decline as a function of tangent effective decline is written in Eq. (17.10).

Monthly nominal tangent hyperbolic:

$$D_{i,\text{tangent}} = -\ln\left[(1-D_{ei})^{\frac{1}{12}}\right] \quad (17.10)$$

$D_{i,\text{tangent}}$ is the monthly nominal tangent hyperbolic (1/time), and D_{ei} is the initial annual effective decline rate from tangent line (1/time).

Decline curve analysis

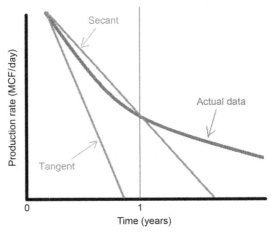

Fig. 17.6 Secant versus tangent decline rates.

Nominal decline as a function of secant effective decline can be written as shown in Eq. (17.11).

Monthly nominal secant hyperbolic:

$$D_{i,\text{secant}} = \left[\frac{1}{12b}\right] \times \left[(1 - D_{\text{eis}})^{-b} - 1\right] \quad (17.11)$$

$D_{i,\text{secant}}$ is the monthly nominal secant hyperbolic (1/time), D_{eis} is the initial annual effective decline rate from secant line (1/time), and b is the hyperbolic exponent.

Secant effective decline rate is calculated from two rates. The first rate is at time 0 and the second rate is exactly after 1 year.

The hyperbolic decline rate equation is shown in Eq. (17.12).

$$\text{Hyperbolic decline rate}: q_{\text{hyperbolic}} = \text{IP} \times (1 + b \times D_i \times t)^{-\frac{1}{b}} \quad (17.12)$$

$q_{\text{hyperbolic}}$ is the hyperbolic decline rate (MSCF/day), IP is the initial production (instantaneous rate) (MSCF/day), D_i is the monthly nominal hyperbolic (1/time), b is the hyperbolic exponent, and t is the time in months.

Hyperbolic example: Calculate the production rate of a hyperbolic decline at the end of 2 years with an IP of 8000 MSCF/day, initial annual secant effective hyperbolic decline (D_{eis}) of 66%, and b value of 1.3.

$$D_i = \left[\frac{1}{12b}\right] \times \left[(1 - D_{eis})^{-b} - 1\right] = \left[\frac{1}{12 \times 1.3}\right] \times \left[(1 - 66\%)^{-1.3} - 1\right]$$

$$= 19.65\%$$

$$q = IP \times (1 + b \times D_i \times t)^{-\frac{1}{b}} = 8000 \times (1 + 1.3 \times 19.65\% \times 24)^{-\frac{1}{1.3}}$$

$$= 1765 \, MSCF/day$$

Monthly hyperbolic cumulative volume:

Monthly hyperbolic cumulative volume can be calculated using Eq. (17.13).

Monthly hyperbolic cumulative volume:

$$N_P = \left\{ \left[\frac{IP}{(1-b) \times \text{Monthly Nominal } Hyp}\right] \right.$$

$$\left. \times \left[1 - (1 + b \times \text{Monthly Nominal } Hyp \times time)^{1-\frac{1}{b}}\right]\right\} \times \frac{365}{12} \quad (17.13)$$

N_P is the monthly hyperbolic cum volume (MSCF), IP is the initial production (MSCF/day), and time is given in months.

Example

Calculate monthly hyperbolic cumulative volume and monthly hyperbolic volume for the first 24 months assuming the following parameters:

IP = 10,500 MSCF/day, $b = 1.5$, Initial annual secant effective decline $(D_{eis}) = 61\%$

Step 1: Calculate monthly nominal secant hyperbolic:

$$D_i = \left[\frac{1}{12b}\right] \times \left[(1 - D_{eis})^{-b} - 1\right] = \left[\frac{1}{12 \times 1.5}\right] \times \left[(1 - 61\%)^{-1.5} - 1\right]$$

$$= 17.25\%$$

To replicate the calculations illustrated below, please use 17.254723% as opposed to 17.25% (rounded). This example was done using excel, and without rounding, the final results should be identical.
Step 2: Calculate hyperbolic cumulative volume for each month starting with month 1:

Decline curve analysis

$$N_p = \left\{ \left[\frac{IP}{(1-b) \times \text{Monthly Nominal } Hyp} \right] \right.$$

$$\left. \times \left[1 - (1 + b \times \text{Monthly Nominal } Hyp \times time)^{1-\frac{1}{b}} \right] \right\} \times \frac{365}{12}$$

$$N_{p,\text{month 1}} = \left\{ \left[\frac{10{,}500}{(1-1.5) \times 17.25\%} \right] \times \left[1 - (1 + 1.5 \times 17.25\% \times 1)^{1-\frac{1}{1.5}} \right] \right\}$$

$$\times \frac{365}{12} = 295{,}208 \, MSCF/1\,\text{month}$$

$$N_{p,\text{month 2}} = \left\{ \left[\frac{10{,}500}{(1-1.5) \times 17.25\%} \right] \times \left[1 - (1 + 1.5 \times 17.25\% \times 2)^{1-\frac{1}{1.5}} \right] \right\}$$

$$\times \frac{365}{12} = 552{,}264 \, MSCF/2\,\text{month}$$

$$N_{p,\text{month 3}} = \left\{ \left[\frac{10{,}500}{(1-1.5) \times 17.25\%} \right] \times \left[1 - (1 + 1.5 \times 17.25\% \times 3)^{1-\frac{1}{1.5}} \right] \right\}$$

$$\times \frac{365}{12} = 781{,}523 \, MSCF/3\,\text{month}$$

$$N_{p,\text{month 4}} = \left\{ \left[\frac{10{,}500}{(1-1.5) \times 17.25\%} \right] \times \left[1 - (1 + 1.5 \times 17.25\% \times 4)^{1-\frac{1}{1.5}} \right] \right\}$$

$$\times \frac{365}{12} = 989{,}466 \, MSCF/4\,\text{month}$$

$$N_{p,\text{month 5}} = \left\{ \left[\frac{10{,}500}{(1-1.5) \times 17.25\%} \right] \times \left[1 - (1 + 1.5 \times 17.25\% \times 5)^{1-\frac{1}{1.5}} \right] \right\}$$

$$\times \frac{365}{12} = 1{,}180{,}447 \, MSCF/5\,\text{month}$$

Step 3: Calculate monthly production volumes simply by subtracting each month's cumulative volume from the previous month:

$$\text{Hyperbolicrate for month 1} = 295{,}208 \, MSCF/\text{month}$$
$$\text{Hyperbolicrate for month 2} = 552{,}264 - 295{,}208$$
$$= 257{,}056 \, MSCF/\text{month}$$
$$\text{Hyperbolicrate for month 3} = 781{,}523 - 552{,}264$$
$$= 229{,}259 \, MSCF/\text{month}$$

Hyperbolic rate for month 4 = 989,466 − 781,523
$$= 207,942\ MSCF/\text{month}$$
Hyperbolic rate for month 5 = 1,180,477 − 989,466
$$= 190,981\ MSCF/\text{month}$$

Cumulative and monthly production volumes for 24 months are summarized in Table 17.2.

Table 17.2 Cumulative and monthly production volumes example

Time Months	CUM volumes MSCF	Monthly rate MSCF/month
1	295,208	295,208
2	552,264	257,056
3	781,523	229,259
4	989,466	207,942
5	1,180,447	190,981
6	1,357,553	177,106
7	1,523,059	165,506
8	1,678,695	155,639
9	1,825,814	147,119
10	1,965,493	139,679
11	2,098,606	133,113
12	2,225,875	127,269
13	2,347,902	122,027
14	2,465,196	117,293
15	2,578,189	112,994
16	2,687,257	109,068
17	2,792,723	105,466
18	2,894,871	102,148
19	2,993,949	99,079
20	3,090,179	96,230
21	3,183,757	93,578
22	3,274,859	91,101
23	3,363,641	88,782
24	3,450,246	86,605

As previously discussed, in modified hyperbolic decline once annual effective decline reaches terminal decline, the decline curve is switched from hyperbolic to exponential. Therefore, annual effective decline must be calculated monthly in an attempt to find the transition point from hyperbolic to exponential decline. Before being able to calculate the annual effective decline, monthly nominal decline must be calculated using Eq. (17.14).

$$\text{Monthly nominal decline}(D) = \frac{\text{Monthly nominal hyperbolic}}{1 + b \times \text{monthly nominal hyperbolic} \times \text{time}} \quad (17.14)$$

Monthly nominal decline is given in 1/time and time is in months.

Afterward, annual hyperbolic effective decline can be calculated using Eq. (17.15).

$$\text{Annual effective decline} = D_e = 1 - (1 + 12 \times b \times D)^{-\frac{1}{b}} \quad (17.15)$$

D is the monthly nominal decline, 1/time.

Example

Calculate annual effective decline after 1, 5, 24, and 50 months subsequently if the initial annual secant effective decline is 85% with a b value of 1.3.

Step 1: Calculate monthly nominal hyperbolic:

$$D_i = \left[\frac{1}{12b}\right] \times \left[(1 - D_{eis})^{-b} - 1\right] = \left[\frac{1}{12 \times 1.3}\right] \times \left[(1 - 85\%)^{-1.3} - 1\right]$$
$$= 69.09\%$$

Step 2: Calculate monthly nominal decline after 1, 5, 24, and 50 months:

$$D_{\text{month 1}} = \frac{\text{Monthly nominal hyperbolic}}{1 + b \times \text{monthly nominal hyperbolic} \times \text{time}}$$

$$= \frac{69.09\%}{1 + 1.3 \times 69.09\% \times 1} = 36.40\%$$

$$D_{\text{month 5}} = \frac{69.09\%}{1 + 1.3 \times 69.09\% \times 5} = 12.6\%$$

$$D_{\text{month 24}} = \frac{69.09\%}{1 + 1.3 \times 69.09\% \times 24} = 3.1\%$$

$$D_{\text{month 50}} = \frac{69.09\%}{1 + 1.3 \times 69.09\% \times 50} = 1.5\%$$

Step 3: Calculate annual effective decline after 1, 5, 24, and 50 months:

$$D_{e,\text{month 1}} = 1 - (1 + 12 \times b \times D)^{-\frac{1}{b}} = 1 - (1 + 12 \times 1.3 \times 36.40\%)^{-\frac{1}{1.3}}$$
$$= 76.8\%$$

$$D_{e,\text{month 5}} = 1 - (1 + 12 \times 1.3 \times 12.6\%)^{-\frac{1}{1.3}} = 56.6\%$$

$$D_{e,\text{month 24}} = 1 - (1 + 12 \times 1.3 \times 3.1\%)^{-\frac{1}{1.3}} = 26.0\%$$

$$D_{e,\text{month 50}} = 1 - (1 + 12 \times 1.3 \times 1.5\%)^{-\frac{1}{1.3}} = 15.0\%$$

Example

A well drilled in the Barnett Shale has a hyperbolic shape with an IP of 6500 MSCF/day, initial annual secant effective decline rate of 55%, and b value of 1.4. Calculate the monthly production rates along with annual effective decline for the first 12 months.

Step 1: Calculate monthly nominal secant hyperbolic:

$$D_i = \left[\frac{1}{12b}\right] \times \left[(1 - D_{eis})^{-b} - 1\right] = \left[\frac{1}{12 \times 1.4}\right] \times \left[(1 - 55\%)^{-1.4} - 1\right]$$

$$= 12.25\%$$

To replicate the calculations illustrated below, please use 12.25272% as opposed to 12.25 (rounded). This example was done using excel, and without rounding, the final results should be identical. Step 2: Calculate hyperbolic cumulative volume for each month starting with month 1:

$$N_p = \left\{\left[\frac{IP}{(1-b) \times \text{Monthly Nominal } Hyp}\right]\right.$$

$$\left. \times \left[1 - (1 + b \times \text{Monthly Nominal } Hyp \times time)^{1-\frac{1}{b}}\right]\right\} \times \frac{365}{12}$$

$$N_{p,\text{month 1}} = \left\{\left[\frac{6,500}{(1 - 1.4) \times 12.25\%}\right] \times \left[1 - (1 + 1.4 \times 12.25\% \times 1)^{1-\frac{1}{1.4}}\right]\right\}$$

$$\times \frac{365}{12} = 186,661 \ MSCF/1\,\text{month}$$

$$N_{p,\text{month 2}} = \left\{\left[\frac{6,500}{(1 - 1.4) \times 12.25\%}\right] \times \left[1 - (1 + 1.4 \times 12.25\% \times 2)^{1-\frac{1}{1.4}}\right]\right\}$$

$$\times \frac{365}{12} = 354,699 \ MSCF/2\,\text{month}$$

$$N_{p,\text{month 3}} = \left\{\left[\frac{6,500}{(1 - 1.4) \times 12.25\%}\right] \times \left[1 - (1 + 1.4 \times 12.25\% \times 3)^{1-\frac{1}{1.4}}\right]\right\}$$

$$\times \frac{365}{12} = 508,034 \ MSCF/3\,\text{month}$$

$$N_{p,\text{month 4}} = \left\{\left[\frac{6,500}{(1 - 1.4) \times 12.25\%}\right] \times \left[1 - (1 + 1.4 \times 12.25\% \times 4)^{1-\frac{1}{1.4}}\right]\right\}$$

$$\times \frac{365}{12} = 649,420 \ MSCF/4\,\text{month}$$

$$N_{p,\text{month 5}} = \left\{ \left[\frac{6,500}{(1-1.4) \times 12.25\%} \right] \times \left[1 - (1 + 1.4 \times 12.25\% \times 5)^{1-\frac{1}{1.4}} \right] \right\}$$
$$\times \frac{365}{12} = 780,873 \, MSCF/5 \, \text{month}$$

$$N_{p,\text{month 6}} = \left\{ \left[\frac{6,500}{(1-1.4) \times 12.25\%} \right] \times \left[1 - (1 + 1.4 \times 12.25\% \times 6)^{1-\frac{1}{1.4}} \right] \right\}$$
$$\times \frac{365}{12} = 903,921 \, MSCF/6 \, \text{month}$$

$$N_{p,\text{month 7}} = \left\{ \left[\frac{6,500}{(1-1.4) \times 12.25\%} \right] \times \left[1 - (1 + 1.4 \times 12.25\% \times 7)^{1-\frac{1}{1.4}} \right] \right\}$$
$$\times \frac{365}{12} = 1,019,747 \, MSCF/7 \, \text{month}$$

$$N_{p,\text{month 8}} = \left\{ \left[\frac{6,500}{(1-1.4) \times 12.25\%} \right] \times \left[1 - (1 + 1.4 \times 12.25\% \times 8)^{1-\frac{1}{1.4}} \right] \right\}$$
$$\times \frac{365}{12} = 1,129,292 \, MSCF/8 \, \text{month}$$

$$N_{p,\text{month 9}} = \left\{ \left[\frac{6,500}{(1-1.4) \times 12.25\%} \right] \times \left[1 - (1 + 1.4 \times 12.25\% \times 9)^{1-\frac{1}{1.4}} \right] \right\}$$
$$\times \frac{365}{12} = 1,233,317 \, MSCF/9 \, \text{month}$$

$$N_{p,\text{month 10}} = \left\{ \left[\frac{6,500}{(1-1.4) \times 12.25\%} \right] \times \left[1 - (1 + 1.4 \times 12.25\% \times 10)^{1-\frac{1}{1.4}} \right] \right\}$$
$$\times \frac{365}{12} = 1,332,445 \, MSCF/10 \, \text{month}$$

$$N_{p,\text{month 11}} = \left\{ \left[\frac{6,500}{(1-1.4) \times 12.25\%} \right] \times \left[1 - (1 + 1.4 \times 12.25\% \times 11)^{1-\frac{1}{1.4}} \right] \right\}$$
$$\times \frac{365}{12} = 1,427,196 \, MSCF/11 \, \text{month}$$

$$N_{p,\text{month 12}} = \left\{ \left[\frac{6,500}{(1-1.4) \times 12.25\%} \right] \times \left[1 - (1 + 1.4 \times 12.25\% \times 12)^{1-\frac{1}{1.4}} \right] \right\}$$
$$\times \frac{365}{12} = 1,518,006 \, MSCF/12 \, \text{month}$$

Step 3: Calculate monthly rates by simply subtracting cumulative volumes from the previous month as shown in Table 17.3:

Table 17.3 Monthly production rate example

Time Months	CUM volumes MSCF	Monthly rate MSCF/month
1	186,661	186,661
2	354,699	168,038
3	508,034	153,335
4	649,420	141,386
5	780,873	131,454
6	903,921	123,047
7	1,019,747	115,826
8	1,129,292	109,545
9	1,233,317	104,025
10	1,332,445	99,128
11	1,427,196	94,751
12	1,518,006	90,810

Step 4: Calculate monthly nominal decline for each month:

$$D_{\text{month 1}} = \frac{\text{Monthly nominal hyperbolic}}{1 + b \times \text{monthly nominal hyperbolic} \times \text{time}}$$

$$= \frac{12.25\%}{1 + 1.4 \times 12.25\% \times 1} = 10.5\%$$

$$D_{\text{month 2}} = \frac{12.25\%}{1 + 1.4 \times 12.25\% \times 2} = 9.1\%$$

$$D_{\text{month 3}} = \frac{12.25\%}{1 + 1.4 \times 12.25\% \times 3} = 8.1\%$$

$$D_{\text{month 4}} = \frac{12.25\%}{1 + 1.4 \times 12.25\% \times 4} = 7.3\%$$

$$D_{\text{month 5}} = \frac{12.25\%}{1 + 1.4 \times 12.25\% \times 5} = 6.6\%$$

$$D_{\text{month 6}} = \frac{12.25\%}{1 + 1.4 \times 12.25\% \times 6} = 6.0\%$$

$$D_{\text{month 7}} = \frac{12.25\%}{1 + 1.4 \times 12.25\% \times 7} = 5.6\%$$

$$D_{\text{month 8}} = \frac{12.25\%}{1 + 1.4 \times 12.25\% \times 8} = 5.2\%$$

$$D_{\text{month 9}} = \frac{12.25\%}{1 + 1.4 \times 12.25\% \times 9} = 4.8\%$$

$$D_{\text{month 10}} = \frac{12.25\%}{1 + 1.4 \times 12.25\% \times 10} = 4.5\%$$

Decline curve analysis

$$D_{\text{month }11} = \frac{12.25\%}{1 + 1.4 \times 12.25\% \times 11} = 4.2\%$$

$$D_{\text{month }12} = \frac{12.25\%}{1 + 1.4 \times 12.25\% \times 12} = 4.0\%$$

Step 5: Calculate annual effective decline for each month:

$$D_{e,\text{month }1} = 1 - (1 + 12 \times b \times D)^{-\frac{1}{b}} = 1 - (1 + 12 \times 1.4 \times 10.5\%)^{-\frac{1}{1.4}}$$

$$= 51.5\%$$

$$D_{e,\text{month }2} = 1 - (1 + 12 \times 1.4 \times 9.1\%)^{-\frac{1}{1.4}} = 48.5\%$$

$$D_{e,\text{month }3} = 1 - (1 + 12 \times 1.4 \times 8.1\%)^{-\frac{1}{1.4}} = 45.8\%$$

$$D_{e,\text{month }4} = 1 - (1 + 12 \times 1.4 \times 7.3\%)^{-\frac{1}{1.4}} = 43.4\%$$

$$D_{e,\text{month }5} = 1 - (1 + 12 \times 1.4 \times 6.6\%)^{-\frac{1}{1.4}} = 41.3\%$$

$$D_{e,\text{month }6} = 1 - (1 + 12 \times 1.4 \times 6.0\%)^{-\frac{1}{1.4}} = 39.4\%$$

$$D_{e,\text{month }7} = 1 - (1 + 12 \times 1.4 \times 5.6\%)^{-\frac{1}{1.4}} = 37.6\%$$

$$D_{e,\text{month }8} = 1 - (1 + 12 \times 1.4 \times 5.2\%)^{-\frac{1}{1.4}} = 36.0\%$$

$$D_{e,\text{month }9} = 1 - (1 + 12 \times 1.4 \times 4.8\%)^{-\frac{1}{1.4}} = 34.5\%$$

$$D_{e,\text{month }10} = 1 - (1 + 12 \times 1.4 \times 4.5\%)^{-\frac{1}{1.4}} = 33.2\%$$

$$D_{e,\text{month }11} = 1 - (1 + 12 \times 1.4 \times 4.2\%)^{-\frac{1}{1.4}} = 31.9\%$$

$$D_{e,\text{month }12} = 1 - (1 + 12 \times 1.4 \times 4.0\%)^{-\frac{1}{1.4}} = 30.8\%$$

A summary of this example is located in Table 17.4.

Table 17.4 Hyperbolic example summary

Time Months	CUM volumes MCF	Monthly rate MCF/month	Monthly nominal %	Annual effective %
1	186,661	186,661	10.5	51.5
2	354,699	168,038	9.1	48.5
3	508,034	153,335	8.1	45.8
4	649,420	141,386	7.3	43.4
5	780,873	131,454	6.6	41.3
6	903,921	123,047	6.0	39.4
7	1,019,747	115,826	5.6	37.6
8	1,129,292	109,545	5.2	36.0
9	1,233,317	104,025	4.8	34.5
10	1,332,445	99,128	4.5	33.2
11	1,427,196	94,751	4.2	31.9
12	1,518,006	90,810	4.0	30.8

Multisegment decline

Multisegment decline is another type of decline curve analysis used amongst various operators. Multisegment decline typically consists of three segments (or more) as follows:

First segment: A hyperbolic decline with a higher b value

Second segment: A hyperbolic decline with a lower b value

Third segment: Transition from the second hyperbolic to an exponential decline

The advantage of using a multisegment decline is that it is customized to fit the production data more precisely. As discussed previously, one constant b value is chosen in modified hyperbolic decline until terminal decline is reached; however, in multisegment decline, the first hyperbolic segment lasts anywhere from a few months to a few years, and the second hyperbolic segment remains until the terminal decline is reached.

Example

Calculate the total EUR and EUR/ft for the following multisegment decline listed below assuming a lateral length of 10,000 ft.

First segment

q_i (MSCF/day)	36,000
D_e	75%
b	2
Time (days) in the first segment (days)	365
Time (months)	12.00

Second segment

D_e	32%
b	1.1
End time	600
Switch to exponential	5%

The equations used in multisegment are the same equations used for modified hyperbolic decline except that, in multisegment decline, two different hyperbolic segments with different parameters must be calculated. Please use the same hyperbolic equations to calculate the first segment using q_i

of 36,000 MSCF/day, annual secant effective decline of 75%, b value of 2 for 365 days. After calculating the hyperbolic volumes (rates) for the first 12 months, the parameters provided in the second segment must be used to calculate the second hyperbolic segment. However, q_i at the start of the second segment is not provided in this example and must be calculated first. Given the info in this example, q_i for the second segment must be found as follows:

Calculate D_i as follows:

$$D_i = \frac{(1-D_{eis})^{(-b)} - 1}{12 \times b} = \frac{(1-75\%)^{(-2)} - 1}{12 \times 2} = 0.625$$

Calculate the daily rate for each day until 365 days is reached (first segment). Below are example calculations for the first 3 days:

$$q_{day\,1} = \frac{q_i}{(1 + b \times D_i \times \text{adjusted time})^{\frac{1}{b}}} = \frac{36,000}{\left(1 + 2 \times 0.625 \times \frac{\frac{1}{365}}{12}\right)^{\frac{1}{2}}}$$

$$= 35,282\,\text{MSCF/day}$$

$$q_{day\,2} = \frac{36,000}{\left(1 + 2 \times 0.625 \times \frac{\frac{2}{365}}{12}\right)^{\frac{1}{2}}} = 34,606\,\text{MSCF/day}$$

$$q_{day\,3} = \frac{36,000}{\left(1 + 2 \times 0.625 \times \frac{\frac{3}{365}}{12}\right)^{\frac{1}{2}}} = 33,967\,\text{MSCF/day}$$

Table 17.5 shows the resulting daily rate for the first 6 days and last 6 days prior to reaching 365 days. As demonstrated, after 1 year of production using the first segment DCA parameters provided in this example, the qi for starting the second hyperbolic segment is 9000 MSCF/day. Therefore, the hyperbolic calculations must be repeated for the second hyperbolic segment with the following parameters:

$q_i = 9000\,\text{MSCF/day}$, $b = 1.1$, $D_e = 32\%$, Terminal decline $= 5\%$

Repeating the hyperbolic calculations for both segments and then switching to an EXP decline at a terminal rate of 5% will yield a total

Table 17.5 Daily rate for 365 days

Day	Daily q (MCSF/day)
0	36,000
1	35,282
2	34,606
3	33,967
4	33,362
5	32,789
6	32,244
...	...
360	9058
361	9047
362	9035
363	9023
364	9012
365	9000

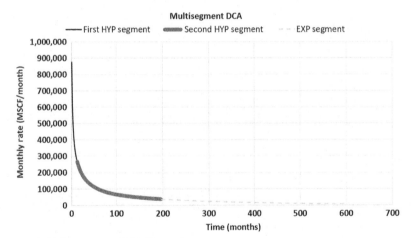

Fig. 17.7 Multi-segment DCA illustration.

EUR of 27,532,616 MSCF or 2.75 BCF/1000′. Fig. 17.7 illustrates the multi-segment hyperbolic DCA and highlights each segment via a varying line style. Table 17.6 shows the rates for each segment (first 12 months, some portions of the second HYP segment, and some portion of the EXP segment).

Table 17.6 Multisegment DCA Monthly rates

Months	Monthly HYP+EXP (MSCF/month)	Decline segment
1	876,000	First_HYP
2	649,692	First_HYP
3	540,704	First_HYP
4	473,111	First_HYP
5	425,898	First_HYP
6	390,509	First_HYP
7	362,704	First_HYP
8	340,108	First_HYP
9	321,273	First_HYP
10	305,259	First_HYP
11	291,425	First_HYP
12	279,316	First_HYP
13	268,420	Second_HYP
14	258,318	Second_HYP
15	248,983	Second_HYP
16	240,330	Second_HYP
...	...	...
197	36,686	EXP
198	36,529	EXP
199	36,374	EXP
...	...	...
596	6665	EXP
597	6637	EXP
598	6608	EXP
599	6580	EXP
600	6552	EXP

Pressure normalized rate

Pressure normalized rate is another approach that can be used for well comparison as well as calculating EUR. Since some of the wells cannot be produced at full capacity due to various operational or technical reasons, it is important to be able to calculate well productivity using a pressure normalized rate approach to account for curtailment issues caused by operational or technical reasons. Pressure normalized rate requires simply plotting $\frac{q}{\Delta P}$/ft of lateral on the y-axis vs CUM/ft of lateral on the x-axis and the intersection of the decline curve to the x-axis yields the EUR/ft. This technique is also very common for quick analysis such as production performance comparison analysis. The following steps can be done to generate pressure normalized rate plot and run a DCA through the plot to calculate EUR:

(1) From actual gas production data to date, calculate CUM/ft of lateral. This calculation takes the CUM value for each row and divides by the lateral length of the well.

(2) Calculate pressure normalized rate and divide by the lateral length as follows:

$$\text{Pressure normalized rate} = \frac{q}{\Delta P} = \frac{q}{P_i - P_{wf}}$$

where q is the production rate, MSCF/day; P_i the initial reservoir pressure, psi; and P_{wf} the flowing bottom hole pressure, psi.

Take the above pressure normalized rate and divide by the lateral length of a well.

(3) Once PNR/ft is calculated, plot PNR/ft on the y-axis vs CUM/ft on the x-axis.

(4) After calculating PNR/ft and CUM/ft for the actual data, the next step is to calculate the forecasted PNR/ft vs CUM/ft using regular DCA equations (hyperbolic and exponential equations). Please note that instead of having rate (q) in traditional DCA, PNR/ft is used in the PNR methodology. In addition, instead of using time in traditional DCA, CUM/ft is used in PNR methodology. The same infamous hyperbolic and exponential decline equations are still being used except they are applied to PNR/ft vs CUM/ft. Once forecasted, the intersection of x-axis will yield EUR/ft for the well.

(5) The next step is to create a forecasted curve to apply to actual data similar to when performing DCA analysis. Calculate forecasted volume for each row as follows:

$$\text{Forecasted volume} = \text{PNR/ft for each day} \times DD \times LL$$

PNR IP is the same as q_i in DCA, Mscf/psi/ft; DD is the Drawdown, reservoir pressure-line pressure converted to BH condition, psi; and LL is the lateral length, ft.

(6) Next, calculate CUM forecasted rate per ft for each row.

(7) One of the most important considerations in PNR is fitting a D_{min} through the data. This action can be performed by fitting the best-fit line through the actual production data and iterating on D_{min} until a fit through the actual production data is reached. That D_{min} can then be used to switch from hyperbolic decline to exponential decline equations as will be illustrated in the example. The D_{min} percentages used in PNR analysis are much higher as compared to the D_{min} percentages used in traditional DCA analysis. The D_{min} percentages of 40%, 50%, or 60% are not unusual and it all depends on the best-fit line through the data. PNR EUR accuracy works the best when a well is not constrained (curtailed). Six months of un-curtailed production data would provide the best PNR EUR accuracy when using this methodology.

Example

Using the following parameters, generate a PNR curve and calculate hyperbolic, exponential, and total EUR.

PNR IP (q_i in DCA) = 0.0018 MSCF/psi/ft

where $b = 0.8$; effective decline = 84%; D_{min} = 50%; drawdown = 4250 psi; and lateral length = 10,000 ft.

Step 1: Convert the effective decline rate to daily nominal hyperbolic as follows:

$$\text{Daily nominal hyperbolic} = \frac{(1 - \text{effective annual decline})^{-b} - 1}{b \times 365}$$

$$= \frac{(1 - 84\%)^{-0.8} - 1}{0.8 \times 365} == 0.011411$$

Step 2: Calculate HYP forecasted volume for day 1:

$$\begin{aligned}\text{HYP forecasted volume for day 1} &= \text{PNR IP} \times DD \times LL \\ &= 0.0018 \times 4250 \times 10,000 \\ &= 76,500 \text{ MSCF}\end{aligned}$$

Step 3: Calculate HYP CUM/ft for day 1:

$$\text{HYP CUM/ft for day 1} = 0$$

Step 4: Calculate effective decline for day 1:

$$\text{Effective decline for day 1} = 100\%$$

Step 5: Calculate HYP CUM/ft for day 2:

$$\begin{aligned}HYP\ CUM/\text{ft for day 2} &= HYP\ CUM/\text{ft for day 1} \\ + \left(\frac{HYP \text{forecasted volume for day 1}}{LL}\right) &= 0 + \left(\frac{76,500}{10,000}\right) = 7.65 \text{ MSCF/ft}\end{aligned}$$

Step 6: Calculate HYP PNR/ft for day 2:

$$HYP\ PNR/ft\ for\ day\ 2 = \frac{PNR\ IP}{(1 + b \times HYP\ CUM/ft\ for\ day\ 2 \times daily\ nominal\ hyperbolic)^{\frac{1}{b}}}$$

$$= \frac{0.0018}{(1 + 0.8 \times 7.65 \times 0.011411)^{\frac{1}{0.8}}}$$

$$= 0.00165434\ MSCF/psi/ft$$

Step 7: Calculate HYP forecasted volume for day 2:

$$HYP\ forecasted\ volume\ for\ day\ 2 = HYP\ PNR/ft\ for\ day\ 2 \times DD \times LL$$
$$= 0.00165434 \times 4250 \times 10{,}000$$
$$= 70{,}309\ MSCF$$

Step 8: Calculate effective decline % for day 2 as follows:

$$\text{Effective decline \% for day 2} = 1 - e^{\frac{\left(\frac{HYP\ PNR/ft\ for\ day\ 2}{HYP\ PNR/ft\ for\ day\ 1} - 1\right) \times 365}{HYP\ CUM/ft\ for\ day\ 2 - HYP\ CUM/ft\ for\ day\ 1}}$$

$$= 1 - e^{\frac{\left(\frac{0.00165434}{0.0018} - 1\right) \times 365}{7.65 - 0}} = 0.9790\ or\ 97.90\%$$

Step 9: Calculate HYP CUM/ft for day 3:

$$HYP\ CUM/ft\ for\ day\ 3 = 7.65 + \left(\frac{70{,}309}{10{,}000}\right) = 14.68\ MSCF/ft$$

Step 10: Calculate PNR/ft for day 3:

$$HYP\ PNR/ft\ for\ day\ 3 = \frac{PNR\ IP}{(1 + b \times HYP\ CUM/ft\ for\ day\ 3 \times daily\ nominal\ hyperbolic)^{\frac{1}{b}}}$$

$$= \frac{0.0018}{(1 + 0.8 \times 14.68 \times 0.011411)^{\frac{1}{0.8}}}$$

$$= 0.00153813\ MSCF/psi/ft$$

Step 11: Calculate HYP forecasted volume for day 3:

$$HYP\ forecasted\ volume\ for\ day\ 3 = HYP\ PNR/ft\ for\ day\ 3 \times DD \times LL$$
$$= 0.00153813 \times 4250 \times 10{,}000$$
$$= 65{,}371\ MSCF$$

Step 12: Calculate effective decline % for day 3 as follows:

$$\text{Effective decline \% for day 2} = 1 - e^{\dfrac{\left(\dfrac{\text{HYP PNR/ft for day 3}}{\text{HYP PNR/ft for day 2}} - 1\right) \times 365}{\text{HYP CUM/ft for day 3} - \text{HYP CUM/ft for day 2}}}$$

$$= 1 - e^{\dfrac{\left(\dfrac{0.00153813}{0.00165434} - 1\right) \times 365}{14.68 - 7.65}} = 0.9739 \text{ or } 97.39\%$$

These steps can be repeated in excel or any programming language for 50 years or 18,250 days (if the curve were to stay in hyperbolic PNR mode for 50 years which is not realistic). Please note that the D_{min} in this question is given to be 50%. Therefore, once effective decline reaches 50%, the decline equations must be switched to exponential decline equations. Continuing with this example, the switch time when effective decline reaches D_{min} of 50% is 353 days and the PNR/ft at switch time is 0.000191406 MSCF/psi/ft. This PNR/ft will be used as the IP for the exponential equation calculations that will be discussed. The HYP CUM/ft at 353 days is 548.45 MSCF/ft or 0.54845 BCF/1000 ft Table 17.7 illustrates all the discussed steps and calculations for the first 12 days and the last 12 days showing the switch over time (highlighted in table). The PNR/ft at switch over time will then be used for applying to the exponential equations.

Step 13: The PNR/ft at 353 days (switch over time) is 0.000191406 MSCF/psi/ft. Calculate the exponential forecasted volume at day 353 as follows:

$$\text{EXP forecasted volume at day 353} = \text{PNR/ft for day 353} \times DD \times LL$$
$$= 0.000191406 \times 4250 \times 10{,}000$$
$$= 8135 \text{ MSCF}$$

Step 14: Calculate the exponential forecasted CUM/ft at 353 days as follows:

$$\text{EXP forecasted CUM/ft at 353 days} = \dfrac{\text{EXP forecasted volume at day 353}}{LL}$$
$$= \dfrac{8135}{10{,}000} = 0.8135 \text{ MSCF/ft}$$

Step 15: Calculate EXP PNR/ft for day 354 as follows:

$\text{EXP PNR/ft for day 354}$
$= \text{PNR/ft @ switch time} \times e^{-\ln(1-D_{min})\left(\frac{1}{365}\right) \times -\text{EXP forecasted CUM/ft at 353 days}}$
$= 0.000191406 \times e^{-\ln\left((1-50\%)^{\left(\frac{1}{365}\right)}\right) \times -0.8135} = 0.000191111 \text{ MSCF/psi/ft}$

Table 17.7 Hyperbolic PNR decline until switch time

Time (days)	HYP PNR/ft (MSCF/psi/ft)	HYP forecasted volume (MSCF)	HYP CUM/ft (MSCF/ft)	Decline (%)	Decline type
1	0.00180000	76,500	0	100.00	HYPERBOLIC
2	0.00165434	70,309	7.65	97.90	HYPERBOLIC
3	0.00153813	65,371	14.68	97.39	HYPERBOLIC
4	0.00144262	61,311	21.22	96.88	HYPERBOLIC
5	0.00136231	57,898	27.35	96.36	HYPERBOLIC
6	0.00129356	54,976	33.14	95.85	HYPERBOLIC
7	0.00123385	52,439	38.64	95.33	HYPERBOLIC
8	0.00118136	50,208	43.88	94.82	HYPERBOLIC
9	0.00113475	48,227	48.90	94.32	HYPERBOLIC
10	0.00109301	46,453	53.72	93.82	HYPERBOLIC
11	0.00105534	44,852	58.37	93.33	HYPERBOLIC
12	0.00102113	43,398	62.85	92.85	HYPERBOLIC
...	...	...	...	...	...
342	0.00019474	8277	539.42	50.49	HYPERBOLIC
343	0.00019443	8263	540.25	50.44	HYPERBOLIC
344	0.00019412	8250	541.08	50.40	HYPERBOLIC
345	0.00019382	8237	541.90	50.35	HYPERBOLIC
346	0.00019351	8224	542.72	50.31	HYPERBOLIC
347	0.00019321	8211	543.55	50.27	HYPERBOLIC
348	0.00019290	8198	544.37	50.22	HYPERBOLIC
349	0.00019260	8186	545.19	50.18	HYPERBOLIC
350	0.00019230	8173	546.01	50.14	HYPERBOLIC
351	0.00019200	8160	546.82	50.09	HYPERBOLIC
352	0.00019170	8147	547.64	50.05	HYPERBOLIC
353	**0.00019141**	**8135**	**548.45**	**50.01**	**EXPONENTIAL**

Step 16: Calculate the EXP forecasted volume at 354 days:

$$\text{EXP forecasted volume at day 354} = \text{PNR/ft for day 354} \times DD \times LL$$
$$= 0.000191111 \times 4250 \times 10,000$$
$$= 8122 \, \text{MSCF}$$

Step 17: Calculate the exponential forecasted CUM/ft at 354 days as follows:

EXP forecasted CUM/ft at 354 days

$$= \text{EXP forecasted CUM/ft at 353 days} + \frac{\text{Exp forecasted volume at day 354}}{LL}$$

$$= 0.813 + \frac{8122}{10,000} = 1.626 \, \text{MSC F/ft}$$

Step 18: Calculate EXP PNR/ft for day 355 as follows:

$EXP\ PNR/ft$ for day 355

$= PNR/ft$ @ switch time $\times e^{-\ln(1-D_{min})\left(\frac{1}{365}\right)} \times -EXP$ forecasted CUM/ft at 354 days

$= 0.000191406 \times e^{-\ln\left((1-50\%)\left(\frac{1}{365}\right)\right) \times -1.626} = 0.000190816\ MSCF/psi/ft$

Step 19: Calculate the EXP forecasted volume at 355 days:

$$EXP\ \text{forecasted volume at day } 355 = PNR/ft \text{ for day } 355 \times DD \times LL$$
$$= 0.000190816 \times 4250 \times 10{,}000$$
$$= 8110\ MSCF$$

Step 17: Calculate the exponential forecasted CUM/ft at 355 days as follows:

EXP forecasted CUM/ft at 355 days

$$= \text{EXP forecasted CUM/ft at 354 days} + \frac{\text{EXP forecasted volume at day 355}}{LL}$$

$$= 1.626 + \frac{8110}{10{,}000} = 2.437\ MSCF/ft$$

Step 18: Continue on with the process until 18,250 days (50 years) is reached and calculate the EXP EUR. Once EXP EUR is calculated, add to HYP EUR to get the total EUR for the well. The EXP CUM/ft at 18,250 days is calculated to be 1766.80 MSCF/ft or 1.77 BCF/1000′. The total EUR per 1000′ is calculated as follows:

Total EUR = HYP CUM/ft @ 353 days + EXP CUM/ft @ 18,250
= 548.45 + 1766.80 = 2315.2 MSCF/ft or 2.3 BCF/1000′

Table 17.8 illustrates the EXP PNR decline for the first 12 days and the last 12 days of the EXP decline period.

Table 17.8 EXP PNR decline until 18,250 days

Time (days)	EXP PNR/ft, MSCF/psi/ft	EXP Forecasted volume, MSCF	EXP CUM/ft, MSCF/ft	Decline type
353	0.000191406	8135	0.81	EXP
354	0.000191111	8122	1.63	EXP
355	0.000190816	8110	2.44	EXP
356	0.000190522	8097	3.25	EXP

Continued

Table 17.8 EXP PNR decline until 18,250 days—cont'd

Time (days)	EXP PNR/ft, MSCF/psi/ft	EXP Forecasted volume, MSCF	EXP CUM/ft, MSCF/ft	Decline type
357	0.000190230	8085	4.05	EXP
358	0.000189938	8072	4.86	EXP
359	0.000189647	8060	5.67	EXP
360	0.000189357	8048	6.47	EXP
361	0.000189068	8035	7.28	EXP
362	0.000188779	8023	8.08	EXP
363	0.000188492	8011	8.88	EXP
364	0.000188205	7999	9.68	EXP
...	...	...	...	...
18,239	0.000006685	284	1766.49	EXP
18,240	0.000006684	284	1766.52	EXP
18,241	0.000006684	284	1766.55	EXP
18,242	0.000006684	284	1766.57	EXP
18,243	0.000006683	284	1766.60	EXP
18,244	0.000006683	284	1766.63	EXP
18,245	0.000006683	284	1766.66	EXP
18,246	0.000006682	284	1766.69	EXP
18,247	0.000006682	284	1766.72	EXP
18,248	0.000006682	284	1766.74	EXP
18,249	0.000006681	284	1766.77	EXP
18,250	0.000006681	284	1766.80	EXP

CHAPTER EIGHTEEN

Economic evaluation

Introduction

There are three commonly used models to determine profit. Each model has a unique way of defining cost. These models are as follows:
- net cash-flow (NCF) model
- financial model
- tax model

Net cash-flow model

From the above three models, the NCF model is the most commonly used model in the oil and gas industry due to the fact that it accounts for time value of money, which is discussed later in this chapter. This model is typically used in any oil and gas property evaluation to calculate profit and net present value (NPV) and other important capital budgeting and financial parameters as needed. One unique feature of the NCF model is the time zero. Time zero refers to the day when the first investment is made. For example, if Company ABC decides to invest roughly $10 million for exploration and development of one well in the Bakken Shale (located in North Dakota), $10 million will be inputted in time zero. Time zero is the point at which the future profits are discounted. If long-term economic analysis is being performed in which wells will be turned in line (TIL) many years from today (e.g., 4 years), all of the future cash flows are typically discounted back to today's dollar to get a comprehensive understanding on the value of an asset today. There are two important concepts in the NCF model. The first one is referred to as *cash outflow*, which is essentially the cash spent (i.e., money coming out of the business) on a project. Examples of cash outflows are investment, operating costs, and income taxes. Companies invest in a particular project to recover the original investment and additionally make a profit on top of what was originally invested. The second important

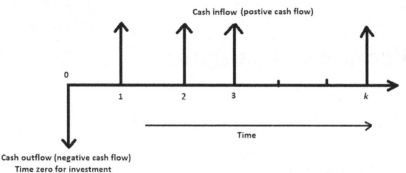

Fig. 18.1 Net cash-flow (NCF) model.

concept in the NCF model is *cash inflow*, which is basically the amount of cash that the company generates from the project. An example of cash inflow is revenue. Fig. 18.1 shows the flow chart of the NCF concept.

Royalty

Royalty is the amount of money paid to the landowners who own the mineral rights. In the oil and gas industry, the first payment is typically the amount of money paid to the landowner per acre to lease the acreage. For example, if a landowner owns 5000 acres in which an oil or gas company is interested (due to the potential or proven reserves), the oil and gas company will end up paying a certain amount of money per acre to the landowner to be able to drill and complete on a particular property. This amount varies from state to state and can be as low as $500/acre to as high as $15,000/acre depending on the formation potential and rate of return of the project. For example, if Mr. Hoss Belyadi owns 5000 acres in Pennsylvania and an operator is interested in leasing his acreage for $2500/acre, Mr. Belyadi will get a big check for $12,500,000 for allowing the company to drill and complete on his property since he owns the mineral rights. Many people became millionaires overnight after signing an agreement with an oil and gas company at the beginning of the shale boom. Aside from getting a big lump sum of money initially, Mr. Belyadi will also get a percentage of the produced hydrocarbon's profit from each well called a *royalty* as soon as the well starts producing hydrocarbon. This percentage varies from contract to contract again and can be anywhere between 12.5% and 20%.

This can add up to a lot of money every year, and many landowners have become rich by earning money after the shale boom. Royalty percentage and other conditions and circumstances of the lease contract can be discussed and agreed upon between the landowner and the operating company when signing the lease and the contract. Some companies' strategy is to buy the land instead of going through the hassle of leasing various properties and renewing leases once expired. Leasing property from mineral-rights owners is more common. One of the biggest disadvantages of leasing a property is that operating companies have a limited period of time to begin operations, which is usually 5 years (can be extended). If the oil and gas market stays healthy by sustaining a profitable oil and gas price, companies would be able to make decisions more easily on the number of wells to be drilled and completed each year. However, when the price of oil or natural gas fluctuates to uneconomic prices, companies that do not have their gas hedged at a certain price will have a hard time staying focused on drilling and completing wells in undeveloped areas where leasing has been signed. Therefore, it is very important for companies to be ready and have strategic plans in place before leasing a property. Companies need to consider all kinds of development plans. If a particular strategic plan does not exist within a particular operating company, renewing expiring leases or losing leases will cost the company millions of dollars each year. Some companies do not have to go through the hassle of leasing and paying royalties to the landowners to a big extent due to the fact that the land and mineral rights have been previously owned from prior activities (coal mining, etc.) in a particular area.

Working interest

Working interest (WI) is fundamentally a percentage of ownership in an oil and gas lease or property that gives the owner of the interest the right to drill, complete, and produce oil and gas on the leased acreage. For example, if an operating company XYZ has an 80% WI in a particular property, this means that this particular company is obligated to pay 80% of any investment and costs incurred. These investments and costs are included but not limited to cost of acquisition, exploration, drilling, completion, operating costs, and so forth. Generally, there are two main considerations in which an operating company elects to obtain the WI percentage. The first consideration is the amount of capital that a company has. If a company has a small amount of capital (i.e., private owner or sometimes family owner), it is very important to only own a small percentage. This is due to the fact that large amounts of capital are required to drill and complete wells. Owning a high percentage of WI means a lot of capital will be needed to pay for all of the previously discussed investments and costs. There are small family business owners in Marcellus and other shale plays that own as low as 1% or less WI. Large and medium developed exploration and production (E&P) companies typically own larger WI percentages but can have small WI in various noncore acreage positions. Companies who have 50% or more WI are normally in charge of the operation.

The second important consideration is the risk and confidence level associated with a project that will determine the WI percentage. For example, let us assume Mr. Hoss Belyadi owns an $8 billion company and he is trying to invest in a particular prospect for the drilling program. The only issue and drawback that he is facing is not being confident in the outcome of that particular prospect upon completion. To split the risk associated with that project, Mr. Belyadi has the option of finding a business partner who is willing to buy 50% of the WI of that prospect under a joint operating agreement (JOA). By doing this, the risk associated with that project will be 50% in the event that particular project does not meet the expectation. This is called a risk mitigation practice and is common among some operators that do not have the confidence level to invest big lump sums of money into a particular project or simply do not have the capital and would like to develop the acreage position faster than their capital budget would allow. There are various methods to obtain third-party acreage and the most common ones are as follows:

- **Exchange**: Exchange is performed by trading leases of equal values, which are typically located in a similar area with similar number of acres. Exchange is recommended when disposing leases that a company does not have any plans to develop. Section 1031 of the IRS tax code of 1986 and treasury regulations permits investors (e.g., companies) to postpone capital gains taxes on any exchange of like-kind properties, which are properties of the same nature, class, or character.
- **Assignment**: Assignment is an agreement assigning leasehold for a certain period of time. In an assignment, the owner leases to operator ABC. Operator ABC assigns all or a portion of those rights to another operator. Assignment is common when a small chunk of acreage in which an operator is trying to develop (within the operator's development plan) is owned or leased by another operator. In this scenario, the operator who has the lease or ownership in that chunk of acreage will have to run economic analysis to determine whether to participate in the well (through a JOA) or simply not participate and assign the acreage in return for a sign-on bonus, overriding royalty interest, well data, etc. (depending on the JOA). Assignments typically include upfront bonus

per acre, overriding royalty interest (ORRI), well data, and site visits. Overriding royalty interest is a royalty in excess of the royalty provided in an oil and gas lease and is usually added through an assignment of the lease.
- **Participation/Joint operating agreement (JOA)**: In a JOA, a third party retains ownership of their leasehold and becomes part WI owner in wells. In addition, the third party will have to pay its proportionate share of well costs (depending on the WI%). Economic analysis and other risk factors associated with participating in a well will determine whether to participate in a well or not. For instance, if company X decides to participate in 200 acres outside of their normal development area, it is extremely important to consider whether the operating company that will be drilling and completing the acreage position has extensive prior experience in the basin. This is to ensure that the amount of capital that will be spent is within a ballpark estimate when Authorization for Expenditure (AFE) is submitted. Let's assume the operating company planning to drill and complete the well lacks experience in that particular basin and sends the total AFE in the amount of $10 MM to the participating company for approval. Let's also assume that the participating company has a WI of 5% in the 200 acre position, meaning the participating company will be responsible for paying $500K of the total $10 MM investment. What if there is low confidence in the operating company due to lack experience in the area? What if the operating company overspends by $5 MM, totaling $15 MM? Now the upfront investment has increased from $500K to $750K. This example demonstrates some of the risks associated with participating in a well with low confidence in the operating company.

Net revenue interest

Net revenue interest (NRI) is the percentage of production that is actually received after all the burdens, such as royalty and overriding royalty, have been deducted from the WI. For example, if a company's WI is 100% but it has agreed to pay 18.5% royalty interest to the landowner, NRI will be a smaller percentage than 100% (in this example 81.5%) since this is the money that the company actually receives after paying off the royalty to the landowner. NRI percentage can be calculated using Eq. (18.1).

$$\text{NRI} = \text{WI} - (\text{WI} \times \text{RI}) \tag{18.1}$$

where WI is the working interest, % and NRI is net revenue interest, %.

Example

Calculate the NRI for the following prospects given the WI and royalty percentages located in Table 18.1.

Table 18.1 Net revenue interest (NRI) example

Prospect	WI (%)	Royalty (%)
A	80	18.5
B	100	12.5
C	76	15.0

Prospect A NRI% = [80% − (80% × 18.5%)] × 100 = 65.2%
Prospect B NRI% = [100% − (100% × 12.5%)] × 100 = 87.5%
Prospect C NRI% = [76% − (76% × 15%)] × 100 = 64.6%

Every company's NRI will vary depending on the royalty percentage agreement, and this has to be taken into account for accurate economic analysis calculations.

British thermal unit content

Another important concept to consider in the oil and gas economic evaluation is BTU content. As previously discussed, BTU stands for British thermal unit and is defined in every textbook as the amount of energy needed to cool or heat 1 pound of water by 1°F. The higher the BTU content, the hotter it will burn. Since gas is sold per MMBTU, it needs to be converted into proper units when performing economic analysis calculation. BTU content can be obtained from the gas composition analysis as shown in Chapter 1. BTU factor is simply BTU divided by 1000. For example, if the BTU of a dry gas well is 1040 (from gas composition analysis), the BTU factor is 1.04. The current and forecasted gas pricing must be adjusted for BTU content by using Eq. (18.2).

$$\text{Adjusted gas price} = \text{Gas price} \times \text{BTU factor} \quad (18.2)$$

where adjusted gas price is $/MSCF and gas price is $/MMBTU.

Shrinkage factor

Shrinkage factor in dry gas areas is typically low (0.5%–3%). Shrinkage in dry gas areas refers to the volumes lost due to possible line losses or field usage (e.g., fueling the compressor station). Since compressor stations are sometimes fueled by natural gas in some areas, a small percentage of produced volume from each well will be used to fuel the compressor station. As a result, a very small percentage of shrinkage must be considered when performing economic analysis calculations in dry gas areas.

Shrinkage factor becomes more important in wet gas areas with much higher BTU. Shrinkage factor is used to convert the produced wet gas into dry sales gas. Hydrocarbons exit the wellhead and hit the separator. Condensate/oil and water exit the bottom of the separator and are metered. Wet gas/dry gas exits the top of the separator and is metered, and that is usually the wellhead volume reported to the state and used for reserve forecasting. At this point, shrinkage can come into play, but it depends on the situation. For example, if wet gas is sold to the market (assuming no processing required), shrinkage factor will only be line losses and field usages. Let us assume the line losses and field usages add up to be roughly 5% of the total volume. The wet gas in this case is reduced by 5% due to the losses. Therefore, a shrinkage factor of 95% must be taken into account when performing the calculations. On the other hand, if gas is processed at a plant and both residue dry gas and natural gas liquid (NGL) revenue are obtained (based on what comes out of the exit of the plant), a much larger shrinkage factor is applied, which will contain field usage, processing shrink, liquid shrinkage, line losses, and so forth. Therefore, it is very important to consider the shrinkage factor when performing economic analysis on any well specifically in wet gas regions where the gas will be processed. Total shrinkage factor for wet gas areas can be calculated using Eq. (18.3). Liquid shrinkage can be calculated based on the inlet gas composition along with the plant removal percentage (varies from plant to plant) for each gas component.

$$\text{Total shrinkage factor}: \begin{aligned} S_T &= [(1 - \text{field usages})(1 - \text{liquid shrinkage})] \\ &\times (1 - \text{processing plant shrinkage})] \times 100 \end{aligned}$$

(18.3)

where S_T is the total shrinkage factor, %; liquid shrinkage is in %; field usages is in %; and processing plant shrinkage is in %.

Example

Calculate the liquid shrinkage assuming the following gas composition and plant removal percentage.

As can be seen from the table below, liquid shrinkage can be calculated as follows:

$$100\% - 88.47\% = 11.53\%$$

	Known		Calculated
Gas component	Inlet gas composition (%)	Plant removal (%)	(1 − plant removal %) × inlet gas composition (%)
Methane (C_1)	77.9731	0.00	77.973
Ethane (C_2)	14.6177	35.00	9.502
Propane (C_3)	4.7239	90.00	0.472
i-Butane (i-C_4)	0.4634	98.00	0.009
n-Butane (n-C_4)	1.0839	99.00	0.011
i-Pentane (i-C_5)	0.2671	99.90	0.000
n-Pentane (n-C_5)	0.1496	99.90	0.000
Hexane$^+$	0.2225	99.90	0.000
Nitrogen (N_2)	0.4379	0.00	0.438
Carbon dioxide (CO_2)	0.0609	0.00	0.061
Sum	100.00		88.47

Operating expense

Operating expense (Opex) or cost is the ongoing cost for running a business. Unfortunately, it is a common mistake among the general public, who thinks once the oil or gas well is produced, there are no more operating costs associated with it. This is a wrong assumption because there are many different operating costs associated with producing a BBL of oil or an MSCF of gas. Some of the most important operating costs associated with operating a well are as follows:

- **Lifting cost**: Lifting cost is the cost of lifting oil or gas out of the ground and bringing it to the surface. Lifting cost typically includes labor cost, cost of supervision, supplies, cost of operating the pumps, electricity, and general maintenance/repairs on the wellhead and surface production equipment. One major part of the lifting cost is the cost of labor and supervision or well-tending cost. Well tenders are the contractors that the operating company hires to go on different well sites, often on a daily basis to perform routine maintenance and make sure the well and surface equipment installed on the well are functioning properly. Well tenders could also be hired on a full-time basis by the operator. The lifting cost is often divided into two categories. The first lifting cost is referred to as the variable lifting cost, which is a function of producing one BBL or MSCF of oil, NGL, condensate, or gas. The second category of the lifting cost is called the fixed lifting cost, which is not a function of the amount of hydrocarbon produced but is a fixed monthly cost associated with the well. It is up to the operating company to classify which costs fall under fixed or variable lifting costs. Every company's categorization can be different. For example, fixed and variable lifting costs on a producing dry gas well can be $650 per month per well and $0.26/MSCF in sequence. If the entire cost is assigned as only fixed lifting cost per month per well, the economics of the well will end prematurely (which is considered to be very conservative). If the entire cost is assigned as only variable lifting cost, the Opex will be too high initially. Reserve auditors do not typically like all costs to be variable lifting since it would be very optimistic from a reserve perspective as there are fixed lifting costs associated with low-producing wells (e.g., 30 MSCF/day) and the reserve life could be prolonged. Therefore, it is crucial to have a combination of both fixed and variable costs to create a balance in the Opex. The categories that fall under the fixed lifting cost do not depend on the production volume over time. For example, snow removal and vegetation control are considered as fixed lifting costs because no matter how much a well produces, snow must be removed from the access road and site to perform routine maintenance on the well. On the other hand, well tending is considered a variable cost because this cost is typically a function of the amount of hydrocarbon produced from a well.

- **Gathering and compression cost (G&C)**: G&C is the cost of gathering gas from the sales line located on every well site and sending it to the compressor station to compress the gas before sending it to the market. In almost every gas field in this country and other countries across the world, compression is an essential operation. Gathering is typically the cost of gathering the gas from the well site to the compressor station, and compression is the cost of compressing the gas per MSCF at the compressor station. Compression is used to increase natural gas pressure to successfully meet various markets' pipeline pressure before injecting gas into the transmission pipeline. For example, if the pipeline in which the gas is being sent has a pressure of 1000 psi, the compressed gas has to be over 1000 psi to send the gas to the transmission pipeline and consequently consumers. Gas always moves from high pressure to low pressure. For the gas to be sent into transmission pipelines, there are minimum requirements such as pressure and vapor percentage that must be met. Examples of G&C costs are leased compression equipment, dehydration, repairs and maintenance of electrical flow meters, etc. The unit for G&C cost is a function of the amount of gas produced and it is in $/MMBTU or $/MSCF. G&C costs also have a small fixed portion associated with each that is considered a fixed cost regardless of the amount of gas compressed.

- **Processing cost**: Processing cost is the cost of processing oil, condensate, wet gas, etc. into more useful products. The petroleum that comes out of the ground has to be processed to obtain products such as gasoline, diesel, heating oil, kerosene, and so forth. Just like all of the other costs discussed, there is a fee associated with processing petroleum. This cost is typically considered in $/BBL for oil fields and $/MMBTU for wet gas and retrograde condensate fields. Processing cost does not apply to fields that only produce dry gas unless the gas has a high percentage of H_2S (hydrogen sulfide).
- **Firm transportation (FT) cost**: Firm transportation (FT) cost is the cost of transporting natural gas from the compressor station to the consumers. FT cost depends on many factors such as the pipeline that the gas flows into and the contract associated with the FT purchased. This cost is typically in $/MMBTU or $/MSCF.
- **General and administrative (G&A) cost**: G&A cost is basically the cost of running the company such as office expenses, employee salaries, professional fees, personal costs, etc. This cost is typically in $/MSCF or $/BBL. This cost is typically not included when performing economic analysis because it is considered to be a sunk cost.
- **Water disposal cost**: Another important cost is the water disposal cost per BBL of produced water. Once a well is TIL, it will most likely produce water for the rest of the life of the well. The wells that have been hydraulically fractured produce more water initially and the water production decreases with time afterward. For example, on average unconventional shale reservoirs are known to typically produce 10%–30% of the total injected fluid throughout the life of the well, depending on many factors such as the amount of water injected, water saturation, target depth, etc. This produced water can be reused on a different frac job or must be disposed of. Water disposal costs the operating companies a lot of money. Water disposal cost is typically in $/BBL. Many operating companies that have continuous operation in a particular basin mix the produced water with freshwater and pump the mix back downhole

on the next hydraulic frac job instead of spending lots of money for disposal. This technique works when there are continuous frac operations in a particular area; otherwise, lots of money must be spent on water disposal. Another alternative when not having a continuous frac operation is to sell or give away the water to a nearby operating company not because of charity, but because it is sometimes cheaper to give away the water than spend more money on the disposal of the water.

Total Opex per month

Total operating costs (Opex) for a dry gas or wet gas well can be calculated. The equation below assumes that the operating company will be responsible for paying all of the operating costs since every Opex is being multiplied by WI%. Depending on the lease and contract, operating companies might be able to deduct some of the operating costs from the landowner (postproduction deduction). If so, WI must be replaced by NRI.

Total Opex per month
$$\begin{aligned} = & [(\text{Gross monthly gas production} \times \text{WI} \times \text{total shrinkage factor} \\ & \times \text{variable lifting cost})] + [(\text{Fixed lifting cost} \times \text{WI})] \\ & + [(\text{Gross monthly gas production} \times \text{WI} \times \text{total shrinkage factor} \\ & \times \text{gathering and compression cost})] \\ & + [(\text{Gross monthly gas production} \times \text{WI} \times \text{total shrinkage factor} \\ & \times \text{processing cost})] \\ & + [(\text{Gross monthly gas production} \times \text{WI} \times \text{total shrinkage factor} \\ & \times \text{FT cost})] + [(\text{Gross monthly NGL production} \times \text{WI} \times \text{NGL OP cost})] \\ & + [(\text{Gross monthly CND production} \times \text{WI} \times \text{CND OP cost})] \\ & + [(\text{Gross monthly water production} \times \text{WI} \times \text{water disposal cost})] \end{aligned}$$
(18.4)

Gross monthly gas production = MSCF per month from decline curve analysis (DCA) or other analyses
WI = working interest, %
Total shrinkage factor = %
Variable lifting cost = $/MSCF
Fixed lifting cost = $/month/well

G&C cost = $/MSCF, must be grossed up and adjusted for applicable shrinkages

FT cost = firm transportation cost, $/MSCF

Processing cost = $/MSCF, must be grossed up and adjusted for applicable shrinkages

Gross monthly NGL production = monthly volumes of NGL in BBLs

NGL OP cost = NGL operating cost, $/BBL

Gross monthly CND production = monthly volumes of condensate in BBLs

CND OP cost = condensate operating cost, $/BBL

Gross monthly water production = monthly volumes of water in BBLs

Water disposal cost = water operating cost, $/BBL

Please note that since G&C and processing costs are being multiplied by total shrinkage, it is very important to gross up the costs based on applicable shrinkages applied to each category.

Example

Calculate total Opex for the first 3 months assuming the production volumes located in Table 18.2 and the following operating costs listed below.

Table 18.2 Gas, CND, and NGL production volumes

Time Month	Gross gas production MSCF	Gross CND production BBL	Gross NGL production BBL
1	350,000	950	19,250
2	330,000	800	18,150
3	300,000	500	16,500

WI = 100%, Inlet BTU = 1240 (Inlet of the plant, BTU factor of 1.240), Outlet BTU = 1100 (Residue gas BTU coming out of the processing plant, BTU factor of 1.1), Compressor burn shrinkage = 1.5%, Liquid shrinkage = 7 %, Plant shrinkage = 0.5%, Variable lifting cost = $0.23/MSCF, Fixed lifting cost = $1600/month/well, Gathering and compression cost = $0.30/MMBTU, Processing cost = $0.28/MMBTU, FT cost = $0.25/MMBTU, NGL fractionation and transportation cost = $7/BBL, CND transportation cost = $11/BBL. Assume water disposal cost of $0/BBL since water will be used on an adjacent frac for the first 3 months.

Step 1: Convert all of the units on operating costs to $/MSCF from $/MMBTU and adjust for shrinkage:

G&C cost is provided in $/MMBTU, therefore it must be converted to $/MSCF by multiplying it by the inlet BTU factor of 1.240 and grossed up since the equation discussed multiplies the G&C cost by the total shrinkage.

Economic evaluation

$$\text{Gathering and compression cost} = \frac{0.30 \times 1.240}{(1-1.5\%)(1-7\%)(1-0.5\%)}$$
$$= \$0.408/\text{MSCF}$$

Processing cost is also provided in $/MMBTU and must be converted to $/MSCF by multiplying it by the inlet plant BTU factor of 1.240 and grossed up for liquid and processing shrinkages.

$$\text{Processing cost} = \frac{0.28 \times 1.240}{(1-7\%)(1-0.5\%)} = \$0.375/\text{MSCF}$$

FT cost is also provided in $/MMBTU and since the residue gas coming out of the processing plant will be sold, the FT cost provided in $/MMBTU must be multiplied by the outlet BTU factor of 1.1.

$$\text{FT cost} = 0.25 \times 1.1 = \$0.275$$

Step 2: Calculate total shrinkage factor:

$$S_T = (1-1.5\%) \times (1-7\%) \times (1-0.5\%) = 91.1\%$$

Step 3: Calculate total Opex for each month using Eq. (18.4).

Total Opex, month 1
$$= [(350,000 \times 100\% \times 91.1\% \times 0.23)]$$
$$+ [(1600 \times 100\%)] + [(350,000 \times 100\% \times 91.1\% \times 0.408)]$$
$$+ [(350,000 \times 100\% \times 91.1\% \times 0.375)]$$
$$+ [(350,000 \times 100\% \times 91.1\% \times 0.275)]$$
$$+ [(19,250 \times 100\% \times 7)]$$
$$+ [(950 \times 100\% \times 11)] = \$557,479$$

Total Opex, month 2
$$= [(330,000 \times 100\% \times 91.1\% \times 0.23)]$$
$$+ [(1600 \times 100\%)] + [(330,000 \times 100\% \times 91.1 \times 0.408)]$$
$$+ [(330,000 \times 100\% \times 91.1\% \times 0.375)]$$
$$+ [(330,000 \times 100\% \times 91.1\% \times 0.275)]$$
$$+ [(18,150 \times 100\% \times 7)] + [(800 \times 100\% \times 11)] = \$524,661$$

Total OPEX, month 3
$$= [(300,000 \times 100\% \times 91.1\% \times 0.23)]$$
$$+ [(1600 \times 100\%)] + [(300,000 \times 100\% \times 91.1 \times 0.408)]$$
$$+ [(300,000 \times 100\% \times 91.1\% \times 0.375)]$$
$$+ [(300,000 \times 100\% \times 91.1\% \times 0.275)]$$
$$+ [(16,500 \times 100\% \times 7)] + [(500 \times 100\% \times 11)] = \$474,610$$

Severance tax

Severance tax is a production tax imposed on operating companies or anyone with a working or royalty interest in certain states. This tax is essentially applied for the removal of nonrenewable resources such as oil, natural gas, condensate, and so forth. The % of severance tax depends on the state. For example, the severance tax in West Virginia is currently 5%, while some states such as Pennsylvania do not have a severance tax yet (Pennsylvania only pays impact fees); however, there is a possibility of such taxes being imposed to the industry in the future. It is very important to deduct the severance tax from the revenue when performing economic analysis calculation using Eq. (18.5).

Severance tax per month
= (Gross monthly gas production × adjusted gas pricing × severance tax %
× NRI × total shrinkage factor)
+ (Gross monthly NGL production × NGL pricing × severane tax % × NRI)
+ (Gross monthly CND production × CND pricing × severance tax % × NRI)

(18.5)

Gross monthly gas production = MSCF per month from DCA or other analyses
Adjusted gas pricing = $/MSCF, gas price must be adjusted for BTU of the sold gas
NRI = net revenue interest, %
Total shrinkage factor = %
Severance tax = %
Gross monthly NGL production = monthly volumes of NGL in BBLs
NGL pricing = sold NGL pricing, $/BBL
Gross monthly CND production = monthly volumes of CND in BBLs
CND pricing = sold CND pricing, $/BBL.

Example

Calculate severance tax for the first month for a dry gas well using the assumptions listed below.

Gross gas production for month 1 = 250,000 MSCF, Gas pricing = $3.5/MMBTU, Severance tax = 5%, BTU = 1070 (no NGL or CND expected since the gas is dry), Total shrinkage factor = 0.98 (2% shrinkage), WI = 80%, RI = 15%

Step 1: Since pricing is provided in MMBTU, calculate adjusted gas pricing at 1070 BTU (1.07 BTU factor)

$$\text{Adjusted gas pricing} = 3.5 \times 1.070 = \$3.745/\text{MSCF}$$

Step 2: Calculate NRI:

$$\text{NRI\%} = [80\% - (80\% \times 15\%)] \times 100 = 68\%$$

Step 3: Use Eq. (18.5) to calculate severance tax for the first month:

$$\text{Severance tax for the first month} = (250,000 \times 3.745 \times 5\% \times 68\% \times 0.98)$$
$$= \$31,196$$

Ad valorem tax

Ad valorem is a Latin phrase meaning *according to value*. This is another form of tax paid when minerals are produced. West Virginia and Texas are examples of states in which ad valorem tax must be paid annually. There are other types of taxes that must be paid in the oil and gas industry in addition to federal income taxes (depending on the state). For example, in Pennsylvania, there is no severance or ad valorem tax (as of the publication date of this book). Instead, there is an impact fee that must be paid. This does not mean that severance or other forms of taxes will not be imposed in the future. As a matter of fact, depending on the person in office for that particular state, such taxes can be added. Ad valorem tax can be calculated using Eq. (18.6).

Ad valorem tax per month
$$= \{[(\text{Gross monthly gas production} \times \text{adjusted gas pricing} \times \text{NRI} \\ \times \text{total shrinkage factor}) \\ + (\text{Gross monthly NGL production} \times \text{NGL pricing} \times \text{NRI}) \\ + (\text{Gross monthly CND production} \times \text{CND pricing} \times \text{NRI})] \\ - \text{Severance tax amount}\} \times \text{Ad valorem tax \%} \qquad (18.6)$$

Gross monthly gas production = MSCF/month from DCA or other analyses

Adjusted gas pricing = $/MSCF, gas price must be adjusted for BTU of the sold gas
NRI = net revenue interest, %
Total shrinkage factor = %
Ad valorem tax = %
Severance tax amount = $/month
Gross monthly NGL production = monthly volumes of NGL in BBLs
NGL pricing = sold NGL pricing, $/BBL
Gross monthly CND production = monthly volumes of CND in BBLs
CND pricing = sold CND pricing, $/BBL.

Example

Calculate ad valorem tax for the first month from a dry gas well using the assumptions listed below.

Gross gas production for month 1 = 300,000 MSCF, Severance tax for the first month = $35,000, Adjusted gas pricing = $2.5/MSCF, Ad valorem tax = 2.5%, Total shrinkage factor = 0.98 (2% shrinkage), NRI = 42%

$$\text{Ad valorem tax}_{\text{month 1}} = [(300{,}000 \times 2.5 \times 42\% \times 0.98) - (35{,}000)] \\ \times 2.5\% \\ = \$6843$$

Net Opex

Net Opex is referred to the total operating costs including severance and ad valorem taxes and can be simply calculated using Eq. (18.7). Production taxes such as severance and ad valorem taxes are taken into account before federal income tax calculations. Although severance and ad valorem taxes are referred to as taxes, these taxes are deducted as production taxes before federal income tax calculations.

Net Opex = Total Opex + severance tax amount + ad valorem tax amount
Net Opex = $/month
Total Opex = $/month
Severance tax amount = $/month
Ad valorem tax amount = $/month
(18.7)

Example

Calculate net Opex for the first 3 months assuming the Opex and production taxes located in Table 18.3.

Table 18.3 Net Opex example

Month	1	2	3
Total Opex ($)	501,564	455,520	401,365
Severance tax ($)	40,250	35,650	30,000
Ad valorem tax ($)	9000	8560	8250

Net Opex$_{\text{month 1}}$ = 501,564 + 40,250 + 9000 = \$550,814
Net Opex$_{\text{month 2}}$ = 455,520 + 35,650 + 8560 = \$499,730
Net Opex$_{\text{month 3}}$ = 401,365 + 30,000 + 8250 = \$439,615

Revenue

In the oil and gas industry, revenue is the amount of money received from normal business activities, services, and products such as selling hydrocarbon, providing various services to the operating companies, or any other activities that generate money. For example, a big portion of an operating company's revenue comes from selling hydrocarbon. It is very important to avoid confusing revenue with profit because revenue is just the gross money that the company earned and does not take the expenses associated with a project into account. A company's gross revenue can be enormous but the profit might actually be negative because of high amounts of expenses associated with performing that project or any other reasons. For natural gas-producing wells, monthly net gas/NGL/CND production can be calculated using Eqs. (18.8)–(18.10).

$$\begin{aligned}\text{Monthly shrunk net gas production} \\ = \text{Monthly unshrunk gross gas production} \\ \times \text{total shrinkage factor} \times \text{NRI}\end{aligned} \quad (18.8)$$

Monthly shrunk net gas production = MSCF/month
Monthly unshrunk gross gas production = MSCF/month, wellhead volumes

Total shrinkage factor = %
NRI = net revenue interest, %

Monthly shrunk net NGL production
= Monthly shrunk gross NGL production × NRI $\qquad$ (18.9)

Monthly shrunk net NGL production = BBLs/month, sold volumes
Monthly shrunk gross NGL production = BBLs/month, sold volumes
NRI = net revenue interest, %

Monthly shrunk net CND production
= Monthly shrunk gross CND production × NRI $\qquad$ (18.10)

Monthly shrunk net CND production = BBLs/month, sold volumes
Monthly shrunk gross CND production = BBLs/month, sold volumes
NRI = net revenue interest, %.

After calculating monthly net gas, NGL, and CND volumes, net revenue can be simply calculated using Eq. (18.11).

Net revenue = (Monthly shrunk net gas production × adjusted gas pricing)
+ (Monthly shrunk net NGL production × NGL sales pricing)
+ (Monthly shrunk net CND production × CND sales pricing)

$\qquad$ (18.11)

Monthly shrunk net gas production = MSCF/month, residue gas
Adjusted gas pricing = $/MSCF
Monthly shrunk net NGL production = BBL/month
NGL sales pricing = $/BBL
Monthly shrunk net CND production = BBL/month
CND sales pricing = $/BBL.

Example

Calculate net revenue from a retrograde condensate well using Table 18.4 for the first 3 months assuming 80% NRI and total shrinkage factor of 90%.

Table 18.4 Net revenue example

Time	Gross volumes			Sales pricing		
Month	Unshrunk gas MSCF	Shrunk NGL BBL	Shrunk CND BBL	Adjusted gas $/MSCF	NGL $/BBL	CND $/BBL
1	450,000	22,500	950	3.5	35	55
2	435,500	21,775	750	3.4	30	56
3	395,400	19,770	720	3.6	33	53

Monthly shrunk net gas production$_{\text{month 1}} = 450,000 \times 80\% \times 90\% = 324,000$ MSCF

Monthly shrunk net gas production$_{\text{month 2}} = 435,500 \times 80\% \times 90\% = 313,560$ MSCF

Monthly shrunk net gas production$_{\text{month 3}} = 395,400 \times 80\% \times 90\% = 284,688$ MSCF

Monthly shrunk net NGL production$_{\text{month 1}} = 22,500 \times 80\% = 18,000$ BBLs

Monthly shrunk net NGL production$_{\text{month 2}} = 21,775 \times 80\% = 17,420$ BBLs

Monthly shrunk net NGL production$_{\text{month 3}} = 19,770 \times 80\% = 15,816$ BBLs

Monthly shrunk net CND production$_{\text{month 1}} = 950 \times 80\% = 760$ BBLs

Monthly shrunk net CND production$_{\text{month 2}} = 750 \times 80\% = 600$ BBLs

Monthly shrunk net CND production$_{\text{month 3}} = 720 \times 80\% = 576$ BBLs

$$\text{Net revenue}_{\text{month 1}} = (324,000 \times 3.5) + (18,000 \times 35) + (760 \times 55)$$
$$= \$1,805,800$$

$$\text{Net revenue}_{\text{month 2}} = (313,560 \times 3.4) + (17,420 \times 30) + (600 \times 56)$$
$$= \$1,622,304$$

$$\text{Net revenue}_{\text{month 3}} = (284,688 \times 3.6) + (15,816 \times 33) + (576 \times 53)$$
$$= \$1,577,333$$

New York Mercantile Exchange (NYMEX)

NYMEX stands for New York Mercantile Exchange and is essentially a commodity exchange located in New York. Trading is conducted into two divisions. The first division is the NYMEX division, which is home to the energy (oil and natural gas), platinum, and palladium markets. The second division is called the COMEX (commodity exchange) division where metals such as gold, copper, and silver are traded (Investopedia).

NYMEX is used among some operating companies to estimate the future price of natural gas for the purpose of economic analysis evaluation. Many companies have developed their own pricing forecast model based on supply and demand and other various factors. On the other hand, some companies prefer to use flat pricing and perform sensitivity analysis instead of using the NYMEX forecast (strip forecast). If NYMEX is used, NYMEX must be corrected for the basis. The basis can have substantial impact on a project's economics.

Henry Hub and basis price

Henry Hub is a natural gas pipeline in Louisiana where onshore and offshore pipelines meet and is the most important natural gas hub in North America. Henry Hub is the pricing point for natural gas futures on NYMEX. The settlement prices at the Henry Hub are used as benchmarks for the entire North American natural gas market. It is very important to understand that when NYMEX is used to estimate the monthly price of natural gas, it is based on the delivery to the Henry Hub. For example, if the price of natural gas at NYMEX is $4/MMBTU in March 2017, this price must be adjusted to represent the price at Henry Hub. The difference between the Henry Hub natural gas price and natural gas price at a specific location is called the *basis differential*. For example, the NYMEX price might be $5/MMBTU; however, a particular pipeline could have a basis differential of −$1.5/MMBTU. Therefore, the price at which the gas is sold is $5/MMBTU plus −$1.5/MMBTU, which yields $3.5/MMBTU. If NYMEX is used for the purpose of economic analysis, NYMEX must be adjusted for the basis by taking the NYMEX forecast for each month and adding the basis forecast for each month to NYMEX.

Basis is a function of NYMEX in that a regional basis is that particular region's differential to NYMEX. In a perfectly balanced market (where supply is equal to demand), the basis is the cost of transportation. For example, if the cost of transporting gas from Appalachia to NYMEX/Henry Hub (Louisiana) was $0.25/MMBTU, in a perfectly balanced market the Appalachian basis would be $0.25/MMBTU. In the Appalachian basin, the basis used to be positive before the development of the Marcellus Shale; however, with the development of the Marcellus Shale and a huge surge in gas supplies, various bases across the basin have become negative. Weather (seasonal variations), geography, natural gas pipeline capacity, product quality, and supply/demand determine the price of natural gas on a particular market (pipeline).

Example

Calculate the actual price of natural gas sold on pipeline ABC for the next 2 years using the provided projected NYMEX forecast and the projected basis forecast using Table 18.5.

Table 18.5 NYMEX and basis forecast example

Date	NYMEX ($/MMBTU)	Basis ($/MMBTU)
Jan-17	3.5	−0.8
Feb-17	3.4	−0.7
Mar-17	3.51	−0.6
Apr-17	3.53	−0.55
May-17	3.6	−0.58
Jun-17	3.9	−0.8
Jul-17	3.87	−0.9
Aug-17	3.88	−0.95
Sep-17	3.6	−0.7
Oct-17	3.9	−0.6
Nov-17	4	−0.5
Dec-17	4.1	−0.55

As can be seen from the table, pipeline ABC's natural gas price can be calculated by taking the NYMEX forecast per month and adding the basis forecast. In this case, the negative basis prices are due to too much supply and not enough demand. For example, the NYMEX price in October 2017 is listed as $3.90/MMBTU and the basis price is listed as −$0.6/MMBTU. Therefore,

$$\text{Pipeline ABC natural gas price on Oct } 2017 = 3.9 + (-0.6)$$
$$= \$3.3/\text{MMBTU}$$

The basis forecast is positive in some places and negative in others depending on the market, and it all boils down to supply and demand at the end. Before the development of unconventional shale plays across the United States, the basis prices used to be positive. However, with the development of unconventional shale plays, the supply has dramatically increased while the demand has not changed in the same proportionality. Therefore, the basis prices have switched from being positive to negative in high-supply markets primarily due to the lack of infrastructure. The basis prices can be easily changed from negative to positive during cold winter months due to too much demand and not much supply. This is one of the main reasons that during the coldest days of winter 2014, the price of natural gas was actually increased to $50/MSCF in Connecticut due to a lack of infrastructure (pipeline) and supply.

Cushing Hub and West Texas Intermediate (WTI)

Cushing Hub is the largest hub for the distribution of crude oil in the world and is located in Oklahoma. Cushing Hub has always been very important for traders because of its role as the delivery point (just like Henry Hub) in the US benchmark oil futures. This hub, just like Henry Hub, is the pricing point for crude oil futures in West Texas Intermediate (WTI). WTI, also known as Texas light sweet, is used as a benchmark in oil pricing. Light sweet crude oil has a low density of approximately 39.6 API gravity and low sulfur content of about 0.24%. Other essential crude oil benchmarks that serve as a reference price for buyers and sellers of crude oil are Brent crude, Dubai crude, Oman crude, and OPEC Reference Basket. Condensate price is typically a function of oil price and as oil price increases or decreases, condensate price will increase or decrease.

Mont Belvieu and Oil Price Information Services (OPIS)

Mont Belvieu is the pricing point for NGL futures. As previously mentioned, the settlement prices at the Henry Hub and Cushing are used as benchmarks for the entire North American natural gas and crude oil market. The settlement prices at the Mont Belvieu are used as benchmarks for the NGL market. Oil Price Information Services (OPIS) has a very similar concept to NYMEX or WTI. OPIS is used as a benchmark for NGLs and is one of the world's biggest sources used for NGL pricing. OPIS can be used just like NYMEX to estimate the future price of NGL for the purpose of economic analysis calculations. NGL price is also typically a function of oil price.

Capital expenditure (CAPEX)

The next important term in oil and gas economics is capital expenditures (also referred to as Capex). Capital expenditures are the money invested upfront to create future benefits. Capex is not a cost, but it is an investment because companies invest money in projects that are expected to create value for the shareholders. Capital expenditures are as follows:
1. **Acquisition**. Acquisition considered to be a Capex and is referred to as the costs when acquiring the rights to develop and produce oil and natural gas. For example, when a company purchases or leases the right to extract the oil and gas from a property not owned by the company, this will be considered as acquisition Capex. Other examples of acquisition Capex are title search, legal expenses, recording costs, and so forth. The land department is typically responsible and is in charge of dealing

with this side of the business. The land department's responsibilities are included but not limited to acquiring/renewing leases, dealing directly and negotiating with landowners, title search, and so on. Before any kind of acquisition is made, reservoir engineers and geologists along with other departments are heavily involved in valuing the asset that is under consideration for acquisition by performing various analyses such as geological potential analysis, type curve analysis, water infrastructure analysis, midstream infrastructure analysis, land analysis, environmental analysis, and finally economic analysis (using the NCF model) for the area. Many acquisition deals occur in a low commodity pricing environment where assets are sold at a discount, which can significantly be cheaper than the intrinsic value of the asset in a regular commodity pricing environment.

2. **Exploration**. Before drilling a well, it is very important to perform some type of seismic to determine the depth of the formation of interest, lithology, formation tops, formation characteristics, directional plan (azimuth, inclination, etc.), and other valuable information. A 2D or 3D seismic is used during the exploration phase to obtain this information. 3D seismic is more accurate while providing better resolution and more information about a particular prospect. 3D seismic is mostly used and preferred over 2D seismic (when capital is available). Therefore, exploration expenditures are charges related to gathering and analysis of geophysical and seismic data.

3. **Development**. These expenditures are associated with constructing the well sites, building or improving the access roads, drilling/completion, gathering, installation of pipelines, and other expenditures incurred during the developmental phase of the operation. For example, it costs on average about $5–14 million to drill and complete a well in Marcellus Shale depending on the true vertical depth (TVD), lateral length, drilling/completions design, and most importantly market condition. The unconventional shale plays are absolutely promising plays across the United States. However, it is very important to understand that developing unconventional shale plays are very capital intensive. This is due to the fact that not only do these shale formations have to be drilled, but also must be properly hydraulically fractured to produce at an economically feasible rate. Therefore, proper economic evaluation/analysis is an important job that reservoir and planning engineers are responsible for performing.

Net Capex is the net capital expenditure based on the WI% of a well. For example, if an operating company has 40% ownership in a gas well

(40% WI), the operating company will only be responsible to pay 40% of the total capital investment on a project. Net Capex can be written as Eq. (18.12).

$$\text{Net Capex} = \text{Gross Capex} \times \text{WI} \qquad (18.12)$$

Net Capex = net capital expenditure, $
Gross Capex = gross capital expenditure, $
WI = working interest, %.

Example

An 8000′ lateral-length well is estimated to be drilled and completed for $7.5 MM. Assuming 40% WI, what is the net Capex?

$$\text{Net Capex} = 7{,}500{,}000 \times 0.4 = \$3{,}000{,}000$$

Opex, Capex, and pricing escalations

When performing economic analysis, escalation is a challenging subject in the oil and gas property evaluation. E&P companies typically apply a percentage of escalation on Opex, Capex, and pricing depending on the company's philosophy. When monthly cash flows are used, the Society of Petroleum Evaluation Engineers' best practices recommend that escalation must take place in a "stair-step" fashion on a monthly basis. For example, if prices are assumed to increase at 3% per year, the monthly increase would be based on an effective annual rate of 3% per year with prices increasing every month.

Example

Perform a stair-step escalation at $3/MMBTU gas price using 3% effective annual rate for the first 12 months:

Month 1 $= 3 \times (1+3\%)^{\frac{1}{12}} = 3.007$ Month 2 $= 3.007 \times (1+3\%)^{\frac{1}{12}} = 3.015$

Month 3 $= 3.015 \times (1+3\%)^{\frac{1}{12}} = 3.022$ Month 4 $= 3.022 \times (1+3\%)^{\frac{1}{12}} = 3.030$

Month 5 $= 3.030 \times (1+3\%)^{\frac{1}{12}} = 3.037$ Month 6 $= 3.037 \times (1+3\%)^{\frac{1}{12}} = 3.045$

Month 7 = $3.045 \times (1 + 3\%)^{\frac{1}{12}} = 3.052$ Month 8 = $3.052 \times (1 + 3\%)^{\frac{1}{12}}$
= 3.060

Month 9 = $3.060 \times (1 + 3\%)^{\frac{1}{12}} = 3.067$ Month 10 = $3.067 \times (1 + 3\%)^{\frac{1}{12}}$
= 3.075

Month 11 = $3.075 \times (1 + 3\%)^{\frac{1}{12}} = 3.082$ Month 12 = $3.082 \times (1 + 3\%)^{\frac{1}{12}}$
= 3.090

Two to four percent is the typical escalation percentage that is assumed among many operating companies. The percentage escalation is directly related to the inflation rate. Using escalation is an attempt to represent inflationary expectations and should be somewhat in line with the historical long-term trend when nominal cash flows are used. From an economic analysis standpoint and NPV calculation (to be discussed), inflation must be treated consistently. When nominal interest rate is used, nominal cash flows must also be used. On the other hand, when real interest rate is used, real cash flows should be used. Nominal interest rate refers to the actual prevailing interest rate, while real interest rate is adjusted for inflation. For example, the return on a particular investment could be 5%, which is referred to as nominal interest. However, after accounting for inflation of 3%, the real interest is only 2%. Nominal interest rate is written in Eq. (18.13):

$$\text{Nominal interest rate} = \text{real rate} + \text{inflation} \qquad (18.13)$$

Profit or net cash flow (NCF)

Profit or NCF is basically revenue minus costs. The most commonly used model in the oil and gas industry to determine profit is the NCF model since this model incorporates the time value of money. Profit in the cash-flow model is also referred to as NCF. As previously mentioned, the NCF model has one unique feature and this unique piece is called time zero. Time zero is the day that the check is written to the contractors to perform a job. Capex is placed in time zero in the NCF model. It is very important that the cash-flow model is used for economic analysis, since it incorporates the time value of money. Profit excluding investment is referred to as operating cash flow and is shown in Eq. (18.14).

$$\text{Profit (excluding investment)} = \text{Net revenue} - \text{net Opex} \qquad (18.14)$$

Profit (excluding investment) = monthly basis, $
Net revenue = monthly basis, $
Net Opex = monthly basis, $.

Before federal income tax monthly undiscounted net cash flow

Before federal income tax (BTAX) monthly cash flows can be calculated by taking the profit in Eq. (18.14) and subtracting net Capex. BTAX monthly undiscounted NCF can be written as Eq. (18.15).

BTAX monthly undiscounted net cash flow = profit − net Capex (18.15)

where profit is in monthly basis in $ and Net Capex is applied at time 0 in $.

Net Capex at time zero is equal to net Capex. However, net Capex for subsequent months is zero unless special activities occur after time zero. Examples of such activities are remedial work, refrac, swabbing, artificial lift, etc.

Before starting the most important and beautiful concept in economic analysis, that is, NPV (net present worth, NPW), it is very important to understand discount rate and its significance.

Discount rate

Discount rate, also known as interest rate, exchange rate, cost of capital, opportunity cost of capital, cost of money, weighted average cost of capital (WACC), or hurdle rate, is used to discount all of the future cash flows to today's dollar. Discount rate is the basis for all the economic analysis performed in any industry. It is basically the cost of doing business. For example, if a company's cost of capital (discount rate) is 10%, it means the return on a particular project must be greater than 10% or the company will not be creating any value for the shareholders. For the purpose of economic analysis, weighted average cost of capital is typically used to discount all of the future cash flows to the present dollar. Weighted average cost of capital accounts for both time value of money and inflation. Time value of money is not the same as inflation, although they are often confused. Time value of money refers to the fact that a dollar today is worth more than a dollar in the future. It is tied back to the fact that people are impatient about their money. If I were to offer you $1000 today versus $1000 a month from today, the chance that you would like to have your $1000 today is very likely because you are impatient about your money. Therefore, time value of money has to deal with the impatience of people regarding their money. Inflation, on the other hand, refers to the reduction in purchasing power of the money. Ten years or so ago, a $5 bill could have bought much more than a $5 bill today because of the inflation with time. Therefore, the purchasing power of the same $5 bill has decreased due to inflation.

What discount rate should be assumed for a project? This is the task for financial people within a corporation. It is very important to understand the concept of discount rate as it is very significant in the determination of NPV. Every company has a cost of capital. The determination of cost of capital can be tricky and complicated. Cost of capital calculation essentially takes three important factors into account. These three factors are debt, common stock (equity), and preferred stock (equity, if exists, some companies do not have preferred stock). Thus, cost of capital is usually the combination of debt and equity since many companies use a combination of debt and equity to finance their business. If a company only uses debt to finance its projects, cost of capital is referred to as cost of debt. On the other hand, if a company only uses equity to finance its projects, cost of capital is called cost of equity. As previously mentioned, many companies' cost of capital consists of both debt and equity. Cost of capital is sometimes referred to as hurdle rate. Hurdle rate is the minimum discount rate or minimum acceptable rate of return that must be overcome for a company to generate value and return for its investors.

- **Debt**. Companies, just like ordinary people, have to incur debt to finance their projects. Debt can be borrowing money through issuing bonds, loans, and other forms of debts from banks or financial institutions.
- **Preferred stock**. Preferred stock is a type of equity or a class of ownership in a corporation that has a higher claim to a company's assets. The reason this type of equity is referred to as "preferred" is because when a company cannot meet its financial obligations as debts become due (insolvency), preferred stockholders get their money before common stockholders. This means that there is a lower risk associated with preferred stocks in addition to lower rates of return (less potential to appreciate in price) in comparison to common stocks. Furthermore, when the company has excess cash and decides to reward the stockholders by distributing cash in the form of dividends, preferred stockholders are paid before common stockholders. The dividends paid to the preferred stockholders are different and typically more than the

dividends paid to common stockholders. Preferred shareholders typically do not have the voting rights, however, under some circumstances these rights can return to shareholders who have not received their dividend.
- **Common stock.** Common stock is also a type of equity or ownership in a corporation in which investors invest their money in a risky stock market. Common stock is riskier than preferred stock due to the fact that those investors will be last in line to get their money back in the event of insolvency. The reward of common stocks would be a higher return compared to preferred stocks and bonds in the long run. Common stockholders have the power of voting in the election of the board of directors and corporate policy.

To summarize, debt and equity (common and preferred stocks) are used in the computation of cost of capital. Since cost of capital consists of cost of debt and equity, both must be combined into an equation referred to as weighted average cost of capital (WACC), shown in Eq. (18.16).

$$\text{Weighted average cost of capital (WACC)} = W_d R_d (1-T) + W_p R_p + W_c R_c \quad (18.16)$$

W_d = weight of debt (% of the company that is debt)
R_d = cost of debt, %
W_p = weight of preferred stock (% of the company that is preferred stock)
R_p = cost of preferred stock, %
W_c = weight of common stock (% of company that is common stock)
R_c = cost of common stock, %
T = Corporate tax rate, %.

Weight of debt and equity

The weight of debt and equity of any company can be obtained using the debt-to-equity ratio, which is publicly available on various financial websites. Weight of debt refers to the percentage of the company that is financed by debt. On the other hand, weight of equity refers to the percentage of the company that is financed by equity. The combination of the two makes up the company's **capital structure**. It is important to have a balance between the amount of debt and equity that a company has. Low-debt companies are typically safer to invest in. Although debt is tax deductible (beneficial from a tax perspective), having too much debt can cause a company to go through so much capital when the stock of that particular company decreases. For example, imagine a house that you are interested in purchasing is worth $250,000. You were able to obtain a loan from your local bank for $200,000. You placed $50,000 as a down payment on the house (20% down payment). Therefore, 80% of the house is financed by debt and only 20% is equity. If the price of the house decreases by 25%, not only have you lost all of the equity that was placed, but also the house is under the market value by an additional 5%. When the debt level is high and the stock of a particular corporation decreases, the company goes through so much capital and they can become history overnight. This concept was very clear when studying some of the financial institutions during the 2008 crash. Many of those institutions had a very high leverage ratio.

Example

Company X currently has a debt-to-equity ratio of 0.65. Calculate the weight of debt and equity of this company.

$$\text{Weight of debt} = \frac{0.65}{0.65 + 1} = 39.4\% \text{ weight of debt}$$

$$\text{Weight of equity} = 1 - 39.4\% = 60.6\% \text{ weight of equity}$$

Cost of debt

Cost of debt can be calculated by taking the weighted average of the percentage interest rate paid on debt. For example, if Matt has a house for $400,000 at 4%, a car for $30,000 at 2.5%, and a small boat for $25,000 at 1.5%, his cost of debt will be the weighted average of the percentage interest rates above, which is 3.76%. When tax season arrives, some of the interests that Matt paid are tax deductible. Companies obtain their cost of debt by taking the weighted average of all their debts.

Example

A bond is issued at $950. After 2 years, your company will pay back $1000 to the investors. What is the cost of debt for this particular bond?

Par value = face value = $950
Maturity value = $1000

$$PV = \frac{FV}{(1+i)^t} \gg 950 = \frac{1000}{(1+i)^2} \gg i = 2.59\%$$

Cost of equity

Cost of equity is the percentage return that shareholders expect from a company. For example, if you decide to invest in a company, you will have some demands from that company in exchange for obtaining the risk of ownership. Let us assume that an investor's required rate of return (demand) in order to be convinced to invest in a company's stock is 20%. This 20% required rate of return is referred to as *cost of equity* for the company. It is important to remember that 20% is not the profit earned by the investors but it is just the demand from the investors for putting their money in a high-risk and volatile stock market. The main reason the cost of equity is considered a cost from a company's perspective is because if the company fails to deliver this kind of return, shareholders will simply sell their stocks (shares) causing the stock price to go down. There are multiple ways to calculate cost of equity but one of the most commonly used models is the capital asset pricing model (CAPM). The general idea about the CAPM model is that investors need to be compensated in two ways:
1. Time value of money ≫ risk-free rate
2. Risk ≫ the amount of compensation for taking additional risk

Cost of equity using CAPM is written in Eq. (18.17).

$$\text{Capital asset pricing model (CAPM)}: K_e = R_f + \beta(R_m - R_f) \quad (18.17)$$

where K_e is the cost of equity, %; R_f is risk-free rate, %; β is beta of security; R_m is expected market return, %; and $R_m - R_f$ is risk premium, %.

Risk-free rate or R_f is the theoretical rate of return obtained from an investment that has no risk associated with it. For example, if a theoretical government asks you to invest in a particular bond and receive 2% in return without any risks, the risk-free rate is 2%. In reality, the risk-free rate does not exist because even the safest investment will have a very small amount of risk. Typically the interest rate of US Treasury bills is used as the risk-free rate among many companies. Governmental treasury bills are called risk-free because the US Government has never defaulted on its debt.

Beta or β, also known as the **beta coefficient**, measures the volatility of a company's share price against the whole market. A beta of 1 means the company moves in line with the market. If beta is more than 1, the security's price will be more volatile than the market. Finally, if beta is less than 1, the security's price will be less volatile than the market. Having a high beta means more risks while offering a possibility of a higher rate of return. Oil and gas companies typically have a beta greater than 1. An example with a beta less than 1 is the US Treasury bill since the price does not change much over time. If market return is 10%, a stock whose beta is 1.5 would return 15% because it would go up 1.5 times as high as the market. Beta considers systematic risk and not idiosyncratic risk. *Systematic risk* refers to the overall market risk. However, *idiosyncratic risk* refers to the risk of change in the price of a security due to the special circumstances of a particular security. Idiosyncratic risk can be eliminated through diversification but systematic risk cannot be eliminated.

Risk premium or $R_m - R_f$ is the return for which investors expect to be compensated for having taken the extra risk in investing in the volatile stock market. It is basically the difference between the risk-free rate and market rate. Risk premium accounts for inflation rate and this is the primary reason escalation is very important to use on Opex, Capex, pricing, etc. when cost of capital is used in economic analysis.

Example

A firm's risk-free rate (R_f) is 6% and the market risk premium ($R_m - R_f$) is 7%. Assuming a beta of 1.5, what is the cost of equity using the CAPM model?

$$K_e = R_f + \beta(R_m - R_f) = 6\% + 1.5 \times 7\% = 16.5\%.$$

Example

A company wants to raise money. The company will sell $15 million shares of common stock with the expected return of 15%. In addition, the company will issue $10 million of debt with the cost of debt of 12%. Assuming a corporate tax rate of 35%, calculate WACC.

$$\text{Total value of the company} = \$15\,\text{MM} + \$10\,\text{MM} = \$25\,\text{MM}$$

$$\text{Weight of equity} = \frac{15}{25} = 0.6 \text{ or } 60\%$$

$$\text{Weight of debt} = \frac{10}{25} = 0.4 \text{ or } 40\%$$

Using Eq. (18.16):

$$\text{WACC} = 40\% \times 12\% \times (1 - 35\%) + (60\% \times 15\%) = 12.12\%$$

Capital budgeting

Capital budgeting is an important part of determining whether to invest in a project such as a drilling/completion program, buying machinery, replacing equipment, and so forth. Capital budgeting defines a firm's strategic direction and planning. Capital budgeting typically involves large capital expenditure (Capex), and making wrong decisions can have serious consequences. Without analyzing capital budgeting parameters and presenting the results, the management committee of any public or private company will not approve projects. This is a very simple concept. The management committee of any company would like to see return as a result of investing money into a project to maximize shareholders' value. Important capital budgeting criteria are NPV, internal rate of return (IRR), modified internal rate of return (MIRR), return on investment (ROI), payback, discounted payback, and profitability index (PI). All of these criteria are very important to understand and comprehend in detail for successful capital budgeting decision making. The capital budgeting decision-making process must be evaluated in great detail before investing big lump sums of money on a project to determine whether the project is worth the investment or not. Now that the concept of capital budgeting is clear, let's discuss the most important capital budgeting criteria involved in the decision-making process.

Net present value

NPV also known as net present worth (NPW) is one way of analyzing the profitability of an investment. NPV is basically the value of specific stream of future cash flows presented in today's dollar. NPV is an essential calculation in petroleum economics due to considering time value of money and inflation. Companies are keen to know what an actual project is worth in today's dollar rather than, say, the dollar of 10 years from now. As a simple example, oil and gas operating companies project the future production rates for each well using various techniques such as decline curve, type curve, reservoir simulation, rate transient analysis, material balance, machine learning, and so forth. Those future production rates that will yield future cash flows must be discounted (using cost of capital) to present value. It really does not make any logical sense for a company to announce that their profit cash flow for doing a project would be $10 million, $8 million, $12 million, and $11 million in subsequently 2, 3, 4, and 5 years, respectively. This is because these cash flows are worth less in today's dollar when discounted back (due to time value of money). Instead, it would make much more sense to use the NPV formula to calculate the present value of all of the future cash flows. A simple insight one can use to think about NPV is that cash flows at different dates are like different currencies. What is the summation of 200 US dollars and 200 euros? Not discounting the future cash flows that are occurring at different points in time to today's dollar is like saying that the answer to the posed question is 400. Therefore, just like currencies that must be converted before being able to perform the summation, cash flows at different points in time must be discounted back to today's dollar. NPV calculation assumes that positive cash flows from a project are reinvested at the cost of capital. NPV can be calculated using Eq. (18.18).

$$\text{NPV} = \sum_{t=0}^{n} \frac{CF_t}{(1+i)^t} \qquad (18.18)$$

where i is the discount rate, %; CF_t is yearly cash flow at time t, $; t is period of time, yearly; and

$$\sum_{t=0}^{n} = \text{Summation sign from time } 0(\text{investment}) \text{ to } n.$$

The reason the term "net" is used in the term *net present value* is that the initial investment is subtracted and taken into account when calculating NPV. NPV is the summation of the present value of all of the future cash flows and initial investment (Capex). Discount rate in NPV calculation is an essential factor that considers the time value of money and inflation. When discount rate increases, the NPV decreases. The discount rate used among some operating companies regardless of cost of capital is typically 10%. Many E&P companies' cost of capital is anywhere from 8% to 12%. Therefore, the industry standard discount rate used in many economic analysis calculations is 10% for simplicity.

The rules of thumb for NPV projects are as follows:
1. Accept independent projects if the NPV is positive.
2. Reject any project that has a negative NPV.
3. Pick the highest positive NPV in mutually exclusive projects that would add the most value.
4. NPV must be considered along with other capital budgeting criteria to make educational decisions.

Table 18.6 NPV example

Year	Profit ($MM)
0 (investment)	−$100.00
1	$20.00
2	$30.00
3	$40.00
4	$80.00
5	$60.00

Although the rule of thumb says to accept any project with positive NPV, would you accept a project that has a NPV of $20,000 after investing $2 billion dollars in that project? Absolutely not, because although the NPV is positive, the project could be so risky that it might end up costing the company more than creating any value for the shareholders. As a result, it is extremely important to comprehend the magnitude of the investment along with other capital budgeting tools before making such decisions (Table 18.6).

Example

Find the NPV for the cash flows (profits) located in Table 18.6 using a 10% discount rate.

$$\text{NPV} = \sum_{t=0}^{n} \frac{\text{CF}t}{(1+i)^t}$$
$$= -100 + \frac{20}{(1+0.1)^1} + \frac{30}{(1+0.1)^2} + \frac{40}{(1+0.1)^3} + \frac{80}{(1+0.1)^4} + \frac{60}{(1+0.1)^5}$$
$$= \$64.92$$

The present value for each year is summarized in Table 18.7.

Table 18.7 Net present value summary

Year	Profit ($MM)	Present value ($MM)
0 (investment)	−$100.00	−$100.00
1	$20.00	$18.18
2	$30.00	$24.79
3	$40.00	$30.05
4	$80.00	$54.64
5	$60.00	$37.26
Summation (NPV at 10%)		$64.92

Example

Imagine you win the million-dollar lottery! Do not get too excited. You will actually get paid $50,000 per year for the next 20 years. If the discount rate is a constant 8% and the first payment will be in year 1, how much have you actually won in present dollars? What are you going to do, take a lump sum or yearly payments for the next 20 years?

This is a classic NPV example. First of all, it is very important to take a lump sum instead of yearly payments for the next 20 years due to the fact that this money can be invested in various projects and make higher returns (as long as the money can be invested and made higher returns). Secondly, based on the time value of money concept, people are impatient about their money and would love to have their money as soon as possible instead of yearly payments for the next 20 years. In addition, taxes will have to be paid on the calculated $490,907 present value of the lottery.

$$\begin{aligned}NPV &= \frac{50,000}{(1+0.08)^1} + \frac{50,000}{(1+0.08)^2} + \frac{50,000}{(1+0.08)^3} + \frac{50,000}{(1+0.08)^4} + \frac{50,000}{(1+0.08)^5} \\ &+ \frac{50,000}{(1+0.08)^6} + \frac{50,000}{(1+0.08)^7} + \frac{50,000}{(1+0.08)^8} + \frac{50,000}{(1+0.08)^9} + \frac{50,000}{(1+0.08)^{10}} \\ &+ \frac{50,000}{(1+0.08)^{11}} + \frac{50,000}{(1+0.08)^{12}} + \frac{50,000}{(1+0.08)^{13}} + \frac{50,000}{(1+0.08)^{14}} + \frac{50,000}{(1+0.08)^{15}} \\ &+ \frac{50,000}{(1+0.08)^{16}} + \frac{50,000}{(1+0.08)^{17}} + \frac{50,000}{(1+0.08)^{18}} + \frac{50,000}{(1+0.08)^{19}} + \frac{50,000}{(1+0.08)^{20}} \\ &= \$490,907\end{aligned}$$

Table 18.8 shows the summary for discounted cash flows for the next 20 years at an 8% interest rate. As can be seen from Table 18.8, the present value of $50,000 at year 20 is only $10,727. This example clearly illustrates the main reason why E&P companies would like to make as much money as possible during the first 5–10 years of the life of a well in order to create the most value for the shareholders. Typically during a well's life (which varies from well to well based on type curve and economic assumptions made), up to 70%–80% of the value is associated with the first 8 years of production with less than 50% of the EUR produced. The next 42 years (if the reserve life is assumed to be 50 years) delivers approximately 20%–30% of present value and over 50% of EUR. A sensitivity analysis on various parameters of this exercise can be made to understand the impact of value creation and EUR throughout the life of a well. As previously mentioned, these percentages (value and EUR) will be different based on type curve and economic parameter assumptions.

Table 18.8 Present value example summary

Year	CF	PV	Year	CF	PV
1	$50,000	$46,296	11	$50,000	$21,444
2	$50,000	$42,867	12	$50,000	$19,856
3	$50,000	$39,692	13	$50,000	$18,385
4	$50,000	$36,751	14	$50,000	$17,023
5	$50,000	$34,029	15	$50,000	$15,762
6	$50,000	$31,508	16	$50,000	$14,595
7	$50,000	$29,175	17	$50,000	$13,513
8	$50,000	$27,013	18	$50,000	$12,512
9	$50,000	$25,012	19	$50,000	$11,586
10	$50,000	$23,160	20	$50,000	$10,727

CF, cash flow; *PV*, present value.

Advantages of NPV:
- NPV accounts for time value of money.
- Cash flows over the economic life of the project are taken into account.
- NPV provides a sense of scale about the value that will be created for the shareholders.
- NPVs can be added. If there are 100 projects with an NPV of $1000 for each project, the total NPV can be easily summed up to be $100,000.
- NPV assumes that all the future cash flows are reinvested at the cost of capital. The same cost of capital does not have to be used for the entire life of the project and different discount rates can be assumed.

Disadvantage of NPV:

NPV does not give any indication on the size of the original investment. For instance, the NPV of a $10 million investment could be $1 million, and the NPV of a $1 billion investment could also be $1 million.

BTAX and ATAX monthly discounted net cash flow

Before tax monthly undiscounted NCF is written in Eq. (18.15). The NPV equation can be used on a monthly basis to calculate before and after tax present value of all the future cash flows as shown in Eq. (18.19). The only difference when discounting monthly cash flows is to divide time in the NPV equation by 12. Calculation of after federal income tax (ATAX) monthly undiscounted NCF is discussed in the tax model.

$$\text{BTAX or ATAX monthly discounted NCF} = \frac{\text{BTAX or ATAX monthly undiscounted NCF}}{(1+\text{WACC})^{\frac{\text{Time}}{12}}} \quad (18.19)$$

where BTAX or ATAX monthly undiscounted NCF is in $; WACC is weighted average cost of capital, $; and time is in month.

1. BTAX NPV is the summation of all BTAX monthly discounted cash flows.
2. ATAX NPV is the summation of ATAX monthly discounted cash flows.

Internal rate of return (IRR)

IRR is known as discounted cash-flow rate of return (DCFROR) or simply rate of return (ROR). IRR is the discount rate when the NPV of particular cash flows is exactly zero. The higher the IRR, the more growth potential a project has. IRR is an important decision metric on any project. IRR is frequently used for project evaluation and profitability of a project. The formula for calculating IRR is basically the same formula as NPV except that the NPV is replaced by zero and the discount rate is replaced by IRR as shown in the following equation. As opposed to NPV, IRR assumes that positive cash flows of a project are reinvested at IRR instead of cost of capital. This is one of the **disadvantages** of using the IRR method since it **defectively** assumes that positive cash flows are reinvested at the IRR.

When the NPV of a particular project is exactly zero, the IRR will yield cost of capital of a project. For example, if the cost of capital of a particular publicly traded company is 9.3% and the NPV of a particular project yields zero, IRR will be 9.3% for that particular project. This means the present value of all the cash inflows is just enough to cover the cost of capital. When NPV is zero, no value will be created for the shareholders. IRR must be higher than the cost of capital of a project to create any value for the shareholders. When IRR is less than the cost of capital, no value will be created for the shareholders.

$$\text{Internal rate of return (IRR)}: 0 = \sum_{t=0}^{n} \frac{CF_t}{(1+\text{IRR})^t} \quad (18.20)$$

IRR rule of thumb:

The rationale behind IRR in an independent project is:

1. If IRR is greater than WACC (IRR > WACC), the project's rate of return will exceed its costs and as a result the project should be accepted.

2. If IRR is less than WACC (IRR < WACC), the project's rate of return will not exceed its costs and as a result the project should be rejected. For example, if a company's cost of capital (WACC) is 12% and IRR for a particular project is calculated to be 11%, the project must be declined because it would cost more to finance the project (through debt and equity) than the actual return of the project. On the other hand, if a company's cost of capital is 12% and the IRR for a specific project is 20%, the project is approved. A lot of companies have a minimum acceptable IRR before investing in a project. This minimum acceptable IRR for one particular company could be 15% while for others could be 20% or 25% depending on many factors, especially market conditions. Typically, in the O&G industry, lower return projects are accepted in a **downturn market condition** as long as better investment alternatives are unavailable.

In mutually exclusive projects, the project with higher IRR must be picked. For example, if IRR on project A is 15% and project B is 20%, project B must be selected.

Example

Calculate the IRR for the cash flows listed in Table 18.9.

Table 18.9 IRR example

Year	Profit ($MM)
0 (investment)	−$500.00
1	−$100.00
2	$20.00
3	$300.00
4	$400.00
5	$500.00
IRR	19.89%

IRR can be calculated using Eq. (18.20):

$$0 = -500 + \frac{-100}{(1+\text{IRR})^1} + \frac{20}{(1+\text{IRR})^2} + \frac{300}{(1+\text{IRR})^3} + \frac{400}{(1+\text{IRR})^4} + \frac{500}{(1+\text{IRR})^5}$$

As can be determined when manually computing IRR, IRR can be solved either using trial and error or linear interpolation methods. Financial calculators or Excel are recommended to perform this calculation. In this example, if various discount rates are inputted into the above equation when the IRR is 19.89% in the denominator of each term, the equation is equal to 0. This means the IRR for this particular project is approximately 20%.

Economic evaluation

Table 18.10 NPV at various discount rates example

Discount rate	10%	15%	18%	25%
Time 0	−500.00	−500.00	−500.00	−500.00
Discounted CF, Year 1	−90.91	−86.96	−84.75	−80.00
Discounted CF, Year 2	16.53	15.12	14.36	12.80
Discounted CF, Year 3	225.39	197.25	182.59	153.60
Discounted CF, Year 4	273.21	228.70	206.32	163.84
Discounted CF, Year 5	310.46	248.59	218.55	163.84
Summation (NPV)	234.68	102.71	37.08	−85.92

CF, cash flow; *NPV*, net present value.

Internal rate of return calculation

As previously discussed, when manually computing IRR, IRR can be calculated using trial and error, which is tedious and time consuming, or linear interpolation. Many commercial economic software packages use linear interpolation, in which the software finds the discount rate when the sign of NPV changes from positive to negative and linearly interpolates between the two discount rates. One of the flaws with this type of calculation is that two users who define different series of discount rates will see different calculated IRR. There are other mathematical methods (not discussed in this book) such as the root finding method, which can be used to perform such calculations. In the above example, let us calculate NPV at different discount rates of 10%, 15%, 18%, and 25%. Afterward, linear interpolation can be used to calculate the discount rate when NPV is zero.

As can be seen from Table 18.10, NPV goes from 37.08 MM at 18% discount rate to −85.92 MM at 25% discount rate. After performing linear interpolation to find the discount rate when NPV is 0, IRR is found to be 20.11%, which is close 19.89%. This difference may be expanded at higher IRRs and widely spaced discount rates. Therefore, users with various series of discount rates will see different calculated IRRs.

Example

Calculate the IRR using Table 18.11 given the NPV for each discount rate.

Table 18.11 IRR example

Discount rate (%)	0	5	10	15	20	25	30	35
NPV ($MM)	200	150	100	20	−6	−11	−16	−21

IRR is the discount rate at which NPV is equal to zero. In this example, NPV at 15% discount rate is $20 MM and NPV at 20% discount rate is −$6 MM. Therefore, NPV is equal to zero when the discount rate is in between 15% and 20%. Linear interpolation can be used to find the discount rate when NPV is 0 given the predefined series of discount rates.

$$Y = Y_a + (Y_b - Y_a) \times \frac{X - X_a}{X_b - X_a}$$
$$= 15 + (20 - 15) \times \frac{0 - 20,000,000}{-6,000,000 - 20,000,000} = 18.85\%$$

From this example, the discount rate when NPV is 0 is equal to 18.85%.

NPV profile

NPV profile is a graphical representation of project's NPV against various discount rates. Discount rates and NPV are subsequently plotted on the x- and y-axis.

Example

Draw the NPV profile for projects A and B and determine which project is better assuming a cost of capital of 5%.

The first task in this problem is to draw the NPV profile by plotting discount rate (x-axis) vs NPVs for projects A and B (y-axis). IRR is the point at which NPV curves cross the x-axis as shown in Fig. 18.2. There is a point referred to as the crossover point (rate) in Fig. 18.2. Crossover point is the discount rate at which the NPV for both projects is equal (Table 18.12).

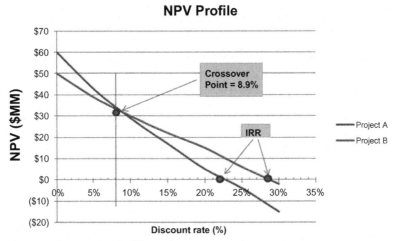

Fig. 18.2 Crossover point illustration.

Table 18.12 Net present value (NPV) profile

Rate (%)	NPV (A)	NPV (B)
0	$60	$50
5	$43	$39
10	$29	$30
15	$17	$22
20	$5	$15
25	($4)	$6
30	($15)	($2)

There are three stages in the following NPV profile. The first stage occurs before the crossover point, and in this phase, the NPV of project A is more than the NPV of project B. In this stage, there is a conflict between IRR and NPV since the NPV of project A is more than B, while the IRR of project B is more than A. The company's cost of capital in this example is given to be 5%. When cost of capital is less than crossover point (rate), a conflict exists. When a conflict exists and the cost of capital is less than the crossover point, the NPV method must be used for decision making. Therefore, project A is superior to project B in this example since the cost of capital is given to be 5%. When the cost of capital is low, delaying cash flows is not penalized as much compared to at a higher cost of capital. When the cost of capital is high (more than the crossover point) delaying cash flows will be penalized.

At the crossover point (second stage), NPV of both projects is equal. Finally, during the third stage, NPV for project B is more than NPV for project A. Please note that if the cost of capital in this problem was given to be 10% instead of 5% (cost of capital > crossover rate), both NPV and IRR methods would have led to the same project selection. It is important to note that it is the difference in timing of cash flows that is causing the crossover between the two projects. The project with faster payback provides more cash flows in the early years for reinvestment. If the interest rate is high, it is vital to get the money back faster because it can be reinvested while if the interest rate is low, there is not such a hurry to get the money back faster.

Advantages of IRR:
- IRR accounts for time value of money.
- Cash flows over the economic life of the project are taken into account.

Disadvantage of IRR:
- IRR does not provide a **sense of scale** about the value created for the shareholders.

- IRRs cannot be added. If there are four projects with IRRs of 15%, 18%, 22%, and 12% the total IRR will not be 67%. Instead, cash flows of all the projects must be combined and IRR can be determined from the combined cash flows.
- IRR assumes that all the future cash flows are reinvested at IRR.
- IRR just like NPV does not give any indication of the size of the original investment.
- IRR cannot be calculated when:
  - cash flows are all negative or positive;
  - total undiscounted revenues are less than the original investment;
  - cumulative cash flow stream changes sign more than once by going positive to negative.

NPV vs IRR

NPV basically measures the dollar benefit (added value) of the project to the shareholders but it does not provide information on the safety margin or the amount of capital at risk. For example, if NPV of a project is calculated to be $2 million, it does not indicate the kind of safety margin that the project has. In contrast, IRR measures the annual rate of return and provides safety margin information. All in all, for mutually exclusive projects and ranking purposes, NPV is always superior to IRR. Unfortunately, in the oil and gas industry, IRR is quite often used for making critical decisions. It is recommended to calculate and understand IRR methodology for each project. However, the ultimate decision whether to perform a project should be determined using NPV calculation.

Modified internal rate of return (MIRR)

MIRR is basically an improved version of IRR and is another tool used in capital budgeting. It is very important to understand the difference between IRR and MIRR. As previously mentioned, IRR defectively assumes that positive cash flows from a particular project are reinvested at IRR. In contrast to IRR, MIRR assumes that cash flows from a project are reinvested at cost of capital or a particular reinvestment rate. In addition to this improvement, MIRR only yields one solution. Consequently MIRR can be defined as the discount rate that causes the present value of a project's terminal value to equal the present value of cost. The MIRR concept is fairly complicated and will only make more sense with examples. This is one of the main reasons that IRR is used more frequently in the real world, that is, since MIRR is not completely understood by a lot of managers. MIRR can be calculated using Eq. (18.21).

$$\text{MIRR} = \sqrt[n]{\frac{\text{Future value (positive cash flows @ reinvestment rate)}}{-\text{Present value (negative cash flows @ cost of capital or finance rate)}}} - 1$$

(18.21)

Example

The cash flows for projects A and B are summarized in Table 18.13. Calculate MIRR assuming a cost of capital of 10% and reinvestment rate of 12%.

Table 18.13 MIRR example

Year	Project A ($MM)	Project B ($MM)
0	($600)	($350)
1	$100	$200
2	$250	$225
3	$320	$250
4	$385	$350
5	$400	$450

The first step is to calculate the present value of negative cash flows at cost of capital for both projects:

$$\text{Project A present value} = \frac{-600}{(1+0.1)^0} = -600$$

$$\text{Project B present value} = \frac{-350}{(1+0.1)^0} = -350$$

Next, future values of positive cash flows at reinvestment rate must be calculated for both projects:

Project A future value

$$= 100 \times (1+12\%)^4 + 250 \times (1+12\%)^3 + 320 \times (1+12\%)^2 + 385 \times (1+12\%)^1 + 400 \times (1+12\%)^0 = \$1741.19$$

Project B future value

$$= 200 \times (1+12\%)^4 + 225 \times (1+12\%)^3 + 250 \times (1+12\%)^2 + 350 \times (1+12\%)^1 + 450 \times (1+12\%)^0 = \$1786.41$$

Using the MIRR equation:

$$\text{Project A MIRR} = \sqrt[5]{\frac{1741.19}{-(-600)}} - 1 = 0.2375 \text{ or } 23.75\%$$

$$\text{Project B MIRR} = \sqrt[5]{\frac{1786.41}{-(-350)}} - 1 = 0.3854 \text{ or } 38.54\%$$

In this example, note that the first-year cash inflow is assumed to be reinvested in 4 years (5 − 1), the second-year cash inflow is assumed to be reinvested in 3 years (5 − 2), the third-year cash inflow is assumed to be reinvested in 2 years (5 − 3), the fourth-year cash inflow is assumed to be reinvested in 1 year (5 − 4), and finally the fifth-year cash inflow is received at the end of the fifth year and is not available for reinvestment since it accords with the end of the project's life.

Payback method

Payback method is another capital budgeting method to determine the quick profitability of an original investment. Payout period is the period of time in which a particular project is expected to recover its initial investment. For example, if $7 MM was initially invested on a particular project for drilling and completing a well and it took 3.5 years to earn $7 MM of profit back, the payback period for this project would be 3.5 years. The payback period when cash inflows per period are even can be calculated using Eq. (18.22).

$$\text{Payback period} = \frac{\text{Initial investment}}{\text{Cash inflow per period}} \qquad (18.22)$$

Example
Calculate the payback period given the following undiscounted cash flows, assuming uneven cash inflows and using Table 18.14.

Table 18.14 Payback period example

Year	Cash flow ($MM)
0	($100)
1	$20
2	$30
3	$40
4	$45
5	$70

Years 1, 2, and 3 are added up to be 90. The transition from year 3 to year 4 is when the cash inflows exceed the original investment of $100 million. Placing this in an algebraic equation,

$$\text{Payback} = 3 + \frac{10}{45} = 3.22 \text{ years}$$

Therefore, it takes 3.22 years to pay back the original investment.

Strengths of payback method:
- easy to calculate and understand
- provides an intuition of project risk and liquidity

Weaknesses of payback method:
- As can be easily determined, payback method ignores the time value of money.
- This method also ignores the cash flows occurring after the payback period.

Discounted payback method:
- Discounted payback method, just like the payback method, is another method used in capital budgeting to determine the profitability of an investment. The difference between payback period and discounted payback period is that in discounted payback period, time value of money is taken into account when calculating the number of years it takes to break even from an initial investment. When calculating discounted payback period, discounted cash flows are used instead of undiscounted cash flows.

Example

Calculate the discounted payback period using the undiscounted cash flows as shown in Table 18.15 by assuming a 10% cost of capital.

Table 18.15 Discounted payback period example problem

Year	Cash flow ($MM)
0	($90)
1	$20
2	$30
3	$40
4	$50
5	$60

Table 18.16 Discounted payback period example answer

Year	Cash flow ($MM)	PV equation	PV of cash flow ($MM)
0	($90)	$-90/1.1^0$	($90)
1	$20	$20/1.1^1$	$18.18
2	$30	$30/1.1^2$	$24.79
3	$40	$40/1.1^3$	$30.05
4	$50	$50/1.1^4$	$34.15
5	$60	$60/1.1^5$	$37.26

The first step is to use the PV equation to discount the future cash flows for each period at 10% discount rate (shown in Table 18.16). Afterward, discounted payback period can be easily calculated using algebra.

$$\text{Discounted payback period} = 3 + \frac{90 - 73.02}{34.15} = 3.5 \text{ years}$$

This method, just like the payback method, ignores the cash flows after the discounted payback period; however, it takes time value of money into account.

Profitability index (PI)

PI is another tool used in capital budgeting to measure the profitability of a project. As previously discussed, NPV yields the total dollar figure of a project (absolute measure), but profitability is a relative measure given by a ratio; the higher the PI the higher the ranking. PI essentially tells us how much money will be gained for every dollar invested. For example, PI of 1.4 of a project tells us that for every dollar invested in the project, an expected return of $1.4 is anticipated. PI is well known among financial managers as representing the bang-per-buck measure. PI can be calculated using Eq. (18.23).

$$\text{Profitability index (PI)} = \frac{\text{PV of future cash flows excluding investment}}{\text{Initial investment}}$$

(18.23)

PI rule of thumb:
- Accept projects that have PI more than 1 (PI > 1)
- Reject projects that have PI of less than 1 (PI < 1)

Example

Calculate the PI assuming 10% discount rate and $200 million investment using Table 18.17.

Table 18.17 Profitability index example problem

Year	Cash flow ($MM)
1	$20
2	$30
3	$55
4	$60
5	$70
6	$55
7	$90

Present value for each cash flow is summarized in Table 18.18 and PI can be calculated as follows:

Table 18.18 Profitability index example answer

Year	Cash flow ($MM)	PV equation	PV ($MM)
1	$20	$20/1.1^1$	$18.18
2	$30	$30/1.1^2$	$24.79
3	$55	$55/1.1^3$	$41.32
4	$60	$60/1.1^4$	$40.98
5	$70	$70/1.1^5$	$43.46
6	$55	$55/1.1^6$	$31.05
7	$90	$90/1.1^7$	$46.18
Summation			$245.97

$$PI = \frac{PV \text{ of future cash flows}}{\text{Initial investment}} = \frac{245.97}{200} = 1.23$$

Tax model (ATAX calculation)

The tax model is used for after-tax calculation in oil and gas property evaluation. This model takes into account depreciation, taxable income, corporation tax rate, and discounting. The discounting equations for the tax model are the same as the NCF model. The primary difference between the two models is that depreciation, taxable income, and corporate tax rate are all taken into account in the tax model.

Table 18.19 ACR2 7-year Depreciation rate

Year	ACR2 7-year (%)
1	14.29
2	24.49
3	17.49
4	12.49
5	8.93
6	8.92
7	8.93
8	4.46

Depreciation

Tangible and intangible capital expenditure must be specified for after-tax calculations. Typically 10%–20% is considered tangible with the remaining percentage being intangible Capex. The Internal Revenue Service (IRS) defines depreciation rates using the accelerated recovery method depreciation for 7 years as shown in Table 18.19. Majority of petroleum investments have a 5 or 7 year guideline life when using the accelerated recovery method depreciation, however, other tables such as ACR 3-year and ACR 10-year can also be used. Most petroleum engineers primarily use ACR2 7-year or ACR2 5-year depreciation tables to perform after tax economic analysis. Monthly depreciation must be calculated for each year using the defined IRS depreciation rate and appropriate tangible Capex. It is important to classify the items that are considered tangible versus items that are intangible in an attempt to accurately account for depreciation of tangible capital using accelerated recovery method depreciation either over 5 or 7 year (depending on the company). Monthly depreciation can be calculated using Eq. (18.24), which assumes depreciation rate for each year is paid equally on a monthly basis. Depreciation occurrence will be different depending on the time of the year that a well is TIL but for the simplicity of the monthly depreciation calculation, yearly depreciation rate is divided by 12.

$$\text{Monthly depreciation} = \frac{\text{Yearly depreciation rate}}{12} \times \text{tangible investment} \times \text{WI} \qquad (18.24)$$

Yearly depreciation rate = IRS-defined accelerated recovery method, 7 years (will vary depending on the company), %

Tangible investment = typically 10%–20% of the total drilling and completions Capex, $
WI = working interest, %.

Example

Calculate depreciation rate for the first month assuming a total drilling and completions Capex of $7 MM with 15% tangible and 65% WI using ACR2 7-year depreciation table.

$$\text{Depreciation}_{\text{month 1}} = \frac{14.29\%}{12} \times (7,000,000 \times 15\%) \times 65\% = \$8127$$

Taxable income

Once depreciation is calculated, taxable income can be calculated using Eqs. (18.25), (18.26).

$$\begin{aligned}\text{Taxable income @ investment date} \\ = -(\text{Intangible investment} \times \text{WI})\end{aligned} \quad (18.25)$$

where intangible investment is typically 80%–90% of the total drilling and completions Capex in $ and WI is working interest in %.

Please note that in Eq. (18.25), profit excluding investment at investment date is 0 and that is why the term (intangible investment × WI) is being multiplied by −1. In addition, depreciation is 0 at investment date. Eq. (18.25) assumes that the entire intangible capital is written off when investment is made. Typically the intangible capital is written off in the first year but for the simplicity of calculating monthly taxable income, intangible capital is written off at investment date in this equation.

$$\begin{aligned}\text{Taxable income after investment} = \text{Profit excluding investment} \\ - \text{depreciation}\end{aligned} \quad (18.26)$$

where taxable income after investment is the monthly taxable income in $ and Profit excluding investment is monthly basis in $.

Example

An 8000′ lateral length well's total Capex is $8.250 MM. Assuming 88% intangible Capex and 100% WI, calculate taxable income at and after initial investment date using Table 18.20 for the first year assuming ACR2, 7-years depreciation schedule:

Table 18.20 Taxable income example problem

Month	Profit excluding investment
0 (investment date)	$0
1	$253,794
2	$207,166
3	$178,231
4	$158,156
5	$143,241
6	$131,633
7	$122,288
8	$114,571
9	$108,068
10	$102,499
11	$97,665
12	$93,421

Depreciation rate for the first year from ACR 2, 7-years schedule is 14.29%.

$$\text{Depreciation}_{\text{month 1 through 12}} = \frac{14.29\%}{12} \times (8{,}250{,}000 \times 12\%) \times 100\%$$
$$= \$11{,}789$$

$$\text{Taxable income @ investment date} = -(8{,}250{,}000 \times 88\% \times 100\%)$$
$$= \$-7{,}260{,}000$$

$$\text{Taxable income}_{\text{month 1}} = 253{,}794 - 11{,}789 = \$242{,}005$$
$$\text{Taxable income}_{\text{month 2}} = 207{,}166 - 11{,}789 = \$195{,}377$$
$$\text{Taxable income}_{\text{month 3}} = 178{,}231 - 11{,}789 = \$166{,}442$$

The remaining taxable incomes for this example are summarized in Table 18.21.

Table 18.21 Taxable income example answer

Question			
	Answer		
Month	Profit excluding investment	Depreciation	Taxable income
0 (investment date)	$0	$0	−$7,260,000
1	$253,794	$11,789	$242,005
2	$207,166	$11,789	$195,377
3	$178,231	$11,789	$166,442
4	$158,156	$11,789	$146,367
5	$143,241	$11,789	$131,452
6	$131,633	$11,789	$119,844
7	$122,288	$11,789	$110,499
8	$114,571	$11,789	$102,782
9	$108,068	$11,789	$96,279
10	$102,499	$11,789	$90,710
11	$97,665	$11,789	$85,876
12	$93,421	$11,789	$81,632

Corporation tax

Corporations, just like individuals and small business owners, have specific tax brackets. Therefore, when performing ATAX calculation corporation tax must also be taken into account. Corporation tax can be calculated using Eq. (18.27).

$$\text{Corporation tax} = \text{Taxable income} \times \text{corporation tax rate} \quad (18.27)$$

where taxable income is in $ and Corporation tax rate is in %.

ATAX monthly undiscounted NCF

Now that depreciation, taxable income, and corporation tax have all been discussed, the last step before discounting the ATAX future cash flows is to calculate ATAX monthly undiscounted NCF. ATAX monthly undiscounted NCF can be calculated using Eq. (18.28). The 2018 tax reform decreased the corporation tax rate from 35% to 21%.

$$\text{ATAX monthly undiscounted NCF} = \text{BTAX monthly undis.NCF} - \text{Corporation tax} \quad (18.28)$$

where BTAX monthly undiscounted NCF is in $ and Corporation tax is in $.

Example

Assuming a corporation tax rate of 35%, calculate monthly corporation tax and ATAX monthly undiscounted NCF for the first year using the assumptions listed in Table 18.22.

Table 18.22 ATAX monthly undiscounted NCF example problem

Month	BTAX monthly undiscounted NCF	Taxable income	Corporation tax rate
0	−$4,379,096	−$3,897,395	35%
1	$253,794	$248,058	35%
2	$207,166	$201,429	35%
3	$178,231	$172,495	35%
4	$158,156	$152,420	35%
5	$143,241	$137,505	35%
6	$131,633	$125,897	35%
7	$122,288	$116,552	35%
8	$114,571	$108,835	35%
9	$108,068	$102,332	35%
10	$102,499	$96,763	35%
11	$97,665	$91,928	35%
12	$93,421	$87,685	35%

Step 1: Calculate corporation tax for each month (sample calculation below):

$$\text{Corporation tax}_{\text{time } 0} = -3,897,395 \times 35\% = -\$1,364,088$$
$$\text{Corporation tax}_{\text{month } 1} = 248,058 \times 35\% = \$86,820$$
$$\text{Corporation tax}_{\text{month } 2} = 201,429 \times 35\% = \$70,500$$

Step 2: Calculate ATAX monthly undiscounted NCF (sample calculation below):

$$\text{ATAX monthly undiscounted NCF}_{\text{time } 0} = -4,379,096 - (-1,364,088)$$
$$= \$-3,015,007$$

$$\text{ATAX monthly undiscounted NCF}_{\text{month } 1} = 253,794 - 86,820$$
$$= \$166,974$$

$$\text{ATAX monthly undiscounted NCF}_{\text{month } 2} = 207,166 - 70,500$$
$$= \$136,665$$

Table 18.23 summarizes the results for this example.

Table 18.23 ATAX monthly undiscounted NCF example answer

	Question			Answer	
Month	BTAX monthly undiscounted NCF	Taxable income	Corporation tax rate (%)	Corporation tax	ATAX monthly undiscounted NCF
0	−4,379,096	−3,897,395	35	−1,364,088	−3,015,007
1	253,794	248,058	35	86,820	166,974
2	207,166	201,429	35	70,500	136,665
3	178,231	172,495	35	60,373	117,858
4	158,156	152,420	35	53,347	104,809
5	143,241	137,505	35	48,127	95,114
6	131,633	125,897	35	44,064	87,569
7	122,288	116,552	35	40,793	81,495
8	114,571	108,835	35	38,092	76,479
9	108,068	102,332	35	35,816	72,252
10	102,499	96,763	35	33,867	68,632
11	97,665	91,928	35	32,175	65,490
12	93,421	87,685	35	30,690	62,732

Example

A type curve is generated from 200 producing dry gas wells from a field with similar reservoir properties. You are to run economic analysis and figure out whether the management should proceed with drilling and completing the well or not. The type curve generated is for an 8000' lateral length well with an IP of 14,500 MSCF/day, annual secant effective decline of 58%, and b value of 1.5. Assuming the following parameters, calculate NPV and IRR for the life of the well (assume 50-year life).

Terminal decline = 5%, WI = 100%, RI = 20%, BTU factor = 1.06 (1060 BTU/SCF), Shrinkage factor = 0.985

Fixed variable and gathering cost = \$426/month/well escalated at 3% to the life of the well, Variable lifting cost = \$0.14/MSCF escalated at 3% to the life of the well, Variable gathering and compression cost = \$0.35/MMBTU escalated at 3% to the life of the well, Firm transportation = \$0.30/MMBTU escalated at 3% to the life of the well, Gas price = assume \$3/MMBTU escalated at 3% to the life of the well, Severance tax = 5%, Ad valorem tax = 2.5%, Tangible investment = \$1,500,000, Intangible investment = \$6,500,000 (assume the entire intangible capital is written off when investment is made), Apply total investment 3 months before start date (TIL date), Discount all of the future cash flows using mid-point discounting to the date (time) the investment is made, Weighted average cost of capital = 8.8%, Corporation tax rate = 40%

The calculations shown below are for the first 6 months only and the remaining time is recommended to be performed using an Excel spreadsheet to compare the final NPV and IRR reported in this problem. This problem should provide step-by-step guidance on how to perform economic analysis on a new well based on the assumptions listed above. Some of the assumptions used in this example (e.g., ATAX calculation method, discounting method, etc.) can greatly vary from company to company.

Step 1: Calculate monthly nominal secant hyperbolic:

$$D_i = \left[\frac{1}{12b}\right] \times \left[(1 - D_{eis})^{-b} - 1\right] = \left[\frac{1}{12 \times 1.5}\right] \times \left[(1 - 58\%)^{-1.5} - 1\right]$$
$$= 14.85\%$$

Step 2: Calculate hyperbolic cumulative rate for each month starting with month 1:

$$N_p = \left\{ \left[\frac{IP}{(1-b) \times \text{Monthly Nominal Hyp}}\right] \right.$$
$$\left. \times \left[1 - (1 + b \times \text{Monthly Nominal Hyp} \times \text{time})^{1-\frac{1}{b}}\right] \right\} \times \frac{365}{12}$$

$$N_{p,\text{month 1}} = \left\{ \left[\frac{14,500}{(1-1.5) \times 14.85\%}\right] \right.$$
$$\left. \times \left[1 - (1 + 1.5 \times 14.85\% \times 1)^{1-\frac{1}{1.5}}\right] \right\}$$
$$\times \frac{365}{12} = 411,820 \, \text{MSCF}/1 \, \text{month}$$

$$N_{p,\text{month 2}} = \left\{ \left[\frac{14,500}{(1-1.5) \times 14.85\%} \right] \right.$$
$$\left. \times \left[1 - (1 + 1.5 \times 14.85\% \times 2)^{1-\frac{1}{1.5}} \right] \right\}$$
$$\times \frac{365}{12} = 776,199 \, \text{MSCF}/2 \, \text{month}$$

$$N_{p,\text{month 3}} = \left\{ \left[\frac{14,500}{(1-1.5) \times 14.85\%} \right] \right.$$
$$\left. \times \left[1 - (1 + 1.5 \times 14.85\% \times 3)^{1-\frac{1}{1.5}} \right] \right\}$$
$$\times \frac{365}{12} = 1,104,815 \, \text{MSCF}/3 \, \text{month}$$

$$N_{p,\text{month 4}} = \left\{ \left[\frac{14,500}{(1-1.5) \times 14.85\%} \right] \right.$$
$$\left. \times \left[1 - (1 + 1.5 \times 14.85\% \times 4)^{1-\frac{1}{1.5}} \right] \right\}$$
$$\times \frac{365}{12} = 1,405,331 \, \text{MSCF}/4 \, \text{month}$$

$$N_{p,\text{month 5}} = \left\{ \left[\frac{14,500}{(1-1.5) \times 14.85\%} \right] \right.$$
$$\left. \times \left[1 - (1 + 1.5 \times 14.85\% \times 5)^{1-\frac{1}{1.5}} \right] \right\}$$
$$\times \frac{365}{12} = 1,683,079 \, \text{MSCF}/5 \, \text{month}$$

$$N_{p,\text{month 6}} = \left\{ \left[\frac{14,500}{(1-1.5) \times 14.85\%} \right] \right.$$
$$\left. \times \left[1 - (1 + 1.5 \times 14.85\% \times 6)^{1-\frac{1}{1.5}} \right] \right\}$$
$$\times \frac{365}{12} = 1,941,935 \, \text{MSCF}/6 \, \text{month}.$$

Step 3: Calculate monthly rate by subtracting cumulative volumes from the previous month:

$q_{\text{hyperbolic, month 1}} = 411,820 \, \text{MSCF/month}$

$q_{\text{hyperbolic, month 2}} = 776,199 - 411,820 = 364,379 \, \text{MSCF/month}$

$q_{\text{hyperbolic, month 3}} = 1,104,815 - 776,199 = 328,616 \, \text{MSCF/month}$

$q_{\text{hyperbolic, month 4}} = 1,405,331 - 1,104,815 = 300,516 \, \text{MSCF/month}$

$q_{\text{hyperbolic, month 5}} = 1,683,079 - 1,405,331 = 277,748 \, \text{MSCF/month}$

$q_{\text{hyperbolic, month 6}} = 1,941,935 - 1,683,079 = 258,856 \, \text{MSCF/month}.$

Step 4: Calculate monthly nominal decline for each month:

$$D_{\text{month 1}} = \frac{\text{Monthly nominal hyperbolic}}{1 + b \times \text{monthly nominal hyperbolic} \times \text{time}}$$

$$= \frac{14.85\%}{1 + 1.5 \times 14.85\% \times 1} = 12.1\%$$

$$D_{\text{month 2}} = \frac{14.85\%}{1 + 1.5 \times 14.85\% \times 2} = 10.3\%$$

$$D_{\text{month 3}} = \frac{14.85\%}{1 + 1.5 \times 14.85\% \times 3} = 8.9\%$$

$$D_{\text{month 4}} = \frac{14.85\%}{1 + 1.5 \times 14.85\% \times 4} = 7.9\%$$

$$D_{\text{month 5}} = \frac{14.85\%}{1 + 1.5 \times 14.85\% \times 5} = 7.0\%$$

$$D_{\text{month 6}} = \frac{14.85\%}{1 + 1.5 \times 14.85\% \times 6} = 6.4\%.$$

Step 5: Calculate annual effective decline for each month:

$$D_{e,\text{month 1}} = 1 - (1 + 12 \times b \times D)^{-\frac{1}{b}} = 1 - (1 + 12 \times 1.5 \times 12.1\%)^{-\frac{1}{1.5}}$$
$$= 53.8\%$$

$$D_{e,\text{month 2}} = 1 - (1 + 12 \times 1.5 \times 10.3\%)^{-\frac{1}{1.5}} = 50.2\%$$

$$D_{e,\text{month 3}} = 1 - (1 + 12 \times 1.5 \times 8.9\%)^{-\frac{1}{1.5}} = 47.1\%$$

$$D_{e,\text{month 4}} = 1 - (1 + 12 \times 1.5 \times 7.9\%)^{-\frac{1}{1.5}} = 44.4\%$$

$$D_{e,\text{month 5}} = 1 - (1 + 12 \times 1.5 \times 7.0\%)^{-\frac{1}{1.5}} = 42.0\%$$

$$D_{e,\text{month 6}} = 1 - (1 + 12 \times 1.5 \times 6.4\%)^{-\frac{1}{1.5}} = 39.9\%$$

After calculating the annual effective decline for the remaining life of the well, it appears that at month 145, the annual effective decline reaches 5% terminal decline. The hyperbolic decline equation must be switched to an exponential decline equation for the life of the well starting with month 145.

Step 6: Calculate monthly nominal exponential decline using the following equation:

$$D = -\ln\left[(1 - D_e)^{\frac{1}{12}}\right] = -\ln\left[(1 - 5\%)^{\frac{1}{12}}\right] = 0.427\%$$

Step 7: Calculate exponential decline rate for each month after reaching 5% terminal decline using the following equation:

$$q_{\text{exponential}} = \left(\text{IP} \times e^{-D \times t}\right) \times \left(\frac{365}{12}\right)$$

The rate at which hyperbolic decline is switched to exponential is 1404 MSCF/day or 42,698 MSCF/month. The first month right after the switch time (month 146) is called month 1 in this particular example, followed by the remaining months for the life of the well.

$$q_{\text{exponential, month 1}} = \left(1404 \times e^{-0.427\% \times 1}\right) \times \left(\frac{365}{12}\right)$$
$$= 42,516 \text{ MSCF/month}$$

$$q_{\text{exponential, month 2}} = \left(1404 \times e^{-0.427\% \times 2}\right) \times \left(\frac{365}{12}\right)$$
$$= 42,334 \text{ MSCF/month}$$

$$q_{\text{exponential, month 3}} = \left(1404 \times e^{-0.427\% \times 3}\right) \times \left(\frac{365}{12}\right)$$
$$= 42,154 \text{ MSCF/month}$$

$$q_{\text{exponential, month 4}} = \left(1404 \times e^{-0.427\% \times 4}\right) \times \left(\frac{365}{12}\right)$$
$$= 41,974 \text{ MSCF/month}$$

$$q_{\text{exponential, month 5}} = \left(1404 \times e^{-0.427\% \times 5}\right) \times \left(\frac{365}{12}\right)$$
$$= 41,795 \text{ MSCF/month}$$

$$q_{\text{exponential, month 6}} = \left(1404 \times e^{-0.427\% \times 6}\right) \times \left(\frac{365}{12}\right)$$
$$= 41,617 \text{ MSCF/month}.$$

Step 8: Calculate net gas production for each month:

Net gas production = Gross gas production × shrinkage factor × NRI%

$$\text{Net gas production}_{\text{month 1}} = 411,820 \times 0.985 \times 80\%$$
$$= 324,514 \text{ MSCF/month}$$

$$\text{Net gas production}_{\text{month 2}} = 364,379 \times 0.985 \times 80\%$$
$$= 287,131 \text{ MSCF/month}$$

$$\text{Net gas production}_{\text{month 3}} = 328,616 \times 0.985 \times 80\%$$
$$= 258,949 \text{ MSCF/month}$$

$$\text{Net gas production}_{\text{month 4}} = 277,748 \times 0.985 \times 80\%$$
$$= 236,806 \text{ MSCF/month}$$

$$\text{Net gas production}_{\text{month 5}} = 258,856 \times 0.985 \times 80\%$$
$$= 218,865 \text{ MSCF/month}$$

$$\text{Net gas production}_{\text{month 6}} = 242,883 \times 0.985 \times 80\%$$
$$= 203,979 \text{ MSCF/month}.$$

Step 9: Calculate gas pricing incorporating an escalation of 3% using a stair-step escalation:

$$\text{Gas price}_{\text{month 1}} = \$3/\text{MMBTU}$$
$$\text{Gas price}_{\text{month 2}} = 3 \times (1+3\%)^{\frac{1}{12}} = \$3.007/\text{MMBTU}$$
$$\text{Gas price}_{\text{month 3}} = 3.007 \times (1+3\%)^{\frac{1}{12}} = \$3.015/\text{MMBTU}$$
$$\text{Gas price}_{\text{month 4}} = 3.015 \times (1+3\%)^{\frac{1}{12}} = \$3.022/\text{MMBTU}$$
$$\text{Gas price}_{\text{month 5}} = 3.022 \times (1+3\%)^{\frac{1}{12}} = \$3.030/\text{MMBTU}$$
$$\text{Gas price}_{\text{month 6}} = 3.030 \times (1+3\%)^{\frac{1}{12}} = \$3.037/\text{MMBTU}.$$

Step 10: Calculate adjusted gas pricing by accounting for 1060 BTU gas:

$$\text{Adjusted gas price} = \text{Gas price} \times \text{BTU factor}$$
$$\text{Adjusted gas price}_{\text{month 1}} = 3 \times 1.06 = \$3.18/\text{MSCF}$$
$$\text{Adjusted gas price}_{\text{month 2}} = 3.007 \times 1.06 = \$3.188/\text{MSCF}$$
$$\text{Adjusted gas price}_{\text{month 3}} = 3.015 \times 1.06 = \$3.196/\text{MSCF}$$
$$\text{Adjusted gas price}_{\text{month 4}} = 3.022 \times 1.06 = \$3.204/\text{MSCF}$$
$$\text{Adjusted gas price}_{\text{month 5}} = 3.030 \times 1.06 = \$3.211/\text{MSCF}$$
$$\text{Adjusted gas price}_{\text{month 6}} = 3.037 \times 1.06 = \$3.219/\text{MSCF}.$$

Step 11: Calculate net revenue for each month:

$$\text{Net revenue} = (\text{Monthly shrunk net gas production} \times \text{adjusted gas pricing})$$

$$\text{Net revenue}_{\text{month 1}} = 324,514 \times 3.18 = \$1,031,955$$
$$\text{Net revenue}_{\text{month 2}} = 287,131 \times 3.188 = \$915,328$$
$$\text{Net revenue}_{\text{month 3}} = 258,949 \times 3.196 = \$827,526$$
$$\text{Net revenue}_{\text{month 4}} = 236,806 \times 3.204 = \$758,630$$
$$\text{Net revenue}_{\text{month 5}} = 218,865 \times 3.211 = \$702,883$$
$$\text{Net revenue}_{\text{month 6}} = 203,979 \times 3.219 = \$656,691.$$

Step 12: Calculate severance tax for each month:

Severance tax per month
$$= (\text{Gross monthly gas production} \times \text{adjusted gas pricing} \times \text{severance tax}$$
$$\times \text{NRI} \times \text{total shrinkage factor}) \text{ OR Net revenue} \times \text{severance tax}$$

$$\text{Severance tax}_{\text{month 1}} = 1,031,955 \times 5\% = \$51,598$$
$$\text{Severance tax}_{\text{month 2}} = 915,328 \times 5\% = \$45,766$$
$$\text{Severance tax}_{\text{month 3}} = 827,526 \times 5\% = \$41,376$$
$$\text{Severance tax}_{\text{month 4}} = 758,630 \times 5\% = \$37,931$$

$$\text{Severance tax}_{\text{month 5}} = 702,883 \times 5\% = \$35,144$$
$$\text{Severance tax}_{\text{month 6}} = 656,691 \times 5\% = \$32,835.$$

Step 13: Calculate ad valorem tax for each month:

Ad valorem tax per month
$$= \{[(\text{Gross monthly gas production} \times \text{adjusted gas pricing} \times \text{NRI}$$
$$\times \text{total shrinkage factor})] - \text{Severance tax amount}\}$$
$$\times \text{Ad valorem tax OR}\,(\text{Net revenue} - \text{severance tax}) \times \text{Ad valorem tax}$$

$$\text{Ad valorem tax}_{\text{month 1}} = (1,031,955 - 51,598) \times 2.5\% = \$24,509$$
$$\text{Ad valorem tax}_{\text{month 2}} = (915,328 - 45,766) \times 2.5\% = \$21,739$$
$$\text{Ad valorem tax}_{\text{month 3}} = (827,526 - 41,376) \times 2.5\% = \$19,654$$
$$\text{Ad valorem tax}_{\text{month 4}} = (758,630 - 37,931) \times 2.5\% = \$18,017$$
$$\text{Ad valorem tax}_{\text{month 5}} = (702,883 - 35,144) \times 2.5\% = \$16,693$$
$$\text{Ad valorem tax}_{\text{month 6}} = (656,691 - 32,835) \times 2.5\% = \$15,596$$

Step 14: First perform escalation on fixed, variable, and FT costs:

$$\text{Fixed cost escalatin}_{\text{month 1}} = \$426$$
$$\text{Fixed cost escalatin}_{\text{month 2}} = 426 \times (1 + 3\%)^{\frac{1}{12}} = \$427.1$$
$$\text{Fixed cost escalatin}_{\text{month 3}} = 427.1 \times (1 + 3\%)^{\frac{1}{12}} = \$428.1$$
$$\text{Fixed cost escalatin}_{\text{month 4}} = 428.1 \times (1 + 3\%)^{\frac{1}{12}} = \$429.2$$
$$\text{Fixed cost escalatin}_{\text{month 5}} = 429.2 \times (1 + 3\%)^{\frac{1}{12}} = \$430.2$$
$$\text{Fixed cost escalatin}_{\text{month 6}} = 430.2 \times (1 + 3\%)^{\frac{1}{12}} = \$431.3$$

Total variable(\$/MSCF)
$$= \text{variable lifting}(\$/\text{MSCF}) + \text{variable gathering}(\$/\text{MSCF})$$
$$+ \text{FT}(\$/\text{MSCF})$$
Total variable cost per MSCF $= \$0.14/\text{MSCF} + (\$0.35/\text{MMBTU} \times 1.06)$
$$+ (\$0.3/\text{MMBTU} \times 1.06) = \$0.829/\text{MSCF}$$

$$\text{Variable cost escalatin}_{\text{month 1}} = \$0.829/\text{MSCF}$$
$$\text{Variable cost escalatin}_{\text{month 2}} = 0.829 \times (1 + 3\%)^{\frac{1}{12}} = \$0.831/\text{MSCF}$$
$$\text{Variable cost escalatin}_{\text{month 3}} = 0.831 \times (1 + 3\%)^{\frac{1}{12}} = \$0.833/\text{MSCF}$$
$$\text{Variable cost escalatin}_{\text{month 4}} = 0.833 \times (1 + 3\%)^{\frac{1}{12}} = \$0.835/\text{MSCF}$$
$$\text{Variable cost escalatin}_{\text{month 5}} = 0.835 \times (1 + 3\%)^{\frac{1}{12}} = \$0.837/\text{MSCF}$$
$$\text{Variable cost escalatin}_{\text{month 6}} = 0.837 \times (1 + 3\%)^{\frac{1}{12}} = \$0.839/\text{MSCF}.$$

Step 15: Calculate total Opex for each month:

Total Opex per month
= [(Gross monthly gas production × WI × total shrinkage factor
  × variable lifting cost)] + [(Fixed lifting cost × WI)]
+ [(Gross monthly gas production × WI × total shrinkage factor
  × gathering and compression cost)]
+ [(Gross monthly gas production × WI
  × total shrinkage factor × FT cost)]

Total Opex$_{month\ 1}$ = $(411820 \times 100\% \times 0.985 \times 0.829) + (426 \times 100\%)$
    = $336,704

Total Opex$_{month\ 2}$ = $(364,379 \times 100\% \times 0.985 \times 0.831) + (427.1 \times 100\%)$
    = $298,700

Total Opex$_{month\ 3}$ = $(328,616 \times 100\% \times 0.985 \times 0.833) + (428.1 \times 100\%)$
    = $270,090

Total Opex$_{month\ 4}$ = $(300,516 \times 100\% \times 0.985 \times 0.835) + (429.2 \times 100\%)$
    = $247,640

Total Opex$_{month\ 5}$ = $(277,748 \times 100\% \times 0.985 \times 0.837) + (430.2 \times 100\%)$
    = $229,475

Total Opex$_{month\ 6}$ = $(258,856 \times 100\% \times 0.985 \times 0.839) + (431.3 \times 100\%)$
    = $214,424.

Step 16: Calculate net Opex for each month:

Net Opex = Total Opex + severance tax amount + ad valorem tax amount

Net Opex$_{month\ 1}$ = $336,704 + 51,598 + 24,509 = \$412,811$
Net Opex$_{month\ 2}$ = $298,700 + 45,766 + 21,739 = \$366,206$
Net Opex$_{month\ 3}$ = $270,090 + 41,376 + 19,654 = \$331,120$
Net Opex$_{month\ 4}$ = $247,640 + 37,931 + 18,017 = \$303,589$
Net Opex$_{month\ 5}$ = $229,475 + 35,144 + 16,693 = \$281,312$
Net Opex$_{month\ 6}$ = $214,424 + 32,835 + 15,596 = \$262,855$.

Step 17: Calculate operating cash flow or profit excluding investment:

Profit(excluding investment) = Net revenue − net Opex
Profit$_{month\ 1}$ = $1,031,955 - 412,811 = \$619,145$
Profit$_{month\ 2}$ = $915,328 - 366,206 = \$549,122$
Profit$_{month\ 3}$ = $827,526 - 331,120 = \$496,406$
Profit$_{month\ 4}$ = $758,630 - 303,589 = \$455,041$
Profit$_{month\ 5}$ = $702,883 - 281,312 = \$421,570$
Profit$_{month\ 6}$ = $656,691 - 262,855 = \$393,836$.

Step 18: Calculate net Capex (since WI is 100%, net Capex is equal to gross Capex):

$$\text{Net Capex} = \text{Gross Capex} \times \text{WI}$$
$$\text{Net Capex} = (1{,}500{,}000 + 6{,}500{,}000) \times 100\% = \$8{,}000{,}000$$

Apply $8,000,000 total net investment 3 months prior to start date (the date where the production begins).

Step 19: Calculate BTAX monthly undiscounted NCF:

$$\text{BTAX monthly undiscounted net cash flow} = \text{profit} - \text{net Capex}$$

Net Capex at time zero is equal to net Capex. However, net Capex for subsequent months is zero.

$$\text{Investment date (time 0)} = 0 - 8{,}000{,}000 = -\$8{,}000{,}000$$
$$\text{BTAX undiscounted NCF}_{\text{month 2}} = \$0$$
$$\text{BTAX undiscounted NCF}_{\text{month 3}} = \$0$$
$$\text{BTAX undiscounted NCF}_{\text{month 4 from investment date}} = 619{,}145 - 0 = \$619{,}145$$
$$\text{BTAX undiscounted NCF}_{\text{month 5 from investment date}} = 549{,}122 - 0 = \$549{,}122$$
$$\text{BTAX undiscounted NCF}_{\text{month 6 from investment date}} = 496{,}406 - 0 = \$496{,}406$$
$$\text{BTAX undiscounted NCF}_{\text{month 7 from investment date}} = 455{,}041 - 0 = \$455{,}041$$
$$\text{BTAX undiscounted NCF}_{\text{month 8 from investment date}} = 421{,}570 - 0 = \$421{,}570$$
$$\text{BTAX undiscounted NCF}_{\text{month 9 from investment date}} = 393{,}836 - 0 = \$393{,}836.$$

Step 20: Calculate BTAX monthly discounted (midpoint) NCF. To perform midpoint discounting, subtract 0.5 from each month as shown in the following equation:

$$\text{BTAX monthly discounted NCF} = \frac{\text{BTAX monthly undiscounted NCF}}{(1 + \text{WACC})^{\frac{\text{Time}-0.5}{12}}}$$

$$\text{BTAX discounted NCF}_{\text{investment date}} = -\$8{,}000{,}000$$
$$\text{BTAX discounted NCF}_{\text{month 2}} = \$0$$
$$\text{BTAX discounted NCF}_{\text{month 3}} = \$0$$

$$\text{BTAX discounted NCF}_{\text{month 4 from investment date}} = \frac{619{,}145}{(1 + 8.8\%)^{\frac{4-0.5}{12}}} = \$604{,}100$$

$$\text{BTAX discounted NCF}_{\text{month 5 from investment date}} = \frac{549{,}122}{(1 + 8.8\%)^{\frac{5-0.5}{12}}} = \$532{,}026$$

$$\text{BTAX discounted NCF}_{\text{month 6 from investment date}} = \frac{496{,}406}{(1 + 8.8\%)^{\frac{6-0.5}{12}}} = \$477{,}583$$

$$\text{BTAX discounted NCF}_{\text{month 7 from investment date}} = \frac{455,041}{(1+8.8\%)^{\frac{7-0.5}{12}}} = \$434,720$$

$$\text{BTAX discounted NCF}_{\text{month 8 from investment date}} = \frac{421,570}{(1+8.8\%)^{\frac{8-0.5}{12}}} = \$399,923$$

$$\text{BTAX discounted NCF}_{\text{month 9 from investment date}} = \frac{393,836}{(1+8.8\%)^{\frac{9-0.5}{12}}} = \$370,997.$$

Step 21: Calculate depreciation for each month starting with production date:

$$\text{Monthly depreciation} = \frac{\text{Yearly depreciation rate}}{12} \times \text{tangible investment} \times \text{WI}$$

$$\text{Depreciation}_{\text{month 1}} = \frac{14.29\%}{12} \times 1,500,000 \times 100\% = \$17,863$$

Depreciation for the next 11 months will be the same using IRS-defined accelerated recovery method.

Step 22: Calculate taxable income starting with when the investment is made:

$$\begin{aligned}\text{Taxable income at investment date} &= -(\text{Intangible investment} \times \text{WI})\\ &= -(6,500,000 \times 100\%)\\ &= -\$6,500,000\end{aligned}$$

$$\text{Taxable income after investment} = \text{Profit excluding investment} - \text{depreciation}$$

$$\text{Taxable income}_{\text{month 2}} = \$0$$
$$\text{Taxable income}_{\text{month 3}} = \$0$$
$$\text{Taxable income}_{\text{month 4}} = 619,145 - 17,863 = \$601,282$$
$$\text{Taxable income}_{\text{month 5}} = 549,122 - 17,863 = \$531,260$$
$$\text{Taxable income}_{\text{month 6}} = 496,406 - 17,863 = \$478,544$$
$$\text{Taxable income}_{\text{month 7}} = 455,041 - 17,863 = \$437,179$$
$$\text{Taxable income}_{\text{month 8}} = 421,570 - 17,863 = \$403,708$$
$$\text{Taxable income}_{\text{month 9}} = 393,836 - 17,863 = \$375,974.$$

Step 23: Calculate corporation tax for each month starting with the investment date:

$$\text{Corporation tax} = \text{Taxable income} \times \text{corporation tax rate}$$
$$\text{Corporation tax}_{\text{investment date}} = -6,500,000 \times 40\% = -\$2,600,000$$
$$\text{Corporation tax}_{\text{month 2}} = \$0$$
$$\text{Corporation tax}_{\text{month 3}} = \$0$$

Corporation tax$_{month\ 4}$ = 601,282 × 40% = $240,513
Corporation tax$_{month\ 5}$ = 531,260 × 40% = $212,504
Corporation tax$_{month\ 6}$ = 478,544 × 40% = $191,417
Corporation tax$_{month\ 7}$ = 437,179 × 40% = $174,871
Corporation tax$_{month\ 8}$ = 403,708 × 40% = $161,483
Corporation tax$_{month\ 9}$ = 375,974 × 40% = $150,389.

Step 24: Calculate ATAX undiscounted NCF for each month starting with the investment date:

$$\text{ATAX monthly undiscounted NCF} = \text{BTAX monthly undis.NCF} - \text{Corporation tax}$$

ATAX monthly undiscounted NCF$_{Invetsment\ date}$
= −8,000,000 − (−2,600,000) = −$5,400,000
ATAX monthly undiscounted NCF$_{month\ 2}$ = $0
ATAX monthly undiscounted NCF$_{month\ 3}$ = $0

ATAX monthly undiscounted NCF$_{month\ 4}$ = 619,145 − 240,513 = $378,632

ATAX monthly undiscounted NCF$_{month\ 5}$ = 549,122 − 212,504 = $336,618

ATAX monthly undiscounted NCF$_{month\ 6}$ = 496,406 − 191,417 = $304,989

ATAX monthly undiscounted NCF$_{month\ 7}$ = 455,041 − 174,871 = $280,170

ATAX monthly undiscounted NCF$_{month\ 8}$ = 421,570 − 161,483 = $260,087

ATAX monthly undiscounted NCF$_{month\ 9}$ = 393,836 − 150,389 = $243,447.

Step 25: Calculate ATAX monthly discounted NCF for each month starting with investment date:

$$\text{ATAX monthly discounted NCF} = \frac{\text{ATAX monthly undiscounted NCF}}{(1+\text{WACC})^{\frac{Time-0.5}{12}}}$$

ATAX discounted NCF$_{investment\ date}$ = −$5,400,000
ATAX discounted NCF$_{month\ 2}$ = $0
ATAX discounted NCF$_{month\ 3}$ = $0

$$\text{ATAX discounted NCF}_{month\ 4\ from\ investment\ date} = \frac{378,632}{(1+8.8\%)^{\frac{4-0.5}{12}}} = \$369,431$$

$$\text{ATAX discounted NCF}_{\text{month 5 from investment date}} = \frac{336{,}618}{(1+8.8\%)^{\frac{5-0.5}{12}}} = \$326{,}138$$

$$\text{ATAX discounted NCF}_{\text{month 6 from investment date}} = \frac{304{,}989}{(1+8.8\%)^{\frac{6-0.5}{12}}} = \$293{,}424$$

$$\text{ATAX discounted NCF}_{\text{month 7 from investment date}} = \frac{280{,}170}{(1+8.8\%)^{\frac{7-0.5}{12}}} = \$267{,}658$$

$$\text{ATAX discounted NCF}_{\text{month 8 from investment date}} = \frac{260{,}087}{(1+8.8\%)^{\frac{8-0.5}{12}}} = \$246{,}732$$

$$\text{ATAX discounted NCF}_{\text{month 9 from investment date}} = \frac{243{,}447}{(1+8.8\%)^{\frac{9-0.5}{12}}} = \$229{,}329$$

BTAX NPV is the summation of all BTAX monthly discounted cash flows for 50 years. ATAX NPV is the summation of ATAX monthly discounted cash flows for 50 years. The summaries of both BTAX and ATAX NPVs are listed in Table 18.24.

Table 18.24 ATAX and BTAX NPV profile example
NPV profile (mid-discounting)

Discount rate (%)	BTAX NPV	ATAX NPV
0	$42,158,086	$25,294,851
5	$17,876,005	$10,637,205
8.8	**$11,250,715**	**$6,608,750**
10	$9,910,182	$5,789,584
15	$6,163,010	$3,487,676
20	$3,962,247	$2,124,255
25	$2,488,747	$1,205,150
30	$1,418,424	$534,014
40	($56,802)	($395,914)
50	($1,043,607)	($1,020,815)
60	($1,760,296)	($1,475,760)
70	($2,309,467)	($1,824,769)
80	($2,746,487)	($2,102,619)
90	($3,104,192)	($2,330,036)
100	($3,403,442)	($2,520,232)

BTAX and ATAX IRR can now be easily calculated by interpolating between 30% and 40% discount rate, in which the BTAX and ATAX NPV switch sign from positive to negative.

$$Y = Y_a + (Y_b - Y_a) \times \frac{X - X_a}{X_b - X_a}$$

$$\text{BTAX IRR} = 30\% + (40\% - 30\%) \times \frac{(0 - 1{,}418{,}424)}{(-56{,}802 - 1{,}418{,}424)} = 39.61\%$$

$$\text{ATAX IRR} = 30\% + (40\% - 30\%) \times \frac{(0 - 534{,}014)}{(-395{,}914 - 534{,}014)} = 35.74\%.$$

CHAPTER NINETEEN

The role of federal reserve

Introduction

The Federal Reserve (The FED) was created in 1913 and is independent from the government regardless of political party in office. In the United States, the FED works toward the improvement of the US economy and people without interest in pleasing any political party. The FED committee meetings occur 8 times a year (~every 6 weeks) to decide whether to increase or decrease the interest rate and money supply. The Federal Open Market Committee (FOMC) members set a target rate for the federal funds rate and depending on various economic condition factors can hold additional meetings outside of its regular schedule. The main objectives of the FED are as follows:
- Maximizing employment
- Minimizing inflation

Interest rate management

The FED regulates the economy by controlling the money supply. The FED's responsibility is to maximize employment while minimizing inflation. A perfect example of the FED's role was during the financial crisis of 2008 when the FED lowered the interest rate to 0% and increased the money supply. In this example, the FED demonstrated that the primary ways to stimulate the economy are lowering interest rate and increasing the money supply. When the FED lowers the bank's interest rate, banks in turn distribute those savings to other borrowers and consumers. When the economy is in a recession or when facing a slowdown (downturn), the FED lowers the interest rate. This phenomenon encourages people to borrow more money. For example, if there is an economic slowdown and the FED drops the interest rate from 10% to 5% to stimulate the economy, they make it possible to borrow more money without raising the monthly payment. To illustrate this concept, let's consider an example.

Let's imagine that you are interested in buying a second house in the amount of $500K using a conventional loan at 10% interest rate, 30-year mortgage, and 20% down payment. Therefore, the amount that you would be borrowing would be 400K at 10% and 30-year amortization schedule. There are two types of cash flows which are referred to as perpetuity and annuity. Perpetuity pays a constant stream of identical cash flow C forever and it is considered a type of annuity. Present value of a perpetuity and present value of a growing perpetuity can be calculated as illustrated in Eqs. (19.1), (19.2).

$$\text{Present value of a perpetuity} = \frac{C}{r} \quad (19.1)$$

$$\text{Present value of a growing perpetuity} = \frac{C}{r-g} \quad (19.2)$$

where C is the cash flow, r is the interest rate, and g is the growth rate.

On the other hand, annuity pays constant cash flow C for T periods and can be calculated as shown in Eqs. (19.3), (19.4).

$$\text{Present value of an annuity} = \frac{C}{r}\left[1 - \frac{1}{(1+r)^T}\right] \quad (19.3)$$

$$\text{Present value of a growing annuity} = \frac{C}{r-g}\left[1 - \frac{(1+g)^T}{(1+r)^T}\right] \quad (19.4)$$

Eq. (19.3) can be rewritten as follows to solve for monthly mortgage payment:

$$C = PV\frac{i(1+r)^T}{(1+r)^T - 1}$$

where C is the monthly payment, PV is the principle amount that is being borrowed ($400K in this example), i is interest rate (divide by 12 to get monthly interest rate), and n is number of months (number of payments). Therefore, the monthly payment for the second house mortgage example can be calculated as follows:

$$C = 400,000 \times \frac{\frac{10\%}{12}\left(1 + \frac{10\%}{12}\right)^{30 \times 12}}{\left(1 + \frac{10\%}{12}\right)^{30 \times 12} - 1} = \$3510$$

Therefore, the monthly mortgage payment would be $3510 for 30 years. Now, let's perform the same calculation to figure out the incremental amount to $400K that one can borrow to equate paying the same principle amount every month at a lower interest rate of 5%.

$$C = 400,000 \times \frac{\frac{5\%}{12}\left(1+\frac{5\%}{12}\right)^{30\times12}}{\left(1+\frac{5\%}{12}\right)^{30\times12}-1} = \$2147$$

As can be seen, the monthly mortgage payment has now dropped by $1363 from **$3510 to $2147**. Next, let's figure out how much more someone can borrow to pay the same amount as $3510 that was calculated initially at 10% interest rate.

$$3510 = X \times \frac{\frac{5\%}{12}\left(1+\frac{5\%}{12}\right)^{30\times12}}{\left(1+\frac{5\%}{12}\right)^{30\times12}-1} \rightarrow X = \$653,848$$

At 5% interest, a borrower can now borrow a whopping $653,848 (~245K incremental) and pay the same monthly payment of $3510 just because the interest rate has dropped from 10% to 5%. This example illustrates the impact of lowering interest rate on a microlevel. Now, the huge effect that the FED has on the US economy on a macrolevel is noticeable. When interest rate is lowered, and more money is borrowed, it encourages people to spend more money simply because they now have more money to spend. As opposed to paying $3510 on the second home, they can now pay $2147 and potentially spend, save, or invest the rest. Spending money will stimulate the economy and create more jobs since consumers primarily drive the economy. The more money that becomes available to consumers, the more they will spend. With more people buying, spending, and working (due to high consumer demand) because of cheap money, it becomes apparent that there is still a **limited amount of supplies**. Therefore, due to limited amount of supply (or reduced supply), cheap money causes **inflation**. For example, if a class E Mercedes is $60,000, and you and 100 more people walk into a Mercedes dealer that only have 10 MB left in the store, the dealer can raise the price of MB if the demand is steady. This causes an inflation which can reduce the purchasing power of money. This impact is of the inherent problems of the FED. Please note that there is a lag from the time the FED increases or decreases interest rate to be felt in the economy. By the time the FED feels the impact of inflation due to lowering the interest rate, they can now change their decision to increase the interest rate to cool down the economy and balance inflation. The increase in interest rate and slowdown in the economy will cause less people to borrow and spend, leading to a drop in prices as the economy cools down. If central bankers significantly increase the interest rate to balance inflation, not many people will have the borrowing ability which could lead to a recession. If the economy slows or encounters a recession, the FED can once again lower the interest rate.

Money supply management

As previously mentioned, the second way that the FED regulates the economy is through the money supply. This process is referred to as quantitative easing (QE) or printing money. A perfect example of QE that the FED has infamously done is the 2008 financial crisis to help the US economy grow out of recession. Increasing the money supply is done through buying bonds (bonds that banks have on their balance sheet) from the banks by placing them in their portfolio and providing the cash to the banks. Those banks no longer have the bonds in their portfolio and now instead have cash on hand. Sitting on cash is not going to bring any profits to the banks; therefore, the banks will now go out and lend out their cash to borrowers and consumers to make profit. So, not only does the FED lowers the interest rate, it provides money supply to the banks through printing money or QE. When the interest rate is low (cheap money period), since banks have excess cash on their balance sheet, they will offer good deals at a low interest rate to encourage borrowers and consumers to spend and stimulate the economy. Therefore, in good times, banks are at much more liberty to lend out money to consumers. The snowball effect of spending will gradually pull the economy out of the recession, a process referred to as a **recovery**. Therefore, the FED works in two cycles such as stimulating spending periods and slowing spending periods. When inflation rises, to cool down the economy, the FED can now pull cash out of the system by offering the banks bonds which will reduce the money supply. They will also increase the interest rate to reduce borrowing ability. As a result, banks have less cash on hand and the interest

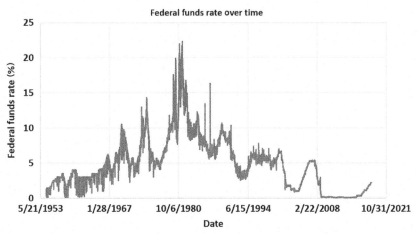

Fig. 19.1 Federal funds rate. *(Data obtained from www.macrotrends.net.)*

rate to borrow money is higher, causing a tighter policy when it comes to lending money to consumers. In President Regan's era, the inflation became so high that the banks had to significantly increase the interest rate to balance the inflation. The interest rate went up so high that a 30-year house mortgage was over 20% as can be seen in Fig. 19.1 which illustrates federal funds rate. After the early 1980s, the FED started lowering interest rates as the economy cooled off and inflation balanced. At the time of writing this book, all central banks have pumped so much money (liquidity) in the system that inflation could potentially get very severe. The only recourse would then be to increase the interest rate and lower the money supply. Therefore, another prominent inherent challenge with the FED is that there is an inherent lag factor between action time and widespread effects of their policy. This issue could very well lead to another recession.

A lower interest rate provides a great opportunity for investors and entrepreneurs to borrow as much money as possible for as long as possible (depending on credit score and lending capability) to invest in various projects if the interest rate is locked in.

In 1971, a lot of us remember the day when President Nixon made the public announcement taking the US dollar off the gold standard which essentially meant that money can be printed from thin air and is only backed by the trusts of other governments in US dollar. In this pivotal moment, our currency devaluation started, reducing the purchasing power of money due to inflation.

Federal funds rate, discount rate, and prime rate

As previously discussed, Fig. 19.1 shows the federal funds rate set by the FED. **The federal funds rate** refers to the bank-to-bank lending in which one bank borrows from another bank (usually overnight). This interaction occurs because banks are required to have a minimum reserve balance. At the end of each day, some banks are below the minimum reserve requirement while others have a surplus. Therefore, the banks that need to boost their funds overnight can borrow from other banks with surplus funds at a federal funds rate. On the other hand, **discount rate** is the interest rate in which the FED offers to banks and financial institutions. The discount rate is usually higher than the federal funds rate. The FED sets both the federal funds and discount rates. The discount rate is used as the last resort for the banks that need to borrow money from the FED for themselves and not for lending to other banks.

Prime rate is another term used in the banking industry, referring to the interest rate offered by the banks to consumers and borrowers. The prime rate is known as the benchmark for lending rates. The prime rate is higher than the federal funds rate, and a margin is usually added to the prime rate depending on the customers' credit score, worthiness, and collateral ability (risk level of the borrower). The prime rate is used as a benchmark when setting credit card, mortgage, small business loans, and home equity lines of credit rates. Business owners that have a track record of success and excellent credit score are known as "prime borrowers" (borrowers who offer the lowest risk) and can quality for the prime rate. The prime rate is generally calculated by adding 3% to the federal funds rate. As of December 2018, the federal funds rate is 2.25% and the prime rate is 5.25% (additional 3% to federal funds rate). Below is a summary of each rate:

Fed funds rate → Bank-to-bank lending
Discount rate → Fed to banks and financial institutions lending
Prime rate → Bank-to-consumer Lending

Historically important financial and political events

Some of the important dates that must be intuitively understood (Investopedia) are given in the following:
1913:
- The Federal Reserve bank was created. The FED was created by the congress to provide flexible and stable monetary and financial system.

- The 16th amendment to the US constitution was passed, it called for the ratification of federal income tax which essentially authorized the government to tax income.

1929:
- The huge stock market crash occurred, leading to the great depression, a worldwide economic crisis and lasting until mid-1930s. Many people who lived through the great depression and experienced the difficulty first hand learned to be more conservative regarding spending and saving habits.[1]

1935:
- The Social Security Act became law and was signed by President Roosevelt.

1943:
- Current tax payment act was passed, allowing the government to take taxes out of employee's paychecks before getting paid.

1944:
- The Bretton Woods Agreement is an important part of the world's financial history which put the world on the dollar standard in which the US agreed to back its dollar with gold. This was considered to be a new international monetary system developed by delegates from 44 countries. The dollar essentially became the reserve currency of the world.[2]

1971:
- President Nixon ended the Bretton Woods Agreement. In 1973, the agreement officially ended. The printing of money began. Today's financial crisis could have potentially been prevented if Nixon had not ended the Bretton Woods Agreement.

1974:
- After the termination of Bretton Woods Agreement by President Nixon, he signed a Petrodollar Agreement with Saudi Arabia to standardize oil price contracts in US dollar. The US dollar is now backed by oil and all countries are required to buy oil in US dollar. This was a paradigm shift from fixed exchange rates and gold-backed currencies to non-backed floating rate.[3]

---

[1] https://www.investopedia.com/terms/s/stock-market-crash-1929.asp
[2] https://www.investopedia.com/terms/b/brettonwoodsagreement.asp
[3] https://www.investopedia.com/terms/p/petrodollars.asp

1978:
- Congress passed The Revenue Act of 1978 in which allowed employees to prevent getting taxed on a portion of their income to be received as deferred compensation. This was the birth of 401K.[4]

1987:
- In the financial world, black Monday refers to the 1978 stock market crash which started in Hong Kong and eventually impacted the US stock market (in some countries it is referred to black Tuesday due to time zone difference). Fed Chairman, Alan Greenspan (1987–2006) passed the Greenspan Put which is referred to relying on the stock market **put option strategy** that could help investors mitigate losses due to a sudden drop in the stock price. Alan Greenspan basically propped up the market by lowering the interest rate and increasing money supply after the stock market crash of 1987. From 1987 to 2000, the Dow gained an almost exponential increase in value when millions of middle class and investors (specifically passive) became very wealthy via inflation through their home, 401K, IRAs, and other company/government pension plans.[5]

21st century:
Three giant crashes in the first 10 years of the 21st century occurred:
- **2000:** The Dot-com Crash after a period of exponential growth in internet embracement and growth.
- **2007:** The Subprime Crash which was triggered by a large decline in home prices.[6]
- **2008:** The financial crisis of 2008 began in 2007 with the Subprime mortgage crash and led to a full-scale international banking crisis. This caused Lehman Brothers which was 150-year-old bank and one of the oldest banks in the United States to file for bankruptcy and close.[7]

As demonstrated by these brief historical facts in the past century, the market experiences cycles and these cycles (including the booms and the busts) have always occurred and are expected to continue.

---

[4] https://www.investopedia.com/ask/answers/100314/why-were-401k-plans-created.asp

[5] https://www.investopedia.com/terms/s/stock-market-crash-1987.asp

[6] https://www.investopedia.com/terms/s/subprime-meltdown.asp

[7] https://www.investopedia.com/articles/economics/09/financial-crisis-review.asp

Example

You just won the lottery and it pays $130,000 a year for 20 years? (Assume 10% discount rate)

Part (a) What is the present value of this annuity?

$$\text{Present value of an annuity} = \frac{C}{r}\left[1 - \frac{1}{(1+r)^T}\right]$$

$$= \frac{130{,}000}{10\%}\left[1 - \frac{1}{(1+10\%)^{20}}\right] = \$1{,}106{,}763$$

Part (b) What if the payments are made for 30 years?

$$\text{Present value of an annuity} = \frac{C}{r}\left[1 - \frac{1}{(1+r)^T}\right]$$

$$= \frac{130{,}000}{10\%}\left[1 - \frac{1}{(1+10\%)^{30}}\right] = \$1{,}225{,}499$$

Part (c) What about the present value of forever perpetuity?

$$\text{Present value of a perpetuity} = \frac{C}{r} = \frac{130{,}000}{10\%} = \$1{,}300{,}000.$$

CHAPTER TWENTY

NGL and condensate handling, calculations, and breakeven analysis

Introduction

Natural gas processing entails separating all the liquid hydrocarbon and water from natural gas to produce a pipeline quality dry natural gas. Natural gas liquids (NGL) obtained after processing and fractionation associated with rich BTU areas have a variety of uses such as petrochemical plants and feedstock, providing raw materials for oil refineries, and enhanced oil recovery (tertiary recovery) in oil wells. Transportation regulations typically impose limitations on the natural gas make up before allowing natural gas transportation through pipelines. These limitations include outlet BTU and Wobbe Index (WI). WI is very often used as a parameter to determine the upper and lower bounds of gas composition specified in gas sales and contracts. Each gas market has been historically isolated from other regions and each market has its own gas quality specifications. Therefore, each transportation company has its own tariff (set of requirements). Typically, 1100 BTU/SCF and 1400 WI are the requirements to meet the pipeline quality specifications. Some pipelines and contracts allow higher gas BTUs of up to 1150 BTU/SCF but 1100 BTU is usually the case for many pipelines and contracts. WI can be calculated as follows:

$$\text{WI} = \frac{\text{Outlet BTU}}{\sqrt{\text{Gas gravity}}} \quad (20.1)$$

One of the most important analyses performed in unconventional reservoirs is referred to as "break even analysis." It identifies the breakeven NGL pricing and breakeven BTU content in wet gas/retrograde CND areas at which processing gas is more economical as compared to blending gas. In wet gas areas, there is a breakeven BTU at which processing gas is justified depending on variables such as gas composition analysis, BTU content,

NGL component pricing (ethane, propane, i-butane, n-butane, C_{5+}), gas pricing, capital expenditure, and operating cost associated with processing gas such as processing fee ($/MMBTU), NGL T&F excluding ethane (transportation, fractionation, loading, and marketing), de-ethanization, ethane transportation and marketing, and electric fuel consumption associated with processing gas. In high BTU areas, the wet gas stream can be blended with low BTU gas (from the same or other formations) to lower the BTU content to pipeline quality specifications which varies but is typically 1100 BTU/SCF. In some areas, the blending capability is not available which could be due to the produced gas having a higher BTU content or not having the proper pipeline infrastructure or low BTU formations to blend the gas to pipeline specifications. Therefore, processing is necessary to achieve the tariff pipeline quality BTU. Unless some money is invested into pipeline infrastructure, the gas must be processed. The blending capability provides a great deal of flexibility for many companies not be pressured to process the gas when NGL and condensate pricing are low. For example, with the oil price crash of 2014, NGL and condensate pricing which are usually in line with oil pricing caused the breakeven BTU processing justification to be much higher than anticipated. For instance, if $77/BBL and $50/BBL were assumed for the price of the CND and NGL, respectively, to justify processing and fractionation fees, a huge drop in price of CND and NGL will change the entire dynamic of the breakeven BTU content. Despite the fact that it was unprofitable, many operators were forced to process their gas because they had no other option. By the time companies shift their focus from wet to dry gas areas, it takes some time to catch up before divesting the assets from liquid rich to dry areas.

The most challenging areas have a BTU content close to the tariff's BTU of 1100. For example, if the BTU in a wet gas area is around 1130 and the Tariff's BTU is 1100, the gas must be processed to be able to sell, unless it can be blended with lower BTU gas or some agreements can be made between some operators to trade their gas (depending on each company's strategic position). This issue could create challenges for some operators that have a lot of their acreage position in the nuisance CND areas where the BTU expected to be very close to the tariff's BTU.

Ideally, it is beneficial to have the capability to process gas when economically defensible (high NGL pricing and low gas pricing environment) and blend the gas at low NGL pricing and high gas pricing. Having the optionality gives the operator the flexibility to choose between one and the other depending on the market condition which is typically unforeseeable from a pricing perspective. Another important analysis is breakeven BTU content at different NGL and gas pricing. At high NGL pricing and low gas pricing, it could be economically justifiable to process lower BTU wet gas such as 1150 BTU/SCF. This is reversed in high gas and low NGL pricing environments where economic analysis does not rationalize processing high wet gas BTU. Therefore, market condition can directly impact whether processing takes place and detailed analysis must be performed to determine the breakeven NGL and BTU content at which processing is necessary.

When gas processing is economically justified and the capability to process a wet gas system exists (wet gas infrastructure), processing will take place in a processing plant (also referred to as cryogenic or stripping plant). A processing plant is a facility where natural gas is cooled in an attempt to condense liquids or NGL. From the processing plant, the total stream (rich gas) will be divided into three streams. NGL, residue gas, and ethane (can be recovered or rejected) will be extracted out of the total gas stream (rich gas stream). After going through a turbo expander or JT (Joule Thompson) plant, there is typically a de-ethanizer (also referred to as de-methanizer) at the cryogenic plant that enables the adjustment of the percentage of ethane needed to be recovered. If a certain % of ethane (e.g., 30%) is recovered, the remaining ethane (in this example 70% of the ethane) will be rejected into the residue gas stream. Theoretically speaking, when 100% of the ethane is recovered and 0% of the ethane is rejected into the residue gas stream, de-ethanizer is referred to as a "de-methanizer." On the other hand, when 0% of the ethane is recovered and 100% of the ethane is being rejected into the residue gas stream, the ethane recovery unit located right after the turbo expander or JT plant is referred to as a "de-ethanizer." It is important to note that these terms are used interchangeably in different parts of the world but refer to the same ethane recovery unit located right after a turbo expander or JT plant. The percentage of ethane recovery will depend on ethane market (pricing) as well as the required BTU needed after processing. Economic analysis can also be performed to calculate the breakeven ethane pricing at which processing a higher percentage of ethane will increase the NPV of the project. At a certain ethane price, recovering a higher percentage of ethane will create value. Below the breakeven ethane pricing, recovery of ethane will result in loss in NPV. For example, if after processing a 1300 BTU wet gas by not recovering any of the ethane (0% ethane recovery due to low ethane pricing), the outlet processing plant BTU is 1155 and pipeline tariff quality is 1100 BTU/SCF, enough percentage of the ethane must be recovered to lower the outlet BTU to pipeline quality BTU of 1100.

This could require the recovery of a certain percentage of ethane at low pricing just to make sure the residue gas BTU meets the pipeline specification. Therefore, the terms and details of each contract must be fully reviewed and analyzed to make sure the most economical decision is made for the shareholders when negotiating the deals with midstream and downstream companies. In some parts of the United States, recovering ethane is not economical due to low ethane pricing but operators sometimes have no other option.

After going through the cryogenic plant, the remaining NGL stream will be directed into a fractionation plant where each gas component such as propane, i-butane, n-butane, i-pentane, n-pentane, and C_{5+} will be separated, transported, marketed, and sold separately. Residue gas is the remaining dry gas (low BTU) after processing takes place and will be transported, marketed, and sold in a dry gas pipeline. After processing, NGL will be sent to a fractionation plant where various NGL components will be fractionated, transported, loaded, and marketed which is typically the case for most of the operators. However, there are other alternatives where NGL can be sold as a y-grade price (which is typically lower) where a certain y-grade price will be offered to sell the NGL. In addition, midstream or upstream companies will handle the unfractionated NGL. Y-grade components include C_{2+}.

NGL yield calculation

NGL yield is one of the most important calculations in wet gas economic analysis. NGL yield calculation is important in order to calculate NGL volume per month when performing economic analysis. NGL yield does not change with time (as long as the gas composition does not change) because all of the NGL components will remain in gaseous phase until fractionated in a fractionation plant.

1. The first step in NGL yield calculation is to calculate entrained NGL in GPM (gallons per MSCF) from a gas composition analysis as shown in Eq. (20.2).

$$\text{Entrained NGL} = \left(\frac{\text{Inlet gas component} \times \frac{\text{gal}}{\text{lb.mol}} \text{ of each component}}{379.49} \right) \times 1000 \quad (20.2)$$

Knowing that 1 lb.mol = 379.4 SCF, entrained NGL can be calculated for each gas component.

Note that gal/lb.mol is available (shown in the following example) for each component.

2. The second step is to calculate mol% remaining based on processing plant % recovery for each gas component. Processing plants typically provide the amount of recovery that the plant can handle for each gas component. Ethane recovery can vary based on the economic analysis of ethane recovery. mol% remaining can be calculated as shown in Eq. (20.3).

$$\text{mol\% remaining per gas component} = (1 - \text{plant removal\% per component}) \times \text{Inlet gas composition} \quad (20.3)$$

The summation of mol% remaining for all gas components will yield the remaining mol% (plant outlet). The difference between **plant inlet** and **outlet** yields **liquid shrinkage**.

3. New mol% in outlet or post extraction residue gas is calculated as shown in Eq. (20.4).

$$\text{Postextration residue gas} = \frac{\text{mol\% remaining per gas component}}{\text{Total\% mol remaining}} \quad (20.4)$$

4. The next step is to calculate plant NGL removal or stripped out GPM (gal/MSCF) to calculate NGL yield. This can be calculated as shown in Eq. (20.5).

$$\text{Plant NGL removal (GPM) per component} = \text{Entrained NGL (GPM) per component} \times \text{plant removal\% per component} \quad (20.5)$$

The summation of plant NGL removal per component will yield NGL yield in gallons per MSCF. The calculated NGL yield in GPM can be converted into BBL/MMSCF by multiplying by 1000 and dividing by 42.

5. Finally, calculate % GPM per component using Eq. (20.6).

$$\%\text{GPM per component} = \frac{\text{Plant NGL removal per component}}{\text{Total plant NGL removal (NGL yield)}} \quad (20.6)$$

6. Weighted average NGL price can be found knowing % GPM per component using Eq. (20.7).

$$\begin{aligned}\text{Weighted avg NGL} =\ & (\text{ethane price} \times \%\text{GPM for ethane}) \\
& + (\text{propane price} \times \%\text{GPM for propane}) \\
& + (i-\text{butane price} \times \%\text{GPM for } i-\text{butane}) \\
& + (n-\text{butane price} \times \%\text{GPM for } n-\text{butane}) \\
& + (C_{5+} \text{ price} \times \%\text{GPM for } C_{5+}) \end{aligned} \quad (20.7)$$

Example

A gas sample with the following properties (inlet gas composition) is provided. Calculate liquid shrinkage and NGL yield as well as weighted average NGL pricing assuming the following pricing per component:

Ethane price = \$0.28/gal, propane price = \$0.66/gal, i-butane price = \$0.87/gal, n-butane = \$0.86/gal, C_{5+} = \$1.22/gal

1. The first step is to calculate the entrained NGL in GPM (gallon per MSCF) from a gas composition analysis. The entrained NGL sample calculation for methane component is given as follows:

$$\text{Entrained NGL} = \left(\frac{\text{Inlet gas component} \times \frac{\text{gal}}{\text{lb.mol}} \text{ of each component}}{379.49} \right) \times 1000$$

$$= \left(\frac{81.7043\% \times 6.417}{379.49} \right) \times 1000 = 13.82 \text{ GPM (for C1)}$$

	Input		Output
Gas component	Inlet gas composition (%)	gal/lb. mol	Entrained NGL(GPM)
Methane (C1)	81.7043	6.417	13.82
Ethane (C2)	12.1416	10.123	3.24
Propane (C3)	3.9237	10.428	1.08
i-Butane (iC4)	0.3849	12.386	0.13
n-Butane (nC4)	0.9003	11.933	0.28
i-Pentane (iC5)	0.2219	13.843	0.08
n-Pentane (nC5)	0.1243	13.721	0.04
Hexane+	0.1848	16.517	0.08
Nitrogen (N_2)	0.3637	4.1643	0.04
Carbon dioxide (CO_2)	0.0506	6.4593	0.01

2. The next step is to calculate liquid shrinkage associated with processing plant and postextraction residue gas. Please note that the processing plant

must provide the % plant removal for each gas component. In the following example, 30% ethane recovery was assumed, but the recovery of this component could be much larger than 30% and various iterations must be run to find the breakeven ethane pricing as is shown later in this chapter. mol% remaining and residue gas % for methane component is given as follows:

$$\text{mol \% remaining per gas component} = (1 - \text{plant removal\%}) \times \text{inlet gas composition}$$
$$= (1 - 0\%) \times 81.7043\% = 81.7043\%$$

$$\text{Postextration residue gas} = \frac{\text{mol\% remaining per gas component}}{\text{Total mol\% remaining}} = \frac{81.7043\%}{91.03\%} = 89.76\%$$

91.03% is the plant outlet BTU % which can be used for liquid shrinkage calculation as follows:

$$\text{Liquid shrinkage} = \text{plant inlet} - \text{plant outlet} = 100\% - 91.03\% = 8.97\%$$

3. The next step is to calculate plant NGL removal for each gas component and sum up the plant NGL removal for all components to obtain the

	Input			Provided	Calculated	
Gas component	Inlet gas composition (%)		gal/lb. mol	Plant removal (%)	(1 − plant removal %) × inlet gas composition (%)	Postextraction (residue gas) (%)
Methane (C1)	81.7043		6.417	0.00	81.70	89.76
Ethane (C2)	12.1416		10.123	30.00	8.50	9.34
Propane (C3)	3.9237		10.428	90.00	0.39	0.43
i-Butane (iC4)	0.3849		12.386	98.00	0.01	0.01
n-Butane (nC4)	0.9003		11.933	99.00	0.01	0.01
i-Pentane(iC5)	0.2219		13.843	99.90	0.00	0.00
n-Pentane(nC5)	0.1243		13.721	99.90	0.00	0.00
Hexane+	0.1848		16.517	99.90	0.00	0.00
Nitrogen (N_2)	0.3637		4.1643	0.00	0.36	0.40
Carbon dioxide (CO_2)	0.0506		6.4598	0.00	0.05	0.06
Sum	UNIX			After processing plant	91.03	

NGL yield. A sample calculation for the ethane component is given as follows:

$$\text{Plant NGL removal (GPM) per component}$$
$$= \text{Entrained NGL (GPM) per component}$$
$$\times \text{plant removal\% per component}$$
$$= 3.24 \times 30\% = 0.972 \text{ GPM}$$

The summation of plant NGL removal (which is 2.552 GPM in this example) gives NGL yield and it can be converted from gal per MSCF to BBL/MMSCF as follows:

$$\text{NGL yield} = 2.552 \frac{\text{gal}}{\text{MSCF}} \text{ or } \frac{2.552 \times 1000}{42} = 60.75 \frac{\text{BBL}}{\text{MMSCF}}$$

	Input	Calculated	Provided	Calculated
Gas component	Inlet gas composition (%)	Entrained NGL(GPM)	Plant removal (%)	Plant NGL removal (GPM)
Methane (C1)	81.7043	13.82	0.00	0.000
Ethane (C2)	12.1416	3.24	30.00	0.972
Propane (C3)	3.9237	1.08	90.00	0.971
i-Butane (iC4)	0.3849	0.13	98.00	0.123
n-Butane (nC4)	0.9003	0.28	99.00	0.280
i-Pentane (iC5)	0.2219	0.08	99.90	0.081
n-Pentane (nC5)	0.1243	0.04	99.90	0.045
Hexane+	0.1848	0.08	99.90	0.080
Nitrogen (N_2)	0.3637	0.04	0.00	0.000
Carbon dioxide (CO_2)	0.0506	0.01	0.00	0.000
Sum	100		NGL Yield	2.552

4. The following step is to calculate % GPM per component for the ethane component:

$$\%\text{GPM per component} = \frac{\text{Plant NGL removal per component}}{\text{Total plant NGL removal (NGL yield)}}$$
$$= \frac{0.972}{2.552} = 38.08\%$$

Gas component	% GPM per component
Methane (C1)	0.00
Ethane (C2)	38.08
Propane (C3)	38.03
i-Butane (iC4)	4.83
n-Butane (nC4)	10.98
i-Pentane (iC5)	3.17
n-Pentane (nC5)	1.76
Hexane+	3.15
Nitrogen (N_2)	0.00
Carbon dioxide (CO_2)	0.00
Total	100.0

5. Using the NGL component pricing provided in this example, weighted average NGL price can be determined as follows:

$$\begin{aligned}\text{Weighted avg NGL} &= (0.28 \times 38.08\%) + (0.66 \times 38.03\%) \\ &\quad + (0.87 \times 4.83\%) + (0.86 \times 10.98\%) \\ &\quad + (1.22 \times (3.17\% + 1.76\% + 3.15\%)) \\ &= \frac{\$0.5926}{\text{gal}} \text{ or } \$24.89/\text{BBL}\end{aligned}$$

NGL yield remains constant whereas CND yield does not because it does not remain in gaseous phase as CND is produced. Therefore, to estimate CND yield and production volumes with time, CND yield for various BTUs is plotted versus time to fit the best equation through the data. The best-fitted equation through the data can be used to estimate CND yield over time. For nonlinear problems, nonlinear machine learning algorithms could also yield more accurate results. Essentially, actual production history can be used to forecast CND yield with time. If production history is not available, numerical simulation and development of equation of state can be the solution to the challenge.

Blending vs processing gas step-by-step guide and calculation

Your manager asks you to calculate breakeven BTU, breakeven NGL, and breakeven ethane pricing to justify processing the gas in a wet gas area. You are also asked to perform sensitivity analysis and create various tables to show the breakeven analysis for each property. The type curve parameters are provided to you as follows:

Gas type curve: IP = 16,000 MSCF/day, D_e = 71%, b = 1.5, Terminal decline = 5%, Lateral length (ft) = 8000'

CND yield type curve: assume 10 BBL/MMSCF for month 1, 9 BBL/MMSCF for month 2, 8 BBL/MMSCF for month 3, 7 BBL/MMSCF from month 4 to life of the well (assume 50-year life)

Assume 1240 BTU/SCF (gas composition data is provided in Table 20.1)

Compressor shrinkage = 2.5%, processing plant shrinkage = 0%, wet gas BTU = 1240, royalty = 15%, WI =100%, NRI =85%, Total Capex = $8,000,000, WACC = 10%, severance tax = 5%, Ad valorem tax = 2.5%, tangible capex =12%, intangible capex =88%, federal tax rate = 21%, variable lifting cost = $0.15/MCF, gathering cost = $0.2/MMBTU, firm transportation = assume it's a sunk cost, fixed operating cost = $1300/month/well, processing (+fuel electric) = $0.32/MMBTU, NGL transportation, fractionation, loading, and marketing (excluding ethane) = $0.09/gal, CND transportation and marketing = $0.025/gal, de-ethanization = $0.03/gal, ethane transportation and marketing = $0.01/gal

Do not escalated OPEX, CAPEX, or pricing for this example.

Pricing info:

CND pricing = $0.92/gal, ethane = $0.15/gal, propane = $0.45/gal, i-butane = $0.71/gal, n-butane = $0.64/gal, C_{5+} = $1.06/gal, realized gas pricing= $2.5/MMBTU (run various sensitivity from $1/MMBTU to $5/MMBTU)

Table 20.1 Gas composition analysis example Provided

Inlet BTU	Inlet gas composition (%)	gal/lb.mol	Plant % removal (%)
Methane	78.812	6.4170	0.00
Ethane	14.255	10.1230	32.00
Propane	4.293	10.4280	92.00
i-Butane	0.519	12.3860	98.00
n-Butane	0.957	11.9330	98.00
i-Pentane	0.232	13.8430	99.00
n-Pentane	0.201	13.7210	99.00
Hexane+	0.269	16.5170	99.00
Inerts	0.463		
Total	100.000		

Step 1: This analysis must be divided into the following sections:
(a) Calculate ATAX NPV at various gas pricing by assuming that the 1240 BTU gas can be sold via blending the gas before sending to pipeline. In this instance, we will receive 24% uplift in gas pricing from the high BTU content (1.24) and pay no operating fees associated with processing or fractionation of gas since the gas can simply be blended and the weighted average gas can be brought down to pipeline quality BTU of 1100 (assuming 1100 is the pipeline BTU quality).
(b) Calculate the ATAX NPV at various gas pricing by assuming the gas must be processed and fractionated incorporating NGL production associated with the gas and processing/fractionation operating costs. In both scenarios since CND will be produced, CND operating fees must be considered.
(c) Assuming blending opportunity exists, determine whether blending yields a higher NPV or processing/fractionation does.

Let's stat the analysis by calculating NGL yield and total shrinkage. As previously shown, the workflow can be repeated for this example to populate Table 20.2. As can be seen from this table, NGL yield and total shrinkage are 3.03 GPM (or 72.24 BBL/MMSCF) and 11.11% (100%–88.89%), respectively. Also, calculate weighted average NGL pricing:

$$\begin{aligned}\text{Weighted avg NGL} &= (0.15 \times 40.10\%) + (0.45 \times 35.77\%) + (0.71 \times 5.47\%) \\ &\quad + (0.64 \times 9.72\%) + (1.06 \times (2.76\% + 2.37\% + 3.81\%)) \\ &= \frac{\$0.4169}{\text{gal}} \text{ or } \$17.51/\text{BBL}\end{aligned}$$

Table 20.2 NGL yield and total shrinkage calculations

Entrained NGL (GPM)	(1 − plant % removal) × inlet gas composition (%)	Calculated Postextraction (residue gas) (%)	Plant NGL removal	% GPM per component
13.3267	78.81	84.45	0	0.00
3.8027	9.69	10.39	1.2168	40.10
1.1797	0.34	0.37	1.0853	35.77
0.1693	0.01	0.01	0.1659	5.47
0.3009	0.02	0.02	0.2949	9.72
0.0845	0.00	0.00	0.0836	2.76
0.0726	0.00	0.00	0.0719	2.37
0.1169	0.00	0.00	0.1157	3.81
	88.89		3.03	100.00

Scenario a: The first step is to calculate ATAX NPV assuming that blending capability exists, and gas can be realized at 1240 BTU/SCF.

All the steps in this example are the same as the steps shown in the previous chapter for the dry gas example except that condensate production and operating cost must be considered. The unshrunk gas rate over time can be obtained based on the modified hyperbolic parameters provided. Following the equations presented in the decline curve analysis chapter, Table 20.3 shows a summary of gas and condensate volumes for the first year. Net sold condensate can be calculated as follows for the first two time-steps:

$$\text{Net sold CND}_{\text{month 1}} = \text{gross unshrunk gas}_{\text{month 1}} \times \text{CND yield}_{\text{month 1}} \times \text{NRI}$$
$$= \frac{427{,}755}{1000} \times 10 \frac{\text{BBL}}{\text{MMSCF}} \times 85\% = 3636 \text{ BBLs}$$

$$\text{Net sold CND}_{\text{month 2}} = \text{gross unshrunk gas}_{\text{month 2}} \times \text{CND yield}_{\text{month 2}} \times \text{NRI}$$
$$= \frac{346{,}163}{1000} \times 9 \frac{\text{BBL}}{\text{MMSCF}} \times 85\% = 2648 \text{ BBLs}$$

Also note that shrunk gas volume can be calculated as follows:

Gross shrunk gas = gross unshrunk gas × (1 − compressor shrinkage)

Please note that in this scenario, net sold NGL will be zero since the gas is being blended. Even though the gas is being blended, CND can be sold separately as it can be trucked off site. The advantage of this scenario can be

Table 20.3 Gas and condensate volumes for the first year

Date	Time	Gross unshrunk gas (MSCF/M)	Gross shrunk gas (MSCF/M)	Net shrunk gas	Net sold CND
01/2018	0	0	0	0	0
02/2018	0.5	427,755	417,061	354,502	3636
03/2018	1.5	346,163	337,509	286,882	2648
04/2018	2.5	294,989	287,614	244,472	2006
05/2018	3.5	259,351	252,868	214,937	1543
06/2018	4.5	232,850	227,029	192,974	1385
07/2018	5.5	212,230	206,925	175,886	1263
08/2018	6.5	195,646	190,755	162,142	1164
09/2018	7.5	181,966	177,417	150,804	1083
10/2018	8.5	170,452	166,191	141,262	1014
11/2018	9.5	160,603	156,588	133,100	956
12/2018	10.5	152,064	148,263	126,023	905
01/2019	11.5	144,577	140,963	119,818	860

observed below where the assumed gas price of $2.5/MMBTU receives a 24% uplift due to the BTU of 1240 as follows:

$$\text{Adjusted realized gas pricing} = \frac{\$2.5}{\text{MMBTU}} \times 1.24 = \frac{\$3.1}{\text{MCF}}$$

In addition, CND operating cost must be considered as follows (assumes operator pays for CND fees):

$$\text{CND Operating cost} = \text{Gross sold CND} \times \text{WI} \times \text{CND fee} \left(\frac{\$}{\text{gal}}\right) \times 42$$

Therefore, calculating the CND operating cost can be calculated for the first couple of months as follows:

$$\text{Month 1} = \left(\frac{3636}{85\%}\right) \times 100\% \times 0.025 \times 42 = \$4492$$

$$\text{Month 2} = \left(\frac{2648}{85\%}\right) \times 100\% \times 0.025 \times 42 = \$3271$$

The rest of the calculations remain the same as in the previous example shown in the economic evaluation chapter. Table 20.4 illustrates the ATAX NPV for this scenario at various gas pricing and discount rates. Please note that we will use the **ATAX NPV at 10% discount rate** when comparing this scenario to the next scenario that will be discussed in part b.

Table 20.5 also illustrates the ATAX NPV at various BTUs assuming a fixed realized gas pricing of $2.5/MMBTU for scenario a. As can be seen, as BTU increases, the ATAX NPV will also increase if the type curve remains the same. This table will be used for comparison purposes with scenario b which is discussed next.

Scenario b evaluation

Step 1: The first step is to calculate outlet BTU based on plant % removal provided for each component. The inlet BTU of 1240 is provided; therefore, it is important to calculate what the outlet BTU would be to get the heating value of the outlet BTU. Postextraction (residue gas) column that was calculated can be used to compute the outlet BTU as follows:

(1) Obtain BTU/SCF for each component from a phase behavior book.
(2) Multiply post extraction (residue gas) by BTU factor (BTU divided by 1000 is called BTU factor).
(3) Add "postextraction*BTU factor" components to obtain uncorrected BTU factor.

Table 20.4 ATAX NPV for unprocessed gas at various realized gas pricing and discount rates

Discount rates (%)	Realized gas pricing, after tax calculation								
	1.00	1.50	2.00	2.50	3.00	3.50	4.00	4.50	5.00
0	2,664,194	8,618,029	14,571,863	20,525,698	26,479,533	32,433,368	38,387,202	44,341,037	50,294,872
5	−362,073	3,481,492	7,325,056	11,168,621	15,012,186	18,855,750	22,699,315	26,542,880	30,386,444
10	**−1,653,428**	**1,334,516**	**4,322,461**	**7,310,405**	**10,298,350**	**13,286,295**	**16,274,239**	**19,262,184**	**22,250,128**
15	−2,371,391	154,527	2,680,446	5,206,365	7,732,284	10,258,202	12,784,121	15,310,040	17,835,959
20	−2,837,329	−606,113	1,625,102	3,856,318	6,087,534	8,318,749	10,549,965	12,781,180	15,012,396
25	−3,169,438	−1,145,938	877,563	2,901,063	4,924,564	6,948,065	8,971,565	10,995,066	13,018,566
30	−3,421,056	−1,553,703	313,649	2,181,001	4,048,354	5,915,706	7,783,058	9,650,411	11,517,763
35	−3,619,948	−1,875,324	−130,700	1,613,924	3,358,548	5,103,172	6,847,796	8,592,420	10,337,043
40	−3,782,121	−2,137,143	−492,164	1,152,814	2,797,793	4,442,771	6,087,750	7,732,728	9,377,707
45	−3,917,524	−2,355,472	−793,420	768,632	2,330,684	3,892,735	5,454,787	7,016,839	8,578,891
50	−4,032,706	−2,541,019	−1,049,331	442,356	1,934,043	3,425,731	4,917,418	6,409,106	7,900,793
55	−4,132,178	−2,701,137	−1,270,096	160,945	1,591,986	3,023,027	4,454,068	5,885,109	7,316,150
60	−4,219,160	−2,841,068	−1,462,975	−84,882	1,293,211	2,671,303	4,049,396	5,427,489	6,805,582
65	−4,296,023	−2,964,659	−1,633,295	−301,932	1,029,432	2,360,796	3,692,160	5,023,523	6,354,887
70	−4,364,553	−3,074,810	−1,785,067	−495,324	794,419	2,084,162	3,373,905	4,663,648	5,953,390
75	−4,426,127	−3,173,750	−1,921,373	−668,995	583,382	1,835,760	3,088,137	4,340,514	5,592,892
80	−4,481,827	−3,263,226	−2,044,626	−826,026	392,575	1,611,175	2,829,775	4,048,375	5,266,976
85	−4,532,511	−3,344,629	−2,156,748	−968,866	219,015	1,406,897	2,594,778	3,782,660	4,970,541
90	−4,578,875	−3,419,081	−2,259,287	−1,099,493	60,300	1,220,094	2,379,888	3,539,682	4,699,476
95	−4,621,487	−3,487,499	−2,353,510	−1,219,521	−85,533	1,048,456	2,182,445	3,316,433	4,450,422
100%	−4,660,818	−3,550,640	−2,440,461	−1,330,283	−220,105	890,074	2,000,252	3,110,430	4,220,609

Table 20.5 ATAX NPV for unprocessed gas at various BTU and fixed realized gas pricing of $2.5/MMBTU
Scenario a (blending gas)

BTU/SCF	ATAX NPV
1100	5,799,450
1125	6,069,263
1150	6,339,077
1175	6,608,890
1200	6,878,704
1225	7,148,517
1250	7,418,331
1275	7,688,144
1300	7,957,958
1325	8,227,771
1350	8,497,585
1375	8,767,398
1400	9,037,212

(4) Calculate the compressibility factor for the gas composition.

(5) Divide uncorrected BTU factor by z factor and multiply by 1000 to obtain the corrected for compressibility factor (z-factor) BTU after processing (postprocessing).

As can be seen in this example, the postprocessing BTU that will be used for economic analysis is 1099 BTU/SCF.

Gas component	Postextraction (residue gas)	BTU/SCF	BTU factor	Postextraction *BTU factor
Methane (C1)	88.21	1012	1.012	0.893
Ethane (C2)	10.85	1774	1.774	0.192
Propane (C3)	0.38	2522	2.522	0.010
i-Butane ($iC4$)	0.01	3259	3.259	0.000
n-Butane ($nC4$)	0.02	3270	3.27	0.001
i-Pentane ($iC5$)	0.00	4010	4.01	0.000
n-Pentane ($nC5$]	0.00	4018	4.018	0.000
Hexane+	0.00	4767	4.767	0.000
Nitrogen (N_2)	0.51	N/A	N/A	N/A
Carbon dioxide (CO_2)	0.00	N/A	N/A	N/A
Total	100.00	Uncorrected BTU factor		1.096
		Compressibility factor (z factor)		0.9975
		Corrected for z factor BTU		1099

Step 2: The following step that is different from previous analysis shown in this book is the calculation of gross shrunk gas which can be done as follows for this scenario since NGL is produced and liquid shrinkage is calculated:

Gross shrunk gas = gross unshrunk gas × (1 − compressor shrinkage) × (1 − liquid shrinkage) × (1 − plant shrinkage)

Note that in addition to compressor shrinkage, liquid shrinkage and processing plant shrinkage must also be considered as shown below. In this example, processing plant shrinkage is provided as 0%, so it can be ignored; however, liquid shrinkage was calculated to be 11.11%.

The gross shrunk gas for the first two time-steps is listed below:

$$\text{Gross shrunk gas}_{\text{month 1}} = 427{,}755 \times (1 - 2.5\%) \times (1 - 11.11\%) \times (1 - 0\%)$$
$$= 370{,}725 \text{ MSCF}$$

$$\text{Gross shrunk gas}_{\text{month 2}} = 346{,}163 \times (1 - 2.5\%) \times (1 - 11.11\%) \times (1 - 0\%)$$
$$= 300{,}011 \text{ MSCF}$$

Net shrunk gas can be obtained by multiplying **gross** shrunk gas by NRI.

$$\text{Net shrunk gas}_{\text{month 1}} = (370{,}725 \times 85\%) = 315{,}117 \text{ MSCF}$$
$$\text{Net shrunk gas}_{\text{month 2}} = (300{,}011 \times 85\%) = 255{,}010 \text{ MSCF}$$

Step 3: Calculate net sold NGL as follows:

$$\text{Net sold NGL} = \frac{\text{Gross unshrunk gas (MSCF)}}{1000} \times \text{NGL yield}\left(\frac{\text{BBL}}{\text{MMSCF}}\right)$$
$$\times (1 - \text{compressor shrinkage}) \times \text{NRI}$$

Net sold NGL for two time-steps can be done as follows:

$$\text{Net sold NGL}_{\text{month 1}} = \frac{427{,}755}{1000} \times 72.24 \times (1 - 2.5\%) \times 85\%$$
$$= 25{,}609 \text{ BBLs}$$

$$\text{Net sold NGL}_{\text{month 2}} = \frac{346{,}163}{1000} \times 72.24 \times (1 - 2.5\%) \times 85\%$$
$$= 20{,}724 \text{ BBLs}$$

Step 4: Realized gas pricing must be calculated based on the residue gas BTU (outlet BTU calculated in step 1) as follows for $2.5/MMBTU:

$$\text{Adjusted realized gas pricing} = 2.5 \times \left(\frac{1099}{1000}\right) = \frac{\$2.7475}{\text{MMBTU}}$$

Step 5: Total revenue for gas stream, condensate stream, and NGL stream for the first two time steps can be calculated as follows:

Total revenue = (Net shrunk gas × adjusted realized gas pricing)
$$+ \left(\text{net sold CND} \times \frac{\text{CND pricing}}{\text{gal}} \times 42\right) + \left(\text{net sold NGL} \times \frac{\text{NGL pricing}}{\text{gal}} \times 42\right)$$

$$\text{Total revenue}_{\text{month 1}} = (315,117 \times 2.7475) + \left(3636 \times \frac{0.92}{\text{gal}} \times 42\right)$$

$$+ \left(25,609 \times \frac{0.4169}{\text{gal}} \times 42\right) = \$1,454,687$$

$$\text{Total revenue}_{\text{month 2}} = (255,010 \times 2.7475) + \left(2648 \times \frac{0.92}{\text{gal}} \times 42\right)$$

$$+ \left(20,724 \times \frac{0.4169}{\text{gal}} \times 42\right) = \$1,165,843$$

Step 6: Severance and ad valorem taxes for the first 2 months can be calculated as follows:

$$\text{Severance tax}_{\text{month 1}} = \text{Total revenue}_{\text{month 1}} \times \text{severance tax\%}$$
$$= \$1,454,687 \times 5\% = \$72,734$$

$$\text{Severance tax}_{\text{month 2}} = \$1,165,843 \times 5\% = \$58,292$$

$$\text{Ad valorem tax}_{\text{month 1}} = (\text{Total revenue}_{\text{month 1}} - \text{Severance tax}_{\text{month 1}})$$
$$\times \text{ad valorem tax\%}$$
$$= (1,454,687 - 72,734) \times 2.5\% = \$34,549$$

$$\text{Ad valorem tax}_{\text{month 2}} = (1,165,843 - 58,292) \times 2.5\% = \$27,689$$

Step 7: Gathering cost can be grossed up as follows:

$$\text{Grossed up gatherign cost} = \frac{\text{gathering cost}}{(1 - \text{compressor shrinkage})(1 - \text{liquid shrinkage})}$$
$$= \frac{\$0.2/\text{MMBTU}}{(1 - 2.5\%)(1 - 11.11\%)} = \frac{\$0.231}{\text{MMBTU}}$$

Step 8: Calculate total OPEX excluding severance and ad valorem taxes as follows:

Total OPEX = (fixed OPEX × WI) + (Gross shrunk gas × Variable OPEX × WI)
$$+ \left(\text{Gross sold } CND \times WI \times CND \text{ fee}\left(\frac{\$}{\text{gal}}\right) \times 42\right)$$

$$+ \left(\text{Gross shrunk gas} \times WI \times \text{Grossed up gathering fee} \times \left(\frac{\text{Inlet } BTU}{1000} \right) \right)$$

$$+ \left(\text{Gross shrunk gas} \times WI \times \frac{\text{Processing fee} \left(\frac{\$}{\text{MMBTU}} \right) \times \left(\frac{\text{Inlet } BTU}{1000} \right)}{(1 - \text{Liquid shrinkage})} \right)$$

$+ \;($Gross sold $NGL \times (De$ ethanization $+$ Ethane $T\&M \times 42) \times WI$

$\times$ Ethane outlet composition$)$

$+ \;($Gross sold $NGL \times$ fractionation fee $\times 42$

$\times \;($propane $+ iso$ butane $+ n -$ butane $+ iso$ pentane $+ n -$ pentane

$+\; C5_+$ outlet compositions$) \times WI)$

As illustrated above, calculating the operating costs for liquid rich regions is more complicated than calculating operating costs in dry gas areas. Please note that outlet compositions referred to in the equation above denotes "% GPM per component" as previously calculated.

Plugging in the numbers in the discussed equation would yield the following total OPEX for the first 2 months:

$$\text{Total OPEX}_{\text{month 1}} = \$421,429$$
$$\text{Total OPEX}_{\text{month 2}} = \$340,928$$

The remaining calculations would remain the same as shown in the economic evaluation chapter of this book.

When ATAX NPV for scenario b (processing the gas) is greater than ATAX NPV for scenario a (blending gas), processing will become economically justified. This was done assuming various gas pricing, NGL pricing, and BTU contents. The summary results of the analysis are listed below:
– Based on the assumptions listed in this example, breakeven NGL pricing to justify processing at $1/MMBTU realized gas pricing is about $14.28/BBL (read from Table 20.6) because at this pricing, the ATAX NPV for processing the gas which is calculated to be $-1,641,522 (as shown in Table 20.6) is **greater** than the ATAX NPV for blending the gas which is obtained from Table 20.4 at 10% discount rate and $1/MMBTU gas pricing to be $-1,653,428. At $2/MMBTU realized gas pricing, breakeven NGL pricing to justify processing is ~$18.06/BBL. Finally, at $3/MMBTU realized gas pricing, the breakeven NGL pricing is ~$21.42/BBL. Therefore, **at a fixed BTU content** of 1240, as gas pricing

Table 20.6 ATAX NPV for scenario b (processing the gas) at various realized gas pricing ($/MMBTU) and NGL pricing ($/BBL)

NGL pricing ($/BBL)	1	1.5	2	2.5	3	3.5	4	5
10.08	−3,113,202	−749,763	1,613,676	3,977,114	6,340,553	8,703,992	11,067,431	15,794,308
10.5	−2,966,034	−602,595	1,760,844	4,124,282	6,487,721	8,851,160	11,214,599	15,941,476
10.92	−2,818,866	−455,427	1,908,012	4,271,450	6,634,889	8,998,328	11,361,767	16,088,644
11.34	−2,671,698	−308,259	2,055,180	4,418,618	6,782,057	9,145,496	11,508,935	16,235,812
11.76	−2,524,530	−161,091	2,202,348	4,565,786	6,929,225	9,292,664	11,656,103	16,382,980
12.18	−2,377,362	−13,923	2,349,516	4,712,954	7,076,393	9,439,832	11,803,271	16,530,148
12.6	−2,230,194	133,245	2,496,684	4,860,122	7,223,561	9,587,000	11,950,439	16,677,316
13.02	−2,083,026	280,413	2,643,852	5,007,290	7,370,729	9,734,168	12,097,607	16,824,484
13.44	−1,935,858	427,581	2,791,020	5,154,458	7,517,897	9,881,336	12,244,775	16,971,652
13.86	−1,788,690	574,749	2,938,188	5,301,626	7,665,065	10,028,504	12,391,943	17,118,820
14.28	−1,641,522	721,917	3,085,356	5,448,794	7,812,233	10,175,672	12,539,111	17,265,988
14.7	−1,494,354	869,085	3,232,524	5,595,962	7,959,401	10,322,840	12,686,279	17,413,156
15.12	−1,347,186	1,016,253	3,379,692	5,743,130	8,106,569	10,470,008	12,833,447	17,560,324
15.54	−1,200,018	1,163,421	3,526,860	5,890,258	8,253,737	10,617,176	12,980,615	17,707,492
15.96	−1,052,850	1,310,589	3,674,028	6,037,466	8,400,905	10,764,344	13,127,783	17,854,660
16.38	−905,682	1,457,757	3,821,196	6,184,634	8,548,073	10,911,512	13,274,951	18,001,828
16.8	−758,514	1,604,925	3,968,363	6,331,802	8,695,241	11,058,680	13,422,119	18,148,996
17.22	−611,346	1,752,093	4,115,531	6,478,570	8,842,409	11,205,848	13,569,287	18,296,164
17.64	−464,178	1,899,261	4,262,699	6,626,138	8,989,577	11,353,016	13,716,455	18,443,332
18.06	−317,010	2,046,429	4,409,867	6,773,306	9,136,745	11,500,184	13,863,623	18,590,500
18.48	−169,842	2,193,597	4,557,035	6,920,474	9,283,913	11,647,352	14,010,791	18,737,668
18.9	−22,674	2,340,765	4,704,203	7,067,642	9,431,081	11,794,520	14,157,959	18,884,836
19.32	124,494	2,487,933	4,851,371	7,214,810	9,578,249	11,941,688	14,305,127	19,032,004

Continued

Table 20.6 ATAX NPV for scenario b (processing the gas) at various realized gas pricing ($/MMBTU) and NGL pricing ($/BBL)—cont'd

NGL pricing ($/BBL)	Realized gas pricing ($/MMBTU)							
	1	1.5	2	2.5	3	3.5	4	5
19.74	271,662	2,635,101	4,998,539	7,361,978	9,725,417	12,088,856	14,452,295	19,179,172
20.16	418,830	2,782,269	5,145,707	7,509,146	9,872,585	12,236,024	14,599,463	19,326,340
20.58	565,998	2,929,437	5,292,875	7,656,314	10,019,753	12,383,152	14,746,631	19,473,508
21	713,166	3,076,605	5,440,043	7,803,482	10,166,921	12,530,360	14,893,799	19,620,676
21.42	860,334	3,223,773	5,587,211	7,950,650	10,314,089	12,677,528	15,040,967	19,767,844
21.84	1,007,502	3,370,941	5,734,379	8,097,818	10,461,257	12,824,656	15,188,135	19,915,012
22.26	1,154,670	3,518,109	5,881,547	8,244,986	10,608,425	12,971,864	15,335,303	20,062,180
22.68	1,301,838	3,665,277	6,028,715	8,392,154	10,755,593	13,119,032	15,482,471	20,209,348
23.1	1,449,006	3,812,445	6,175,883	8,539,322	10,902,761	13,266,200	15,629,639	20,356,516
23.52	1,596,174	3,959,613	6,323,051	8,686,490	11,049,929	13,413,368	15,776,807	20,503,684
23.94	1,743,342	4,106,781	6,470,219	8,833,658	11,197,097	13,560,536	15,923,975	20,650,852

increases, breakeven NGL pricing to justify processing the gas also **increases**. As a result, lower gas pricing and higher NGL pricing is ideal for processing gas because most of the value is created from the sale of NGL components. Table 20.6 shows the summary of this scenario. Please note that cell numbers for each scenario represent ATAX NPV in all the tables discussed in this chapter.

- The next analysis identifies the breakeven NGL pricing at **a fixed realized gas pricing** of $2.5/MMBTU and various BTU contents. At a fixed realized pricing of $2.5/MMBTU, breakeven NGL pricing at 1200 BTU is ~$21.84/BBL. At the same fixed realized gas pricing of $2.5/MMBTU and 1300 BTU/SCF, the breakeven NGL pricing to justify processing the gas is decreased to ~$17.64/BBL. Finally, at the same fixed realized gas pricing of $2.5/MMBTU and 1400 BTU/SCF, the breakeven NGL pricing to justify processing the gas is ~$15.96/BBL. Therefore, at a fixed realized gas pricing, as BTU content increases, breakeven NGL pricing will decrease to justify processing the gas. The summary of this analysis is shown in Table 20.7.
- Table 20.7 can be used to create the next table with the two following conditions:

    (a) If ATAX NPV for processing the gas (scenario b) at each BTU content and NGL pricing is bigger than ATAX NPV of blending the gas (scenario a), then proceed with processing.

    (b) If the above scenario is not true, don't process and blend the gas (of course if the capability is available)

- As can be seen from Table 20.8, at a fixed NGL pricing of $15.96/BBL and fixed gas pricing of $2.5/MMBTU, the breakeven BTU content to justify processing is 1400 BTU/SCF. At a fixed NGL pricing of $17.64/BBL and fixed gas pricing of $2.5/MMBTU, the breakeven BTU content to justify processing is 1300 BTU/SCF. Finally, at $21.84/BBL NGL pricing and a fixed gas pricing of $2.5/MMBTU, the breakeven BTU content to justify processing is only 1200 BTU/SCF. This analysis illustrates that, at a fixed gas pricing, as NGL pricing increases, the breakeven BTU content to justify processing decreases. This example demonstrates that a high NGL pricing is necessary to justify processing gas at a lower BTU content because at high NGL pricing, it is justified to extract every barrel of NGL. For example, if the price of NGL is $75/BBL, very low BTU content would be required to process the gas and still make the whole process economical.

Table 20.7 Breakeven NGL pricing at various BTU contents and fixed realized gas pricing of $2.5/MMBTU ATAX NPV at various BTUs and a fixed realized gas pricing of $2.5/MMBTU

NGL pricing ($/BBL)	1150	1175	1200	1225	1250	1300	1350	1400
10.08	3,108,067	3,349,579	3,591,152	3,832,445	4,073,509	4,554,717	5,034,230	5,511,352
10.5	3,193,952	3,452,074	3,710,453	3,969,045	4,227,802	4,745,578	5,263,238	5,780,082
10.92	3,279,836	3,554,469	3,829,753	4,105,644	4,382,094	4,936,440	5,492,245	6,048,811
11.34	3,365,720	3,656,864	3,949,053	4,242,243	4,536,387	5,127,302	5,721,252	6,317,540
11.76	3,451,504	3,759,259	4,068,353	4,378,843	4,690,679	5,318,163	5,950,259	6,586,270
12.18	3,537,488	3,861,554	4,187,653	4,515,442	4,844,972	5,509,025	6,179,267	6,854,999
12.6	3,623,372	3,964,050	4,306,953	4,652,041	4,999,265	5,699,887	6,408,274	7,123,729
13.02	3,709,257	4,066,445	4,426,253	4,788,640	5,153,557	5,890,748	6,637,281	7,392,458
13.44	3,795,141	4,168,840	4,545,553	4,925,240	5,307,850	6,081,610	6,866,239	7,661,188
13.86	3,881,025	4,271,235	4,664,854	5,061,839	5,462,142	6,272,472	7,095,296	7,929,917
14.28	3,966,909	4,373,630	4,784,154	5,198,438	5,616,435	6,463,333	7,324,303	8,198,546
14.7	4,052,793	4,476,025	4,903,454	5,335,038	5,770,728	6,654,195	7,553,310	8,467,376
15.12	4,138,678	4,578,420	5,022,754	5,471,637	5,925,020	6,845,057	7,782,318	8,736,105
15.54	4,224,562	4,680,815	5,142,054	5,608,236	6,079,313	7,035,918	8,011,325	9,004,835
15.96	4,310,446	4,783,210	5,261,354	5,744,836	6,233,606	7,225,780	8,240,332	9,273,564
16.38	4,396,330	4,885,605	5,380,654	5,881,435	6,387,898	7,417,642	8,469,339	9,542,294
16.8	4,482,214	4,988,000	5,499,954	6,018,034	6,542,191	7,608,503	8,698,347	9,811,023
17.22	4,568,098	5,090,395	5,619,254	6,154,633	6,696,483	7,799,365	8,927,354	10,079,752
17.64	4,653,983	5,192,790	5,738,555	6,291,233	6,850,776	7,990,227	9,156,361	10,348,482
18.06	4,739,867	5,295,186	5,857,855	6,427,832	7,005,069	8,181,088	9,385,363	10,617,211
18.48	4,825,751	5,397,581	5,977,155	6,564,431	7,159,361	8,371,950	9,614,376	10,885,941
18.9	4,911,635	5,499,976	6,096,455	6,701,031	7,313,654	8,562,812	9,843,333	11,154,670
19.32	4,997,519	5,602,371	6,215,755	6,837,630	7,467,946	8,753,673	10,072,390	11,423,399
19.74	5,083,403	5,704,766	6,335,055	6,974,229	7,622,239	8,944,535	10,301,398	11,692,129

20.16	5,169,288	5,807,161	6,454,355	7,110,828	7,776,532	9,135,397	10,530,405	11,960,858
20.58	5,255,172	5,909,556	6,573,655	7,247,428	7,930,824	9,326,258	10,759,412	12,229,583
21	5,341,056	6,011,951	6,692,956	7,384,027	8,085,117	9,517,120	10,988,419	12,498,317
21.42	5,426,940	6,114,346	6,812,256	7,520,626	8,239,410	9,707,932	11,217,427	12,767,047
21.84	5,512,824	6,216,741	6,931,556	7,657,226	8,393,702	9,898,343	11,446,434	13,035,776
22.26	5,598,709	6,319,136	7,050,856	7,793,825	8,547,995	10,089,705	11,675,441	13,304,505
22.68	5,684,593	6,421,531	7,170,156	7,930,424	8,702,287	10,280,567	11,904,448	13,573,235
23.1	5,770,477	6,523,926	7,289,456	8,067,024	8,856,580	10,471,428	12,133,456	13,841,964
23.52	5,856,361	6,626,322	7,408,756	8,203,623	9,010,373	10,662,290	12,362,463	14,110,694

Table 20.8 Breakeven BTU content at a fixed realized gas pricing of $2.5/MMBTU and various NGL pricing

Process or don't process

NGL pricing ($/BBL)	1150	1175	1200	1225	1250	1300	1350	1400
10.08	Don't process	Don't process	Don't process	Don't process	Don't process	Don't process	Don't process	Don't process
10.5	Don't process	Don't process	Don't process	Don't process	Don't process	Don't process	Don't process	Don't process
10.92	Don't process	Don't process	Don't process	Don't process	Don't process	Don't process	Don't process	Don't process
11.34	Don't process	Don't process	Don't process	Don't process	Don't process	Don't process	Don't process	Don't process
11.76	Don't process	Don't process	Don't process	Don't process	Don't process	Don't process	Don't process	Don't process
12.18	Don't process	Don't process	Don't process	Don't process	Don't process	Don't process	Don't process	Don't process
12.6	Don't process	Don't process	Don't process	Don't process	Don't process	Don't process	Don't process	Don't process
13.02	Don't process	Don't process	Don't process	Don't process	Don't process	Don't process	Don't process	Don't process
13.44	Don't process	Don't process	Don't process	Don't process	Don't process	Don't process	Don't process	Don't process
13.86	Don't process	Don't process	Don't process	Don't process	Don't process	Don't process	Don't process	Don't process
14.28	Don't process	process	process	process	process	process	process	process

14.7	Don't process	Don't process	Don't process	Don't process	Don't process	Don't process	Don't process	Don't process
15.12	Don't process	Don't process	Don't process	Don't process	Don't process	Don't process	Don't process	Don't process
15.54	Don't process	Don't process	Don't process	Don't process	Don't process	Don't process	Don't process	Don't process
15.96	Don't process	Don't process	Don't process	Don't process	Don't process	Don't process	Don't process	Process
16.38	Don't process	Don't process	Don't process	Don't process	Don't process	Don't process	Don't process	Process
16.8	Don't process	Don't process	Don't process	Don't process	Don't process	Don't process	Process	Process
17.22	Don't process	Don't process	Don't process	Don't process	Don't process	Process	Process	Process
17.64	Don't process	Don't process	Don't process	Don't process	Don't process	Process	Process	Process
18.06	Don't process	Don't process	Don't process	Don't process	Process	Process	Process	Process
18.48	Don't process	Don't process	Don't process	Don't process	Process	Process	Process	Process
18.9	Don't process	Don't process	Don't process	Process	Process	Process	Process	Process
19.32	Don't process	Don't process	Don't process	Process	Process	Process	Process	Process
19.74	Don't process	Don't process	Process	Process	Process	Process	Process	Process
20.16	Don't process	Don't process	Process	Process	Process	Process	Process	Process

Continued

Table 20.8 Breakeven BTU content at a fixed realized gas pricing of $2.5/MMBTU and various NGL pricing—cont'd

NGL pricing ($/BBL)	1150	1175	1200	1225	1250	1300	1350	1400
				Process or don't process				
20.58	Don't process	Don't process	Don't process	Process	Process	Process	Process	Process
21	Don't process	Don't process	Don't process	Process	Process	Process	Process	Process
21.42	Don't process	Don't process	Don't process	Process	Process	Process	Process	Process
21.84	Don't process	Don't process	Process	Process	Process	Process	Process	Process
22.26	Don't process	Don't process	Process	Process	Process	Process	Process	Process
22.68	Don't process	Don't process	Process	Process	Process	Process	Process	Process
23.1	Don't process	Don't process	Process	Process	Process	Process	Process	Process
23.52	Don't process	Process	Process	Process	Process	Process	Process	Process

The last analysis reveals breakeven ethane pricing where increasing recovery of the ethane results in a higher value for the shareholders. Therefore, different % ethane recovery vs ethane pricing was constructed to identify the breakeven ethane pricing at which recovering more of the ethane will create value. As discussed in this example, some companies are forced to recover enough ethane to achieve pipeline quality despite low pricing. Those companies recover ethane even though they know that recovering any ethane will just deteriorate value (NPV), but due to lack of optionality, they still recover enough ethane to bring the residue gas BTU to pipeline spec. At $2.5/MMBTU gas pricing as well as all the other pricing assumptions listed in this example, the breakeven ethane pricing at which recovering more and more of the ethane will create value for the shareholders is ~$0.12/gal. If ethane pricing is less than $0.11/gal, as ethane recovery increases, ATAX NPV will decrease. On the other hand, if ethane pricing is higher than $0.12/gal, as ethane recovery increases, ATAX NPV will also increase as shown in Table 20.9.

Table 20.9 Breakeven ethane pricing at fixed gas pricing of $2.5/MMBTU as well as other pricing assumptions listed in the pricing assumption table (inlet BTU of 1240)

1240 BTU

Ethane pricing ($/gal)	\multicolumn{8}{c}{Ethane recovery}							
	0%	10%	20%	30%	40%	50%	60%	70%
0.03	6,327,463	6,181,968	6,033,121	5,880,764	5,724,725	5,564,823	5,400,864	5,232,643
0.04	6,327,463	6,200,452	6,070,090	5,936,217	5,798,662	5,657,245	5,511,771	5,362,034
0.05	6,327,463	6,218,937	6,107,059	5,991,670	5,872,600	5,749,667	5,622,678	5,491,425
0.06	6,327,463	6,237,421	6,144,028	6,047,124	5,946,538	5,842,089	5,733,584	5,620,816
0.07	6,327,463	6,255,905	6,180,997	6,102,577	6,020,476	5,934,511	5,844,491	5,750,207
0.08	6,327,463	6,274,390	6,217,966	6,158,030	6,094,413	6,026,934	5,955,397	5,879,598
0.09	6,327,463	6,292,874	6,254,934	6,213,483	6,168,351	6,119,356	6,066,304	6,008,989
0.1	6,327,463	6,311,359	6,291,903	6,268,937	6,242,289	6,211,778	6,177,210	6,138,380
0.11	6,327,463	6,329,843	6,328,872	6,324,390	6,316,226	6,304,200	6,288,117	6,267,771
0.12	6,327,463	6,348,328	6,365,841	6,379,843	6,390,164	6,396,622	6,399,024	6,397,162
0.13	6,327,463	6,366,812	6,402,810	6,435,297	6,464,102	6,489,044	6,509,930	6,526,553
0.14	6,327,463	6,385,296	6,439,779	6,490,750	6,538,040	6,581,466	6,620,837	6,655,944
0.15	6,327,463	6,403,781	6,476,748	6,546,203	6,611,977	6,673,889	6,731,743	6,785,335
0.16	6,327,463	6,422,265	6,513,716	6,601,656	6,685,915	6,766,311	6,842,650	6,914,726
0.17	6,327,463	6,440,750	6,550,685	6,657,110	6,759,853	6,858,733	6,953,556	7,044,117
0.18	6,327,463	6,459,234	6,587,654	6,712,563	6,833,790	6,951,155	7,064,463	7,173,508
0.19	6,327,463	6,477,719	6,624,623	6,768,016	6,907,723	7,043,577	7,175,370	7,302,899
0.2	6,327,463	6,496,203	6,661,592	6,823,470	6,981,666	7,135,999	7,286,276	7,432,290
0.21	6,327,463	6,514,687	6,698,561	6,878,923	7,055,604	7,228,421	7,397,183	7,561,681
0.22	6,327,463	6,533,172	6,735,530	6,934,376	7,129,541	7,320,844	7,508,039	7,691,072
0.23	6,327,463	6,551,656	6,772,498	6,989,829	7,203,479	7,413,266	7,618,996	7,820,463
0.24	6,327,463	6,570,141	6,809,467	7,045,283	7,277,417	7,505,688	7,729,902	7,949,854
0.25	6,327,463	6,588,625	6,846,436	7,100,736	7,351,354	7,598,110	7,840,809	8,079,245
0.26	6,327,463	6,607,110	6,883,405	7,156,189	7,425,292	7,690,532	7,951,716	8,208,636
0.27	6,327,463	6,625,594	6,920,374	7,211,643	7,499,230	7,782,954	8,062,622	8,338,027

CHAPTER TWENTY-ONE

Well spacing and completions optimization

Introduction

One of the most critical aims regarding field optimization within a corporation is to optimize well spacing and completions design **simultaneously**. Unfortunately, this optimization problem was evaluated separately; the reservoir engineering would propose optimum well spacing while completions engineering would propose completions design. While there might have been some merit in this approach, more companies are realizing that, to produce the best economic outcome for the shareholders, both completions design and well spacing must be considered together, on the same level. Optimizing completions design and well spacing is a direct function of various economic parameters. Any tornado chart would show that commodity pricing has the highest impact on the economics of wells. Therefore, as discussed in the economic evaluation chapter of this book, **commodity pricing** has the highest impact on the completions design and well spacing analysis. Therefore, this combo must be evaluated at various commodity pricing based on a corporation's outlook on pricing. In addition to pricing, as highlighted by Belyadi et al. (2016a,b), CAPEX, OPEX, NRI, and lateral lengths also have significant impact on the economic outlook of well spacing and completions design. Curtailment must also be considered. As gas pricing increases, tighter spacing and more expensive completions design become desirable. Inversely, as gas pricing decreases, wider spacing and less expensive completions design become necessary. If a field started developing in 2010 when oil pricing was above $100/BBL and used 500-ft well spacing (interlateral spacing) and very expensive completions design, would the same well spacing and completions design still stand when the price of oil is only $50/BBL? The answer is no, and this concept will be illustrated with an example later in this chapter. At higher commodity pricing, every molecule of gas or drop of oil should be produced as quickly as possible. Therefore, it is recommended to place the wells closer together and pump a very expensive completions design to **accelerate production volumes**. In contrast, at lower commodity pricing, there is no need to accelerate production volumes as much as a higher commodity pricing by placing the wells at a tighter spacing and pumping expensive frac designs.

Gas pricing and CAPEX influence

In addition to gas pricing, CAPEX is another important factor involved in well spacing and completions design optimization. The energy sector is one of the most volatile industries. Service companies within the oil and gas industry continuously change their service costs for providing various services based on increasing and decreasing commodity pricing as well as supply and demand. If all the other economic assumptions remain the same, increasing total CAPEX spent on a well will increase well spacing. On the other hand, lowering CAPEX will dictate a tighter well spacing. Like CAPEX, change in OPEX could also have an impact on the optimum economic well spacing design. Increasing OPEX (if all other assumptions remain the same) will result in wider well spacing because it is more expensive to produce the commodity per MMBTU or barrel.

Lateral length influence

Another important consideration is lateral length. As lateral length increases, total $/ft decreases. This relationship results from **drilling CAPEX/ft** decreasing while **completions CAPEX/ft** remains the same or slightly increases (depending on casing design and higher friction pressure mitigation strategy during frac jobs). Therefore, as lateral length increases, total Capex/ft decreases which indicates placing the wells closer together. Therefore, a sensitivity analysis will show that drilling longer lateral will result in tighter well spacing (assuming all other parameters remain constant). To complicate the problem, let's take curtailment into account. First, let's define curtailment. Production curtailment refers to not having enough surface equipment to produce from the well at its max capacity. In addition, production curtailment could be due to **system back offs**, occurring when pipeline sizing and underestimation in production type curves lead to system curtailment or back offs. If a system is closer to its max capacity, bringing more wells on line will result in an increase in line pressure and a resulting decrease in production rate from other well pads on the same system. This response is a common problem in some systems within various organizations in different basins. Careful planning and production type curve estimation can certainly help mitigate this problem. When curtailment is an issue due to surface equipment limitation or system curtailment issues, well spacing is directly impacted. As curtailment increases within a system, a tighter well spacing would be more cost-effective. For example, if a well with 15,000 ft lateral length that could produce 60, 40, 30, 20 MMSCF/day in months 1, 2, 3, and 4, respectively, is curtailed back to 15 MMSCF/day until casing or tubing pressure reaches line pressure (at which production rate will start declining), this curtailment or back off will indicate a tighter well spacing. This is a one well example scenario and as will be illustrated, if curtailment is an issue, it needs to be evaluated on a field level to determine the optimum well spacing.

Inventory influence

Considering a lower commodity pricing environment would dictate a wider well spacing, why would some companies choose not to increase well spacing? The answer could depend on **inventory**. How many years of inventory does a company have? For example, if well spacing is increased by 200', how does this increase impact the number of remaining undeveloped locations that companies report in their analyst presentations? As evident, the answer could become complicated especially when some corporations are limited in terms of their inventory. If the remaining inventory is not a problem, at a low commodity pricing, increasing well spacing should be most cost-effective. As demonstrated, determining optimum well spacing and completions design can get very complicated based on various factors. Therefore, special care must be taken to ensure the best economic decision is made for the company.

Simultaneous optimization of well spacing and completions design

Next, a workflow for simultaneously optimizing completions design and well spacing will be illustrated. One of the first matters of investigation is the impact of various completions design (frac designs) on frac geometry. Does increasing sand and water per ft or sand and water per cluster result in increasing total fracture surface area and as a result fracture half-length? If increasing sand and water per ft increases frac half-length, would it be more cost effective to increase sand and water amount and contact more SRV at a larger well spacing (assuming it has not been optimized yet)? The answer to this question depends on the economic parameters discussed as well as the fracture surface area increase from one design to another. Before pursuing larger well spacing, it is important to consider optimizing the design at current well spacing by pumping various designs (big and small) and observing the impact. If all the alternatives have failed to increase the production performance by retaining the same well spacing, increasing well spacing should be considered next. The idea is to ensure nothing gets left behind by increasing well spacing. For instance, if after increasing the well spacing, it turns out that the previously designed well spacing could have produced more by tweaking the frac design (e.g., adjusting cluster spacing, pumping diversion, etc.), it would lose millions or even billions (on some occasions) for the shareholders. Therefore, as engineers, it is our due diligence to make sure all other alternatives have truly failed before increasing well spacing.

Outer and standalone wells' influence on optimization design

As previously discussed in the flow back chapter, when unbounded wells (outer wells) perform better than bounded wells (inner wells), their performance indicates that the well spacing could potentially be considered for enlargement. After analyzing more than 120 wells in Marcellus Shale located in Washington and Greene Counties (Pennsylvania), Belyadi et al. (2015) showed that the best wells are either unbounded or standalone based on performing rate transient analysis and obtaining $A\sqrt{K}$ for each well. In addition, many other unconventional fields across the United States have shown similar behavior. As previously indicated, this comparison should be one of the indications on a field that is developed at a tight well spacing and could be considered for increasing well spacing. This relationship shows the importance of understanding the performance in each field in a **full-field development** as opposed to one well's performance. For example, the performance of one standalone well without any other wells nearby might not represent the potential value of an asset. This understanding becomes more important especially when standalone or unbounded wells are known to have exceptional production outcome. Can the performance of a standalone well be replicated over and over in a full-field development at a tighter well spacing that is being proposed? Or should the well spacing be increased to replicate such performance? This question is probably the most important one that a company should answer before making a conclusive decision on the performance of a new exploration area in unconventional reservoirs. In general, most publicly traded companies do a good job in this regard to provide the most accurate information to investors and shareholders.

Frac hit and its influence on optimization design

As discussed earlier, companies might decide to use tighter well spacing to access more reservoir area and increase production from shale reservoirs. They may also base such a decision on high commodity gas pricing, lower CAPEX required, drilling longer lateral lengths, or an increase in curtailment due to system back off (assuming all other parameters remain constant). However, decreasing the well distances with tighter well spacing may result in cross-well communication during hydraulic fracturing. This effect is known as fracture hit or frac hit and must be considered when designing the well spacing. Frac hit is not only a short distance phenomenon affecting nearby wells in the same pad and adjacent pads and has also been reported in wells with more than 2000 ft spacing. When frac hit impacts more widely spaced wells, it may be due to the presence of natural fractures that encourage pressure and fluid communication between the producing well (parent well) and the well that is hydraulically fractured (child well). When a child well is hydraulically fractured, the fracture extension from the child well could reach the producing (parent) well and become detrimental to its production performance. In addition to the reduced production of the parent well, the child well usually underperforms by 20%–40% (Klenner et al. 2018), depending on the length of time that the parent well has been in production. The longer the parent well has been in line usually translates to a more severe fracture hit on the parent well due to more pressure depletion in the region.

The well interference could simply be pressure communication between the wells or, in more severe cases, pressure and frac fluid communication between the wells. In some cases, frac hit has a neutral effect on production and, in other cases, it improves a well's performance. However, the majority of frac hits appear to be detrimental to the production performance of parent wells. A child well's performance could also be impacted by pressure depletion from the parent well. The severity of the frac hit can be used to quantify the impact of frac hit on parent and child wells. Since the child and parent wells production are of higher importance to the oil and gas industry, the change in production resulting from frac hit can be used to quantify the severity of the frac hit. Weichun et al. (2018) also introduced a new technique for quantifying pressure interference in fractured horizontal shale wells. They illustrated the use of Chaw Pressure Group (CPG) shown in Eq. (21.1) to quantify the frac hit impact. This analysis can be performed by plotting ΔP, $\Delta P'$, and $\frac{\Delta P}{2\Delta P'}$ vs material balance time on log-log plot. When the slope of CPG stabilizes, it can be recorded as the magnitude of pressure interference (MPI) and can be used to perform various analyses. A BHP gauge might be necessary in multiphase flow to ensure that this analysis can be performed properly.

$$\text{Chow Pressure Group} = \frac{\Delta P}{2\Delta P'} \qquad (21.1)$$

Fig. 21.1 shows an example of a "Frac Hit" in Marcellus Shale. The parent well normalized flow rate has been severely impacted by frac hit after 1000 days of gas production at the time the child well has been hydraulically fractured in the nearby pad. Different conditions can promote the occurrence of frac hit in the field including: (a) the presence of pressure sink (low-pressure region) due to the production from parent wells or pressure source due to injecting high-pressure fluid in the child well during hydraulic fracturing. Introducing the pressure sink or source due to parent well production or child well stimulation will change the effective stress of the field

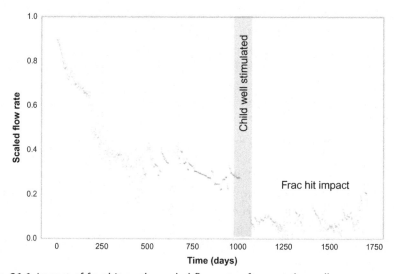

Fig. 21.1 Impact of frac hit on the scaled flow rate of parent the well.

and will locally change the direction of maximum and minimum stresses that will result in deviation in hydraulic fracture propagation path toward the parent well. (b) Fault, sealed natural fractures, or discontinuities activation due to stimulation of the child well during hydraulic fracturing. As a result, new pathways will be generated for frac fluid to approach the parent well from activated faults, natural fracture or any activated discontinuities. (c) Tight well spacing between the parent and child well that results in an overlap of their stimulated reservoir volumes that increases the risk of pressure and fluid communications.

Fracture hit detection and mitigation strategies

The conventional approach to identifying frac hit is through monitoring the pressure and flow rates in the parent wells during the production and while the child well in the area is being stimulated. Noticing sharp changes or sudden fluctuations in pressure and flow rate measurements around the time of child well stimulation enables one to detect the frac hit in parent wells. Using this method, one can efficiently and directly detect frac hits when few wells and frac jobs are involved. However, as the number of wells and frac jobs increases, resulting in more complicated pressure and rate responses, it becomes extremely difficult to distinguish each parent-child well communication. There are also other techniques that can be used to detect frac hit such as measuring the density of the produced hydrocarbon. In severe frac hit conditions, the frac fluid from the stimulation of the child well will enter the parent well and change the density of produced hydrocarbon. Frac hit can also be detected indirectly by investigating the micro seismic events or tracking the tracer injected during hydraulic fracturing that show the extension of the fractures and possible communication between stimulated and adjacent wells. While different techniques have been used to detect the frac hit after it occurs in the field, only a few efforts have been made to predict the frac hit and detect it in real time. Recently, Chamberlain (2018) used artificial intelligence (AI) technique to predict and detect the frac hit in real time, based on actual field data. Unlike previous studies in which numerical or analytical solutions are used to identify the well interference using pressure build up test, or rate transient analysis, this approach uses actual field data and a model is developed that can locate and determine well-to-well hydraulic fracture interference (frac-hit) in shale plays. In this technique, a combination of adaptive moment estimation (ADAM) neural networks with designated parameters and target outputs in conjunction with graphs of gas flow rate, tubing pressure, and cumulative gas prediction of more than 200 wells in Marcellus shale have been used to identify the well interference effects. Their model shows a great premise that the AI can be used for more robust and accurate predictions of frac hit and developing mitigation strategies in real time using actual field data.

Depending on the type of interference observed in a field development, various strategies can be used to mitigate fracture hits. The mitigation strategy could be (a) to increase well spacing on the parent and child wells, (b) to alter completions design by increasing the number of clusters per stage or pumping less sand and water per ft, (c) to well stagger or stack frac the first few stages of a child well (near a parent well) to provide a pressure barrier from other frac wells (near or next to child well), (d) to shut in the parent wells around the child well to build up pressure in the region, (e) to use a tank development approach in which one well or a few wells are shut-in after fracing the wells (pad A) and finish fracing the next adjacent pad (pad B) before flowing back pad A to keep the high pressure in the region. This process positions a pressure buffer (also referred to as pressure wall) in between the previously fraced well and new wells that are being fraced (Thompson et al., 2018).

Even though frac hit prediction, detection and remediation is still in its early development stage, and a unique protocol has not been developed to quantify its impact on hydrocarbon production from shale reservoirs, it is extremely important to consider these concepts when designing well spacing, performing completion design optimization, and optimizing the drill schedule.

A dynamic workflow for design and well spacing optimization

As previously discussed, well spacing and completions design optimization must be done together. The objective function in any field development is typically to maximize NPV/acre. However, some companies decide based on IRR. Belyadi et al. (2016a,b) illustrated how NPV and IRR metric can yield separate results, analyzing managed pressure drawdown wells. They basically showed that if the objective is to maximize NPV, a smaller percentage uplift in EUR is needed to economically justify curtailing the wells back as compared to the IRR metric. Similarly, an objective function of maximizing NPV would indicate placing the wells at a tighter spacing as compared to the IRR. Therefore, the first task in well spacing and completions design optimization is to choose the objective function. The objective function is typically guided by the executive team depending on their vivid vision for the company. Throughout this chapter, NPV metric will be used for well spacing and completions optimization.

There are different approaches to optimizing well spacing and completions design. The most promising approach is calibrating a model to actual production history and using that calibrated model to run various sensitivity analysis. Belyadi and Smith (2018) illustrated a dynamic workflow for completions design and well spacing optimizing. In their paper, they performed the following analyses:

- Upscaling analysis using Residual optimization algorithm: this was done to create a model with multiple layers as opposed to a single layer. This optimization algorithm essentially determines the optimum grouping of layers from all possible layer groupings in the original fine grid model by minimizing the difference between the upscaled and original model, called "Residual." This algorithm retains high permeability streaks, porosity, flow barriers, etc., thereby minimizing the loss of reservoir heterogeneity from the original geological/fine grid model. Li and Beckner published a paper in 1999 defining the Residual function as

$$R = \sum_{k=1}^{n_z} \sum_{j=1}^{n_y} \sum_{i=1}^{n_x} \frac{\left(P_{ijk}^c - P_{ijk}^f\right)^2}{n_x n_y n_z}$$

where R is the residual; P_{ijk}^c is the coarse-layer property mapped (downscaled) at the location (i,j,k) of the fine-layer model; P_{ijk}^f is the fine layer property at the location (i,j,k) of the fine layer model; n_x and n_y are number of cells in the x and y directions of the fine-layer model, and n_z is the number of layers of the fine-layer model. P in the Residual equation can be any variable that is acceptable to characterize reservoir heterogeneity.
- Created a base mode with a normal distribution of fracture half-length as opposed to a fixed fracture half-length.
- Used some of the most uncertain parameters such as matrix perm, frac perm and width (frac conductivity), and fracture half-length multiplier (to be applied to the distribution of fracture half-lengths) to perform history matching using Markov Chain Monte Carlo as well as other algorithms. Other parameters such as fracture height, compaction/dilation table, NFZ perm (enhanced perm region), and relative perm can also be used as needed for HM analysis.

- Performed sensitivity analysis by applying the history matched completions parameters to various well spacing scenarios and obtaining a production profile for each case.
- Finally, performed economic sensitivity analysis at various gas pricing.

The advantage of this approach is that actual production history of various well spacing and completions design will be used for decision making. In addition, the test result would have to be a controlled test to understand the impact of each parameter on production outcome. The disadvantage is the lack of data for those designs. An example to go through field-level economic analysis after obtaining a forecast from the HM is given below.

Full-field completions design and well-spacing optimization example

Nine scenarios of production type curves (flow rates over time in monthly volumes) for various completions designs and well spacing have been provided. You are tasked by your manager to find the **optimum well spacing and completions design** for full-field development as shown in Fig. 21.2. The production type curves are provided for the following scenarios:
- 1000' well spacing for 2000 #/ft, 2750 #/ft, and 3500 #/ft (sand/ft design)
- 850' well spacing for 2000 #/ft, 2750 #/ft, and 3500 #/ft (sand/ft design)
- 700' well spacing for 2000 #/ft, 2750 #/ft, and 3500 #/ft (sand/ft design)

All the economic assumptions are provided in Table 21.1.

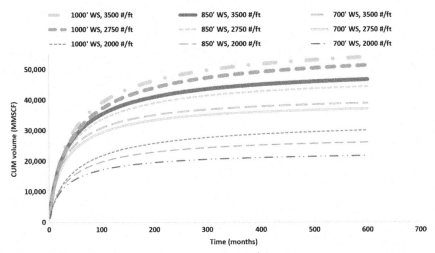

Fig. 21.2 Production CUM volume vs time (example).

Table 21.1 Example economic assumptions

Economic assumptions	
Compressor shrinkage, %	2.00%
BTU/SCF	1065
WI, %	100%
Royalty, %	20%
NRI, %	80%
Severance tax, %	5%
Ad valorem tax, %	2.50%
CAPEX, tangible %	11%
CAPEX, intangible %	89%
Federal tax rate, %	21%
WACC, %	10.00%
Number of wells TIL per month	12
Total acreage position, acres	300,000
Avg field LL per well (ft)	14,000
Variable OPEX, $/MCF	0.15
Gathering OPEX, $/MM3TJ	0.350
Firm transportation, $/MMBTU	0.25
Fixed operating costs, $/month/well	1000
CAPEX for 3500 #/ft Design, $	14,095,157
CAPEX for 2750 #/ft Design, $	11,095,157
CAPEX for 2000 #/ft Design, $	8,595,157

Part A (BASE CASE): Calculate **full-field ATAX NPV** (based on an avg LL of 14,000′, total acreage position of 300,000 acres, and turning 12 wells in line per month as shown in the assumption sheet) for each scenario and recommend the optimum field design going forward. Perform this analysis at the following fixed gas pricing (Do NOT escalate gas pricing, OPEX, or Capex for this analysis):

$1.5/MMBTU, $2/MMBTU, $2.5/MMBTU, $3/MMBTU, $3.5/MMBTU, and $4/MMBTU

Part B: Repeat the analysis using the same assumptions as the BASE CASE but **increasing** AND **decreasing** the total CAPEX by **30%**. Report the optimal design and well spacing based on these two scenarios (30% decrease in total CAPEX AND 30% increase in total CAPEX).

Part C: Repeat the analysis by assuming a 10% royalty interest as opposed to the given 20% (provided in the BASE CASE) and report the optimal design and well spacing based on 10% royalty interest.

Part D: The provided volumes are **uncurtailed volume** (unconstraint assuming full system capacity). Repeat the analysis by curtailing the given production rates to **10** and **20 MMSCF/day** and repeat the analysis by finding the optimal completions design and well spacing.

Part E: Repeat the analysis by assuming 50,000 remaining acres as opposed to the original 300,000 acres.

Part F: Repeat the analysis by assuming a 20% WACC as opposed to the given 10% (provided in the BASE CASE) and report the optimal design and well spacing based on 20% WACC.

Part G: Repeat the analysis by assuming turning 36 wells per month in line as opposed to the given 12 wells per month (provided in the BASE CASE) and report the optimal design and well spacing based on turning 36 wells per month in line (faster acreage development).

Note: Make your final decision based on **field ATAX NPV** as you were told by your manager that your company's focus is on creating long-term value for the shareholders.

Solution Part A. One of the most important considerations is what economic metric to choose for your final decision. As was discussed, choosing various economic metric will have a direct impact on your result. For this problem, field NPV will be chosen to make the final design and well-spacing decision. One of the easier approaches to solve this problem is to calculate the ATAX NPV per well assuming that capital was spent at time 0 (effective date = start date). All the future cash flows are discounted back to effective date. After calculating the ATAX NPV per well, field NPV can simply be calculated by scheduling the wells (turning 12 wells on line per month) until the whole field is drilled. The steps that can be taken to perform such analysis are given as follows:

Step 1: Calculate the drainage radius/well for each scenario:

$$\text{Drainage radius at 700 ft WS} = \frac{\text{Lateral length} \times \text{well spacing}}{43,560}$$

$$= \frac{14,000 \text{ ft} \times 700 \text{ ft}}{43,560} = \frac{224.9 \text{ acres}}{\text{well}}$$

Step 2: Calculate the total number of wells that can be fitted into 300,000 net acres:

$$\text{Total wells assuming 300,000 net acres at 700 ft well spacing} = \frac{300,000}{224.9}$$
$$= 1333.5 \text{ wells}$$

Step 3: Following the same step-by-step calculation as was shown in Chapter 18, the ATAX NPV per well for each well can be calculated. After calculating the ATAX NPV which assumes start date = effective date, schedule 12 wells per month until all the wells are drilled up. The ATAX NPV per well for 700′ WS and 3500 #/ft design was calculated to be **$5,120,653 @ $2/MMBTU** gas pricing. Table 21.2 can be created by scheduling 5121 M$ by assuming turning 12 wells on line. Please note that this table only shows the first and last 12 months. Field undiscounted NPV is simply taking number of wells to be TIL multiplied by the ATAX NPV per well. For this example, taking 12 and multiplying by $5,120,653 yields 61,448 M$. Field discounted NPV can then simply be calculated as follows:

Table 21.2 Field discounted NPV analysis

Time (months)	Number of wells per month	Total wells	NPV per well (M$) (start date = effective date)	Field undiscounted NPV (M$)	Field discounted NPV (M$)
1	12	12	5121	61,448	60,962
2	12	24	5121	61,448	60,479
3	12	36	5121	61,448	60,001
4	12	48	5121	61,448	59,526
5	12	60	5121	61,448	59,055
6	12	72	5121	61,448	58,588
7	12	84	5121	61,448	58,125
8	12	96	5121	61,448	57,665
9	12	108	5121	61,448	57,209
10	12	120	5121	61,448	56,756
11	12	132	5121	61,448	56,307
12	12	144	5121	61,448	55,862
...	...	...	...	...	...
100	12	1200	5121	61,448	27,769
101	12	1212	5121	61,448	27,550
102	12	1224	5121	61,448	27,332
103	12	1236	5121	61,448	27,116
104	12	1248	5121	61,448	26,901
105	12	1260	5121	61,448	26,688
106	12	1272	5121	61,448	26,477
107	12	1284	5121	61,448	26,268
108	12	1296	5121	61,448	26,060
109	12	1308	5121	61,448	25,854
110	12	1320	5121	61,448	25,649
111	12	1332	5121	61,448	25,446
112	1.5	1333.5	5121	7524	3091

$$\text{Field discounted NPV for month 1} = \frac{61,448}{(1+10\%)^{\frac{1}{12}}} = 60,962 \text{ M\$}$$

$$\text{Field discounted NPV for month 1} = \frac{61,448}{(1+10\%)^{\frac{2}{12}}} = 60,479 \text{ M\$}$$

$$\text{Field discounted NPV for month 1} = \frac{61,448}{(1+10\%)^{\frac{3}{12}}} = 60,001 \text{ M\$}$$

Step 4: Repeat the process for various gas pricing, completions design, and well spacing to create Table 21.3. Please note that these are synthetic cases that were created for the illustration of the concept and these are not the optimum designs in any of the Shale plays across the United States or around the world. The last chapter of this book will use actual publicly available data to go through a full analysis. As shown in Table 21.3, the highest ATAX NPV for these 9 scenarios is **2750 #/ft and 1000 ft WS** up to $3.25/MMBTU gas pricing. The optimum design changes to **3500 #/ft and 1000 ft WS** at $3.5/MMBTU gas pricing and above. These figures indicate that higher gas pricing justifies more expensive completions design. Please note that this example uses only nine scenarios for simplicity of the analysis; however, other scenarios can also be used in addition to those illustrated to fully capture more completions design and well spacing that are within operational limits. Though, as previously discussed and shown in this example, higher gas pricing will justify more expensive completions design.

Solution part B
30% increase in total CAPEX:

CAPEX is another economic uncertainty that must be sensitized to interpret its impact on completions design and well spacing. The solution to part A will be used as the **base case scenario** going forward in this example. As displayed in Table 21.4, after applying a 30% increase to Total CAPEX (for the three designs provided in this example) and rerunning the analysis, the optimum design stays at 2750 #/ft and 1000 ft WS at different gas pricing. It is important to note that in the base case (Part A), the optimum design changed from 2750 #/ft to 3500 #/ft starting at $3.5/MMBTU gas pricing; however, after applying a 30% increase to total CAPEX, higher gas pricing of up to $4/MMBTU does not yet justify pumping a more expensive completions design of 3500 #/ft. Of course, if this example was shown for gas pricing higher than $4/MMBTU, it would have eventually crossed over and justified a more expensive completions

Table 21.3 Field ATAX NPV (MM$) for various gas pricing, completions design, and well spacing

Gas pricing ($/MMBTU)	1.5	1.75	2	2.25	2.5	2.75	3	3.25	3.5	3.75	4
3500 #/ft, 1000 ft WS	−755	2910	6575	10,241	13,906	17,571	21,237	24,902	**28,567**	**32,232**	**35,898**
3500 #/ft, 850 ft WS	−1614	2097	5808	9519	13,231	16,942	20,653	24,364	28,075	31,787	35,498
3500 #/ft, 700 ft WS	−2852	833	4518	8203	11,888	15,573	19,258	22,943	26,628	30,313	33,998
2750 #/ft, 1000 ft WS	**552**	**4038**	**7523**	**11,008**	**14,494**	**17,979**	**21,464**	**24,950**	28,435	31,920	35,406
2750 #/ft, 850 ft WS	−111	3418	6947	10,476	14,005	17,534	21,063	24,592	28,121	31,650	35,179
2750 #/ft, 700 ft WS	−1096	2408	5912	9416	12,920	16,424	19,929	23,433	26,937	30,441	33,945
2000 #/ft, 1000 ft WS	−848	1201	3249	5298	7346	9395	11,443	13,492	15,540	17,589	19,637
2000 #/ft, 850 ft WS	−1380	695	2769	4843	6917	8991	11,066	13,140	15,214	17,288	19,362
2000 #/ft, 700 ft WS	−2136	−76	1983	4043	6102	8162	10,221	12,281	14,341	16,400	18,460

Note: The bold values show the optimum sand per foot based on maximum Field ATAX NPV for each predicted gas pricing.

Table 21.4 Field ATAX NPV (MM$) assuming 30% increase in total CAPEX

Gas pricing	1.5	1.75	2	2.25	2.5	2.75	3	3.25	3.5	3.75	4
3500 #/ft, 1000 ft WS	−3089	577	4242	7907	11,572	15,238	18,903	22,568	26,234	29,899	33,564
3500 #/ft, 850 ft WS	−4230	−519	3193	6904	10,615	14,326	18,038	21,749	25,460	29,171	32,882
3500 #/ft, 700 ft WS	−5821	−2136	1549	5234	8919	12,604	16,289	19,975	23,660	27,345	31,030
2750 #/ft, 1000 ft WS	**−1285**	**2201**	**5686**	**9171**	**12,657**	**16,142**	**19,627**	**23,113**	**26,598**	**30,084**	**33,569**
2750 #/ft, 850 ft WS	−2170	1359	4888	8417	11,946	15,475	19,004	22,533	26,062	29,591	33,120
2750 #/ft, 700 ft WS	−3433	71	3575	7080	10,584	14,088	17,592	21,096	24,600	28,104	31,608
2000 #/ft, 1000 ft WS	−2271	−222	1826	3875	5923	7972	10,020	12,069	14,117	16,166	18,214
2000 #/ft, 850 ft WS	−2975	−900	1174	3248	5322	7396	9471	11,545	13,619	15,693	17,767
2000 #/ft, 700 ft WS	−3946	−1887	173	2233	4292	6352	8411	10,471	12,530	14,590	16,650

Note: The bold values show the optimum sand per foot based on maximum Field ATAX NPV for each predicted gas pricing.

design. Therefore, this illustration shows that as CAPEX increases, more expensive designs may not be the optimum solution. Sensitivity analysis must then be performed to make sure that the selected base design based on the original CAPEX assumption still holds true. This analysis is applied when vendors decide to dramatically increase their pricing for various services based on the **market condition**. When this phenomenon happens, optimization models must be repeated to confirm the optimum design that was selected for the project.

30% decrease in total CAPEX:

This analysis was also repeated when a 30% decrease in total CAPEX was applied (to all three completions design scenarios). As demonstrated in Table 21.5, the breakeven gas pricing to justify more expensive completions design decreased when compared to the base scenario. The optimum design is 2750 #/ft and 1000 ft WS up to $2.5/MMBTU gas pricing. However, at $2.75/MMBTU gas pricing (as opposed to $3.5/MMBTU gas pricing on the base design) the optimum design changed to 3500 #/ft and 1000 ft WS. This change illustrates that reduction in CAPEX will justify more expensive completions design sooner than the base case scenario. In addition, reduction in CAPEX will justify tighter well spacing. In this example, the minimum and maximum provided well spacing are 700 and 1000 ft and other well-spacing scenarios such as 500, 600, 1100, 1200, etc. could be included for a more conclusive analysis.

Solution part C: Another factor to include when it comes to sensitivity analysis is royalty interest. The base case was repeated using 10% RI as opposed to the original 20% royalty interest to see the impact on completions design and well spacing. As can be seen from Table 21.6 and expected, the breakeven gas pricing to justify more expensive completions design decreased as compared to the base case. Starting around $3/MMBTU gas pricing (as opposed to $3.5/MMBTU for the base case), 3500 #/ft completions design is justified. Therefore, as RI decreases, it is justifiable to pump more expensive completions design.

Solution part D: As the lateral lengths increase, some systems will lack the capabilities of producing at full capacity. Therefore, it is important to take this issue into account when performing various sensitivity analysis to find the optimum design and well spacing because it will have a direct impact on the optimum solution. Applying 10 and 20 MMSCF/day curtailment changes the shape of the production curve for the first few months to few years (depending on the amount of production from each well). Figs. 21.3 and 21.4 illustrate CUM production over time for the nine

Table 21.5 Field ATAX NPV (MM$) assuming 30% decrease in total CAPEX

Gas pricing	1.5	1.75	2	2.25	2.5	2.75	3	3.25	3.5	3.75	4
3500 #/ft, 1000 ft WS	1578	5244	8909	12,574	16,240	**19,905**	23,570	27,235	30,901	34,566	**38,231**
3500 #/ft, 850 ft WS	1001	4712	8424	12,135	15,846	19,557	23,268	26,980	30,691	34,402	38,113
3500 #/ft, 700 ft WS	116	3801	7486	11,171	14,856	18,542	22,227	25,912	29,597	33,282	36,967
2750 #/ft, 1000 ft WS	**2389**	**5874**	**9360**	**12,845**	**16,330**	19,816	23,301	26,787	30,272	33,757	37,243
2750 #/ft, 850 ft WS	1948	5477	9006	12,535	16,064	19,593	23,122	26,651	30,180	33,709	37,238
2750 #/ft, 700 ft WS	1241	4745	8249	11,753	15,257	18,761	22,265	25,769	29,274	32,778	36,282
2000 #/ft, 1000 ft WS	575	2624	4672	6721	8769	10,818	12,866	14,915	16,963	19,012	21,060
2000 #/ft, 850 ft WS	215	2289	4364	6438	8512	10,586	12,660	14,735	16,809	18,883	20,957
2000 #/ft, 700 ft WS	−326	1734	3793	5853	7913	9972	12,032	14,091	16,151	18,210	20,270

Note: The bold values show the optimum sand per foot based on maximum Field ATAX NPV for each predicted gas pricing.

Table 21.6 Field ATAX NPV (MM$) assuming 10% royalty interest as opposed to 20% RI

Gas pricing	1.5	1.75	2	2.25	2.5	2.75	3	3.25	3.5	3.75	4
3500 #/ft, 1000 ft WS	1994	6117	10,241	14,364	18,488	22,611	**26,735**	**30,858**	**34,981**	**39,105**	**43,228**
3500 #/ft, 850 ft WS	1169	5344	9519	13,694	17,870	22,045	26,220	30,395	34,570	38,745	42,920
3500 #/ft, 700 ft WS	−88	4057	8203	12,349	16,494	20,640	24,786	28,931	33,077	37,223	41,368
2750 #/ft, 1000 ft WS	**3166**	**7087**	**11,008**	**14,929**	**18,850**	**22,771**	26,692	30,613	34,534	38,455	42,376
2750 #/ft, 850 ft WS	2535	6506	10,476	14,446	18,416	22,386	26,356	30,327	34,297	38,267	42,237
2750 #/ft, 700 ft WS	1532	5474	9416	13,358	17,301	21,243	25,185	29,127	33,069	37,011	40,953
2000 #/ft, 1000 ft WS	688	2993	5298	7602	9907	12,211	14,516	16,821	19,125	21,430	23,734
2000 #/ft, 850 ft WS	176	2509	4843	7176	9510	11,843	14,177	16,510	18,844	21,177	23,511
2000 #/ft, 700 ft WS	−591	1726	4043	6360	8677	10,994	13,311	15,628	17,945	20,262	22,579

Note: The bold values show the optimum sand per foot based on maximum Field ATAX NPV for each predicted gas pricing.

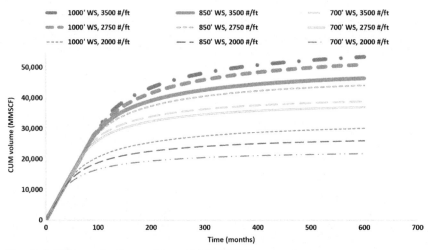

Fig. 21.3 CUM production vs time at 10 MMSCF/day curtailment.

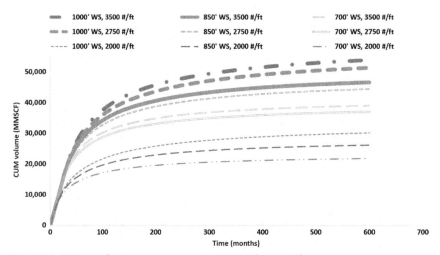

Fig. 21.4 CUM production vs time at 20 MMSCF/day curtailment.

discussed cases at 10 and 20 MMSCF/day curtailed rates, respectively. Please spend some time building a curtailment function into excel and compare the shape of the curves at various curtailment rates.

Curtailment essentially prevents a well from producing at its full capacity by delaying production into the future. This function has a direct impact on the total field NPV of the project by causing the optimum design to change to tighter well spacing and smaller designs. As displayed in Table 21.7, the optimum design changed from 2750 #/ft and 1000 ft WS to 2750 #/ft and 850 ft WS at ~$3/MMBTU gas pricing when assuming 10 MMSCF/day/well curtailment. Although a larger spacing and bigger completions design were profitable as shown in the base starting at $3.5/MMBTU gas pricing, curtailing production volume changed the whole dynamic of this exercise and simply indicates smaller designs and tighter well spacing since those volumes cannot be produced. This example simply indicates that spending the same CAPEX while constrained to producing due to system or pad level curtailment causes the optimum design to change to tighter well spacing and smaller designs. Table 21.8 shows the repeated analysis at 20 MMSCF/day/well curtailment (less curtailment as compared to 10 MMSCF/day/well). Although the directional guidance is the same for this scenario, this table illustrates that quantifying curtailment rate also matters. In addition to knowing that a field is going to be curtailed, it is very important to quantify the exact curtailment rate per well to make the right well spacing and completions design decision for the shareholders. It is also important to note that quality areas with severe curtailment should be addressed or potentially avoided because the economic analysis of other core areas could exceed the economic analysis of severely curtailed areas with quality reservoirs. Although the quality of reservoir could be tier 1 asset, severe curtailment could significantly deteriorate the value of the tier 1 asset and, as a result, move lower tiered assets to exceed economically. In those situations, core assets with heavy curtailment should be reevaluated, addressed, or improved to make sure the highest net asset value is generated for the shareholders.

Solution part E: Another important consideration when it comes to completions design and well spacing is the number of remaining acres. The whole dynamic of well spacing and completions design optimization changes when a large public company is compared with a smaller public or private company. If a small company has only 50,000 acres left in the core of a certain play while another operator has 300,000 acres in the same core play, the optimum well spacing differs for the two companies due to **inventory difference**. Therefore, inventory is crucial to consider when designing well spacing and completions design. The same analysis was repeated for 50,000 acres (as opposed to the original 300,000 acres) and the result is summarized in Table 21.9. As can be seen from this table, the optimum design is 2750 #/ft and 1000 ft WS until a gas pricing of $2.25/MMBTU. The optimum design changes to 850 ft WS and 2750 #/ft between $2.5/MMBTU and $3/MMBTU. The optimum design changes again to 700 ft WS and

Table 21.7 Field ATAX NPV (MM$) assuming 10 MMSCF/day/well curtailment

Gas pricing	1.5	1.75	2	2.25	2.5	2.75	3	3.25	3.5	3.75	4
3500 S/ft, 1000 ft WS	−2472	306	3084	5862	8640	11,418	14,196	16,974	19,752	22,530	25,307
3500 #/ft, 850 ft WS	−3198	−306	2586	5479	8371	11,264	14,156	17,049	19,941	22,834	25,726
3500 #/ft, 700 ft WS	−4237	−1267	1702	4671	7641	10,610	13,580	16,549	19,518	22,488	25,457
2750 #/ft, 1000 ft WS	**−972**	**1725**	**4422**	**7120**	**9817**	**12,514**	**15,211**	**17,909**	**20,606**	**23,303**	**26,001**
2750 #/ft, 850 ft WS	−1518	1285	4087	6889	9691	12,493	15,295	18,098	20,900	23,702	26,504
2750 #/ft, 700 ft WS	−2325	545	3414	6283	9152	12,022	14,891	17,760	20,629	23,498	26,368
2000 #/ft, 1000 ft WS	−1219	638	2494	4351	6208	8064	9921	11,778	13,634	15,491	17,348
2000 #/ft, 850 ft WS	−1723	173	2070	3967	5863	7760	9657	11,553	13,450	15,346	17,243
2000 #/ft, 700 ft WS	−2436	−532	1373	3277	5181	7086	8990	10,894	12,799	14,703	16,608

Note: The bold values show the optimum sand per foot based on maximum Field ATAX NPV for each predicted gas pricing.

Table 21.8 Field ATAX NPV (MM$) assuming 20 MMSCF/day/well curtailment

Gas pricing	1.5	1.75	2	2.25	2.5	2.75	3	3.25	3.5	3.75	4
3500 #/ft, 1000 ft WS	−1282	2111	5503	8896	12,289	15,682	19,075	22,468	25,860	29,253	32,646
3500 #/ft, 850 ft WS	−2104	1354	4812	8270	11,728	15,186	18,644	22,102	25,560	29,018	32,476
3500 #/ft, 700 ft WS	−3279	185	3650	7114	10,578	14,043	17,507	20,971	24,436	27,900	31,364
2750 #/ft, 1000 ft WS	**104**	**3357**	**6611**	**9864**	**13,118**	**16,371**	**19,625**	**22,878**	**26,131**	**29,385**	**32,638**
2750 #/ft, 850 ft WS	−528	2786	6100	9414	12,728	16,041	19,355	22,669	25,983	29,297	32,610
2750 #/ft, 700 ft WS	−1464	1850	5164	8478	11,792	15,106	18,420	21,734	25,049	28,363	31,677
2000 B/ft, 1000 ft WS	−908	1109	3126	5143	7161	9178	11,195	13,213	15,230	17,247	19,264
2000 #/ft, 850 ft WS	−1439	604	2648	4691	6734	8778	10,821	12,865	14,908	16,952	18,995
2000 #/ft, 700 ft WS	−2191	−159	1872	3903	5935	7966	9997	12,028	14,060	16,091	18,122

Note: The bold values show the optimum sand per foot based on maximum Field ATAX NPV for each predicted gas pricing.

Table 21.9 Field ATAX NPV (MM$) assuming 50,000 acres as opposed to 300,000 acres

Gas pricing	1.5	1.75	2	2.25	2.5	2.75	3	3.25	3.5	3.75	4
3500 #/ft, 1000 ft WS	−160	618	1396	2174	2952	3730	4509	5287	6065	6843	7621
3500 #/ft, 850 ft WS	−357	463	1283	2102	2922	3742	4561	5381	6201	7020	7840
3500 #/ft, 700 ft WS	−665	194	1054	1914	2774	3633	4493	5353	6212	7072	**7932**
2750 #/ft, 1000 ft WS	**117**	**857**	**1597**	**2337**	3077	3817	4557	5297	6037	6777	7517
2750 #/ft, 850 ft WS	−25	755	1534	2314	**3093**	**3872**	**4652**	5431	6211	6990	7769
2750 #/ft, 700 ft WS	−256	562	1379	2197	3014	3832	4649	**5467**	**6285**	**7102**	7920
2000 #/ft, 1000 ft WS	−180	255	690	1125	1560	1994	2429	2864	3299	3734	4169
2000 #/ft, 850 ft WS	−305	153	611	1070	1528	1986	2444	2902	3360	3818	4276
2000 #/ft, 700 ft WS	−498	−18	463	943	1424	1904	2385	2865	3346	3826	4307

Note: The bold values show the optimum sand per foot based on maximum Field ATAX NPV for each predicted gas pricing.

2750 #/ft between $3.25/MMBTU and $3.75/MMBTU. Finally, at $4/MMBTU gas pricing, the optimum design changes to 3500 #/ft and 700 ft WS. Please spend some time comparing this scenario as well as the other scenarios discussed with the base case (Part A).

Solution part F: Increasing the WACC from 10% to 20% will also have a direct impact on the optimum design because the cost of doing business is much more expensive now as illustrated in Table 21.10. As opposed to the base case scenario in which the optimum design changed to 3500 #/ft and 1000 ft WS at $3.5/MMBTU, the optimum design change occurs at $4/MMBTU in this scenario. This change is due to cost of borrowing money as well as the increased market expectation which would result in a less expensive completions design for a longer gas pricing period.

Solution part G: As shown in Table 21.11, the optimum design changes from 2750 #/ft and 1000 ft WS to 2750 #/ft and 850 ft WS at $2.75/MMBTU due to faster acreage development as compared to the base case scenario. The optimum design changes again at $3.75/MMBTU to 3500 #/ft while remaining at 850' well spacing. The outcome from this part illustrates that faster acreage development will result in tighter well spacing. The result of this section analysis is in line with the reduced inventory example in **part E**.

Well spacing and completions optimization 465

Table 21.10 Field ATAX NPV (MM$) assuming 20% WACC as opposed to 10% WACC

Gas pricing	1.5	1.75	2	2.25	2.5	2.75	3	3.25	3.5	3.75	4
3500 tt/ft, 1000 ft WS	−1689	616	2922	5227	7533	9838	12,144	14,450	16,755	19,061	**21,366**
3500 it/ft, 850 ft WS	−2220	76	2372	4667	6963	9259	11,555	13,851	16,147	18,442	20,738
3500 #/ft, 700 ft WS	−2932	−710	1512	3734	5956	8178	10,400	12,622	14,844	17,065	19,287
2750 #/ft, 1000 ft WS	**−605**	**1587**	**3779**	**5972**	**8164**	**10,356**	**12,549**	**14,741**	**16,934**	**19,126**	21,318
2750 #/ft, 850 ft WS	−1027	1156	3340	5523	7706	9889	12,072	14,255	16,438	18,621	20,804
2750 #/ft, 700 ft WS	−1611	502	2615	4727	6840	8953	11,066	13,179	15,292	17,405	19,518
2000 ft/ft, 1000 ft WS	−1268	21	1309	2598	3886	5175	6463	7752	9040	10,329	11,618
2000 #/ft, 850 ft WS	−1592	−309	975	2258	3541	4824	6107	7390	8673	9957	11,240
2000 #/ft, 700 ft WS	−2020	−778	464	1706	2948	4190	5431	6673	7915	9157	10,399

Note: The bold values show the optimum sand per foot based on maximum Field ATAX NPV for each predicted gas pricing.

Table 21.11 Field ATAX NPV (MM$) assuming 36 wells per month TIL as opposed to 12

Gas pricing	1.5	1.75	2	2.25	2.5	2.75	3	3.25	3.5	3.75	4
3500 #/ft, 1000 ft WS	−915	3526	7966	12,406	16,847	21,287	25,728	30,168	34,609	39,049	43,490
3500 #/ft, 850 ft WS	−2017	2620	7258	11,895	16,532	21,170	25,807	30,444	35,082	**39,719**	**44,357**
3500 #/ft, 700 ft WS	−3720	1086	5892	10,698	15,504	20,309	25,115	29,921	34,727	39,533	44,339
2750 #/ft, 1000 ft WS	**669**	**4891**	**9114**	**13,336**	**17,559**	21,781	26,004	30,226	34,449	38,671	42,894
2750 #/ft, 850 ft WS	−139	4271	8680	13,090	17,500	**21,910**	**26,319**	**30,729**	**35,139**	39,548	43,958
2750 #/ft, 700 ft WS	−1430	3140	7710	12,280	16,850	21,420	25,990	30,560	35,130	39,700	44,269
2000 #/ft, 1000 ft WS	−1027	1454	3936	6418	8900	11,381	13,863	16,345	18,827	21,309	23,790
2000 #/ft, 850 ft WS	−1724	868	3460	6052	8643	11,235	13,827	16,419	19,011	21,603	24,194
2000 #/ft, 700 ft WS	−2785	−100	2586	5272	7958	10,644	13,330	16,016	18,702	21,388	24,074

Note: The bold values show the optimum sand per foot based on maximum Field ATAX NPV for each predicted gas pricing.

CHAPTER TWENTY-TWO

Stacked pay development

Introduction

Stacked pay codevelopment is another important concern when developing a field. If more than one viable formation is present, stacked pay codevelopment becomes an important strategic and economic matter of contention. In this situation, one must decide between developing both formations or only one, resulting in the loss of the opportunity cost of an additional formation. The factors that contribute to this economic and strategic decision include gas/oil pricing, capital expenditure to develop each formation, lateral length of development wells, cycle timing, NRI (how much royalty is being paid to the landowners), OPEX, etc. In a high commodity pricing, one might develop both formations and generate net asset value for the shareholders. However, in lower commodity pricing, it could be more challenging to justify codevelopment, especially when the production performance of one formation is lower than the other.

As the lateral length of a well increases, $/ft CAPEX decreases, helping to build a stronger case for codevelopment due to the reduction in capital. In addition, if codevelopment results in an increase in cycle timing to the drill schedule by pushing the turn in line (TIL) dates to a further date in the future, it will negatively affect the net present value of the codeveloped scenario. For example, if the spud date for completing six wells in formation X is 01/2020, and it takes about 7 months to turn all those six wells in line (TIL date of 8/2020), codeveloping 12 wells in formations X and Y (six wells in formation X and six wells in formation Y) could potentially cause the TIL date to be delayed by X number of months. This delay would result in a lower net present value for codeveloped scenario than would have resulted from the development of just one formation. This relationship is only applicable if codeveloping does cause a delay in TIL date. When wellheads are buried underground, turning a limited number of wells in line on a specific date and turning the remaining wells later is not a concern. In those scenarios, codeveloping does not adversely affect the net asset value of the field when it comes to delaying the TIL dates.

NRI is another important factor that contributes to the economic feasibility of codevelopment of the field. In areas where most of the acreage is fee acreage, meaning 0% royalty is paid out to the landowners due to owning the mineral rights, the economic feasibility of codevelopment of the field is significantly improved. As NRI increases, the economic feasibility of codevelopment will also increase. In addition to NRI, OPEX could have some impact on codevelopment of the field. As OPEX decreases, the economic feasibility of codevelopment increases.

Another important aspect of codevelopment of the field is referred to as "fracture communication" between the two codeveloped formations. Does developing one formation negatively impact the production performance of the base prolific formation or both? Does the formation of interest have a strong barrier to avoid fracture communications between the two codeveloped formations? If yes, the production performance of neither formation should be impacted by codevelopment. On the other hand, if fracture communication does exist between the two formations of interest, the production performance of one or both formations would deteriorate. For example, if an expected type curve from formation X has an IP of 14,000 MSCF/day, b value of 1.2, secant annual effective decline of 60%, and terminal decline of 5%, does developing the overlying or underlying formation negatively affect the mentioned type curve for the area? If yes, how much degradation in production performance of the base type curve can be tolerated before codevelopment becomes uneconomic? Again, the answer to these questions is a function of gas pricing, capital expenditure, lateral length, cycle timing, NRI, OPEX, etc.

Step-by-step workflow for codevelopment analysis

The strategic capital budgeting metric for a company is another important influence on the decision whether to codevelop a field. Is a company's metric NPV or IRR? The answer to this question will depend on the company's level of constraint regarding capital expenditure. Are there short-term debts and liabilities that the company must address immediately or are debts and liabilities less of a pressing issue? Typically, the smaller private or public companies tend to choose IRR over NPV because they are capital constrained and prefer a fast return on their investment. However, bigger and more established corporations tend to choose NPV over IRR to create long-term value for the shareholders rather than focusing on short-term goals. This rule of thumb is not set in stone but that is the consensus when it comes to choosing between NPV and IRR. Therefore, if a company's metric is NPV, codevelopment is favored even when the production performance of one formation is less than the other assuming the two formations are not communicating and gas pricing as well as other economic parameters discussed are accommodating. However, if a company's metric is IRR, codeveloping is most likely unfavorable depending on various economic factors. IRR metric can be justified at the right commodity pricing and other variables, but it is less favorable as compared to the NPV metric. As shown, codevelopment decision making is complex and involves many factors. Therefore, a detailed reservoir and economic analysis must be performed to make the right decision for the company. Below are a few guidelines to follow when performing such an analysis:

(1) Understand the geology by answering the following questions: Is there a competent barrier between the two formations that can prohibit fracture communication when hydraulic fracturing occurs? The answer to this question can be achieved by investigating the petrophysical and geo-mechanical properties across the two formations such as minimum horizontal stress, Young's Modulus, Poisson's ratio, etc. Hydraulic fracture numerical simulation can also be performed to estimate the fracture height growth during the stimulation. Tracer analysis as well as microseismic analyses can also be used to identify whether the fracture height growth in one formation enters the other codeveloped formation.

(2) Of course, modeling can be done before performing the actual test. However, field-testing is essential to making a multimillion-dollar decision for the company. This testing can be done through developing one pad (e.g., six wells) using only one formation of interest, called X. Then, develop another pad with 12 codeveloped wells (six wells in formation X and six wells in formation Y). Next, collect at least 6 months of production data (preferably longer) before analyzing the well, that is, 6 months of production data is typically the minimum time required to perform any type of meaningful rate transient analysis as well as other reservoir analyses such as numerical simulation and history matching.

(3) Develop a type curve for the six stand-alone wells that were completed in formation X. Afterward, develop the second type curve for the six codeveloped wells in formation X and the other six codeveloped wells in formation Y.

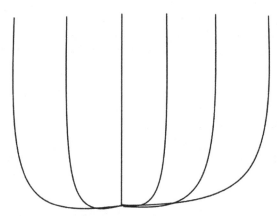

Fig. 22.1 Formation X stand-alone scenario.

(4) Once the testing has been completed, the following three type curves should be developed:
   - **Formation X stand-alone** type curve for an average lateral length well for the development area. Fig. 22.1 illustrates a stand-alone development with six formation X wells. By standalone, it is referred to when the wells are not codeveloped with any overlying or underlying formations.
   - **Formation X codeveloped** type curve for an average lateral length well for the development area
   - **Formation Y codeveloped** type curve for an average lateral length well for the development area. Fig. 22.2 illustrates a codevelopment scenario with six formation X (solid line) and six formation Y (dotted line) wells.

(5) Record any detrimental impact observed on the codeveloped type curve as compared to stand-alone type curve which could indicate hydraulic fracture communication. Using various diagnostic plots such as pressure normalized rate (normalized $q/\Delta P$) vs CUM/ft, P_{wf} vs CUM/ft, and superposition plot (normalized $\Delta P/q$) vs square root of time to measure the detrimental impact of codevelopment (if any). Also use numerical simulation to history match the production data for each scenario and obtain basic fracture geometry for each scenario. Did the fracture geometry on the stand-alone type curve differ from the fracture geometry for codeveloped scenario? In theory, if the two formations do not communicate, the fracture geometry should be close to one another and no significant difference should be observed.

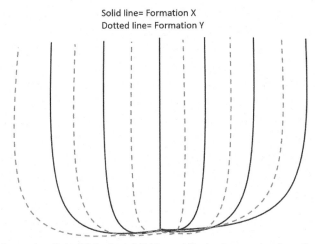

Fig. 22.2 Formation X (solid line) and Y (dotted line) when codeveloped.

(6) Gather all the economic parameters, especially capital expenditure. It is important to note that one of the biggest advantages of codevelopment is shared infrastructure capital. Therefore, in codevelopment scenario, the capital expenditure that will be applied to formation Y in codeveloped scenario will not include some of the capital expenditures such as pad construction, gathering infrastructure, water infrastructure, etc. which of course will depend on each area of development and investigation. The reason that some of these expenditures should not be included when performing the codevelopment economic analysis scenario is because of shared capital which is considered sunk because those infrastructures will be needed when developing the stand-alone formation scenario.

(7) Run a six well economic analysis scenario for formation X using formation X stand-alone type curve and record the NPV and IRR for this scenario.

(8) Run a 12 well economic analysis scenario using formation X codeveloped type curve and formation Y codeveloped type curve and record the NPV and IRR for this scenario. Essentially six stand-alone formation X is being compared to 12 codeveloped formations X and Y to define the net asset value per acre of each scenario.

(9) Once done, one scenario should be superior depending on the capital budgeting metric of NPV or IRR that has been chosen. It will be clear that the NPV metric will lean more toward codeveloping than the IRR metric since the goal is creating long-term shareholder's value.

The decision will depend on many factors discussed and must be evaluated on an area-to-area basis. In some areas, the geologic feature could indicate fracture communication between the two formations; however, other geologic areas and competent fracture barriers will have a completely different story when it comes to codevelopment.

(10) Call these two scenarios the base case and start performing a sensitivity analysis on commodity pricing, capital expenditure, cycle timing, lateral length, OPEX, NRI, etc. as recommended below:
 - If developing a gas field, rerun the analysis at various flat or escalated gas pricing of $1.5/MMBTU, $2/MMBTU, ..., $6/MMBTU depending on the market condition and being as close as possible to a realistic realized gas pricing (current market condition). For instance, if current realized gas pricing is $3/MMBTU, perform various flat or escalated gas pricing sensitivities around this pricing. If developing oil or any other type of formation, follow the same methodology. It will become apparent that as commodity pricing increases, **the justification for codevelopment is reinforced.**
 - Perform economic analysis at different lateral lengths such as 8000', 9000', ..., 20,000' depending on the average achievable lateral length for the field. It will be obvious that, as lateral length increases, codeveloped scenario will become favorable due to a decrease in capital expenditure per ft.
 - If cycle timing is affected by codevelopment, find out the TIL date delays for each lateral length and rerun the same analysis with the delayed TIL dates to find out the impact of cycle timing delay. This sensitivity is only applied if and only if cycle timing is affected or delayed in any shape or form when codeveloping.
 - If there is any room for improving the NRI from a land perspective, run a sensitivity analysis on various feasible NRIs to show the impact.
 - OPEX is generally improved over time, and it is recommended to run various sensitivities on the impact of lower or higher OPEX on codevelopment.

(11) Next, find out the breakeven commodity pricing for different lateral lengths in which codevelopment becomes economical. There is a breakeven pricing at which codevelopment becomes economic that is a function of multiple variables such as pricing, lateral length, capital expenditure, OPEX, NRI, and type curve assumptions. Type curve assumption will also have a significant impact on the analysis.

For instance, if formation Y can produce similar to formation X when codeveloped without having any type of fracture communication or detrimental impact caused by one formation or another, codeveloping is the right solution to field development. However, the decision becomes more challenging when formation Y produces 1/2 or 1/3 of formation X. In this case, codevelopment will be a function of many discussed variables and higher gas pricing, longer lateral length wells, higher NRI wells, lower OPEX, and significant reduction in capital are needed to justify codevelopment.

(12) Finally, find out the percentage loss in production that can be tolerated and still yield a higher ATAX NPV as compared to developing only formation X on a stand-alone basis. Perform this analysis at different commodity pricing.

When deciding whether to codevelop, one must consider the **lost opportunity cost of not codeveloping** if and only if returning later and developing the skipped formation is geologically and practically prevented by pressure sink and depletion in the base formation. The reason is because, if only one formation is developed, the pressure sink or depletion will potentially negatively impact the producing wells when coming back after so many years to develop the skipped formation right above or below the base formation. Essentially, it is referred to as use it or lose it investment since not using that formation could not be able to be developed later. This discussion is valid when there is potential for fracture communication. Stacked pay can occur in formations that are completely far away from each other as well. For example, if one formation is located at 9000′ TVD and the other is at 15,000′ TVD, there is no concern about fracture communication or pressure sink. In these instances, stacked pay development can potentially have significant upside to the field if both formations are economically viable.

Sometimes, stacked pay for some companies is more feasible as compared to other companies depending on the amount of inventory that each company has. If company X has various assets across the world, deciding on developing one formation vs. stacked pay development becomes a completely different argument because that company has the flexibility of investing in higher valued projects as compared to codevelopment. On the other hand, if company X has only acreage position in one basin and not codeveloping causes running out of Tier A asset, codevelopment may be the ideal solution depending on a lot of other factors that were discussed. Another consideration when codeveloping is gathering infrastructure and liquid handling. For example, if formation X is expected to have 1010

BTU/SCF and formation Y is expected to have 1200 BTU/SCF and only dry gathering infrastructure exists in the area with an allowed maximum pipeline quality BTU of 1100, codeveloping and blending the gas could add value by preventing the cost of investing millions or billions of dollars into a new wet infrastructure system to handle liquids and lowering the BTU to pipeline quality. The gas can be simply blended to obtain pipeline quality BTU gas by having the right development strategy and timing for codeveloping each formation. This can add a lot of value to a company's long-term shareholder's value and companies do take advantage of this gift in certain areas.

CHAPTER TWENTY-THREE

Production analysis and wellhead design

Introduction

Gas well deliverability indicates the rate at which a gas well can produce at a given bottom hole or wellhead pressure. Nodal analysis plays an essential role in determining the size of tubing and whether tubing is necessary. Nodal analysis is essential in determining the flow constraint at various operating conditions. Without performing proper nodal analysis, it is impossible to optimize the production volume per well. The lack of a proper tubing size such as a lower ID tubing size (than needed) can unwontedly curtail production rates and as a result, losing the time value of money. On the other hand, having a larger than the necessary tubing size can cause a faster need for using artificial lift, such as plunger lift, gas lift, etc. to lift the water and condensate out of the wellbore and to surface. With an increasing number of longer lateral developments exceeding 10,000 ft and, in some areas, getting closer to 20,000 ft, many operators do not run tubing from the start of the flow back to make sure the necessary volumes can be produced by taking advantage of time value of money.

Nodal analysis

Nodal analysis can be run to determine when tubing will be needed depending on the area. Inflow performance relationship (IPR) is the ability of the reservoir to deliver fluid to the wellbore against a given bottom hole pressure. Inflow performance relationship is also sometimes referred to as sandface deliverability curve. On the other hand, outflow performance relationship also known as tubing performance curve (TPC) reveals the capacity of the tubing to deliver fluid to the surface for a given wellhead pressure. Combining the IPR and TPC curve reveals the operating point at which a well can produce at a given pressure and rate.

There are different models that can be used to obtain the parameters needed to construct an IPR curve. Economides, 1994 developed a general solution to pseudo-steady-state flow in a radial flow gas reservoir as shown in Eq. 23.1.

General solution to pseudo steady equation:

$$q = \frac{kh\left[m(P) - m(P_{wf})\right]}{1424\,T\left[\ln\left(\dfrac{0.472 r_e}{r_w}\right) + s + Dq\right]} \quad (23.1)$$

where q = gas production rate in MSCF/day, k = effective permeability, md, h = pay zone thickness, ft, $m(P)$ = real gas pseudo pressure at reservoir pressure, psi^2/cp, $m(P_{wf})$ = real gas pseudo pressure at flowing bottom hole pressure, psi^2/cp, T = reservoir temperature, °R, r_e = radius of drainage area in ft, r_w = wellbore radius, ft, s = skin factor, and D = non-Darcy coefficient.

Eq. (23.1) can be simplified using pressure squared approach as follows:

$$q = \frac{kh\left[P^2 - P_{wf}^2\right]}{1424\,\mu z T\left[\ln\left(\dfrac{0.472 r_e}{r_w}\right) + s + Dq\right]}$$

At pressures higher than 3000 psi, highly compressible gases behave like liquids, Eq. (23.1) can be approximated using pressure approach as follows:

$$q = \frac{kh\left[P - P_{wf}\right]}{141.2 \times 10^3\, B_g \mu \left[\ln\left(\dfrac{0.472 r_e}{r_w}\right) + s + Dq\right]}$$

B_g is the average formation volume factor.

Pseudo-pressure concept and calculation

Before discussing the empirical models, it is very important to understand the concept of pseudo-pressure and its use in production analysis in unconventional reservoirs. Pseudo-pressure is basically normalizing pressure for gas viscosity and compressibility as gas viscosity and compressibility varies at different pressures. Gas pseudo-pressure is defined in Eq. (23.2).

$$\text{Gas pseudo-pressure}: m(p) = 2\int_{P_1}^{P} \frac{P}{\mu z}\,dp \quad (23.2)$$

In this equation, P_1 is some arbitrary base pressure, μ is gas viscosity, and z is gas compressibility. **Trapezoidal method** can be used to calculate pseudo-pressure as shown in Eq. (23.3).

Pseudo-pressure calculation using trapezoidal method:

$$m(p) = \sum_0^P \left(\frac{2P}{\mu z}\right) \times \Delta p \tag{23.3}$$

The area of a trapezoid is given by the following equation:

$$\text{Area} = h\left(\frac{b1 + b2}{2}\right)$$

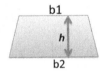

Pseudo-pressure concept and calculations are essential calculations in the majority of the diagnostic plots used in rate transient analysis. Therefore, it is very important to understand the underlying concept behind pseudo-pressure calculation.

Example
Calculate pseudo-pressure given gas viscosity and compressibility at each pressure in Table 23.1.

At 0 psi pressure:

$$\frac{2 \times P}{\mu \times z} = 0 \, \text{psia/cp}$$

At 150 psi pressure:

$$\frac{2 \times P}{\mu \times z} = \frac{2 \times 150}{0.01238 \times 0.9856} = 24{,}587 \, \text{psia/cp}$$

At 300 psi pressure:

$$\frac{2 \times P}{\mu \times z} = \frac{2 \times 300}{0.01254 \times 0.9717} = 49{,}240 \, \text{psia/cp}$$

Between 0 psi and 150 psi pressure:

Table 23.1 Pressure, viscosity, and compressibility Input

P (psia)	Gas viscosity (cp)	z (compressibility)
0	0	0
150	0.01238	0.9856
300	0.01254	0.9717
450	0.01274	0.9582
600	0.01303	0.9453
750	0.01329	0.9332
900	0.0136	0.9218
1050	0.01387	0.9112
1200	0.01428	0.9016
1350	0.01451	0.8931
1500	0.01485	0.8857
1650	0.0152	0.8795
1800	0.01554	0.8745
1950	0.01589	0.8708
2100	0.0163	0.8684
2250	0.01676	0.8671
2400	0.01721	0.8671
2550	0.01767	0.8683
2700	0.01813	0.8705
2850	0.01862	0.8738
3000	0.01911	0.878
3150	0.01961	0.883

$$\text{Trapezoid area} = \frac{\left(\frac{2P}{\mu z} \text{ at } 150\,\text{psi} + \frac{2P}{\mu z} \text{ at } 0\,\text{psi}\right) \times (150 - 0)}{2}$$

$$= \frac{(24{,}587 + 0) \times (150 - 0)}{2} = 1{,}844{,}001 \text{ psia}^2/\text{cp}$$

Pseudo-pressure $= 1{,}844{,}001 \text{ psia}^2/\text{cp}$

Between 300 psi and 150 psi pressure:

$$\text{Trapezoid area} = \frac{\left(\frac{2P}{\mu z} \text{ at } 300\,\text{psi} + \frac{2P}{\mu z} \text{ at } 150\,\text{psi}\right) \times (300 - 150)}{2}$$

$$= \frac{(49{,}240 + 24{,}587) \times (300 - 150)}{2} = 5{,}537{,}030 \text{ psia}^2/\text{cp}$$

Pseudo-pressure $= 1{,}844{,}001 + 5{,}537{,}030 = 7{,}381{,}032 \text{ psia}^2/\text{cp}$

Table 23.2 Pseudo-pressure example result

Input			Output		
P (psia)	Gas viscosity (cp)	z (compressibility)	2p/viscosity × psia/cp	Trapezoid area psia²/cp	Pseudo-pressure, psia²/cp
0	0	0	0		
150	0.01238	0.9856	24,587	1,844,001	1,844,001
300	0.01254	0.9717	49,240	5,537,031	7,381,032
450	0.01274	0.9582	73,725	9,222,432	16,603,463
600	0.01303	0.9453	97,424	12,836,223	29,439,686
750	0.01329	0.9332	120,946	16,377,771	45,817,457
900	0.0136	0.9218	143,581	19,839,524	65,656,981
1050	0.01387	0.9112	166,161	23,230,649	88,887,630
1200	0.01428	0.9016	186,410	26,442,823	115,330,453
1350	0.01451	0.8931	208,351	29,607,097	144,937,550
1500	0.01485	0.8857	228,091	32,733,175	177,670,724
1650	0.0152	0.8795	246,851	35,620,634	213,291,358
1800	0.01554	0.8745	264,906	38,381,753	251,673,111
1950	0.01589	0.8708	281,853	41,006,901	292,680,013
2100	0.0163	0.8684	296,717	43,392,703	336,072,716
2250	0.01676	0.8671	309,649	45,477,402	381,550,118
2400	0.01721	0.8671	321,656	47,347,829	428,897,946
2550	0.01767	0.8683	332,402	49,054,335	477,952,281
2700	0.01813	0.8705	342,158	50,592,040	528,544,321
2850	0.01862	0.8738	350,335	51,936,980	580,481,301
3000	0.01911	0.878	357,599	53,095,011	633,576,312
3150	0.01961	0.883	363,833	54,107,395	687,683,707

Table 23.2 illustrates the output of the example including the trapezoid area and pseudo pressure.

Rawlins and Schellhardt (back pressure model)

In field applications, empirical methods are mostly used for nodal analysis as it is expensive to get all the parameters in the analytical equations. There are different types of empirical models, but the most commonly used model is referred to as back pressure model. Rawlins and Schellhardt (1935) equation can be used to construct the IPR curve. In Rawlins and Schellhardt equation, a multirate test is required to obtain C and n values. At least two stabilized pressures must be measured at different rates to estimate the constants. This method is one of the most commonly used approaches in various commercial softwares.

$$\text{Back pressure model}: q = C\left(P_R^2 - P_{wf}^2\right)^n \quad (23.4)$$

where q is the gas production rate, MSCF/day, P_R is the reservoir pressure, psi, P_{wf} is the flowing bottom hole pressure, psi, n is the turbulence factor (represents non-Darcy effects), dimensionless, and C is the constant, MSCF/psi^{2n}.

In Eq. (23.4), n is the non-Darcy effect or the measure of the effect of turbulence in the near wellbore region. Values of n are only valid between 0.5 and 1. A value of 1 represents fully laminar flow so it has the least amount of pressure losses near the wellbore while a value of 0.5 represents fully turbulent flow with significant pressure losses near the wellbore region. n values closer to 1 represents an aging or low deliverability wells and n values closer to 0.5 shows high drawdown and high deliverability potential. The C value of a well denotes the magnitude of a wells' rate response to changes in pressure. It is important to note that larger C values indicate higher deliverability wells (high sensitivity to changes in C values) and there are no limits on the range of allowed C values.

Back pressure model can be used in the three steps shown below:
Step 1: Calculate n value from a multirate test as follows:

$$n \text{ value calculation from a multirate test}: n = \frac{\log\left(\dfrac{q_1}{q_2}\right)}{\log\left(\dfrac{P_R^2 - P_{wf1}^2}{P_R^2 - P_{wf2}^2}\right)} \quad (23.5)$$

Step 2: Calculate C by rearranging Rawlins and Schellhardt equation:

$$C = \frac{q_1}{\left(P_R^2 - P_{wf1}^2\right)^n}$$

Step 3: Calculate the gas rate (deliverability rate) at the desired flowing bottom hole pressure using Rawlins and Schellhardt equation:

$$q = C\left(P_R^2 - P_{wf}^2\right)^n$$

Production analysis and wellhead design

Example
A gas well was tested at the following two points listed below. Average reservoir pressure is 4000 psi. Estimate the deliverability of the gas well under a pseudo-steady-state flow condition at a flowing bottom hole pressure of 900 psi. Use the back pressure model.

Test point 1: q of 1200 MSCF/day at 3000 psi
Test point 2: q of 1700 MSCF/day at 1600 psi

Step 1: Calculate n value:

$$n = \frac{\log\left(\frac{q_1}{q_2}\right)}{\log\left(\frac{P_R^2 - P_{wf1}^2}{P_R^2 - P_{wf2}^2}\right)} = \frac{\log\left(\frac{1200}{1700}\right)}{\log\left(\frac{4000^2 - 3000^2}{4000^2 - 1600^2}\right)} = 0.534$$

Step 2: Calculate C by rearranging Rawlins and Schellhardt equation:

$$C = \frac{q_1}{\left(P_R^2 - P_{wf1}^2\right)^n} = \frac{1200}{\left(4000^2 - 3000^2\right)^{0.534}} = 0.2654$$

Step 3: Calculate the gas rate (deliverability rate) at 900 psi flowing bottom hole pressure:

$$q = C\left(P_R^2 - P_{wf}^2\right)^n = 0.2654\left(4000^2 - 900^2\right)^{0.534} = 1814 \, \text{MSCF/day}$$

Now that the basic concept behind the back pressure model is understood, the steps below can be used to construct an IPR curve:

Step 1: Solve for P_{wf} by rearranging the Rawlins and Schellhardt equation (Eq. 23.1).

$$P_{wf} = \sqrt{P_R^2 - \left(\frac{q}{C}\right)^{\frac{1}{n}}}$$

Step 2: Solve for C by rearranging the Rawlins and Schellhardt equation.

Step 3: Solve for sandface AOF (absolute open flow) to understand the rate at which the well would produce against zero sandface pressure. It is used as a measure of gas well performance because it quantifies the ability of the reservoir to deliver gas to the wellbore. AOF essentially quantifies the absolute open flow rate with 100% drawdown.

$$\text{SF AOF (sandface absolute open flow)} = C\left(P_R^2 - 0\right)^n = C\left(P_R^2\right)^n$$

Step 4: Divide the SF AOF rate into 10 equal segments (rates) and calculate the corresponding flowing bottom hole pressure at each rate.

Step 5: Plot flowing bottom hole pressure (y-axis) vs gas production rate (x-axis) to construct the IPR curve.

Example

Construct an IPR curve given the following info:

Reservoir pressure = 8550 psi, sandface flowing pressure (FBHP) = 4000 psi, gas production rate = 20,000 MSCF/day, n value = 0.5

Step 1: Solve for P_{wf} by rearranging the Rawlins and Schellhardt equation.

$$P_{wf} = \sqrt{P_R^2 - \left(\frac{q}{C}\right)^{\frac{1}{n}}}$$

Step 2: Solve for C:

$$C = \frac{q}{\left(P_R^2 - P_{wf}^2\right)^n} = \frac{20,000}{\left(8550^2 - 4000^2\right)^{0.5}} = 2.647 \text{ MSCF/day psi}^{2n}$$

Step 3: Solve for sandface AOF:

$$\text{SF AOF} = C\left(P_R^2\right)^n = 2.647\left(8550^2\right)^{0.5} = 22,629 \text{ MSCF/day}$$

Step 4: Divide the SF AOF rate into 10 equal segments (rates) and calculate the corresponding flowing bottom hole pressure at each rate using step 1.

$$\text{Example at 2263 MSCF/day} \rightarrow P_{wf} = \sqrt{P_R^2 - \left(\frac{q}{C}\right)^{\frac{1}{n}}}$$

$$= \sqrt{8550^2 - \left(\frac{2263}{2.647}\right)^{\frac{1}{0.5}}} = 8507 \text{ psi}$$

$$\text{Example at 4526 MSCF/day} \rightarrow P_{wf} = \sqrt{P_R^2 - \left(\frac{q}{C}\right)^{\frac{1}{n}}}$$

$$= \sqrt{8550^2 - \left(\frac{4526}{2.647}\right)^{\frac{1}{0.5}}} = 8377 \text{ psi}$$

Table 23.3 and Fig. 23.1 illustrates rate vs inflow performance curve (or P_{wf}) in a table and graphical format.

Determining the tubing size and timing of its installation in a well is a function of the area of investigation, the estimated flow rate from a well, reservoir pressure, capacity constraint, erosional velocity limitations, etc. If it is determined that a well can only produce through 5½″ casing at a given flow rate based on nodal analysis and erosional velocity calculation (will be discussed), tubing should not be run in the well. However, if the

Production analysis and wellhead design

Table 23.3 Inflow performance relationship example

q (MSCF/day)	Flowing bottom hole pressure (psi)
0	8550
2263	8507
4526	8377
6789	8156
9052	7836
11,315	7405
13,577	6840
15,840	6106
18,103	5130
20,366	3727
22,629	0

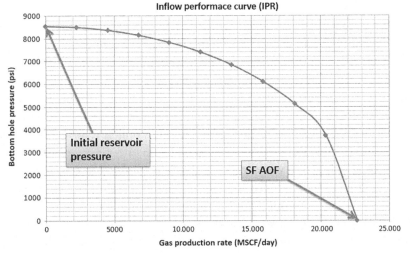

Fig. 23.1 Inflow performance relationship example.

pipeline capacity is limited due to the system being undersized when developed in the area causing production curtailment to a certain rate, it might make more sense to run tubing from the beginning because volume curtailment will not allow the well to produce at its maximum capacity. Across various basins, some systems are undersized when it comes to production volume forecasting while others are oversized as compared to what was originally estimated. This difference can be a serious challenge when it comes to creating the maximum present value for the shareholders. If a system is undersized due to the expectation of lower productivity from wells (e.g., type curve underestimation from reservoir engineers), some wells will not have the opportunity to produce at their peak productivity and will be curtailed at a certain rate until the wells go on natural decline. While curtailment can be beneficial in some areas where managed pressure drawdown is highly recommended due to various reasons discussed in this book (e.g., overpressure reservoirs), this can also deteriorate value if the area is not sensitive to high-pressure drawdown. Therefore, there will be two options in this scenario. The first option is to spend more capital on system infrastructure to expand the capacity that can produce at full capacity to take advantage of the time value of money. The second option is to invest in other prolific areas without any system constraints to take advantage of timing and create the maximum value for the shareholders. These options must be evaluated on a case-by-case basis and economic analysis should be the solution to the problem.

If a system is oversized due to overestimating the type curve for the area or initial strategic development acreage, it can easily deteriorate value unless some other arrangements can be made to take full advantage of the system. This topic indicates the importance of accurately forecasting production performance of wells and having a solid strategy in place from the get-go (type curve for each area) to maximize the net asset value of the field. Infrastructure decision is highly complicated. For instance, let's assume a dry gas field was supposed to be fully developed in 5 years by turning in line 500 wells over time and the production performance of each well was accurately forecasted. Suddenly, the gas price significantly drops to an uneconomic level in that field while the oil price is high. Higher oil price and lower gas price will cause the operator to shift its focus to wet gas and oil reservoirs because that operator has higher return oil projects to focus on. Of course, for this example, we assume that the operator has a diversified portfolio of oil and gas assets. This diverse portfolio will cause a system generated to handle large gas volumes to produce far below its target because of the market condition. At this point, the concept of hedging commodity pricing against risks comes into play. Therefore, system capacity and under/oversizing could be challenging, and careful effort must be taken to maximize the shareholder's value.

Tubing performance relationship

As previously discussed, reservoir properties control the inflow performance of wells. However, the achievable gas production rate is determined by wellhead pressure and flow performance of production string which could be tubing, casing, or both. Using the average temperature and compressibility factor method, the following equations can be used to construct the TPR curve (Katz et al., 1959):

TPR relationship construction:

$$P_{wf}^2 = e^s P_{hf}^2 + \frac{6.67 \times 10^{-4}[e^s - 1]f q_{sc}^2 z^2 T^2}{d_i^5 \cos\theta} \quad (23.6)$$

where P_{wf} is bottom hole flowing pressure, psia, P_{hf} is the wellhead flowing pressure, psia, f is the moody friction factor, q_{sc} is the gas production rate, MSCF/day, z is the average gas compressibility, T is the average temperature, °Rankin, d_i is the tubing ID, in., and θ is the inclination angle.

Where "s" and "fanny friction factor" can be calculated as follows:

$$s = \frac{0.0375 \gamma_g L \cos\theta}{z T}$$

where γ_g is gas-specific gravity, L the measured depth of tubing, ft, θ the inclination angle, z the average compressibility factor, and T the average temperature, °Ranking.

For fully turbulent flow, which is applied to most gas wells, use a simple empirical relation (Katz and Lee, 1990):

$$\text{For } d_i < 4.227 \text{ in.} \rightarrow f = \frac{0.01750}{d_i^{0.224}}$$

$$\text{For } d_i > 4.227 \text{ in.} \rightarrow f = \frac{0.01603}{d_i^{0.164}}$$

Please note that this method ignores water production.

Example

Construct and Plot the TPR curve for the previous IPR curve example with the following assumptions:

Gas specific gravity = 0.57, tubing ID = 1.995 in., tubing MD = 9825 ft, inclination angle = 70, reservoir pressure = 8550 psi, wellhead pressure = 3000 psia, flowing bottom hole pressure (sandface flowing pressure) = 4000 psi, wellhead temperature = 100°F, bottom hole temperature = 180°F, n value = 0.5, gas rate = 20,000 MSCF/day, TVD = 9527 ft, C = 2.647 MSCF/day psi^{2n}, and assume water production of zero

Step 1: Calculate average pressure and temperature and compressibility factor (z factor) at average pressure and temperature.

$$T_{avg} = \left(\frac{\text{Wellhead temp} + \text{BHT}}{2}\right) + 460 = \left(\frac{100 + 180}{2}\right) + 460 = 600°R$$

$$P_{avg} = \frac{\text{Wellhead pressure} + \text{reservoir pressure}}{2} = \frac{3000 + 8550}{2} = 5775\,\text{psi}$$

At 600°Rankin and 5775 psi, the compressibility factor (z factor) is 1.0311 (assumes zero N_2, CO_2, and H_2S mol%)

Step 2: Calculate "s" as follows:

$$s = \frac{0.0375\,\gamma_g\,L\,\cos\theta}{z\,T} = \frac{0.0375 \times 0.57 \times 9825 \times \cos(70)}{1.0311 \times 600} = 0.1161$$

Step 3: Calculate fanning friction factor. Since the tubing ID is less than 4.227″, use the following equation:

$$f = \frac{0.01750}{d_i^{0.224}} = \frac{0.01750}{1.995^{0.224}} = 0.0149$$

Step 4: Calculate the flowing bottom hole pressure to construct the TPC at each rate (from the previous example). Sample calculations are 2263 and 4526 MSCF/day:

At 2263 MSCF/day:

$$P_{wf}^2 = e^s P_{hf}^2 + \frac{6.67 \times 10^{-4}[e^s - 1]f q_{sc}^2 z^2 T^2}{d_i^5 \cos\theta} = e^{0.1161} \times 3000^2$$

$$+ \frac{6.67 \times 10^{-4}[e^{0.1161} - 1] \times 0.0149 \times 2263^2 \times 1.0311^2 \times 600^2}{1.995^5 \cos(70)}$$

$$= 10{,}329{,}847 \rightarrow P_{wf} = 3214\,\text{psi}$$

At 4526 MSCF/day:

$$= e^{0.1161} \times 3000^2 + \frac{6.67 \times 10^{-4}[e^{0.1161} - 1] \times 0.0149 \times 4526^2 \times 1.0311^2 \times 600^2}{1.995^5 \cos(70)}$$

$$= P_{wf} = 3316\,\text{psi}$$

Table 23.4 Tubing performance relationship example Output

q (MSCF/day)	Flowing bottom hole pressure; Inflow performance curve (IPFt) psi	Flowing bottom hole pressure; Tubing performance curve (TPR) psi
0	8550	3179
2263	8507	3214
4526	8377	3316
6789	8156	3480
9052	7836	3697
11,315	7405	3958
13,577	6840	4256
15,840	6106	4583
18,103	5130	4934
20,366	3727	5303
22,629	0	5688

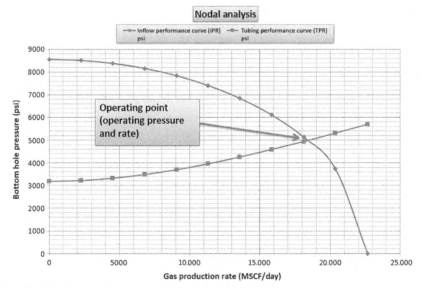

Fig. 23.2 Tubing performance relationship example.

Table 23.4 and Fig. 23.2 illustrate IPR and TPR outputs. As can be seen from Fig. 23.2, the operating point is 18.2 MMSCF/day at 5000 psi. The figure illustrates that at a 5000 psi BHP, the rate achievable by the well based on the assumption provided in this example is 18.2 MM/day.

This analysis can be run at different tubing and casing sizes to determine the size of tubing needed for the anticipated flow rates. For example, if anticipated flow rate from a well is 12 MMSCF/day at 3500 psi at initial production and the operating point from nodal analysis indicates that the well can be flowed 15 MMSCF/day at 3500 psi using 2 3/8″ tubing, running 2 3/8″, tubing in the well will be justified initially instead of flowing through larger tubing or casing sizes as long as it does not exceed the erosional velocity of the tubing which will be discussed next. Nodal analysis is used for tubing size and timing optimization.

Erosional velocity calculation

It is crucial to calculate the erosional velocity of casing and tubing to make sure the rate flowing through pipe does not exceed the erosional velocity of the pipe. Erosional velocity calculation is an essential part of tubing, piping, and system designs. If the tubing is incorrectly sized or if the maximum allowable production rate through tubing or casing is exceeded, erosion of the pipe can occur, resulting in major operational issues when it comes to washing and parting pipe in the wellbore. Therefore, erosional velocity calculations are crucial to a proper understanding of pipe and system constraints and capacity. American Petroleum Institute Recommended Practice 14E (API RP 14E, 1991) proposed correlation for erosional velocity. The following steps can be taken to calculate the erosional velocity of the pipe:

Step 1: Calculate gas-to-liquid ratio as follows:

$$\frac{Gas}{Liquid\ ratio} = \frac{Gas\ rate \times 1{,}000{,}000}{Oil\ rate + water\ rate}$$

where gas/liquid ratio is measured in ft^3/BBL, gas rate is in MMSCF/day, oil rate is in BBL/day, and water rate is in BBL/day.

Step 2: Calculate liquid SG (specific gravity):

$$Liquid\ SG = \frac{Oil\ rate \times oil\ SG + water\ rate \times water\ SG}{Oil\ rate + water\ rate}$$

Step 3: Calculate liquid/1000 BBLs:

$$\frac{Liquid}{1000\ BBL} = \frac{Oil\ rate + water\ rate}{1000}$$

Step 4: Calculate gas/liquid mix density (P is operating pressure in psia):

$$= \frac{\dfrac{\text{Gas}}{\text{Liquid mix density}} (\text{lbs/ft}^3)}{(12{,}409 \times \text{liquid SG} \times P) + \left(2.7 \times \dfrac{\text{Gas}}{\text{Liquid ratio}} \times \gamma_g \times P\right)}{(198.7 \times P) + \left(\dfrac{\text{Gas}}{\text{Liquid ratio}} \times T \times z\right)}$$

Step 5: Calculate fluid erosional velocity:

$$\text{Fluid erosional velocity} = \frac{C}{\sqrt{\dfrac{\text{Gas}}{\text{Liquid mix density}}}}$$

Step 6: Calculate the minimum cross-sectional area required:

$$\text{Min cross} - \text{sectional area required} = \left(\frac{9.35 + \dfrac{z \times T \times \dfrac{\text{Gas}}{\text{liquid ratio}}}{21.25 \times P}}{\text{Fluid erosional velocity}}\right) \times \frac{\text{Liquid}}{1000\,\text{BBL}}$$

Step 7: Calculate pipe cross-sectional area for the desired pipe size.
Step 8:

If pipe cross − sectional area > minimum cross − sectional area required
→ Not erosional

If pipe cross − sectional area < minimum cross − sectional area required
→ Erosional

Example

A new well in dry gas area is being TIL (turned in line). Your boss asks you to perform the erosional velocity calculation and determine whether flowing at the following conditions will exceed the erosional velocity of the tubing or not.

Gas rate = 15 MMSCF/day, oil rate = 0 BBL/day, water rate = 2000 BBL/day, gas SG = 0.61, water SG = 1, operating pressure = 3500 psig or 3514.7 psia, operating temperature = 100°F, z = 0.94, C factor = 125, Tubing ID = 2.441 in.

Step 1: Calculate gas-to-liquid ratio

$$\frac{\text{Gas}}{\text{Liquid ratio}} = \frac{\text{Gas rate} \times 1{,}000{,}000}{\text{Oil rate} + \text{water rate}} = \frac{15 \times 1{,}000{,}000}{0 + 2000} = 7500\,\text{ft}^3/\text{BBL}$$

Step 2: Calculate liquid SG:

$$\text{Liquid SG} = \frac{\text{Oil rate} \times \text{oil SG} + \text{water rate} \times \text{water SG}}{\text{Oil rate} + \text{water rate}} = \frac{0 \times 0 + 2000 \times 1}{0 + 2000}$$
$$= 1.0000$$

Step 3: Calculate liquid/1000 BBL

$$\frac{\text{Liquid}}{1000\,\text{BBL}} = \frac{\text{Oil rate} + \text{water rate}}{1000} = \frac{0 + 2000}{1000} = 2$$

Step 4: Calculate gas/liquid mix density:

$$\frac{\text{Gas}}{\text{Liquid mix density}} = \frac{(12{,}409 \times \text{liquid SG} \times P) + \left(2.7 \times \dfrac{\text{Gas}}{\text{Liquid ratio}} \times \gamma_g \times P\right)}{(198.7 \times P) + \left(\dfrac{\text{Gas}}{\text{Liquid ratio}} \times T \times z\right)}$$

$$= \frac{(12{,}409 \times 1.0000 \times 3514.7) + (2.7 \times 7500 \times 0.61 \times 3514.7)}{(198.7 \times 3514.7) + (7500 \times (100 + 460) \times 0.94)}$$

$$= 18.73\,\text{lb}/\text{ft}^3$$

Step 5: Calculate fluid erosional velocity:

$$\text{FEV} = \frac{C}{\sqrt{\dfrac{\text{Gas}}{\text{Liquid mix density}}}} = \frac{125}{\sqrt{18.73}} = 28.88\,\text{ft/s}$$

Step 6: Calculate the min cross-sectional area required

$$\text{Min CSA} = \left(\frac{9.35 + \dfrac{z \times T \times \dfrac{\text{Gas}}{\text{liquid ratio}}}{21.25 \times P}}{\text{Fluid erosional velocity}}\right) \times \frac{\text{Liquid}}{1000\,\text{BBL}}$$

$$= \left(\frac{9.35 + \dfrac{0.94 \times (100 + 460) \times 7500}{21.25 \times 3514.7}}{28.88}\right) \times 2 = 4.31\,\text{in.}^2$$

Step 7: Calculate the pipe cross-sectional area:

$$\text{Pipe cross} - \text{sectional area} = \frac{\pi}{4} \times (2.441^2) = 4.68\,\text{in.}^2$$

Step 8: Since pipe cross-sectional area is bigger than minimum cross-sectional area (4.68 > 4.31) → **NOT EROSIONAL.**

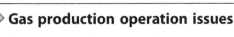

Gas production operation issues

The Three of the biggest challenges in gas production wells are as follows:
- Managed pressure drawdown
- Liquid loading
- Blockage of gas hydrates of pipeline and equipment

Managed pressure drawdown is discussed in Chapter 6 of this book. Liquid loading is a phenomenon that occurs when the hydrostatic pressure of a liquid column is higher than the flowing bottom hole pressure. Liquid loading occurs when gas flow rate (velocity) is not sufficient to carry all fluids out of the wellbore. As a reservoir depletes with time, the energy required to lift the fluid will be reduced. It is important to note that liquid loading could happen on new, old, depleted, energized, vertical, or horizontal wells. The liquid loading phenomenon can happen on new wells (which can be regarded as counterintuitive). For example, let's assume a new Haynesville shale well was expected to produce at 25 MMSCF/day initially (IP). From the nodal analysis, it was recommended to delay running tubing in the well for the first 6 months to get closer to the critical velocity of the casing. This delay would prevent exceeding the erosional velocity of the tubing and being forced to be curtailed back due to tubing limitations. However, after turning the well in line (producing from the well), the well production falls below expectations at 7 MMSCF/day. This result is far below the critical velocity of the casing to lift fluid out of the wellbore. Consequently, the well will be liquid loaded from the beginning even though it is brand new. In this scenario, tubing will be run in the well as soon as possible to efficiently lift fluid out of the wellbore and solve the liquid loading issue. There are different types of liquid loading flow regimes as gas velocity decreases which can be categorized as below:

(1) Mist flow—liquids are typically produced as a mist entrained in the gas stream.
(2) Annular flow—liquids are no longer in the mist and condense on the walls of the production tubing. The liquids have enough energy to move up and out of the well, but annular flow is regarded as less efficient than mist flow regime.
(3) Slugging—is a common phenomenon in gas wells since as gas rates and velocities continue to decrease, gravity impact on the liquids increases. This increase will result in the liquids on the tubing walls that were

moving upward to cease, and gas can go through the center of the liquid. When enough liquids accumulate, liquid "slugs" are created which constrain gas flow.

(4) Bubble—loaded, no upward movement of liquids, which essentially means the well is dead.

The idea is to predict liquid loading in each well in advance to prevent liquid loading issues. Liquid loading can cost any operator much value due to inefficient production from the well (if not identified and handled properly on time). In today's day and age, liquid loading calculation on each well can be done easily and quickly. It determines when tubing or some type of artificial lift will be needed in each well. Production engineers are typically responsible for identifying liquid loading and remedial actions associated with liquid loading.

There are various indications as to when a well is loaded. Practically speaking, it is extremely easy to identify liquid loading at the surface as the well will be constantly slugging fluid at surface and well tenders that are responsible for maintaining the well can easily detect this as soon as it happens. Below are the most common liquid loading indications that can be observed:

- Start of liquid slugs at the surface of the well
- Increasing differential between tubing and casing pressures with time
- Sharp changes in flowing bottom hole pressure (increase in flowing bottom hole pressure)
- Sharp drop in a well's production rate

It is important to note that the accumulation of liquid increases bottom hole pressure that reduces gas production rate. If a well is liquid loaded, flowing bottom hole pressure calculation from surface pressure data would be erroneous. Therefore, performing any type of valid analysis that incorporates flowing bottom hole pressure wouldn't be accurate. Various methodologies have been developed to predict the critical velocity needed to lift fluid out of the wellbore and prevent liquid loading. For operators that frequently perform rate transient analysis, it is crucial to make sure the analyzed wells are not liquid loading since almost all the RTA calculations rely on using a calculated flowing bottom hole pressure.

Turner rate

The method presented by Turner et al. (1969) is referred to as Turner's method. Turner's terminal velocity equation is expressed as

Turner's terminal velocity equation: $V_{sl} = 1.92 \left(\dfrac{\sigma^{\frac{1}{4}} \left(\rho_L - \rho_g \right)^{\frac{1}{4}}}{\rho_g^{\frac{1}{2}}} \right)$ (23.7)

where V_{sl} is the terminal settling velocity, ft/s, σ is the interfacial tension, dyn/cm.

Note that:
water/gas interfacial tension is 60 dyn/cm, condensate/gas interfacial tension is 20 dyn/cm, ρ_L is the liquid density, lb_m/ft^3.

Note that:
water density of ~8.7 ppg is 65 lb_m/ft^3, condensate density can be assumed to be 45 lb_m/ft^3, and ρ_g is the gas density, lb_m/ft^3, function of pressure and temperature.

The minimum required gas flow rate for liquid removal can be calculated using Eq. (23.8).

Minimum required gas flow rate for liquid removal:

$$Q_{gm} = \dfrac{3.06 P V_{sl} A}{Tz}$$ (23.8)

where Q_{gm} is the minimum required gas flow for liquid removal, MMSCF/day, P is the pressure at the depth of interest (BHP), psia, A is the cross-sectional area of conduit, ft^2, T is the temperature, °R, and z is the gas compressibility factor.

Coleman rate

One of the assumptions that Turner and Coleman correlations make is free-flowing liquid in the wellbore form droplets that are suspended in the gas stream. The two forces that act on the droplets are gravity and drag forces in which one pulls the droplets down (gravity) while the other pushes the droplets upward (drag forces). Turner correlation was developed from droplet theory followed by comparing the theoretical calculations to empirical field data and applying a 20% fudge factor. The Coleman equation basically removed the 20% factor that Turner had added after finding out that this provided a better prediction for low-pressure wells (less than 500 psi). Besides this difference, Coleman's methodology is consistent with Turner. Coleman's critical velocity is as follows:

Coleman's terminal velocity equation:

$$V_{sl} = 1.59 \left(\frac{\sigma^{\frac{1}{4}}(\rho_L - \rho_g)^{\frac{1}{4}}}{\rho_g^{\frac{1}{2}}} \right) \qquad (23.9)$$

As will be discussed in detail in the machine learning chapter, Ansari et al. (2018) used an unsupervised ML algorithm called K-means clustering to detect liquid loading and automate the process of liquid loading detection in Marcellus Shale.

Example
Calculate the **critical velocity** and **rate** using both Turner and Coleman methodologies (Report your answers in ft/s and MSCF/day) using the following data:

BHP = 700 psi, Temperature = 140°F, Gas gravity = 0.59, Water/gas interfacial tension = 60 dyn/cm, Water density = 65 lb$_m$/ft^3, Tubing = 2 3/8″ (ID = 1.995″), Mole fraction N$_2$, CO$_2$, and H$_2$S = 0

Step 1: Calculate the gas compressibility factor at 700 psi and 140°F using any of the methodologies in phase behavior.
  Z factor = 0.9402

Step 2: Calculate gas gravity using the following equation:

$$\text{Gas density} = \frac{2.699 \gamma_g P}{Tz} = \frac{2.699 \times 0.59 \times 700}{(140 + 460) \times 0.9402} = 1.976 \, lb_m/ft^3$$

Step 3: Calculate Turner and Coleman's critical velocity:
  Turner's critical velocity:

$$V_{sl} = 1.92 \left(\frac{\sigma^{\frac{1}{4}}(\rho_L - \rho_g)^{\frac{1}{4}}}{\rho_g^{\frac{1}{2}}} \right) = 1.92 \left(\frac{60^{\frac{1}{4}} \times (65 - 1.976)^{\frac{1}{4}}}{1.976^{\frac{1}{2}}} \right) = 10.71 \, ft/s$$

Coleman's critical velocity:

$$V_{sl} = 1.59 \left(\frac{\sigma^{\frac{1}{4}}(\rho_L - \rho_g)^{\frac{1}{4}}}{\rho_g^{\frac{1}{2}}} \right) = 1.59 \left(\frac{60^{\frac{1}{4}} \times (65 - 1.976)^{\frac{1}{4}}}{1.976^{\frac{1}{2}}} \right) = 8.87 \, ft/s$$

Step 4: Calculate the cross-sectional area of 2 3/8″ production tubing (ID = 1.995″) in ft:

$$A = \frac{\pi}{4} D^2 = \frac{\pi}{4} \left(\frac{1.995}{12} \right)^2 = 0.0217 \, ft^2$$

Step 5: Critical flow rate for both Turner and Coleman:

$$Q_{gm, Turner} = \frac{3.06 P V_{sl, Turner} A}{Tz} = \frac{3.06 \times 700 \times 10.71 \times 0.0217}{(140 + 460) \times 0.9402}$$
$$= 0.883 \text{ MMSCF/day or } 883 \text{ MSCF/day}$$

$$Q_{gm, Coleman} = \frac{3.06 P V_{sl, Coleman} A}{Tz} = \frac{3.06 \times 700 \times 8.87 \times 0.0217}{(140 + 460) \times 0.9402}$$
$$= 0.731 \text{ MMSCF/day or } 731 \text{ MSCF/day}$$

Example

Construct a data table and show the Turner and Coleman critical velocity rates at each pressure and gas gravity (Report your answer in MSCF/day in the table).

BHP = Varying BHP based on Table 23.5
Temperature = 150°F
Gas gravity = Varying gas gravity based on Table 23.5
Water/gas interfacial tension = 60 dyn/cm
Water density = 65 lb_m/ft^3
Tubing = 2 7/8″ ≫ ID = 2.441″
Mole fraction N_2, CO_2, and H_2S = 0.

(1) First, gas compressibility factor at various BHP and gas gravity must be calculated. For this analysis, Brill and Beggs (1974) can be used to calculate z values at various BHP and gas gravities. Step-by-step equations for performing z factor calculation are listed below. Please create a spreadsheet in excel where z factor can be calculated by changing gas gravity (γ_g), BHP, BHT, and mole fractions of N_2, CO_2, and H_2S. Once this spreadsheet is built, a data table in excel can be used to calculate the z-factor by changing BHP and gas gravity shown in Table 23.5 while keeping temperature and mole fractions of N_2, CO_2, and H_2S constant.

$$P_{pc} = 678 - 50\left(\gamma_g - 0.5\right) - 206.7 y_{N_2} + 440 y_{CO_2} + 606.7 y_{H_2S}$$

$$T_{pc} = 326 + 315.7\left(\gamma_g - 0.5\right) - 240\, y_{N_2} - 83.3\, y_{CO_2} + 133.3\, y_{H_2S}$$

$$P_{pr} = \frac{P}{P_{pc}}$$

$$T_{pr} = \frac{T}{T_{pc}}$$

$$A = 1.39\left(T_{pr} - 0.92\right)^{0.5} - 0.36 T_{pr} - 0.10$$

Table 23.5 BHP and gas gravity data table

BHP (psi)	Gas gravity									
	0.55	0.56	0.57	0.58	0.59	0.6	0.61	0.62	0.63	0.64
5000										
4800										
4600										
4400										
4200										
4000										
3800										
3600										
3400										
3200										
3000										
2800										
2600										
2400										
2200										
2000										
1800										
1600										
1400										
1200										
1000										
800										
600										
400										
200										

$$B = (0.62 - 0.23 T_{pr}) P_{pr} + \left(\frac{0.066}{T_{pr} - 0.86} - 0.037 \right) P_{pr}^2 + \frac{0.32 P_{pr}^6}{10^E}$$

$$C = 0.132 - 0.32 \log(T_{pr})$$

$$D = 10^F$$

$$E = 9(T_{pr} - 1)$$

$$F = 0.3106 - 0.49 T_{pr} + 0.1824 T_{pr}^2$$

$$z = A + \frac{1-A}{e^B} + C P_{pr}^D$$

(2) Follow the steps discussed in the previous example and use the calculated z factor at each BHP and gas gravity (step 1) to calculate Turner and Coleman rates (MSCF/day) as illustrated in Tables 23.6 and 23.7.

Table 23.6 Turner rate at various BHP and gas gravity

	0.55	0.56	0.57	0.58	0.59	0.6	0.61	0.62	0.63	0.64
5000	3400	3370	3340	3310	3281	3253	3225	3197	3170	3144
4800	3361	3331	3302	3273	3245	3218	3191	3164	3138	3112
4600	3318	3289	3261	3234	3206	3180	3154	3128	3103	3078
4400	3272	3244	3217	3191	3165	3139	3114	3089	3065	3041
4200	3222	3195	3170	3144	3119	3095	3071	3047	3024	3001
4000	3168	3143	3118	3094	3070	3046	3024	3001	2979	2957
3800	3110	3086	3062	3039	3016	2994	2972	2951	2930	2909
3600	3047	3024	3001	2979	2958	2937	2916	2896	2876	2857
3400	2978	2957	2935	2914	2894	2874	2855	2836	2817	2799
3200	2905	2884	2864	2844	2825	2806	2788	2770	2752	2735
3000	2825	2805	2786	2768	2749	2732	2714	2698	2681	2665
2800	2739	2720	2702	2685	2667	2651	2635	2619	2603	2588
2600	2647	2629	2612	2595	2579	2563	2548	2533	2518	2504
2400	2548	2531	2514	2498	2483	2468	2453	2439	2426	2413
2200	2441	2425	2409	2394	2379	2365	2351	2338	2325	2313
2000	2327	2312	2297	2282	2268	2255	2242	2229	2217	2205
1800	2206	2191	2176	2162	2149	2136	2123	2111	2100	2088
1600	2076	2061	2047	2034	2021	2009	1996	1985	1974	1963
1400	1936	1923	1909	1896	1884	1872	1860	1849	1838	1828
1200	1787	1773	1761	1748	1737	1725	1714	1703	1693	1683
1000	1624	1612	1600	1588	1577	1566	1556	1546	1536	1526
800	1446	1435	1424	1413	1403	1393	1383	1373	1364	1355
600	1247	1236	1227	1217	1208	1198	1190	1181	1173	1165
400	1013	1004	996	988	980	972	965	957	950	943
200	713	706	700	694	689	683	678	672	667	662

Table 23.7 Coleman rate at various BHP and gas gravity

	0.55	0.56	0.57	0.58	0.59	0.6	0.61	0.62	0.63	0.64
5000	2816	2791	2766	2741	2717	2694	2670	2648	2625	2603
4800	2783	2758	2734	2711	2687	2665	2642	2620	2598	2577
4600	2748	2724	2701	2678	2655	2633	2612	2590	2570	2549
4400	2709	2687	2664	2642	2621	2599	2579	2558	2538	2519
4200	2668	2646	2625	2604	2583	2563	2543	2523	2504	2485
4000	2624	2603	2582	2562	2542	2523	2504	2485	2467	2449
3800	2575	2555	2536	2517	2498	2479	2461	2444	2426	2409
3600	2523	2504	2485	2467	2449	2432	2415	2398	2382	2366
3400	2467	2448	2431	2414	2397	2380	2364	2348	2333	2318
3200	2405	2388	2372	2355	2339	2324	2309	2294	2279	2265

Continued

Table 23.7 Coleman rate at various BHP and gas gravity—cont'd

	0.55	0.56	0.57	0.58	0.59	0.6	0.61	0.62	0.63	0.64
3000	2340	2323	2307	2292	2277	2262	2248	2234	2220	2207
2800	2268	2253	2238	2223	2209	2195	2182	2169	2156	2143
2600	2192	2177	2163	2149	2136	2122	2110	2097	2086	2074
2400	2110	2096	2082	2069	2056	2044	2032	2020	2009	1998
2200	2022	2008	1995	1983	1970	1959	1947	1936	1926	1915
2000	1927	1914	1902	1890	1878	1867	1856	1846	1836	1826
1800	1827	1814	1802	1791	1780	1769	1758	1748	1739	1729
1600	1719	1707	1695	1684	1674	1663	1653	1644	1634	1625
1400	1604	1592	1581	1570	1560	1550	1541	1531	1522	1514
1200	1480	1469	1458	1448	1438	1429	1419	1411	1402	1394
1000	1345	1335	1325	1315	1306	1297	1288	1280	1272	1264
800	1198	1188	1179	1170	1162	1153	1145	1137	1130	1122
600	1032	1024	1016	1008	1000	992	985	978	971	964
400	839	832	825	818	811	805	799	793	787	781
200	590	585	580	575	570	566	561	557	552	548

As shown in both tables, at a fixed gas gravity, BHP decreases, so Turner and Coleman rates decrease. In addition, at a fixed BHP, Turner and Coleman rates decrease because gas gravity increases.

CHAPTER TWENTY-FOUR

Application of machine learning in hydraulic fracture optimization

Introduction

The subject of artificial intelligence (AI) in general and application of machine learning (ML) has gained lots of popularity in the past few years throughout various industries. This rise in popularity is due to new technologies such as sensors and high-performance computing services (e.g., Apache Hadoop, NoSQL, etc.) that enable big-data acquisition and storage in different fields of study. Big data refers to a quantity of data that is too large to be handled (i.e., gathered, stored, and analyzed) using common tools and techniques, for example, Terabytes of data. In the oil and gas industry, in addition to pressure, rate, and surface and downhole seismic measurements, we are now able to collect information using fiber optics that provide high-resolution temperature and acoustic measurements in time and space. The oil and gas industry has also collected large amounts of data corresponding to evaluation, drilling, completion, stimulation, and operation of the wells. This valuable and expensive data has not been studied and analyzed in detail, simply due to the lack of knowledge and the complexity of the data collected. The application of AI in the oil and gas industry, using different data mining and ML techniques, has enabled us to use this information not only to optimize drilling, completions, stimulation, and operation procedures but also to make real-time decisions to avoid any failure or malfunction, that is, real-time operation center or RTOC. The application of AI will empower our industry to take advantage of new technologies developed in industrial monitoring systems such as sensor technologies, high-performance computing, and use our current and previously collected data to increase the NPV of different projects.

This chapter focuses on basic definitions of AI and different ML algorithms. This chapter also provides some practical examples of the application of AI in completions and stimulation optimization in unconventional reservoirs. As the amount and volume of data increases, human cognition is no longer able to decipher important information from data. Therefore, data mining and ML techniques are used to derive inferences from the raw data and find hidden patterns in the data. Before discussing various applications of ML in different industries, it is important to review the basic definitions that are used in this chapter.

Theory

Artificial intelligence (AI)

AI is machine intelligence as opposed to natural human intelligence. It is essentially a branch of computer science that studies the simulation of human intelligence processes such as learning, reasoning, and self-correction by computers.

Data mining (DM)

Data mining is defined as the process of extracting specific information from a database that was hidden and not explicitly available for the user, using a set of different techniques such as ML. Data mining is used by ML algorithms to find links between various linear and nonlinear relationships. Companies often use data mining to help **collect data** on various aspects of the business such as sales trend, production performance, completions data, stock market key indicators and information, etc. Data mining can also be used to go through websites, online platforms, and social media to collect and compile information.

Machine learning (ML)

ML is a subset of AI. It is defined as the collection of various techniques used to teach computers to find patterns in data to be used for future prediction and forecasting or as a quality check for performance optimization. Patterns discovered using ML can be used to make important business decisions to add value to the shareholders of any corporation. ML and data mining are closely related; however, data mining deals with searching specific information and is more open to interpretation while ML focuses on performing a certain task by building accurate and high precision models. There are three main ML types, which are as follows:

Supervised learning

In this type of ML, set of M number of inputs (x_i) and output (y_i) pairs are available and used as a training set to train models to find patterns existing in the training data with high accuracy. x is a $M \times N$ matrix where M is the frequency of each feature *and* N is the number of input features, attributes or features. y_i is a vector of response feature. Both x and y could appear as numbers, text, image, etc. A response feature (y_i) in the form of a number indicates a regression problem. Otherwise, it is a classification or pattern recognition problem. Let's assume that X_i is a matrix of five features (SWR, GIP, Cluster Spacing, Proppant Loading/Cluster, Average Rate/Cluster) for 100 wells drilled and completed in the Marcellus shale reservoir and the y_i is the 360 days of cumulative gas production per foot of lateral from these 100 wells. If the question is to find the pattern between these features and cumulative gas production, then we are looking at a regression problem. However, if instead of 360 days of cumulative gas production per foot of lateral, the y_i indicates having well interference or not having well interference during stimulation of these 100 wells, then the problem is a classification problem. In both cases, we are using a supervised learning algorithm since pairs of input X_i and output y_i are available. Examples of supervised ML algorithms include artificial neural network (ANN), decision tree, random forest (RF), linear regression (LR), multi-linear regression (MLR), logistic regression, K-nearest neighbor (KNN), and support vector machine (SVM).

Unsupervised ML

In this type of ML, only sets of input features x_i are available, and we seek to decipher existing patterns in the input data. The major difference between this technique and supervised technique is that here, we do not have any output to compare our predictions, while in supervised learning, the model predictions could always be compared with actual available output. Unsupervised learning reduces waste and labor within an organization because unsupervised learning algorithms can be used in place of human efforts to filter through large sets of data for clustering purposes. Examples of unsupervised ML algorithms are *K*-means clustering, hierarchical clustering, DBSCAN clustering, and a priori algorithm.

Reinforcement learning

Reinforcement learning is a learning technique that directs the action to maximize the reward of an immediate action and those following. In this type of ML algorithm, the machine trains itself continuously by using a computational approach to learning from action. Imagine that you are a curious child in a kitchen watching your parents using a knife to cut vegetables and fruits into pieces. You somehow manage to get a hold of the knife and use it to cut an apple into pieces. You have learned that the knife can be used for the positive action of cutting vegetables and fruits. Now, you try to mess around with it and manage to cut yourself. You then realize that it can be used for the negative action of hurting yourself when used inappropriately. This learning procedure helps the kid to learn proper use of a knife for positive and not negative action. Humans learn by interaction and reinforcement learning. Learning through trial and error with prize and punishment is the most important feature of reinforcement learning.

ML case usages in various industries

Most companies, especially in the tech sector, have spent a significant amount of money toward ML to predict attributes that can generate billions of dollars in revenue each year. When using an App such as Zillow or any other house hunting applications, the price of a house is estimated using various types of ML algorithms. For example, if information on 100 houses in an area based on the number of bedrooms, baths, square footage, year built, location, type of house, parking condition, cooling/heating, house condition, crime rate, etc., is collected in a data base, it can be used to predict the price of a new house once it becomes available on the market. However, the accuracy of the prediction is highly dependent on the technique used and expertise of the person performing the prediction. If the person performing this prediction is a real-estate expert who has been trained in data science, one would expect their prediction to be highly accurate. What about stock market prediction? Can a ML algorithm be used to predict how a stock will perform based on all the important financial information about a company's stock? Again, various hedge fund firms use various ML algorithms to predict the price of a stock.

ML also has many applications in Human Resources. One of the applications can be used in large organizations to find high-valued employees that are more susceptible to quitting. If these star employees can be identified in advance, promotion, raises, or other types of benefits can be offered to retain these employees. Furthermore, ML is currently being heavily used for marketing personalization. Amazon often uses ML to propose products to customers. I am sure that you have previously made some purchases on Amazon and have received emails or notifications about other products that might interest you for days afterward or the digital ads simply pop up across the web. Another important use for ML is fraud protection. For instance, PayPal uses various ML algorithms to fight money laundering by comparing millions of transactions, accurately distinguishing legitimate and fraudulent activities. Online search ML algorithms is another area that Google uses extensively by constantly improving the algorithm. If a certain key word such as "Hydraulic fracturing simulation" is searched on Google engine and users constantly go to the second or third page to get information about the search, the search engine surmises that the search result was not good enough and will try to redeem the mistake and yield a better result the next time "Hydraulic fracturing simulation" is searched. A strong ML algorithm can distinguish between best search engines such as Google and other competitors.

Natural language processing (NLP) is another area in which ML is used extensively to quickly route customers to specific information. Self-driving cars would not be possible today without the use of ML and reinforcement learning. Facebook uses ML extensively to predict its content. If you simply pause on a specific topic, Facebook notices that that topic caught your interest and, as a result, will propose related topics when you get back on Facebook next time. This process is extremely powerful since Facebook makes most of its profits by advertising. Providing a pleasant user experience and targeting the right audience are essential to their business strategy. Market basket analysis, where stores use unsupervised ML to place items that sell together in the same isle is another important application of ML in many retail businesses. As can be seen, ML is improving our lives, and it is important for us to embrace this new technology to create value for the shareholders across various industries.

In the O&G industry, ML also has various applications and big corporations have embraced ML and data science by dedicating full teams to its advancements. Areas that ML can be applied for prediction purposes include but are not limited to bit failure diagnosis and prevention, equipment failure and detection, well screen-out prediction, frac hit and liquid-loading detection, gas and liquid leakage detection, plunger lift optimization, reservoir simulation, reservoir performance forecasting, well indexing and ranking, completions and reservoir optimization, electric submersible pump (ESP optimization), oil and gas pricing prediction, acreage and asset evaluation prediction, creating synthetic well logs, predicting petrophysical and geomechanical properties, and EUR maps as well as many other ideas, which will be developed as the O&G industry are in the infancy stage of using ML in various departments.

Knowledge of data science and domain expertise

Domain expertise plays a big role in performing and implementing a successful ML project. Unfortunately, when a data scientist without domain expertise tries to tackle a ML project, the outcome might not be as favorable as compared to combining a knowledgeable data scientist and expert in the topic of interest with years of experience. Combining domain expertise and knowledge of data science will most likely result in the best outcome when it comes to using and implementing ML within an organization. There have been numerous cases within various organizations where ML has been misused, resulting in erroneous results at the shareholders' expense. This misuse stems from either a lack of statistical knowledge and limitations of various ML algorithms or lack of domain expertise. Therefore, combining the two is the key in a successful ML project. There are far too many companies that are starting their own consulting firms and providing AI services, but they fail to meet certain standards for a project. Therefore, it is imperative for corporations to perform a full investigation of various consultations that are being offered before rushing into projects. It might also be much more economical to dedicate a full team of data scientists and ML experts and outsource the work as needed.

Another concern relevant to ML is data gathering and preprocessing. Data preprocessing is necessary to the successful implementation of a ML project. Without proper application of this step, the entire analysis could easily fail not because of failure in ML algorithm but due to lack of preprocessing the data in a manner that would yield the best outcome. Data preprocessing includes but is not limited to data cleaning, outlier detection, data normalization or standardization, and input/feature selection. In ML, typically 80% of the time is spent preprocessing and only 20% is spent performing the actual analysis. Therefore, it is essential for corporations to spend capital on data infrastructure and storage creation. Many companies are moving toward cloud server to accommodate the increase in data. This progression is a big step toward automation and the future of ML. Currently, corporations with long legacies have access to large unstructured data sets. On one hand, it is great to have access to a large amount of data, but on the other hand, the differences in structure, format, and ease of accessibility to the data results in inefficiency when performing various ML projects. The goal is to reduce the 80% preprocessing time to the minimum percentage possible for a successful future of efficient use of ML and automation. The authors foresee that the processing time will significantly reduce as companies embrace on their ML journey.

Methodology
Workflow for building ML models

Different companies have their own best practices and workflows to perform AI-related projects. However, the following steps are usually used in ML model developments:

Step 1:

1-1. Data gathering: gathering the related data for specific ML project is the first step in any type of ML workflow.

1-2. Data integration: since data related to ML projects usually comes from different sources, data needs to be integrated to ensure that their format is appropriate for further analysis.

Step 2:

2-1. Data cleaning: the next step is to obtain the basic statistics of the data including frequency, minimum, maximum, average, median, standard deviation, histogram, probability density function (PDF), and cumulative density function (CDF) of each feature. This step will help us to identify the missing values, errors, typos, etc.

2-2. Outlier detection: different techniques such as cross plots, heat maps, and Z-score tests (will be discussed later in this chapter) can be used to obtain the possible outliers in the data.

2-3. Data imputation: if some wells are missing information regarding specific features, before moving forward, we either need to remove those features or wells or impute the missing data to proceed to the next step. There are different imputation techniques available such as imputation mean value (IMV), Soft Impute, and the KNN model. Among all the techniques used in our studies, the KNN was the most promising with the highest accuracy.

Step 3:

3-1. Data analysis: when analyzing data, one must ensure that measurements and units are consistent. For example, if the flow rate is measured daily and pressures are measured in minutes, then some averaging technique (e.g., arithmetic, Harmonic, moving average, etc.) is required to make the average daily pressure values consistent with rates. One must ensure that all the features have the same resolution and units.

3-2. Feature selection: before proceeding to model development, one important step is to reduce the number of features that will be used. This reduction can be achieved by using different techniques to quantify the impact of each feature x_i on output y_i and select the N number of the most important parameters. We would also like to see some features in the database that we believe are highly related to the physics of the problem or found out to be important by experience. Different techniques such as fuzzy pattern recognition, RF, support vector regression, F-regression test, and ANN can serve this purpose.

Step 4:

4-1. Scaling, normalization, and standardization: to make sure the learning algorithm is not biased to the magnitude of the data, the data (input and outputs) need to be scaled. This can also speed up the optimization algorithms such as gradient descent that will be used in model development by having each of our input values in roughly the same range. This process can be performed using Eq. (24.1):

$$\text{Feature normalization}: X' = \frac{X - \min(x)}{\max(x) - \min(x)} \qquad (24.1)$$

where X' is the normalized data point and X is the input data point.

Data normalization guarantees that each feature would be rescaled to a range of [0, 1]. Since most of the techniques used in ML are based on multivariate Gaussian distribution, normalization technique is used to transfer data distribution of each feature to Gaussian or bell-shaped distribution. Standardization transforms each feature with Gaussian distribution to Gaussian distribution with a zero mean and a variance of one. Some learning algorithms, such as SVM, assume data is distributed around zero with the same order of variance. If this condition is not met, then the algorithm will be biased toward features with a larger variance. Eq. (24.2) can be used to standardize the data:

$$\text{Feature standardization}: X' = \frac{X - \mu}{\sigma} \qquad (24.2)$$

where X' is the standardized data point, μ is the mean of the data set, and σ is the standard deviation of the data set.

Scaling to unit length is another widely used technique when it comes to feature scaling in ML. In this technique, each component gets divided by the Euclidean length of the vector.

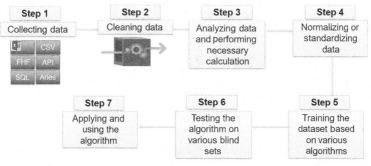

Fig. 24.1 Workflow for building ML models.

$$\text{Feature scaling to unit length}: X' = \frac{X}{\|X\|} \qquad (24.3)$$

Step 5:

5-1. Model development: once the data has been rescaled and important features are selected, the next step is to use various ML algorithms to develop an intelligent model. There are different supervised, unsupervised, and reinforcement learning algorithms available that can be implemented depending on the nature of the problem under investigation.

5-2. Supervised learning algorithm: if the problem requires a supervised learning algorithm, then the database will be divided into three subsets including *training, validation, and testing*.

    5-2-1. Sampling: when dividing the database into training, validation, and testing sets, it is essential to pick subsets randomly with special attention to the distribution of each subset and comparing that with original distribution of the database. Ideally training, validation, and testing subsets should have very similar statistical descriptions when compared to the original data set.

Step 6:

6-1. Blind set testing: after model training and validation, the next step is to apply the training model to a blind set to see the predictive and generalization capability of the model.

Step 7:

Save and apply the model: once satisfied with the training model, the model is ready to be deployed or integrated into a real-time operations center. Fig. 24.1 illustrates all the steps discussed.

Artificial neural network (ANN)

ANN (also known as artificial neural network) is one of the most commonly used supervised ML algorithms throughout various industries. The idea of ANN came from brain neurons and acts as an artificial version of the human nervous system in data receiving, processing, and transportation. In ANN, neurons are placed in a well-defined structure of input, output, and hidden layers. The three main elements in a neural network are as follows:

(1) Input layer: this is the layer where all input features are imported into the model. The number of neurons in the input layer is equal to the input features, which is a function of the problem statement and objective. There is no rule of thumb on the number of input features to include in the model.

(2) Hidden layer: an ANN model can consist of one or multiple hidden layer(s) with multiple hidden neurons depending on the complexity of the problem. The higher the complexity, the more hidden neurons and layers are needed to process the input data. Hidden neurons and layers are basically used to process the inputs received from input layer(s). These hidden layers reveal the existing pattern between input and output layers.

(3) Output layer: after processing the data via the hidden layer(s), the data becomes available at the output layer. The number of neurons in the output layer is equal to the number of target parameters. ANN model can be built for cases with a single output or multiple outputs (objectives).

Input and output layers are selected based on domain expertise. Hidden neurons and layer(s) can be optimized by iterating on various numbers of neurons and hidden layers that would lead to the best model with the highest accuracy on the blind data set. Typically, 1½ or 2 times the input parameters would be ideal for selecting the number of neurons in the first hidden layer. However, it is recommended to start with the same number of input parameters as the number of hidden neurons and increase the number of input parameters until the highest model accuracy on the blind data set is obtained. More hidden layers can also be added to see whether the training and blind testing set's model accuracy increases or not. If problems can be solved with one hidden layer, it is recommended to avoid having too many hidden layers in an attempt to avoid overfitting. This statement is also valid for selecting the number of neurons. If a model's accuracy is high with a limited number of neurons and increasing the number of neurons does not improve a model's accuracy, just stick to a simpler model to avoid overfitting. Fig. 24.2 illustrates an ANN model with nine input features including total GIP, sand/cluster, water/cluster, number of perforations, stage spacing, cluster spacing, landing zone, well spacing, and well boundedness (input layer), 1 hidden layer with 18 hidden neurons, and an objective parameter EUR/1000' (output layer).

ANN can be divided into two categories including supervised and unsupervised learning ANN. Both algorithms have been used extensively in the oil and gas industry. We have used supervised ANN for different projects such as completion design optimization to enhance overall NPV of Marcellus Shale reservoir and diagnosis and prevention of bit balling in drilling. We have also used unsupervised ANN for real-time liquid loading diagnostics in Marcellus Shale and prediction and quantification of well interference "Frac Hit" in Marcellus Shale. However, supervised learning is the most common algorithm used in the oil and gas industry.

In supervised learning as discussed earlier, both input features and output objectives are available. This availability can help the ANN to use both inputs and outputs to train a model by minimizing the cost function. A cost

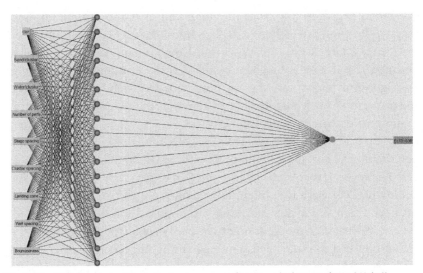

Fig. 24.2 ANN Example (Waikato Environment for Knowledge Analysis (Weka)).

function is the measure of accuracy and performance of the model in terms of its relationship between input and output features. Minimizing the cost function can be done by iteratively running a model to minimize predicted vs. actual values. The objective here is to minimize the cost function as shown in Eq. (24.4). Cost function is also referred to as loss function or error function. In simple words, cost function can be defined by identifying the difference between actual data point (y) and model's prediction ($\hat{y}$), squaring the difference, adding them up, and finally taking the average. Instead of dividing by n, it is divided by $2n$ for simplicity in calculations. Gradient descent technique is one of the well-known optimization algorithms used to minimize the cost function. Gradient descent technique allows a model to find the direction (gradient) necessary to minimize the cost function.

$$\text{Cost function}: J(\theta_0, \theta_1) = \frac{1}{2n}\sum_{i=1}^{n}(\hat{y} - y_i)^2 = \frac{1}{2n}\sum_{i=1}^{n}(h_\theta(X_i) - y_i)^2 \quad (24.4)$$

In this case, gradient descent algorithm changes θ values multiple times to minimize the cost function. Learning rate defines how much change needs to be applied to θ each time to reach the mimimum cost function. In other words, learning rate determines how fast the weights (coefficients in case of linear or logistic regression) change. The algorithm can reach the minimum loss function by choosing large learning rate. However, it is also possible to overshoot, i.e., diverging from the minimum. It is also possible to use an adaptive learning rate where a higher learning rate is initially used, and the learning rate is lowered down until the cost function is minimized. Learning rate is usually between 0 and 1. If the learning rate is too high, the algorithm might miss out on important steps (and fail to converge) and if it is too low, it might take too long to reach to its destination which is minimizing the cost function.

When using gradient descent in an ANN model, ideally, the loss function should reach its global minimum without encountering any local minima. However, it is possible for loss function to get stuck in one of the local minimums. The algorithm might mistakenly assume that a global minimum has been reached which could lead to suboptimal solution. To avoid this issue, a coefficient of momentum can be used in gradient descent algorithms that as opposed to using each step gradient, it also retains a percentage of gradients of previous steps. The coefficient of momentum is between 0 and 1 as well. A small value of momentum could potentially cause the model to fall in the local minima and not be able to jump over it. In addition, it would take longer for the model to converge. A larger momentum could lead to faster convergence. There is no one-size fits all learning rate and momentum for all problems. These are important parameters when using gradient descent in an ANN model that can be used in an iterative algorithm to find the optimal solution to a problem. Finding the correct learning rate and momentum is a function of the number of expected local minima and smoothness of the function, as well as other factors. If a model keeps falling in a local minimum, it might be necessary to choose a higher momentum to avoid this problem.

As Fig. 24.2 illustrates, in ANN structure, each neuron in the hidden layer is connected to all neurons in the input layer. ANN assigns weights to each connection that multiply by values of each feature and sum up at each neuron in the hidden layer. Each neuron can have multiple input connections but only provides one output. An activation function transfers the input of the neuron to the output. Different activation functions are introduced but the most common activation functions are identity, logistic, sigmoid, and tangent hyperbolic activation function. Table 24.1 shows the list of the most common activation functions and their characteristics. In ANN training process, weight assigned to each connection can be initialized randomly or assigned by user. These weights will evolve as ANN structure minimizes the loss function during the training process. The final weights assigned to each feature determine the importance of that feature in the output objective (target).

To train an ANN model, the largest portion of the database is usually used to teach the machine to learn from the database. It is important to note that a model might have the highest accuracy in the training set obtained from multiple attempts, but it still might not perform the best on a blind data set when it comes to prediction and generalization. This problem occurs when the machine memorizes the training set in an iterative process (e.g., back propagation algorithm in ANN), and therefore, loses the prediction and generalization capability. In ML, this problem is called overfitting, where the model has memorized the trend, noise, and detail in the training set instead of intuitively understanding the trend in the data set. In order to avoid the overfitting problem, stoppage criteria for learning should be set where the model tests its predictive capability on a validation set and continues retraining if high validation accuracy is not achieved.

Weight and bias in ANN:

Before providing an example that shows the application of weights and biases in an ANN model, it is important to intuitively understand the

Table 24.1 Example of activation functions can be used for ANN training

Activation function	Equation	Shape
Identity	$F(x) = x$	
Logistic	$F(x) = 1/(1 + \exp(-x))$	
Tanh	$F(x) = \tanh(x)$	
ArcTan	$F(x) = \tan^{-1}(x)$	
Binary step	$F(x) = 0$ for $x < 0$ $F(x) = 1$ for $x > 0$	

concept behind weight and bias. The output of a network is calculated by multiplying input times weight and passing the result through some type of activation function (depending on the nature of the problem). Assuming a logistic activation function, changing the weight basically changes the steepness of the function. What if changing steepness is not sufficient to converge the model? Bias allows a model to shift the entire curve to the left or right. A simpler way to understand bias is to think of it as constant b (y-intercept) in a linear equation ($y = mx + b$). In Fig. 24.3, X_1, X_2 and X_3 are the input parameters (assuming an ANN model with three input parameters), θ_1, θ_2, θ_3, θ_5, θ_6, θ_7, θ_9, θ_{10}, θ_{11} are weights of each neuron and input

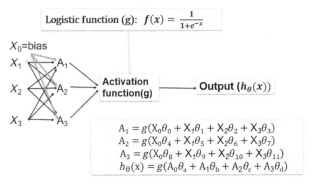

Fig. 24.3 Weight and bias concept/calculation.

parameter, X_0 is the input bias which is 1. θ_0, θ_4, and θ_8 are the biases for each neuron, θ_a is the last bias, θ_b, θ_c, θ_d are the final set of weights (also obtained from the output of an ANN model). Finally, $h_\theta(x)$ is the output for the analysis. In Fig. 24.3, the logistic function is used for illustration purposes only and the activation function can vary depending on the model.

Example

An ANN classification model with five input features and one hidden layer that includes 11 hidden neurons have been trained with high accuracy. The resulting weights and biases have been summarized for each feature and neuron as shown in Tables 24.2 and 24.3. Apply this trained model to the input matrix provided in Table 24.4 (the min and max of each input feature is also provided in Table 24.5). Use logistic function for transformation and classify

Table 24.2 Weights and biases of each input parameter and neuron (model output)

Parameters	Bias	Parameter A	Parameter B	Parameter C	Parameter D	Parameter E
Neuron 1	1.86	7.54	−1.83	−9.59	2.35	11.28
Neuron 2	3.23	−2.36	−2.88	3.85	−5.18	−10.06
Neuron 3	−0.33	0.13	0.46	0.06	0.44	1.24
Neuron 4	−1.19	−5.10	5.42	−10.42	2.36	−5.68
Neuron 5	−0.29	1.40	−1.62	−16.96	1.01	3.73
Neuron 6	4.05	−13.85	−2.71	10.65	−9.33	9.42
Neuron 7	−0.75	−2.38	1.01	0.98	0.67	13.81
Neuron 3	−0.21	0.08	0.34	−0.07	0.01	1.08
Neuron 9	−0.77	0.17	0.47	0.17	0.03	8.08
Neuron 10	1.24	−3.34	2.51	8.28	−2.07	−4.04
Neuron 11	6.38	−5.61	1.36	−6.95	−0.02	−9.05

Table 24.3 Last bias and second set of weights
Last bias and second set of weights

Last bias	1.24
Theta 2, 1	10.83
Theta 2, 2	−12.33
Theta 2, 3	1.37
Theta 2, 4	−8.95
Theta 2, 5	12.24
Theta 2, 6	10.67
Theta 2, 7	13.14
Theta 2, 8	1.07
Theta 2, 9	7.21
Theta 2, 10	−7.63
Theta 2, 11	−12.94

Table 24.4 Input parameters
Input parameters

Parameter A	7890
Parameter B	94
Parameter C	0
Parameter D	0.006
Parameter E	−0.09771

Table 24.5 Minimum and maximum from training data set for normalization of blind data set
Min and max from training data set

Parameters	Min	Max
Parameter A	6447	13827
Parameter E	10	104
Parameter C	0	4.12
Parameter D	0.006	0.022562
Parameter E	−1.07086	19.92199

the result less than 0.5 as GOOD and more than 0.5 as BAD since logistic functions falls between 0 and 1.

Step 1: The first step in applying the trained model to a blind data set model is to normalize the input data (blind data) using minimum and maximum values from the training data set. Since the training data used normalization before training the model, normalization must also be applied to the blind data (input data):

$$\text{Parameter A normalization} = \frac{7890 - 6447}{13,827 - 6447} = 0.1955$$

$$\text{Parameter B normalization} = \frac{94 - 10}{104 - 10} = 0.8936$$

$$\text{Parameter C normalization} = \frac{0 - 0}{4.12 - 0} = 0$$

$$\text{Parameter D normalization} = \frac{0.006 - 0.006}{0.022562 - 0.006} = 0$$

$$\text{Parameter E normalization} = \frac{-0.09771 - (-1.07086)}{19.92199 - (-1.07086)} = 0.04636$$

Step 2: The next step is the matrix multiplication of normalized input parameters from step 1 by bias and weights for each neuron (1 is used for input parameter bias to have the same number of matrix multiplication):

$$\begin{array}{cc} 1 & 1.86 \\ 0.1955 & 7.54 \\ 0.8936 & -1.83 \\ 0 & -9.59 \\ 0 & 2.35 \\ 0.04636 & 11.28 \end{array}$$

Matrix multiplication for neuron 1 $= (1 \times 1.86) + (0.1955 \times 7.54)$
$+ (0.8936 \times (-1.83)) + (0 \times (-9.59))$
$+ (0 \times 2.35) + (0.04636 \times 11.28) = 2.22$

A summary of the result when applying to the 10 remaining neurons following the same approach is given in the following:

Result	
Neuron 1	2.22
Neuron 2	−0.27
Neuron 3	0.16
Neuron 4	2.39
Neuron 5	−1.28
Neuron 6	−0.64
Neuron 7	0.33
Neuron 8	0.16
Neuron 9	0.06
Neuron 10	2.64
Neuron 11	6.09

Step 3: In this step, the logistic function (since training data set used logistic function) is applied to the matrix multiplication result obtained in step 2.

$$\text{Sigmoidal function for neuron } 1 = \frac{1}{1+e^{-2.22}} = 0.9019$$

The sigmoidal function for the remaining neurons are summarized in the following:

Result	
Sigmoidal #1	0.9019
Sigmoidal #2	0.4321
Sigmoidal #3	0.5403
Sigmoidal #4	0.9163
Sigmoidal #5	0.2171
Sigmoidal #6	0.3452
Sigmoidal #7	0.5825
Sigmoidal #3	0.5398
Sigmoidal #9	0.5139
Sigmoidal #10	0.9334
Sigmoidal #11	0.9977

Step 4: The next step is to perform another matrix multiplication of last bias and second set of weights by sigmoidal function results obtained in step 3:

$$\begin{aligned}\text{Output} =\ & (1.24 \times 1) + (10.83 \times 0.9019) + (-12.33 \times 0.4321) \\ & + (1.37 \times 0.5403) + (-8.95 \times 0.9163) \\ & + (12.24 \times 0.2171) + (10.67 \times 0.3452) \\ & + (13.14 \times 0.5825) + (1.07 \times 0.5398) \\ & + (7.21 \times 0.5139) + (-7.63 \times 0.9334) \\ & + (-12.94 \times 0.9977) = -3.5439\end{aligned}$$

Step 5: The last step is to apply the logistic function to the output obtained from step 4 as follows:

$$\text{Sigmoidal function applied to output} = \frac{1}{1+e^{-3.5439}} = 0.02808$$

Since 0.02808 is less than 0.5, the classification for this input set of data would be "GOOD." The same workflow can be applied to any data point as it comes in to determine the classification for each data row. The step-by-step calculation can be very useful when necessary to be applied to an already trained model on a real-time basis. In addition, the same workflow can be applied for regression models.

Example #1: Marcellus shale completion and stimulation optimization using supervised ML

Identifying the best completion and stimulation design (CSD) to achieve the maximum ultimate recovery from shale reservoirs has been one of the major topics of research in the oil and gas industry. Most companies are using experience obtained from their previous completion and stimulation practices to improve current CSD's. This development is due to the fact that the CSD optimization to enhance hydrocarbon production is a complicated problem. In addition to completion and stimulation features that can highly impact the efficiency of hydrocarbon production from shale reservoirs, a significant number of features such as reservoir quality, drilling, operation, field development, and history of field production can also play an important role. Therefore, there is no guarantee that a successful CSD in one area results in a similar success in another area.

Generally, the oil and gas industry uses cross plots in which different completion and stimulation features are plotted against cumulative hydrocarbon production per foot of lateral to study the behavior and quantify the impact of each feature. Next, for CSD optimization, different numerical reservoir simulations are performed using variety of completion and stimulation features. Then, expected well production obtained from each simulation will be plotted against type well of the field to obtain the best CSD. However, these conventional approaches have not been successful in hydrocarbon production from shale reservoirs due to the extremely complex nature of the problem and highly nonlinear correlations between different features involved in hydrocarbon production from these tight formations.

In this example, we briefly review the steps required to apply a supervised ANN model to quantify the impact of each parameter on cumulative gas production from a Marcellus shale reservoir and obtain the best completions and stimulation design. For this purpose, we first implemented the conventional rate transient analysis to gain basic knowledge of the field under investigation. This step has been accomplished using (i) flow regime identification and (ii) flow capacity $A\sqrt{k}$ analysis.

It is extremely important to identify the flow regime of each well to make sure we are not mixing data obtained from a well in transient flow regime with that of boundary dominated flow regime. Transient flow regime can be observed during the early time of production and in extremely low permeability formations. In transient flow regime, flow occurs while a pressure response is moving out in an infinite acting reservoir. At late time and depending on the matrix permeability in unconventional reservoirs, flow experiences a boundary dominated flow regime where a reservoir is in a state of pseudo-equilibrium and informations such as original oil or gas in place can be obtained. For this example, we studied 123 wells in Marcellus where almost all the wells showed transient flow regime. In some cases, signs of a transition period between transient and boundary dominated were observed. However, it is safe to assume that all wells are under the transient flow regime. For this purpose, the normalized rate, that is, flow rate divided by the difference between initial and flowing bottom hole pseudo-pressures is plotted against material balance time, which is the ratio of cumulative gas production over instantaneous rate, on a log-log scale. Data on the left side of the type curve with a half slope indicates the transient flow regime as shown in Fig. 24.4. Next, we used the flow capacity $A\sqrt{k}$ analysis, in which A is the contacted surface area and k is the effective permeability of the contacted rock, to tie the completions design and landing zone to production

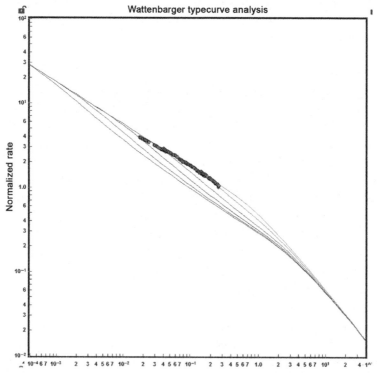

Fig. 24.4 Transient flow regime signature for well #1 in Marcellus Shale reservoir.

performance. The flow capacity $A\sqrt{k}$ can be obtained using the slope of the line passing through normalized pressure vs squared root of time. As long as the flow regime is transient, the $A\sqrt{k}$ will stay constant. We have found great correlation between $A\sqrt{k}$ and EUR (expected ultimate recovery) of the wells under investigation. Using $A\sqrt{k}$ to rank the wells, higher sand per foot loading and lower stage spacing leads to higher performance of the wells in the area under investigation. Followings are the lessons learned through RTA as listed in paper prepared by Belyadi et al. (2015), flow capacity, and diagnostic plots study of 123 wells in Marcellus shale reservoir:

- The 10 top performers in the field are located in the Middle Marcellus.
- Outer wells using $300'$ stage spacing design appears to be outperforming the inner wells with $150'$ stage spacing.
- Pumping higher percentage of 100 mesh sand might allow more surface area to be contacted, improving production.
- Many of the top performing wells in the field have higher lateral lengths.

We have used the lessons learned from RTA and flow capacity analysis to develop a physics based AI model. Next, we have followed the workflow discussed earlier in Fig. 24.1 to build a smart model for completion and stimulation optimization of Marcellus shale. Data preprocessing comprised of the first three steps of the workflow, which included data gathering, cleaning, and analysis. In this example, well information, reservoir properties, drilling, completions, stimulation parameters, and cumulative gas and condensate productions are collected and integrated in an excel file. Table 24.6 shows the list of parameters used for developing smart models for 123 wells in Marcellus shale. To calculate the cumulative gas and condensate production, we used the active dates of the well and removed days where the wells were shut in or not producing due to work overs. Next, average reservoir properties, drilling, completions, stimulation, and production data are assigned to each well. We have performed the data cleaning to identify outliers and missing information. For outlier detection Z-score test and cross plots are used. As an example, one well with shot density equal to 24 appeared to be an outlier using Z-score test and cross plots. All other wells have shot density of 40 and higher as shown in Fig. 24.5. However, by contacting the operator, it was found that this specific well was an old well, with old completion design, so it was not an outlier and should be kept in the dataset.

Some missing data are also identified in the data set. Not all wells have all the features reported. However, for application of AI, we need to have data for all entries of the database, so all the missing data needs to be filled in

Table 24.6 List of features used for smart model development

Reservoir features	Drilling features	Stimulation features	Completion features	Production data
Gas content	Offset	Avg, Tr, pressure (psi)	Cluster spacing (ft)	90 days cum gas/ft (MCF/ft)
Porosity%	Distance (well spacing)	Avg. Tr. rate (BBL-min)	Number of stages	90 days cum condensate/ft (BBLs/ft)
Gross thickness (ft)	Bounded/ unbounded	Fluid volume per ft of lateral (BBL/ft)	Clusters per stage	180 days cum gas/ft (MCF/ft)
Density		100 mesh	Shot density (shots-ft)	180 days cum condensate/ft (BBLs/ft)
TOC%		Proppant per ft of lateral (lb/ft)		360 days-cum gas/ft (MCF/ft)
				360 days cum condensate/ft (BBLs/ft)

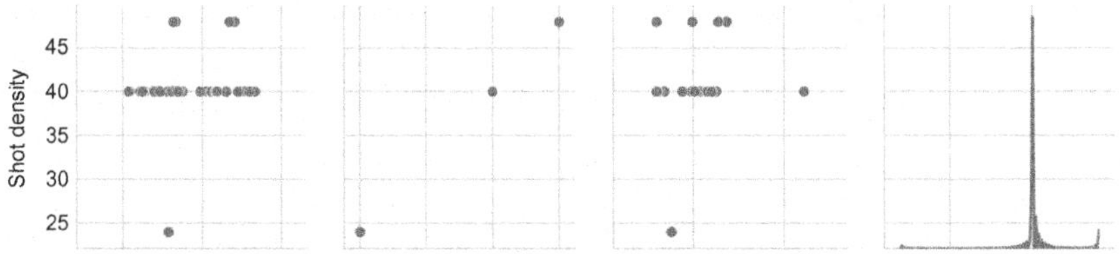

Fig. 24.5 Shot density outlier detection using cross-plot.

before we can start the analysis. In our dataset of 123 wells, there were 17 wells missing one or more feature values. These values were filled/imputed using different techniques such as IMV, soft impute, and the KNN model. Among all the techniques used, the KNN was more promising with higher accuracy that has been used for the rest of this example.

Generally, the way KNN works is that it takes one column at a time to predict values based on # of neighbors selected. However, by utilizing a python package called *Fancyimpute,* all the samples containing missing data are imputed at once. No specific selection was made to fill the data except that the data sets were divided separately based on the target value, that is, 90, 180, and 360 days cumulative gas production per foot of lateral. In order to confirm the accuracy of KNN technique, known values from the dataset are removed and then imputed using the KNN algorithm. For this purpose, 200 random measurements were removed from the original dataset, and regenerated and compared with actual data. The error percentage was calculated using Eq. (24.5). The results of the data filling determined the mean error between parameters obtained with an average error of 9.7%, which is acceptable for our example.

$$\text{Mean error percentage calculation}: \%\text{Error} = \left| \left(\frac{X_{\text{fill}} - X_{\text{org}}}{X_{\text{org}}} \right) \right| \quad (24.5)$$

For feature selection and dimensionality reduction, different techniques such as linear support vector regression, linear least squares, and F-regression test are performed using open source Python packages. Python "sklearn" package provides support vector regression implemented in terms of "liblinear." It has more flexibility in the choice of penalties and loss functions for large number of samples. Given data with n-dimensional features and one target variable (real number), the objective is to find a function, $f(x)$, with the most deviation from the target y assuming that the relationship between X and y is approximately linear. The technique applied for 90, 180, and 360 days cumulative gas production per foot of lateral, consistently identified reservoir and completion/stimulation features as most influential features for all targets as shown in Fig. 24.6.

We have also performed Ridge regression, which is a powerful technique generally used for creating models in the presence of a large number of features. It works by penalizing the magnitude of coefficients of features along with minimizing the error between predicted and actual observations. These are called *regularization* techniques. Ridge regression performs L2 regularization, that is, adds penalty equivalent to the square of the magnitude of coefficients. This model solves a regression model where the loss function is the linear least squares function and regularization is given by the l2-norm. This estimator has built-in support for multivariate regression. For all target features of 90, 180, and 360 days cumulative gas production per foot of lateral, cluster spacing showed the highest impact on production, followed by completions, and stimulation features such as proppant per foot injected. It is

Features (90 days CUM/ft)	Normalized_linearSVR	Features (180 days CUM/ft)	Normalized_linearSVR	Features (360 days CUM/ft)	Normalized_linearSVR
Cluster spacing	21.0	Cluster spacing	22.8	Cluster spacing	18.7
Cluster per stage	19.1	Proppant/ft	18.7	Gas content	15.6
100 mesh	11.8	100 mesh	12.4	Proppant/ft	10.5
Proppant/ft	10.4	Gas content	8.4	Bounded	7.7
Fluid volume/ft	7.9	Cluster per stage	5.3	100 mesh	6.1
Gas content	6.0	Bounded	5.0	Cluster per stage	5.0
Thickness	4.1	Fluid volume/ft	4.1	Fluid volume/ft	4.3
Porosity	3.5	Distance	3.6	Distance	3.7
Avg. Tr. rate	3.4	Thickness	2.9	Number of stages	3.3
Avg. Tr. pressure	2.9	Number of stages	2.8	Porosity	3.0
Number of stages	2.6	Porosity	2.4	TOC	3.0
Shot density	2.4	TOC	2.2	Density	2.7
Density	1.5	Shot density	2.1	Thickness	2.0
Bounded	1.2	Avg. Tr. rate	2.0	Offset	1.9
Distance	0.9	Avg. Tr. pressure	1.9	Avg. Tr. rate	1.8
Offset	0.8	Density	1.8	Avg. Tr. pressure	1.8
TOC	0.5	Offset	1.6	Shot density	1.7

Fig. 24.6 Feature selection and ranking using Linear SVR.

important to note that some parameters such as gas content and boundedness are gaining importance. Early on (e.g., 90 days), they do not show high impact on cumulative gas production per foot of lateral because mostly free gas in natural fractures and matrix will be produced and a well is not impacted by well interference; however, as time progresses (e.g., after 360 days), the gas content and boundedness become more important as a well starts producing from gas stored as adsorbed gas. Under these circumstances, well interference could potentially occur. In addition, field curtailment could also have a direct impact on the first 6 months of the wells. Therefore, the safest assumption is to use CUM360 per foot as the output of all of these models used for feature ranking. Ridge regression has also been applied to our data set; however, this technique did not result in consistent rankings among all the target features. F-regression tests have also been performed for this example. F-regression tests indicate that whether any of the independent features in a MLR model are significant. Since F-regressor only considers one variable at a time and our features are highly correlated, it did not result in consistent rankings among all the target features.

RF is another powerful supervised ML algorithms that can be used for feature ranking. RF can be applied to both regression and classification problems. For classification problems, RF uses Gini importance or mean decrease in impurity (MDI) to calculate the importance of each input feature in relation to output feature. Please note that Gini importance is also known as the total decrease in a node impurity. Essentially, it measures how much a model's accuracy is decreased when a variable is dropped. The larger the decrease, the more significant the variable is. We also applied RF algorithm to this data set to make sure the feature ranking that was obtained from LSVR was consistent with RF. For this analysis, it was consistent and therefore, it provided more confidence in the feature ranking analysis. Please note that the input and output data would need to be either normalized or standardized before applying any of the discussed algorithms. In addition, if the shape of the data is not Gaussian or normal, it could be beneficial to apply a log transform in an attempt to change the distribution to become close to normal.

First, 10 most important features, cross plots, and heat maps were generated as shown in Figs. 24.7 and 24.8. Both figures are generated using "seaborn" package in python. The correlation between features are defined using the **Pearson correlation coefficient** that is a measure of the linear correlation between two features X and Y as shown in Eq. (24.6). The

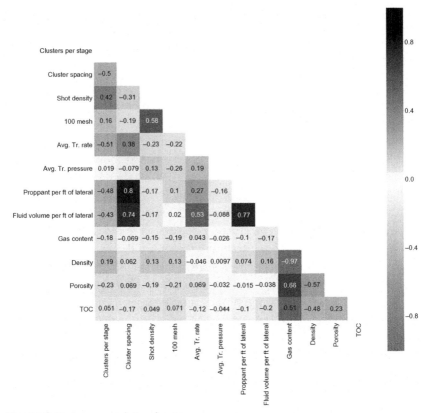

Fig. 24.7 Features correlation heat-map.

Pearson correlation coefficient has a value between −1 and 1, where 1 is total positive linear correlation, 0 is no linear correlation, and −1 is total negative linear correlation. In Figs. 24.7 and 24.8, positive correlation is shown in black and negative correlation between features is shown in gray. The magnitude of the correlations are quantified and presented by the intensity of the colors. For this study, any parameter above a |90|% correlation is considered highly correlated. Fig. 24.7 shows that the gas content has 97% negative correlation with bulk density, indicating that the bulk density has been used to calculate the gas content. As bulk density increases, the gas content decreases. Therefore, bulk density can be removed from the feature list since it carries similar information as gas content. Fig. 24.8 shows the correlation cross plot in which each feature in the database will be shared in the y-axis across a single row and in the x-axis across a single column. The main

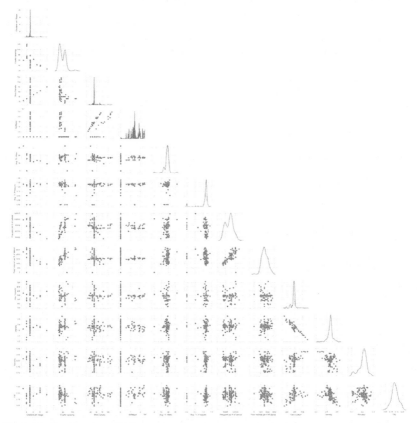

Fig. 24.8 Features correlation cross plot.

diagonal is treated differently, drawing a plot to show the univariate distribution of the data for the feature in that column. Cross plots also provide visual representation of the correlations between features and can also be used to identify diversity of CSD in the field.

$$\text{Pearson correlation coefficient}: P_{X,Y} = \frac{\text{cov}(X, Y)}{\sigma_X \sigma_Y} \quad (24.6)$$

where $cov(X, Y)$ is the covariance between parameters X and Y, σ_X is *Standard deviation of parameter X*, and σ_Y is Standard deviation of parameter Y.

Fig. 24.8 also shows high positive correlation between proppant per ft of lateral and fluid volume per ft of lateral. This figure can also be used for outlier detection as long as it was not based on an old design.

Data partitioning is a practice used to separate portions of a dataset for modeling and training using ANN technique. In this example for data partitioning, we are using Latin hypercube sampling (LHS). LHS is a technique that generates samples based on the provided distribution of a dataset. In order to properly and accurately sample a dataset, partitioning of the data needs to be performed with special care. The samples must be capable of preserving the statistical description of the original data. Training a model on a dataset that focuses mainly on the higher production wells and tested on the lower production wells will lead to inaccurate results and a biased model. An example of the 180-days production distributions along with the training and testing partitioning obtained using LHS algorithm can be found in Fig. 24.9. As shown in Fig. 24.9, both training and test data subsets preserved the original distribution characteristics with high accuracy. This preservation guarantees an unbiased and accurate model development procedure in the next phase of predictive model development. The training set will be separated into training and validation subsets, while the test will be used as a blind set.

Neural network training, validation, and testing

Once the data had been sampled for the training and testing, we started the model development. For this example, multilayer perceptron regressor called "MLPRegressor" in Python is used for model development. This function requires parameters such as: hidden layer sizes, number of layers, activation, solver, learning rate, maximum number of iteration, and momentum. The number of layers and neurons in each layer is highly dependent on the number of features and complexity of the problem. For activation function, different options are available such as logistic sigmoid

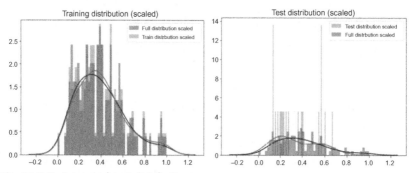

Fig. 24.9 Training and test distributions.

function "logistic," tangent hyperbolic function "tanh," "relu" which is a rectified linear unit function and "identity" function of $f(x) = x$. Solver includes different options such as "lbfgs," which uses quasi-Newton technique, "sgd" that uses stochastic gradient descent technique, and "adam," a stochastic gradient-based technique to optimize the squared loss defined earlier in this chapter. The learning rate could be constant, gradually decreasing, or adaptive and momentum is a value between 0 and 1. Some of these parameters are only available for certain solvers. For example, momentum and learning rate are used if the solver is defined as "sgd."

Finding the correct set of parameters for MLPRegressor that will result in the most accurate model is a key factor in model development. For this example, we have used the full factorial design where different models based on different combinations of parameters in "MLPRegressor" function are built and the accuracy of the model in the testing set is calculated and compared. For accuracy calculation, each model has been run 25 times and the average accuracy is calculated as a cross-validation test. This process leads to the most accurate model and is sometimes referred to as "grid search" in other literatures. For this specific example, it turns out that the choice of a single layer with 30 neurons and a logistic solver will result in the most accurate model for both training and testing sets, with 92% and 84% accuracy, respectively. In Fig. 24.10, the loss values for each run (i.e., 25 runs) are calculated and sorted to get the loss value for the optimum model. Fig. 24.11 also shows the test data set and model predictions obtained using ANN model developed.

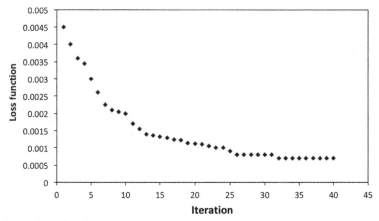

Fig. 24.10 Loss function.

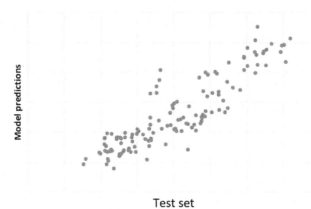

Test set
Fig. 24.11 Blind test prediction using ANN model.

Type curves

An ANN model that has been validated with a test data set can be used to generate different type curves. Type curves can be used for uncertainty quantification and design optimization in which the impact of each input feature on the target can be quantified as a function of another input feature. In this case, one feature will be changed in the range of feature variations while other features remain field average values. For example, the impact of proppant injected per foot of lateral on 360 days cumulative gas production per lateral foot can be obtained while other features are kept as field average values. As shown in Fig. 24.12, injecting a higher amount of proppant per foot of lateral while other parameters are fixed to average field values, resulting in a higher cumulative gas production per foot. However, to obtain the optimum proppant per foot for any special case, the economic analysis reveals whether the extra capital expenditure on higher proppant/ft can be justified by incremental production increase. Fig. 24.13 also shows the impact of cluster spacing on cumulative gas production. In this specific field example, decreasing the cluster spacing results in a higher cumulative gas production per ft. This type curve analysis suggests having more stages with tighter cluster spacing; however, economic analysis must be the deciding factor as discussed in Chapter 18. Fig. 24.14 shows the impact of gas content on cumulative gas production per ft, and as expected, a higher gas content will result in a higher cumulative gas production per ft. Other type curves can also be generated, but it is crucial to investigate the frequency of the specific values used for each parameter based on the distribution of each parameter obtained in Fig. 24.8. Histogram

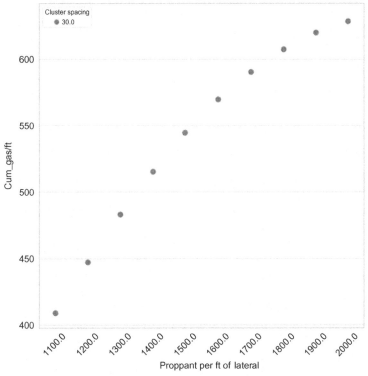

Fig. 24.12 Type curve of proppant per foot of lateral on 360 cumulative gas production per foot of lateral length.

distribution should also be used for each parameter to ensure sufficient data is available at the higher or lower ends of the distribution.

Field completion design optimization (Monte-Carlo analysis)

For completions optimization of the field, an ANN model is used with Monte-Carlo simulation. For this purpose, completions parameters are randomly selected from their corresponding distributions. Other parameters such as reservoir and drilling were kept as field average parameters. A 1000 different simulation scenarios are run and P10, P50, and P90 values of the field cumulative 360 gas production is obtained. Average actual field 360 cumulative gas production is compared with P10, P50, and P90 values and a set of completions parameters leading to maximum production output is obtained. This set becomes optimum completions and stimulation design parameters. Fig. 24.15 shows the Monte-Carlo simulation results and comparison of P10, P50, and P90 of cumulative gas production of the field with actual field production. As

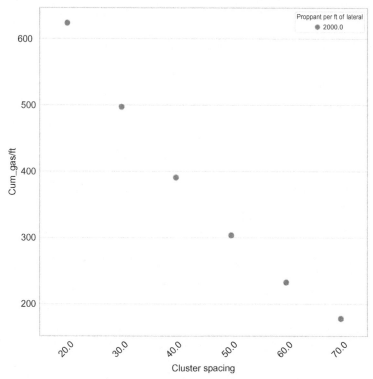

Fig. 24.13 Type curve of cluster spacing on 360 cumulative gas production per foot of lateral length.

shown in Fig. 24.15, the actual field production is slightly lower than P50 field potential, indicating poor quality completions and stimulation design across this specific field. The good news is that by changing the completions and stimulation design for future wells, there is a good chance that the current gas production volume of the field will significantly increase.

K-means clustering

K-means clustering is one of the most commonly used unsupervised ML algorithms for clustering analysis. One of the applications of K-means is to cluster the data into various groups instead of dividing them manually. Another application of K-means is outlier detection. Although other outlier detection algorithms such as local outlier factor (LOF) can be used, K-means is also useful in detecting outliers in a data set. K-means is a simple and easy technique used to classify a data set into a certain number of clusters (K

Application of machine learning in hydraulic fracture optimization 531

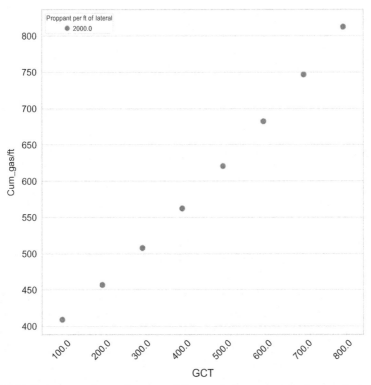

Fig. 24.14 Type curve of gas content on 360 cumulative gas production per foot of lateral length.

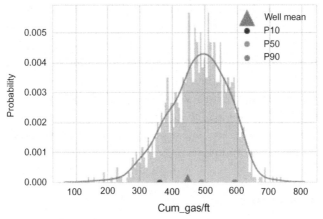

Fig. 24.15 Monte-Carlo simulation.

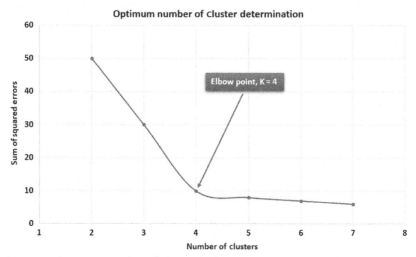

Fig. 24.16 Optimum number of cluster determination.

clusters). Determining the number of clusters can be the most challenging part of the K-means algorithm. Domain expertise plays a big role in determining the number of clusters; however, another approach to discover the number of clusters needed when using this algorithm is to run K-means algorithm multiple times at various clusters (2 clusters, 3 clusters, ..., etc.) and plot "number of clusters" on the x-axis vs "sum of squared errors" on the y-axis until an elbow point is observed. Increasing the number of clusters beyond the elbow point will not tangibly improve the result of the K-means algorithm. Therefore, that point can be selected as the optimum number of clusters for a problem as shown in Fig. 24.16.

When a data set has no label (no output), K-means is a powerful technique to partition data into K number of clusters. Some projects require the data to be clustered (partitioned) first before importing into a supervised ML algorithm. K-means can be used to perform the first task, portioning and labeling data. After partitioning and clustering the data (assigning an output for each row of data), the data set is now ready to apply a supervised ML algorithm. When a combination of both unsupervised and supervised ML is used, it is referred to as semisupervised ML which can be powerful in some applications in various industries.

How does K-means work?

The following steps are typical for performing K-means clustering analysis:

(1) The first step in K-means algorithm is determining number of centroids (clusters) which was discussed.
(2) Next is the initialization of centroids within a data set which can be either initialized randomly or selected purposefully. The default initialization method for most open source ML software is random initialization. If random initialization does not work, carefully selecting the initial centroids could potentially help the model.
(3) Next, the model will find the distance between each data point (instance) and the randomly selected (or carefully selected) cluster centroids. The model will then assign each data point to each cluster centroid based on the distance calculation presented below. For example, if two centroids have been randomly selected within a data set, the model will calculate the distance from each data point to centroid #1 and #2. In this case, each data point will be clustered under either centroid #1 or #2 based on the distance from each data point to the randomly initialized cluster centroids. There are various ways to calculate the distance. The following functions are commonly used techniques, the most common being the Euclidean distance function:

$$\text{Euclidean distance function} = \sqrt{\sum_{i=1}^{k}(x_i - y_i)^2} \quad (24.7)$$

$$\text{Manhattan distance function} = \sum_{i=1}^{k}|x_i - y_i| \quad (24.8)$$

$$\text{Minkowski distance function} = \left(\sum_{i=1}^{k}(|x_i - y_i|)^q\right)^{1/q} \quad (24.9)$$

(4) The next step is to average the values of data points (instances) within each cluster centroid (assigned in step 3) and calculate a new centroid for each cluster (moving the cluster centroid).
(5) Since new centroids have been created in step 4, in step 5, each data point will be **reassigned** to the newly generated centroids based on one of the distance calculation functions.
(6) Steps 4 and 5 are repeated until the model converges which means additional iteration will not lead to significant change in final centroid selection. In other words, cluster centroids will not move any further.

Fig. 24.17 illustrates the *K*-means clustering step by step for a small data set in two-dimensional space. Fig. 24.18 also shows a data set getting divided into two and three clusters. The plus sign represents the final centroid for each cluster.

Distance equation calculation example:

Using the Euclidean distance function, calculate the distance between the following two points:

(2, 5, −1) and (5, −16, 9)

$$\sqrt{\sum_{i=1}^{k}(x_i - y_i)^2} = \sqrt{(5-2)^2 + (-16-5)^2 + (9-(-1))^2} = \sqrt{550} = 23.4$$

After the convergence of the final centroids, they can simply be applied to any new data set to determine which cluster each new instance will fall under. This application can be done for static or dynamic data. One of the simple advantages of *K*-means clustering as compared to other algorithms is that it can be used in real-time scenarios as it can be easily

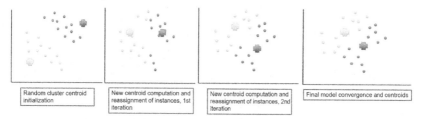

Fig. 24.17 *K*-means clustering illustration.

Fig. 24.18 Data set division into two and three clusters.

Table 24.7 Final centroid cluster example
Final cluster centroids

Attributes	Cluster 1	Cluster 2
Gas rate	4323	3923
Casing pressure	660	2073
Tubing pressure	571	600
Line pressure	377	479

Table 24.8 Example data set

Gas rate	Casing pressure	Tubing pressure	Line pressure
4500	1000	850	500
4460	960	810	500
4420	920	770	500
4380	880	730	500
4340	840	690	500

implemented computationally (will be illustrated). To illustrate this concept, let's go over an example.

K-means application example:

The production data such as gas rate, casing pressure, tubing pressure, and line pressure from 400 wells were gathered into a single data source. After applying K-means clustering using random initialization, Euclidean distance function using two clusters, the data were converged into the two following cluster centroids as shown in Table 24.7. Use the data provided in Table 24.8 to determine which cluster (1 or 2) each row in the data set will fall under.

Step 1: Calculate distance function for cluster 1:

Cluster 1 distance for row 1

$$= \sqrt{(4500-4323)^2 + (1000-660)^2 + (850-571)^2 + (500-377)^2}$$
$$= 490$$

Cluster 1 distance for row 2

$$= \sqrt{(4460-4323)^2 + (960-660)^2 + (810-571)^2 + (500-377)^2}$$
$$= 425$$

Cluster 1 distance for row 3

$$= \sqrt{(4420-4323)^2 + (920-660)^2 + (770-571)^2 + (500-377)^2}$$
$$= 363$$

Cluster 1 distance for row 4

$$= \sqrt{(4380-4323)^2 + (880-660)^2 + (730-571)^2 + (500-377)^2}$$
$$= 303$$

Cluster 1 distance for row 5

$$= \sqrt{(4340-4323)^2 + (840-660)^2 + (690-571)^2 + (500-377)^2}$$
$$= 249$$

Step 2: Calculate distance function for cluster 2:

Cluster 1 distance for row 1

$$= \sqrt{(4500-3923)^2 + (1000-2073)^2 + (850-600)^2 + (500-479)^2}$$
$$= 3170$$

Cluster 1 distance for row 2

$$= \sqrt{(4460-3923)^2 + (960-2073)^2 + (810-600)^2 + (500-479)^2}$$
$$= 3223$$

Cluster 1 distance for row 3

$$= \sqrt{(4420-3923)^2 + (920-2073)^2 + (770-600)^2 + (500-479)^2}$$
$$= 3277$$

Cluster 1 distance for row 4

$$= \sqrt{(4380-3923)^2 + (880-2073)^2 + (730-600)^2 + (500-479)^2}$$
$$= 3332$$

Cluster 1 distance for row 5

$$= \sqrt{(4340-3923)^2 + (840-2073)^2 + (690-600)^2 + (500-479)^2}$$
$$= 3388$$

Step 3: If the calculated distance for each row is smallest, assign the row to that cluster. In this example, the calculated distance for cluster 1 (step 1) in each row is smaller than the calculated distance for cluster 2 (step 2) in each row. Therefore, all the new data rows will be **clustered** as cluster 1.

K-means clustering for liquid-loading detection

Ansari et al. (2018) illustrated how *K*-means clustering can be used for detecting liquid loading. They used basic production data such as gas rate, casing pressure, tubing pressure, line pressure, and water rate as the inputs of the model. The first trial involved using a supervised ML algorithm such as ANN to classify each row of data as "Loaded" or "unloaded" using Turner and Coleman as the main criteria. While this technique was successful in predicting the loading status and condition of a well, the idea was to develop a model that is completely **independent** of Turner and Coleman techniques. This effort was primarily because the Turner and Coleman techniques were developed many decades ago empirically and the output result from Turner and Coleman calculations could erroneously classify a well as loaded when the well is not or vice versa. Therefore, an unsupervised ML algorithm, *K*-means clustering was used to determine the loading status of the well. To prove the concept, they applied *K*-means clustering to two wells located on the same pad. After obtaining the final centroids, these centroids were successfully applied to multiple wells on various pads located within a 10-mile radius. As can be seen below, the predictive capability of *K*-means clustering when applied to a complete blind data set is very high. Therefore, *K*-means clustering can be applied in real time for optimizing production performance of wells and avoid losing gas volumes by getting notified as soon a well's status becomes loaded. Fig. 24.19 illustrates the prediction capability of *K*-means clustering by classifying the points that appear to be unloaded as "unloaded" and points that are loaded (erratic points) as "loaded." Turner rate is shown for illustration purposes only and was not used in the model. Such unsupervised technique can also be used as an outlier tool when performing various reservoir or completions analyses.

Other application of unsupervised ML algorithms

Anomaly detection is another application of unsupervised ML algorithms. DBSCAN is a powerful unsupervised ML algorithm that is often used for anomaly detection especially in production curve auto-fitting. Another application of unsupervised ML such as *K*-means is in type curve

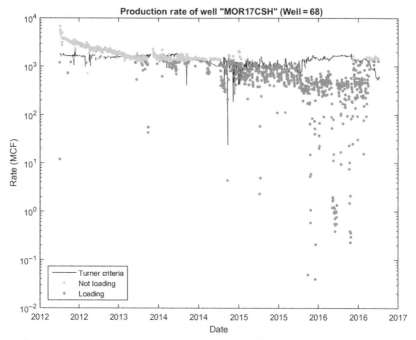

Fig. 24.19 *K*-means clustering applied on a blind well.

clustering analysis. In the past, operators have used manual analysis and intuition to define their type curve boundaries; however, unsupervised ML algorithms can be very powerful for type curve clustering. *K*-means algorithms is one of the most used unsupervised ML algorithms across various industries, and it is a powerful technique to cluster various input parameters into different clusters and find the centroid for each cluster. The final deliverables from this analysis will be providing number of clusters with each cluster centroid. As a new data point becomes available, the process can be automated in a fashion so that each new data point can be assigned to the predefined cluster centroids. This automated activity can be used to **draw type curve boundaries** accurately across a company's acreage position as compared to using a human bias to draw those boundaries. To perform such analysis, it is recommended to use important geologic, completions, and production parameters such as total GIP or OIL for the entire target interval, BTU, geologic complexity, EUR, as well as some completions parameters (if deemed to be necessary). This analysis can be performed using different combinations of parameters to see which combination of parameters would provide the best clustering output of a company's acreage position based on already existing knowledge of the area. That is where domain expertise plays a big role in choosing the right number of clusters and the correct combination of parameters.

In addition to type curve clustering, lithologic classification is another powerful use of unsupervised K-means algorithm.

CHAPTER TWENTY-FIVE

Numerical simulation of real field Marcellus shale reservoir development and stimulation

Introduction

The objective of this chapter is to provide real field Marcellus Shale reservoir development and stimulation examples including completions design optimization and economic analysis. Marcellus formation is black shale that extends from New York to Pennsylvania, Ohio, Maryland, and West Virginia. For this study, MIP-3H multistage hydraulic fractured horizontal well located in Monongalia County, Morgantown, West Virginia is selected. The MIP-3H well is drilled above the Cherry Valley Limestone in the Upper Marcellus Shale. The data regarding the MIP-3H well is available to the public at http://mseel.org/. Field data available for the MIP-3H well includes well logs in vertical and horizontal sections, completions, microseismic, fiber optics, and production data as presented in Table 25.1 and available at http://mseel.org/research/research.html.

Problem statement

For this example, let's assume you are a project manager of Marcellus North division, and you have been directed to optimize your field by defining the optimum hydraulic fracture sand schedule followed by obtaining the net present value of the well in order to provide guidance to the analyst on the true intrinsic value of the well. The purpose of this project is as follows:
1. Find the optimum economic sand design schedule.
2. Find the BTAX and ATAX NPV of the well using various pricing provided at the end of this project.

Below are the steps that must be taken to complete this project:
(a) Perform a literature review and create a summary report on current knowledge of best practices for reservoir modeling, completion design,

Table 25.1 MIP-3H well available data

MIP 3H available data

Well logs		Completions	Micro seismic	Fiber optics	Production
Directional survey (depth, inclination, azimuth) MWD logs (MD) Wireline logs (TVD)	Sonic, density, neutron porosity Resistivity Caliper Geomechanical Gamma	Lithology, TOC, elements FMI NMR Production log	Treatment reports: Stage design: (spacing perforations shots/ phasing) Injection design: (rate time, fluid/ material)	Time Position Magnitude (159–1383 samples/ stage, stages 7–28)	Temperature DTS (during production—8 samples/day, during fracturing—2 samples per minute) Acoustic DAS	Production log (gas rate/cluster, stages 1–28) Daily rates of gas and water Well is restricted to 10 MMCF/day

and forecasting based on published information and your field engineers' experience.
(b) Perform petrophysical analysis on your well to find the basic reservoir properties.
(c) Build a shale gas reservoir model using CMG GEM and perform horizontal well drilling and multistage hydraulic fracturing.
(d) Perform sensitivity analysis and uncertainty quantification (CMG CMOST).
(e) Perform history-matching to match the field production data (CMG CMOST).
(f) Perform well performance analysis (IHS HARMONY).
(g) Evaluate the basic reservoir and fracture properties using a diagnostic fracture injection test (DFIT) conducted in the field. Obtain pore pressure and fracture closure pressure and use these values for running your model. (Refer to Chapter 14 of this book)
(h) The formation of focus is brittle with high Young's modulus and low Poisson's ratio and is ideal for slick water frac job. Therefore, use the workflow for designing a slick water frac job for various sand schedules of 1000, 1500, 2000, 2500, 3000, 3500, and 4000 lb/ft using 40% 100 mesh, and 60% 40/70 mesh sand size. The type of 40/70 mesh sand type must be determined using the closure pressure obtained from the DFIT. When creating these sand/ft schedules, keep your sand concentrations the same across all sand stages (e.g., 0.25 ppg, 0.5 ppg, 0.75 ppg, etc.) and only change the amount of clean water volume for each sand stage to obtain the aforementioned sand/ft loadings. (refer to Chapter 10 Fracture Treatment Design).
(i) Once each sand schedule is created, enter these sand schedules into preferred commercial fracture software, such as FracPro and run these sand schedules in the fracture software to obtain fracture geometry for each design (propped half-length, propped width, and propped conductivity). After obtaining fracture geometry for each design, write down a summary of the fracture geometry for each. Finally, take the fracture geometry of each design and obtain production rate vs time for each scenario that was obtained numerically using CMG GEM compositional simulator. The propped half-length obtained from each design must provide some guidance on well spacing. For the sake of time, run each numerical simulation (seven cases) for 10 years only, because the first 10 years are the most important value creation during the life of a well.
(j) Next, take the production rate vs time for each scenario and perform BTAX and ATAX economic analysis following the step-by-step

Table 25.2 Approximate total capital cost for each sand/ft design

Sand/ft (lb/ft)	Total Capex ($MM)
1000	$ 5.2
1500	$ 5.6
2000	$ 5.9
2500	$ 6.3
3000	$ 6.7
3500	$ 7.0
4000	$ 7.4

workflow in Chapter 18. Plotting ATAX NPV vs various sand schedules of 1000, 1500, 2000, 2500, 3000, 3500, and 4000 lb/ft, there is an optimum sand schedule based on the highest NPV of all sand designs (use $3/MMBTU gas price and the assumptions listed at the end of this project Table 25.2).

(k) Next, run a modified hyperbolic decline curve fit through the optimum sand schedule production volume vs time that was obtained numerically to obtain the production rate vs time for 50 years instead of the first 10 years that was obtained numerically. Assume a terminal decline rate of 5% and use a hyperbolic exponent (b value) of between 1.0 and 1.7 to fit your decline curve to the numerical decline curve that was obtained for the optimum sand design schedule. Record the modified decline curve parameters and use the monthly volumes generated from this decline curve to run all your economic analysis going forward.

(l) Obtain the BTAX and ATAX NPV at each pricing below from the modified hyperbolic decline fit and obtain the well NPV at various pricing listed below:
  – $2.5/MMBTU gas pricing escalated using a monthly stair-step fashion at 3%/year
  – $3.0/MMBTU gas pricing escalated using a monthly stair-step fashion at 3%/year
  – $3.5/MMBTU gas pricing escalated using a monthly stair-step fashion at 3%/year
  – $4/MMBTU gas pricing escalated using a monthly stair-step fashion at 3%/year
  – Using NYMEX provided with this project followed by 3%/year monthly stair-step fashion at the end of the last provided NYMEX pricing

Economic assumptions below can be used for the above tasks:
1. BTU = 1060 BTU/SCF (dry gas)
2. Shrinkage = 1% (or 0.99 shrinkage factor due to line losses and fuel at the compressor station). Use the numerical 10-year decline curve (rate vs time and use the modified hyperbolic decline curve obtained by fitting through the numerical data for 50 years.
3. Variable lifting cost = $0.22/MCF escalated using a monthly stair-step fashion at 3%/year
4. Fixed lifting cost = $100/month/well escalated using a monthly stair-step fashion at 3%/year
5. Variable gathering cost = $0.3/MMBTU escalated using a monthly stair-step fashion at 3%/year
6. Firm transportation (FT) cost = $0.2/MMBTU escalated using a monthly stair-step fashion at 3%/year
7. WI = 100%
8. NRI = 85%

Phase I: Petrophysics and well log analysis

Petrophysics and well log analyses have been performed given the MIP-3H well logs listed in Table 25.1. The actual well logs in. LAS and .pdf format are available to the public on the MSEEL website at http://mseel.org/Data/Wells_Datasets/MIP_3H_Pilot/GandG_and_Reservoir_Engineering/. For this purpose, Petra (geological interpretation software) is used for plotting, well log construction, and cross plot studies. Detailed well log interpretation is performed using Fig. 25.1 to identify the top and bottom of each formation (presented in Table 25.3) and petrophysical properties, such as porosity, permeability, and saturation of the target formations. Gamma ray as well as resistivity logs are used as major logs to identify the top and bottom of the formations. The first track, from left in Fig. 25.1, shows three logs of high-resolution gamma ray, Uranium Concentration (HURA), and total organic content (TOC) plotted on top of one another. HURA indicates fractures or the presence of organic matter. A large spike in uranium log is observed from 7450 to 7560 ft indicating the upper and lower Marcellus Shale due to an increase in TOC. The TOC log also shows high values in the range of upper and lower Marcellus. The next track is the deep resistivity log, RT_HRLT also known as the true formation resistivity, and the shallow depth resistivity log, that is, the micro-cylindrically focused log (R_{XOZ}). The formation resistivity log shows high resistivity in upper and lower Marcellus, indicating the presence of hydrocarbon in these

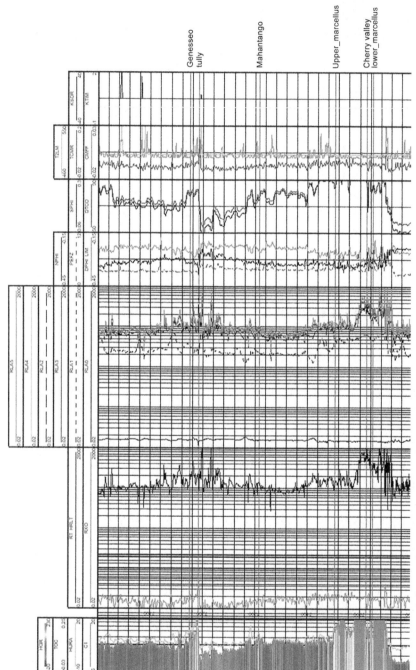

Fig. 25.1 Sample log analysis to identify the top formations.

Table 25.3 Summary of top, bottom, and thickness of different formations in well MIP-3H

Formation name	Formation top (ft)	Formation bottom (ft)	Formation thickness (ft)
Middlesex	6602	7175	573
Genesseo	7175	7196	21
Tully	7196	7283	87
Hamilton	7283	7454	171
Upper Marcellus	7454	7520	66
Cherry Valley	7520	7528	8
Lower Marcellus	7528	7553	25
Onondaga	7553	7583	30
Huntersville	7583	7849	266

layers. The next track is a set of different depth resistivity logs from 0 to 5, in which 0 corresponds to the mud resistivity at the borehole, and 1–5 shows medium depth resistivity values between the resistivity of the flushed zone and the true formation. RLA5 is matching with RT_HRLT log in the second track showing high resistivity of the formation and indicating the presence of hydrocarbon in upper and lower Marcellus Shale. The next track is the photoelectric absorption log (PEFZ) in which z is the average atomic number of the formation together with the density porosity and neutron porosity logs. The combination of these logs will help to identify the mineralogy of the formation and presence of light hydrocarbon (gas) bearing formations. In the presence of gas in the formation, the density porosity log shows higher values as compared to the neutron porosity log observed in both interest zones of upper and lower Marcellus Shale. The next track is the sonic porosity (SPHI) and Delta-T compressional logs (DTCO) in which both show the same trend. The sonic porosity with the density and neutron porosity logs can help in lithology identification, formation of mechanical properties, and abnormal formation pressure identifications. Delta-T compressional logs can also be used to calculate the Poisson's ratio. The next track includes Total Magnetic Resonance Porosity (TCMR), Free Fluid Volume (CMFF), CMRT-B Log, and T_2 distribution log (T_{2LM}). TCMR log is used to compute clay bound water volumes, which helps to calculate more accurate hydrocarbon saturations and provides lithology independent porosity values, which is extremely helpful in complex structures. CMFF presents the standard free-fluid porosity (CMFF) and T_{2Lm} is the log mean T_2 relaxation time that will be used in SDR equation to obtain NMR permeability as presented in Eq. (25.4). The last track includes Timur-Coates and SDR permeability values calculated using Eqs. (25.4), (25.7).

In this stratigraphic column, the Marcellus Shale (upper and lower Marcellus) is targeted as interest zone for gas production. Detailed well log analysis of the upper and lower Marcellus resulted in obtaining the average petrophysical parameters of Marcellus Shale, such as porosity, water saturation, and permeability. The upper and lower Marcellus is identified by high gamma ray >250, having NPHI (neutron porosity) > DPHI (density porosity), Photoelectric log (PE) value of 3.5, high TOC, and low water saturation. The porosity and water saturation are obtained using average porosity and Archie's equation using Eqs. (25.1), (25.2) as follows:

Average porosity obtained from NPHI and DPHI logs : $\phi_{average}$

$$= \sqrt{\frac{DPHI^2 + NPHI^2}{2}} \qquad (25.1)$$

$$\text{Archie's equation} : S_w = \sqrt[n]{\frac{a}{\Phi^m} \frac{R_w}{R_t}} \qquad (25.2)$$

In Archie's equation, the values of $a = 1$, $m = 1.7$, and $n = 1.7$ are used, where a is a constant, m is a cementation factor that varies around 2 and n is a saturation exponent. Cementation factor is directly related to the pore connectivity of the formation. As the pore connectivity decreases, the m value increases. In carbonate reservoirs, the m value can get values as high as 5 (Tiab and Donaldson, 2015). The water resistivity was found using Eq. (25.3). The R_{mf} value (mud filtrate resistivity) was reported to be 0.02 ohm.m for the entire log. The R_{xo} (mud filtrate invaded zone resistivity) and R_t (true formation resistivity) logs were used to create a R_w log in Petra for the entire log interval, where R_w is:

$$\text{Water resistivity} : R_w = R_{mf} \times \frac{R_t}{R_{xo}} \qquad (25.3)$$

Using Eqs. (25.1), (25.2), the average porosity and water saturation of upper and lower Marcellus Shale is obtained to be ($\phi = 8\%$, $S_w = 15\%$) and ($\phi = 5.5\%$, $S_w = 17\%$), respectively.

The permeability of the target formation is obtained using nuclear magnetic resonance (NMR) logs. Different equations, such as the Timur-Coates and SDR equations are suggested to calculate the permeability of the formation using NMR log as follows:

$$\text{Timur-Coates equation} : K_{TIM} = a\varnothing_{NMR}^m \left(\frac{FFV}{BFV}\right)^n \qquad (25.4)$$

$$\text{Bound-fluid volume}: \text{BFV} = \varnothing S_{\text{wirr}} \quad (25.5)$$

$$\text{Free-fluid volume}: \text{FFV} = \varnothing(1 - S_{\text{wirr}}) \quad (25.6)$$

$$\text{SDR equation}: K_{\text{SDR}} = b\varnothing_{\text{NMR}}^{m}(T_{2\text{LM}})^{n} \quad (25.7)$$

where $\varnothing_{\text{NMR}}$ is total porosity obtained from NMR log, S_{wirr} is the irreducible water saturation, and $T_{2\text{LM}}$ is logarithmic mean T_2. The exponents of m, and n together with a and b constants need to be determined for any specific formation. The permeability obtained from these logs for upper and lower Marcellus ranges from 0.0001 to 0.001 md. Recently, NMR high-resolution indicator has been used to obtain permeability of the formation from NMR logging; however, state core sample permeability measurements under reservoir effective stress conditions are still unsteady and provide more accurate and robust information.

Geomechanical log

Geomechanical logs are calculated based on different logs, such as sonic, density, and porosity logs. They are used to obtain the geomechanical properties of different formations, such as bulk modulus, shear modulus, Young's modulus, Poisson's Ratio, unconfined compressive strength, and the closure stress. The log of geomechanical rock properties will help the hydraulic fracture design engineer to investigate the possibility of engineering design of a hydraulic fracturing job and compare that with traditional geometric design. As discussed earlier in Chapter 5, slick water hydraulic fracturing is used in shale gas reservoir to generate a complex fracture system with maximum contacted surface area, while cross-linked gel fluid system is used in shale oil reservoirs to generate the bi-wing fracture system. However, in addition to the hydraulic fracturing fluid system, the geomechanical properties of the formation highly impact the type of fracture system being generated through hydraulic fracturing. As discussed earlier in Chapter 13, hydraulic fracture propagates in the direction of the maximum horizontal stress and perpendicular to the minimum horizontal stress. The hydraulic fracture propagation is also a function of brittleness and fracability ratios of the formations. These parameters are functions of static Young's modulus and Poisson's ratio as presented in Eqs. (13.8), (13.11). The idea in engineering hydraulic fracturing design is to find the optimum stage length, number of clusters and the best locations in each stage to place the perforations for each cluster, that is, high Young's modulus, low Poisson's ratio, and low anisotropic closure stress areas. Lower closure stress areas in the lateral section of the well indicates less energy to initiate and propagate the fracture. Therefore, longer stage spacing and more clusters per stage can be used. However, the high anisotropic closure stress areas require more energy for fracture propagation, and therefore, shorter stage length and fewer clusters are recommended. In areas with high anisotropic closure stress, stage length shorter than 200 ft is advised. For optimum perforation locations the high Young's modulus and low Poisson's ratio areas which indicate higher brittleness index are selected using the criteria obtained from the distributions of these parameters in the lateral section of the well. Fig. 25.2 shows the anisotropic closure stress measured along each stage of the MIP3H well. As shown in this figure, stages 17, 18, 19, and 20 show higher anisotropic stress indicating the need for shorter stage lengths in comparison to stages 13, 14 15, and 16 with lower anisotropic closure stress.

Fig. 25.3 shows the geomechanical log performed in the lateral section of the MIP-3H well. From the left, the first track shows the values of different stresses including C11, C12, C13, C33, C44, C55, and C66. They measure the compression stress (C11), elastic constants (C12 and C13), and shear modulus constants (C33, C44, C55, and C66). These variables can then be used to obtain shear modulus, unconfined compressive strength, and closure stress. The next tracks are the vertical static Young's modulus, horizontal static Young's modulus, vertical Poisson's ratio, and horizontal Poisson's ratio, respectively. To calculate these parameters, the shear modulus is first obtained using bulk density log and the values of the Δt_s from shear wave as follows:

$$\text{Shear modulus}: G = 1.34 \times 10^{10} \times \frac{\rho_b}{\Delta t_s} \quad (25.8)$$

The shear modulus is then used to calculate the Dynamic Young's Modulus as follows:

$$\text{Dynamic Young's modulus}: E = 2G(1 + \nu) \quad (25.9)$$

Here, ν is derived from Poisson's ratio log. Having the dynamic Young's modulus, Static Young's Modulus can be obtained as follows:

$$\text{Static Young's modulus}: E_{stat} = 0.835 \times E_{dyna} - 0.424 \quad (25.10)$$

The Poisson's ratio log can also be used to obtain the fracture closure pressure, which is approximately equal to minimum horizontal stress that can be obtained using Eq. (13.14) in Chapter 13.

We have used the geomechanical logs to locate perforations for each stage to increase the efficiency of the hydraulic fracturing job, that is,

Marcellus shale reservoir development and stimulation 551

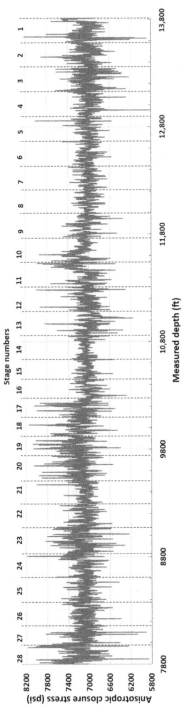

Fig. 25.2 Anisotropic closure stress.

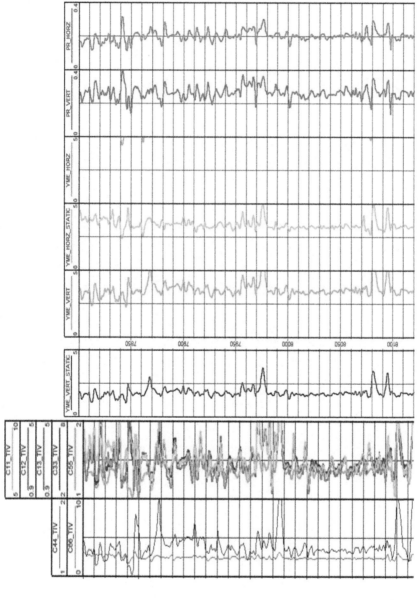

Fig. 25.3 Geomechanical logging in lateral section of the well.

engineering design. Here, we seek the locations with high Young's modules and low Poisson's ratio. For this purpose, first the distribution of the Young's modulus and Poisson's ratio in lateral section of the well is obtained, any value of Young's modulus and Poisson's ratio greater than their mean plus 3 standard deviation $(E \geq \overline{E} + 3\sigma_E, \nu \geq \overline{\nu} + 3\sigma_\nu)$ is considered high, any value between mean minus three standard deviation and mean plus three standard deviation is considered average $(\overline{E} - 3\sigma_E < E < \overline{E} + 3\sigma_E, \overline{\nu} - 3\sigma_\nu < \nu < \overline{\nu} + 3\sigma_\nu)$, and any value less than mean minus three standard deviation, $(E \leq \overline{E} - 3\sigma_E, \nu \leq \overline{\nu} - 3\sigma_\nu)$ is considered low. Fig. 25.4 shows the recommended perforation locations in 5 clusters per stage of 222 feet hydraulic fracture stage. The y-axis of one corresponds to the location for perforation and zero corresponds to the undesirable perforation locations. Blue stands for excellent location for perforation, that is, high Young's modulus and low Poisson's ratio as defined earlier, and red stands for good perforation location, that is, either high Young's modulus and average Poisson's ratio or low Poisson's ratio and average Young's modulus. True measured depth of the top and bottom of each cluster has also been shown in the top left of Fig. 25.4. Table 25.4 shows the actual overview of the MIP-3H completions design. A variety of design parameters are practiced in different hydraulic fracture

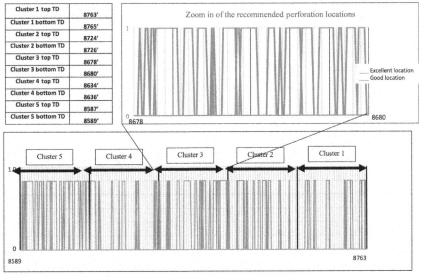

Fig. 25.4 Recommended perforation locations in five clusters per stage of 300 feet hydraulic fracture stage.

Table 25.4 MIP-3H completions overview
MIP 3H completions overview

No. of stages:	28	Cluster spacing:	34–60 ft
No. of clusters:	133	No. of shots per cluster:	6, 8 or 10
Cluster length:	3, 4, or 5 ft	Perforated lateral section:	7750 ft–13815 ft MD

stages of this well so that the best design can be obtained for consecutive wells being drilled in the area.

Phase II: Shale gas reservoir base model development

The CMG GEM compositional and unconventional simulator is used to develop a dual permeability Marcellus Shale gas reservoir model. The process of model development includes:

1. Reservoir Data Collection/Preparation: (i.e., reservoir dimensions, numerical grid type and dimensions, formations of structural and isopac maps, formations and natural fracture porosity and permeability, adsorption characteristics of the formations, formation compressibility, PVT information/fluid properties, relative permeability curves, etc.)
2. Initial Reservoir Conditions: (i.e., initial reservoir pressure, temperature, and water gas contact)
3. Well Information: (well type, trajectories, constraints, perforations, operation, and production)
4. Hydraulic Fracturing: (i.e., Fracture properties, Grid refinement, non-Darcy effect, fracture placement)

The base Marcellus Shale gas reservoir model is developed including 5 layers of Hamilton, Upper Marcellus, Cherry Valley, Lower Marcellus, and Onondaga formations. The top and bottom of each formation are obtained using Table 25.3. The petrophysical properties of each layer are obtained using well log analysis discussed earlier and core analysis reports available at http://mseel.org/research/research.html and assumed to be homogeneous across the formation. Upper Marcellus is selected as a target zone since it has higher thickness, higher porosity, lower water saturation, and higher TOC. Even though our focus is on well MIP-3H, however, MIP4H, MIP5H, and MIP6H are also included in the model to investigate the possibility of well interference in production from MIP-3H well. The physical dimension of the model is defined as $18,000' \times 4500' \times 300'$ (x,y,z). For target formation, Upper Marcellus orthogonal corner point fine grid is selected

Table 25.5 Base model Upper Marcellus shale properties

Permeability	200	nd
Matrix porosity	8	%
Fracture 1/2 length	300	ft
Water saturation	15	%
Fracture height	325	ft
Fracture width	0.027	ft
Sg	85	%
Fracture porosity	0.1	%
Fracture permeability	300	md
Langmuir adsorption constant	0.0002	1/psi
Max adsorbed mass (CH_4)	269.86	(SCF/ton)

Table 25.6 Average porosity, initial water saturation and matrix permeability obtained for all the layers

Formation	Avg. porosity (%)	Avg. initial water saturation (%)	Avg. permeability (nD)
Hamilton	4.5	27	180
Upper Marcellus	8	15	200
Cherry Valley	6	13	330
Lower Marcellus	6	16	380
Onondaga	3	46	37

with $50' \times 50' \times 10'$ dimensions and course grids are used to reset the model $500' \times 500' \times 100'$ dimensions. Table 25.5 shows the base model parameters obtained from our well log analysis, literature review, and core analysis reported for upper Marcellus Shale and Table 25.6 shows the average porosity, initial water saturation and matrix permeability obtained for all the layers. For relative permeability curves, model presented by Osholake (2010) is used.

Fig. 25.5 shows the 3D numerical model built using CMG GEM as our base model, and Table 25.7 presents the average gas composition used for this model. The reservoir model includes 4 producing wells including MIP-3H, MIP-4H, MIP-5H, and MIP-6H. However, MIP-3H is investigated for this study and the reservoir model has been built and history-matched with MIP-3H production and pressure behaviors. The lateral section of the MIP-3H is divided equally into 28 stages with 4–5 clusters per stage resulting in 43 ft cluster spacing. Since the fine grids in upper

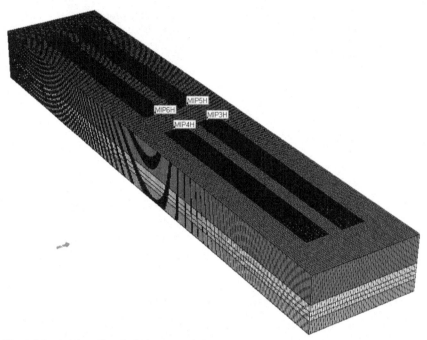

Fig. 25.5 3D Marcellus shale base model.

Table 25.7 Average gas compositions of the Marcellus shale wells

Well	C_1	C_2	C_3	CO_2	N_2
Average	0.852	0.113	0.029	0.004	0.003

Marcellus base mode are 50 × 50 × 10, every grid in the lateral section will be perforated to account for 133 clusters in MIP-3H. For the sake of simplicity, we assumed the geometric completions design and did not use the engineering design suggested by mechanical logs. Figs. 25.6–25.8 show the gas and water rate, cumulative production, casing, and tubing pressures as a function of time for well MIP-3H, respectively. For operating conditions managed pressure drawdown has been implemented to resemble the field production.

Phase III: Sensitivity analysis and history-matching

After the Marcellus Shale reservoir base model is built, it must be able to reproduce the gas, water rate, and cumulative productions as well as pressure profiles corresponding to the actual production of MIP-3H well. This

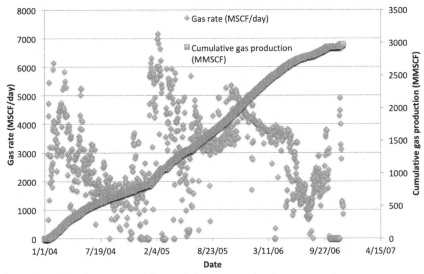

Fig. 25.6 MIP-3H gas rate and cumulative gas production.

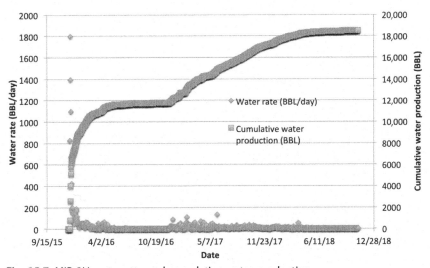

Fig. 25.7 MIP-3H water rate and cumulative water production.

capacity will require more attention since the base model is usually unable to accurately predict these parameters. This inability is mainly because the average rock and fluid properties obtained using well log and core analyses are not representative of these parameters in the field scale. They have been obtained in small scales, such as crushed samples and core plugs in the

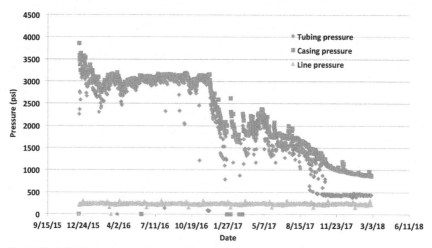

Fig. 25.8 MIP-3H casing, tubing, and line pressures.

laboratory or limited by the resolution and radius of investigation of the well logs. There are different analytical and numerical techniques proposed to upscale these properties from the laboratory to field scale average values. The most commonly used approaches are integral transform methods, stochastic-convective approaches, moment methods, central limit or martingale methods, projection operators, mathematical homogenization, mixture and hybrid mixture theory, renormalization group techniques, space transformational methods, and continuous time random walks. These techniques are developed to capture the effect of heterogeneity and anisotropy of the reservoir and the presence of features that might not be considered while developing the base model, such as the presence of high permeability layers or channels "Thief zones" on fluid flow and storage in porous media. However, in the oil and gas industry, it is more common to use numerical reservoir simulations to history-match the actual field production by performing sensitivity analysis and nonlinear optimization to minimize the error between simulation results and actual field values. This technique treats the problem as an underdetermined inverse problem with multiple solutions where the most probable estimate of the parameters will be used as the optimum solution.

To perform the sensitivity analysis and history-matching studies for this project, we are using the commercial software called CMG CMOST from computer modeling group LTD. CMOST provides the modules for sensitivity analysis, history-matching, and optimization and uncertainty analysis.

There are many parameters, such as rock and fluid properties, well stimulation, and operations that can impact the reservoir simulation results. Therefore, to optimize the history-matching process, it is essential to first perform the sensitivity analysis and reduce the number of parameters. The main objective in sensitivity analysis is to obtain the most important parameters impacting the reservoir simulation results, such as gas and water rate, cumulative production or pressures and their correlations. Tornado and Pareto charts are usually used to show the results of the sensitivity analysis. The Pareto chart ranks all the parameters in descending order of their impact on the simulation results. The standardized Pareto chart displays the values of the t-test in which the negative coefficients imply inverse and positive coefficients imply direct relationships with the simulation results. If the parameter falls to the left of the "p-level" line, it means that we cannot make "statistical significance" claims about the importance of this parameter at the specified p-level. For engineering purposes, the 95% confidence interval or a $p = 0.05$ is considered.

For sensitivity analysis, different experimental design techniques, such as two-level Plackett-Burman and three level Full Factorial designs are used in this study. The Plackett-Burman "PB" design is the most compact two-level design usually used for screening studies. The PB design captures all the main effects of the parameters but cannot fully capture the effect of interactions between parameters on the simulation results. The PB design requires $N+1$ simulations, where N is the number of parameters, but PB design is only available in multiples of 4. For example, for 10 parameters, the number of simulation runs in a PB design is equal to 12. If the parameter takes a deterministic value, it is a good practice to multiply the base value by 1.2 (i.e., 20% increase) and 0.8 (i.e., 20% decrease) to define the upper and lower levels. If the parameter has a normal distribution, mean value plus/minus three standard deviations can be used to define the upper and lower levels. If the parameter is not normally distributed, first convert the distribution of the parameter to normal distribution and then use mean value plus/minus three standard deviations. One can also define two different tables for relative permeability or maps for structures and assign them to upper and lower levels.

Table 25.8 shows an example of PB design table for 10 parameters randomly placed in P_1 to P_10, in which -1 corresponds to the lower and $+1$ to the upper levels of each parameter. To reduce the confounding effect in the PB table, a second round of simulations is suggested in which the same number of simulations as the original design will be added to the table;

Table 25.8 Plackett-Burman design for 10 variables

Run	P_1	P_2	P_3	P_4	P_5	P_6	P_7	P_8	P_9	P_10
1	1	1	1	1	1	1	1	1	1	1
2	−1	1	−1	1	1	1	−1	−1	−1	1
3	−1	−1	1	−1	1	1	1	−1	−1	−1
4	1	−1	−1	1	−1	1	1	1	−1	−1
5	−1	1	−1	−1	1	−1	1	1	1	−1
6	−1	−1	1	−1	−1	1	−1	1	1	1
7	−1	−1	−1	1	−1	−1	1	−1	1	1
8	1	−1	−1	−1	1	−1	−1	1	−1	1
9	1	1	−1	−1	−1	1	−1	−1	1	−1
10	1	1	1	−1	−1	−1	1	−1	−1	1
11	−1	1	1	1	−1	−1	−1	1	−1	−1
12	1	−1	1	1	1	−1	−1	−1	1	−1
13	−1	−1	−1	−1	−1	−1	−1	−1	−1	−1
14	1	−1	1	−1	−1	−1	1	1	1	−1
15	1	1	−1	1	−1	−1	−1	1	1	1
16	−1	1	1	−1	1	−1	−1	−1	1	1
17	1	−1	1	1	−1	1	−1	−1	−1	1
18	1	1	−1	1	1	−1	1	−1	−1	−1
19	1	1	1	−1	1	1	−1	1	−1	−1
20	−1	1	1	1	−1	1	1	−1	1	−1
21	−1	−1	1	1	1	−1	1	1	−1	1
22	−1	−1	−1	1	1	1	−1	1	1	−1
23	1	−1	−1	−1	1	1	1	−1	1	1
24	−1	1	−1	−1	−1	1	1	1	−1	1

however, this time, the signs will be reversed as compared to the original runs. Table 25.8 shows 12 original simulation runs (i.e., 1–12) and 12 folded simulation runs (i.e., 13–24). After running all the experiments, the main effects will be calculated and presented as Pareto and Tornado charts to identify the most important parameters and the magnitude of their impact on the simulation results. Table 25.9 shows the reservoir, completions, and operation parameters and their base, upper, and lower values obtained for sensitivity analysis in this study.

Fig. 25.9 shows the Pareto chart of the standardized effects for $a = 0.05$ or 95% confidence interval. The dashed line shows the corresponding standardized effect for $a = 0.05$. Parameters to the right side of the dashed line, that is, hydraulic fracture permeability, hydraulic fracture half-length, matrix permeability, fracture width, and maximum flow rate constraint, show an important impact on the simulation results while other parameters did

Table 25.9 Sensitivity analysis parameters and their level of changes

Parameter	Lower level (−1)	Base	Upper level (+1)
Matrix porosity, ϕ	0.02	0.08	0.12
Matrix permeability, k (nD)	10	200	1000
Frac. half length, X_f (ft)	100	300	500
Hyd. frac. permeability, k_f (mD)	1600	3000	7000
Equivalent fracture width, w (ft)	0.011	0.015	0.019
Water saturation, S_w	0.12	0.15	0.18
Maximum adsorbed mass, V_L (gmole/lb)	0.111	0.1617	0.197
Langmuir adsorption constant, P_L (1/psi)	0.0001	0.0002	0.0045
Max flow rate (MMCF/day)	5	7	10
Methane diffusion, D	1.0E-5 cm^2/s	5.0E-5 cm^2/s	1.0E-4 cm^2/s

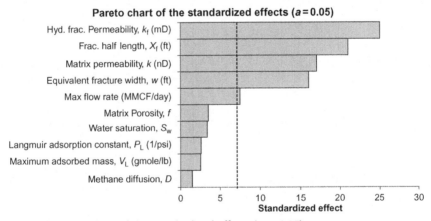

Fig. 25.9 Pareto chart of the standardized effects ($\alpha = 0.05$).

not show statistical significance in 95% confidence interval. This does not mean that these parameters are not important. Fig. 25.10 shows the Tornado chart obtained from PB design sensitivity analysis. As shown in Fig. 25.10, the stimulation parameters including fracture permeability and half-length show the greatest impact in cumulative gas production from MIP-3H, and gas adsorption and diffusion parameters show the lowest impact on cumulative gas production. The results obtained using Pareto chart and Tornado chart agree and can be used for further analysis during the history-matching procedure.

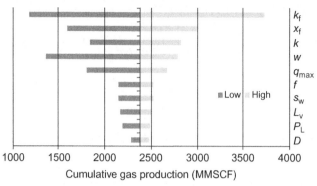

Fig. 25.10 MIP-3H Tornado chart of cum gas production.

To capture the full range of uncertainty and the effect of correlations between different parameters on simulation results, one needs to perform the full factorial design. The full factorial design (FFD) quantifies the impact of all the possible combinations of parameters and their correlations on simulation results. The total of simulation runs required for N number of parameters with L level of changes is equal to L^N.

For the sensitivity analysis, we have taken the following five steps:

1. Determine the parameters of interest and their statistical descriptions (distributions, mean, standard deviation, and upper and lower levels)
2. Perform a linear screening analysis using Plackett-Burman Analysis to find the most important parameters
3. Perform a comprehensive analysis using Full factorial design to quantify the impact of each parameter as well as their interactions on simulation results
4. Generate a Response Surface
5. Perform Monte Carlo simulations on the Response Surface to gauge the uncertainty in reservoir performance

After the most important parameters impacting reservoir simulation results (gas and water rate and cumulative production and pressures) are obtained, these parameters will be carried to CMG CMOST history-matching algorithm for a more comprehensive analysis to obtain the optimized parameters that can lead in reservoir simulation results matching the actual field history.

Before performing the history-matching, it is always a good practice to make sure that there is no error in input data and that all initial and boundary conditions are valid. Next, the three-step history-matching procedure can be followed by history-matching the average reservoir pressure first, gas

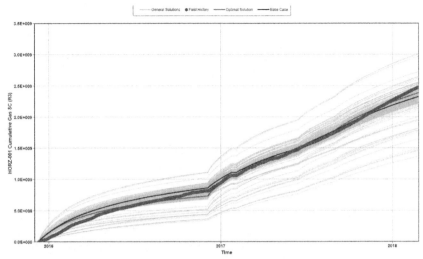

Fig. 25.11 History-matched cumulative gas production of MIP-3H.

and water rates second, and finally bottom-hole flowing pressures. It's a good practice to match the average field parameters first and then adjust the model to match the individual wells.

In this study, we used CMG CMOST optimization algorithm called design **exploration controlled evolution (DECE)** in which continuous or discrete parameters can be used. Fig. 25.11 shows the actual MIP-3H well cumulative gas production in dark gray, base case reservoir simulation results in black, different realization in light gray and the optimum history-matched solution in light black. As shown in Fig. 25.11, a good match is found between field history and optimum reservoir simulation results. Fig. 25.12 shows the actual MIP-3H well daily gas rate in dark gray dots, base case simulation results in black, and optimum history-matched gas rates in black. As shown in Fig. 25.12, a good agreement is found between MIP-3H well daily gas rate and the optimum solution. Finally, Fig. 25.13 shows the MIP-3H well bottom-hole pressure (BHP) in dark gray dots, simulation base case in black, and the optimum history-matched solution in light black. The higher error in gas rate and bottom-hole pressure calculations are due to a lack of sufficient information on well choking schedule, that is, only shut in information was available for MIP-3H well as presented in Table 25.10. Table 25.11 shows the global error of 8.36% with 1.93% error in cumulative gas production, 15.8% error in gas rate calculations and 10.74% error in bottom-hole pressure calculations.

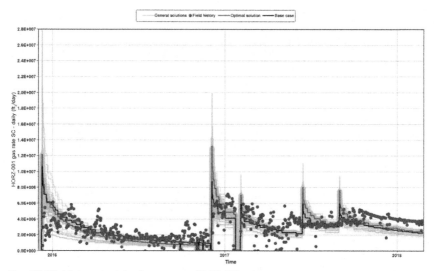

Fig. 25.12 History-matched gas rate of MIP-3H.

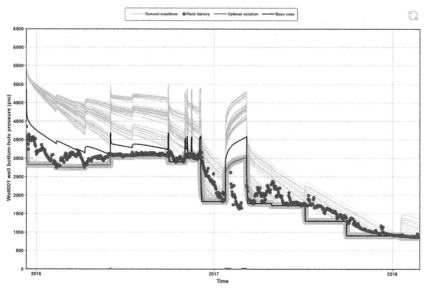

Fig. 25.13 History-matched bottom-hole pressure of MIP-3H.

Table 25.10 MIP-3H shut-In schedules
MIP-3H Shut-in dates

7/18/2016 9/30/2016 11/3/2016 11/16/2016 1/24–31/2017 2/1–3/2017 3/5–7/2017

Table 25.11 History-matching percent errors for MIP-3H

Global Hm error	Cumulative gas production	Gas rate	BHP
8.36%	1.93%	15.8%	10.74%

Table 25.12 Optimum values obtained using history-matching of MIP-3H

Parameter	Base	Optimum value
Matrix porosity, ϕ	0.08	0.04
Matrix permeability, k (nD)	200	400
Frac. half length, X_f (ft)	300	280
Hyd. frac. permeability, k_f (mD)	3000	3206.937
Equivalent fracture width, w (ft)	0.015	0.02095
Water saturation, S_w	0.15	0.10
Langmuir adsorption constant (1/psi)	0.0002	0.00489
Max adsorbed mass (CH_4) (SCF/ton)	269.86	210
Max flow rate (MMCF/day)	7	10
Methane diffusion, D	$5.0\text{E-}5 \text{ cm}^2/\text{s}$	$1.0\text{E-}4 \text{ cm}^2/\text{s}$

The optimum reservoir rock, stimulation and operation parameters obtained from history-matching the cumulative gas production, gas rate, and bottom hole pressure (BHP) is presented in Table 25.12.

After the Marcellus Shale reservoir model is history-matched with actual field production history, we can use the model to forecast the gas production for the next 10 years as requested by the project description.

Phase IV: Well performance analysis

Well performance analysis forecasting the production rates and expected ultimate recovery (EUR) of the well is of special interest in the oil and gas industry. The traditional well performance analysis is based on production rates in which the empirical equations and curve fitting are used to obtain the production forecast and EUR. These techniques are easy to apply and can be used for complex flow behaviors; however, these techniques assume that the operation condition will remain constant during the production and might result in nonunique solutions. Arps in 1945 introduced his empirical decline curve analysis (DCA) for EUR calculations, followed by Fetkovich in 1980 and others. Originally, Arps technique is used for well performance analysis; however, due to uncertainties in forecasting and the assumptions in Arps equation, different rate transient analysis (RTA) techniques were developed to increase the forecasting accuracy by releasing the assumptions of Arps equation and considering more realistic operation conditions. The so-called "modern well performance analysis" uses both production rates and pressures, and instead of empirical equations, it is based on the physics of fluid flow and storage governed by material balance equations. In addition to forecasting the production and estimating the ultimate hydrocarbon recovery, when the pressure and production history of the well/field is available, RTA can also be used to obtain important well/field information, such as permeability and skin and reservoir shape and boundaries. This information can be used to reduce the uncertainties in completions design and enhance hydrocarbon recovery. RTA analysis is very reliable in homogeneous/isotropic reservoir with boundary dominated well flow behavior. However, none of these conditions are valid when dealing with unconventional reservoirs, such as shale gas/oil reservoirs. Shale reservoirs are highly complex and heterogeneous in rock properties, and due to ultra-low permeability of these formations, they show long-term transient flow behavior. Therefore, using RTA for shale reservoirs requires more attention. Usually, a combination of traditional and modern well performance analysis using different techniques will be used and the results obtained will be compared to derive more reliable well performance analysis. For RTA, different type curves and unconventional methods available in IHS harmony commercial software have been used for this study.

Agarwal-Gardner type curve for fractured gas well

The first type curve analysis performed in this study is the Agarwal-Gardner type curve developed in 1999 (Agarwal et al., 1999). This type curve can be used to obtain the gas in place and reservoir parameters, such as permeability, fracture half-length, and skin factor in horizontal and naturally fractured reservoirs. The main application of this type curve is for radial flow regime. To obtain the gas in place using this technique, reservoir data, fluid properties, and operation conditions are required. In this technique, the normalized rate ($\hat{q}$), Eq. (25.11) on y-axis is plotted against material balanced pseudo-time (t_{mp}), Eq. (25.13), on x-axis in a log-log plot. The best type curve that matches the data while the axes of the two plots are kept parallel can be used to calculate the permeability, reservoir radius, skin, fracture half-length, and original gas in place. From the matched type curve, $\left(\frac{r_e}{x_f}\right)$, that is, reservoir radius over fracture half-length, and from the matching point, $(\hat{q}t_{mp})$ of the data and (q_D, t_D) of the type curve will be noted for further calculations.

$$\text{Normalized rate}: \hat{q} = \frac{q}{P_{pi} - P_{pwf}} \quad (25.11)$$

where q is the gas rate and P_{pi} and P_{pwf} are pseudo-initial pressure and pseudo-flowing BHPs. The pseudo-pressure is defined as follows:

$$\text{Pseudo pressure}: P_p(p) = 2\int_0^p \frac{p}{\mu z} dp \quad (25.12)$$

where μ is gas viscosity and z is a gas compressibility factor.

$$\text{Material balance pseudo time}: t_{mp} = \frac{\mu c_t}{q(t)} \int_0^t \frac{q(t)}{\mu(\bar{p})c_t(\bar{p})} dt \quad (25.13)$$

where c_t is total gas compressibility and $\bar{p}$ is the average pressure. Fig. 25.14 shows the Agarwal-Gardner type curve analysis of the MIP-3H well. The hydraulic fracture half-length is found to be 212 ft with a r_e/x_f value of 5.0.

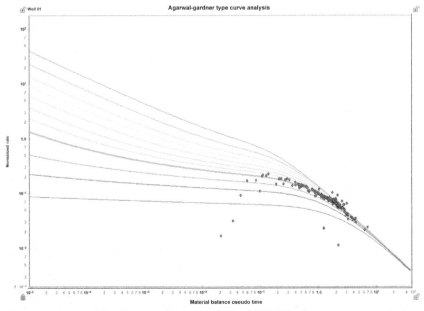

Fig. 25.14 Agarwal-Gardner type curve analysis of MIP-3H.

Blasingame type curve

The next type curve used in this study is the Blasingame type curve developed in 1994 (Doublet et al., 1994). This technique is suitable for open-hole horizontal wells and can provide the formation permeability, original hydrocarbon in place, and reservoir drainage area. Blasingame introduced a new integral function for dimensionless variables that can be used for different flow regimes, such as linear, bilinear, and radial flow. Blasingame decline curve is based on constant rate solutions in which only harmonic decline is plotted. Like the Agrawal-Gardner type curve, data plotted in Blasingame type curve is normalized rate vs material balance pseudo-time for gas wells. The information obtained from fitting the Blasingame type curve to MIP-3H well was consistent with Agrawal-Gardner type curve analysis. Fig. 25.15 shows the match obtained using Blasingame type curve.

Wattenbarger type curve

The Wattenbarger type curve analysis is more suitable for extended linear flow, so it can be used for shale reservoirs with ultra-low permeability (Kanfar and Wattenbarger, 2012). Like Agrawal-Gardner type curve matching the transient part of the Wattenbarger type curve provides

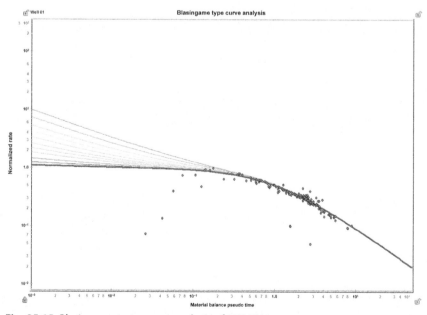

Fig. 25.15 Blasingame type curve analysis of MIP-3H.

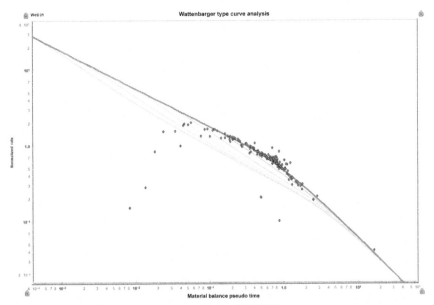

Fig. 25.16 Wattenbarger type curve analysis of MIP-3H.

information regarding the fracture half-length and reservoir boundaries. Wattenbarger type curve assumes hydraulic fracturing in a center of a rectangular reservoir where the initial transient flow is perpendicular to the hydraulic fracture. The Wattenbarger type curve also plots the normalized rate vs material balance pseudo-time, and from the match, the reservoir area and original gas in place can be calculated as shown in Fig. 25.16.

Transient type curves

The transient type curves are also used for analyzing the early time or transient data (Fig. 25.17). This technique is commonly used for cases with long transient time. Like other type curves discussed earlier, they also use normalized rate vs material balance pseudo-time for gas wells. The transient match is used to estimate the permeability and skin, OGIP, and EUR. EUR is dependent on the reservoir size; therefore, it's not recommended to use the transient part for these calculations.

These methodologies are primarily used for flow regime diagnostics and are not typically used for EUR and OGIP predictions. Therefore, caution must be used when determining EUR from these techniques.

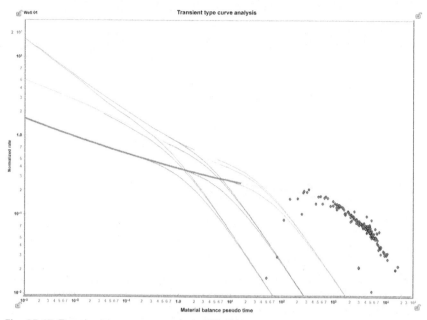

Fig. 25.17 Transient type curve analysis MIP-3H.

Variable flowing pressure

IHS harmony unconventional reservoir analysis and rate transient methods have also been used for this study. First, we have used the variable flowing pressure technique in which the normalized pressure $\hat{P}$, that is, Eq. (25.14), is plotted against the square root of time. $A\sqrt{K}$ is a popular parameter used for well-to-well comparisons since it is a signature for unconventional reservoirs. Please note that $A\sqrt{K}$ must be normalized for lateral length of wells before being used for well-to-well comparison. $A\sqrt{K}/\text{ft}$ can also be used as an output parameter when building a ML model as it signifies the flow capacity of each well. Many operators use $A\sqrt{K}/\text{ft}$ as an output for their ML analysis due to the correlation between $A\sqrt{K}/\text{ft}$ and EUR/ft as illustrated by Belyadi et al. (2015).

$$\text{Normalized pressure}: \hat{P} = \frac{\Delta P_p}{q} \qquad (25.14)$$

The slope of the straight-line "m" passing through the data can be used to obtain the $A\sqrt{K}$ as follows:

$$\text{Flow capacity}: A\sqrt{K} = \frac{1262}{\sqrt{\varphi\mu C_t}} \frac{T}{m} \qquad (25.15)$$

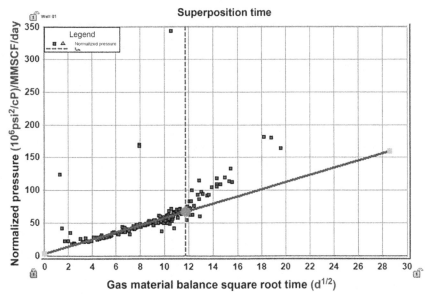

Fig. 25.18 Superposition time plot for MIP-3H.

where T is the reservoir temperature, φ is the porosity, μ is the gas viscosity, and C_t is total compressibility. The flow capacity, $A\sqrt{K}$, is defined as contacted surface area times the square root of permeability. The straight-line behavior indicates the presence of linear flow. Fig. 25.18 shows the superposition time plot for MIP-3H well and the straight line passed through data. The slope of the straight line leads to $A\sqrt{K}$ value equal to 101,850 mD$^{1/2}$ ft^2. Please note that this value must be normalized to lateral length before using in any type of comparison analysis. Fig. 25.18 cannot be used for original gas in place (OGIP) calculations since the data does not show the end of linear flow, and therefore, the boundary-dominated flow has not been reached. However, the minimum OGIP can be estimated using Eq. (25.16).

$$\text{OGIP using square-root of time plot}: \text{OGIP} = \frac{200.8 T S_{gi}}{(\mu C_t B_g)_i} \left(\frac{\sqrt{t_{elf}}}{m} \right) \quad (25.16)$$

Fig. 25.19 shows the flowing material balance plot for MIP-3H well where the gas normalized rate is plotted against the normalized gas cumulative production. The straight-line fit extension indicates 5552 MMCF OGIP. As previously mentioned, since the end of transient flow has not been observed

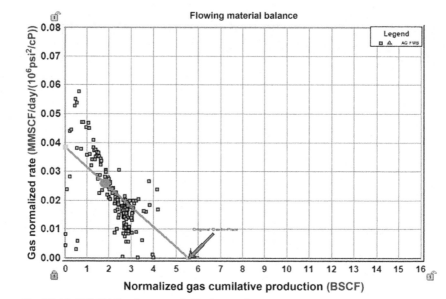

Fig. 25.19 MIP-3H flowing material balance plot.

yet, this OGIP is the minimum OGIP from this analysis. Once the end of transient flow is observed, a more accurate estimate of OGIP can be determined from this analysis.

Next, the type curve is used in which the normalized gas rate is plotted against gas material balance time, that is, the ratio of cumulative gas production to instantaneous rate, on a log-log plot to identify the flow regimes. In Fig. 25.20, the ½ slope line in dark gray corresponds to linear flow and the unit slope line in gray indicates the boundary dominated flow. MIP-3H well data is falling on a ½ slope line in dark gray which indicates the presence of linear flow.

The well performance analysis of the MIP-3H well was performed using the available decline curve, material balance, type curve, and unconventional methods available within the IHS Harmony software and the results are summarized in Table 25.13.

The basic parameters obtained from superposition time analysis can then be used as a starting point when performing history-matching using various analytical and hybrid models within the HIS harmony package. Once a desirable HM is obtained, a forecast can be run, and a reliable EUR can be predicted for each well. The most important parameter that is used from superposition time analysis is the normalized $A\sqrt{k}$ for well to well comparison

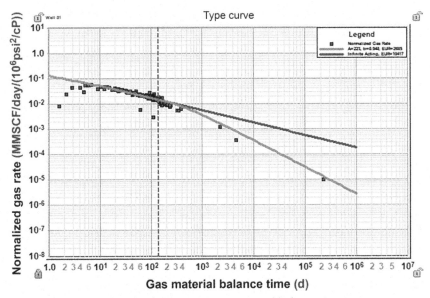

Fig. 25.20 MIP-3H normalized gas rate vs gas material balance time.

Table 25.13 Estimated parameters obtained using well performance analysis

Parameter	Estimated value from RTA
$A\sqrt{k}$ (md$^{1/2}$ ft^2)	101,850
Permeability (md)	1.24E-4 (16.566E-4 through analysis)
Fracture half length (ft)	212

and the remaining parameters, such as fracture half-length, effective permeability, etc. can be used as starting parameters when performing HM using analytical or hybrid models within the HIS harmony package.

Liquid loading

Liquid loading is an accumulation of water, gas condensate or both in the tubing that can impair gas production and, if not diagnosed in a timely manner, can kill the well. The major cause of liquid loading is low gas flow rate or gas velocity. If gas velocity drops below the critical velocity required to carry liquid to the surface, the liquid starts accumulating in the down-hole of a vertical well, lateral section of the horizontal well and even in the hydraulic fractures. Different models, such as droplet, film or transient multiphase flow models are used in unconventional gas reservoirs to predict the liquid loading. Turner et al. (1969) and Coleman et al. (1991) terminal velocity equations are the most common models used in the oil and gas industry to predict the liquid loading. They introduced a minimum gas velocity required to prevent liquid loading using the droplet movement model (see chapter production analysis and wellhead design). Another indication of the liquid loading is the high casing over tubing pressure. Due to liquid accumulation in the down-hole, the flowing bottom-hole pressure goes up leading to high casing pressure. Fig. 25.21 shows the MIP-3H well gas rate plotted with Coleman and Turner rates. It shows that the gas rate (in dark gray) has been greater than both the Coleman and Turner rates, indicating that liquid loading has not happened in this well. The gray line shows the water production from MIP-3H well. Fig. 25.8 presented earlier also shows the casing and tubing pressures of the MIP-3H, indicates that liquid loading has not occurred in MIP-3H well. If liquid loading occurs, the well performance analysis needs to be done with more attention. This is because the calculated flowing bottom-hole pressure will be inaccurate when the well is loaded when a bottom hole gauge is not used.

Diagnostic fracture injection test (DFIT)

Before performing the hydraulic fracturing stimulation optimization using FracPro, the diagnostic fracture injection test performed in MIP-3H well reveals the instantaneous shut-in pressure (ISIP), fracture gradient, net extension pressure, fluid leak-off mechanism, time to closure, closure pressure (minimum horizontal stress), approximation of maximum horizontal stress, anisotropy, fluid efficiency, effective permeability, transmissibility, and pore pressure (for more details on the DFIT analysis please refer to Chapter 14).

As discussed in the DFIT chapter, G-function plot, square root plot, and log-log plot can be used together or separately to identify closure pressure. For this analysis, it is recommended to use all three plots in conjunction with one another to determine closure pressure. As shown in Fig. 25.22 (G-function plot), closure pressure of 7402 psi is obtained, which is the point at which the second derivative curve starts deviating from the extrapolated line going through the origin (highlighted on the graph). In addition, PDL signature is observed from this plot which is the concave down shape above the extrapolated line going through the origin. This indicates the presence of natural fractures and possibly the use of smaller sand sizes, such as 100 mesh to plug off natural fractures and micro-fractures during hydraulic fracture treatments.

Marcellus shale reservoir development and stimulation

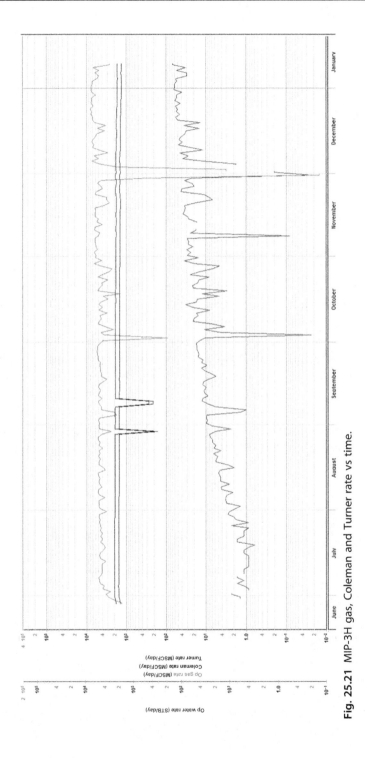

Fig. 25.21 MIP-3H gas, Coleman and Turner rate vs time.

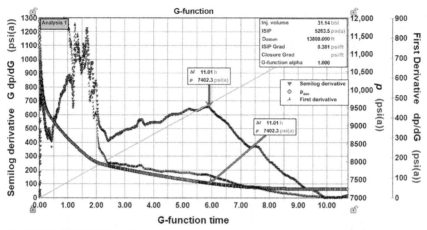

Fig. 25.22 MIP-3H G-function plot.

Another form of before closure analysis (BCA) is called square root analysis in which BHP is plotted against the square root of time. This graph can also be used to find closure pressure. In this plot, pressure, first derivative, and second derivative are plotted. To identify fracture closure, a linear extrapolated line from the origin is drawn on the second derivative curve. Fracture closure can be approximated when the second derivative curve deviates from the linear line. Fig. 25.23 shows the MIP-3H well square root of time plot where the second derivative curve deviates from a straight line and fracture closure pressure is approximated to be **7402 psi** in line with the G-function analysis. Fig. 25.24 shows the MIP-3H well log-log plot. This

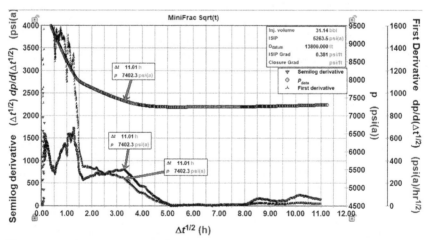

Fig. 25.23 MIP-3H square root of time plot.

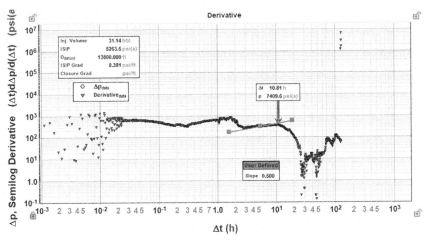

Fig. 25.24 MIP-3H log-log plot.

plot sufficiently identifies closure and various flow regimes before and after closure. Various flow regimes on the second derivative of the log-log plot can be determined. In Fig. 25.24, half-slope line (1/2 slope) before closure and negative half-slope line after closure shows the linear flow regime. In this figure, closure pressure is determined to be approximately ~7400 psi (in line with G-function and square root analyses). The closure pressure is the point where the data starts deviating from positive ½ slope line as shown on the plot. The negative unit slope corresponding to pseudo-radial flow has not been observed in the log-log plot. Therefore, the Horner plot cannot be used to obtain the reservoir pore pressure and approximation of reservoir pressure and transmissibility becomes challenging. Due to the quality of the data on this DFIT, after closure analysis (ACA), such as ACA (Nolte, 1997) and (Soliman et al., 2005) were not performed.

Phase V: Frac job sand schedule optimization

Slick water frac job is selected for Marcellus Shale reservoir. The actual frac job schedule for MIP-3H can be obtained at http://mseel.org/Data/Wells_Datasets/MIP_3H/Completions/. Here, the idea is to optimize the frac job sand schedule to maximize the NPV of the project. For this purpose, first, different sand schedules including 1000, 1500, 2000, 2500, 3000, 3500, and 4000 lb/ft using 40% 100 mesh, and 60% 40/70 mesh sand size are designed by keeping the sand concentrations the same across all sand stages (e.g., 0.25, 0.5, 0.75 ppg, etc.) and changing the amount of clean water volume for each sand stage. Next, each sand schedule is used in commercial fracture software, such as FracPro to obtain the fracture geometry, that is, propped half-length, propped width, and propped conductivity. Finally, the fracture geometry of each design is imported in history-matched Marcellus Shale reservoir model built using CMG GEM to obtain production rate vs time for each scenario that will be used for economic analysis to obtain the optimum frac job sand schedule.

Hydraulic fracture treatment design

FracPro commercial software has been used for our hydraulic fracture treatment design. The plug-and-perf completion technique for multiple fracture treatments is used as a common practice in unconventional shale plays. Four stages including acidizing, pad, proppant, and flush are completed for each fracture treatment design. In acidizing stage, HCL (hydrochloric acid) or HF (hydrofluoric acid) is pumped downhole to clean the perforations, in this case, we have used 15% HCL acid with 85 barrel per minute pump rate. Next, pad, which is a combination of only water and some chemicals, is pumped to initiate the hydraulic fracture and generate the required fracture length, width, and height. Most of the hydraulic fracture is generated through the pad injection, so it's extremely important to inject the designed pad volume during this stage. After pumping the calculated pad volume, the proppant stage can be started. The proppant stage is the stage during which combinations of proppant, water, and chemicals (called slurry) are pumped downhole. In a slick water frac, it is imperative to establish enough flow rate before starting the proppant stage. After pumping the designed pump schedule, proppant is cut and the well is flushed. Flushing means water and chemicals are only pumped downhole to clear the inside of the production casing of sand until all the remaining proppant in the casing has been removed/flushed to the formation. Flush volume can be calculated given the casing size, grade, weight, and bottom perforation. In this study, the concentrations and pump rates are kept constant and absolute volume factor (AVF), slurry volume, total clean volume, stage proppant, total proppant, water per foot, sand water ratio, and time is calculated for each sand schedule design. For more details and sample calculations, please refer to Chapter 10 of this book that is, Fracture Treatment Design. The amount of clean water volume for each sand stage is changed using excel solver to make sure the desired sand per foot with 40%, 100 mesh and 60%, 40/70 mesh is achieved. Table 25.14 shows the slick water schedule for 2000 lb/ft of sand and 40% 100 Mesh and 60% 40/70 obtained using excel solver.

Table 25.14 Slick water schedule for 2000 lb/ft of sand and 40% 100 mesh and 60% 40/70

Stage name	Pump rate bpm	Fluid name	Stage fluid clean vol. BBLs	Stage fluid slurry vol. BBLs	% of total clean vol. %	Prop. conc. ppg	Stage proppant lbs	% of total prop. %	Cumulative prop. lbs	Stage time min
Pump ball	15	Slick water	300	300	3.40%	0	0	0.0%	0	20.00
5% HCL acid	85	Acid	60	60	0.68%	0	0	0.0%	0	0.71
Pad	85	Slick water	350	350	3.97%	0	0	0.0%	0	4.12
100 mesh	85	Slick water	400	405	4.54%	0.25	4200	0.9%	4200	4.76
100 mesh	85	Slick water	400	409	4.54%	0.5	8400	1.9%	12,600	4.81
100 mesh	85	Slick water	400	414	4.54%	0.75	12,600	2.8%	25,200	4.87
100 mesh	85	Slick water	410	429	4.65%	1	17,220	3.9%	42,420	5.04
100 mesh	85	Slick water	500	528	5.67%	1.25	26,250	5.9%	68,670	6.22
100 mesh	85	Slick water	500	534	5.67%	1.5	31,500	7.1%	100,170	6.28
100 mesh	85	Slick water	500	540	5.67%	1.75	36,750	8.3%	136,920	6.35
100 mesh	85	Slick water	500	545	5.67%	2	42,000	9.5%	178,920	6.42
40/70 mesh	85	Slick water	450	460	5.10%	0.5	9450	2.1%	188,370	5.41
40/70 mesh	85	Slick water	450	465	5.10%	0.75	14,175	3.2%	202,545	5.47
40/70 mesh	85	Slick water	450	470	5.10%	1	18,900	4.3%	221,445	5.53
40/70 mesh	85	Slick water	450	475	5.10%	1.25	23,625	5.3%	245,070	5.59
40/70 mesh	85	Slick water	450	481	5.10%	1.5	28,350	6.4%	273,420	5.65
40/70 mesh	85	Slick water	450	486	5.10%	1.75	33,075	7.4%	306,495	5.71
40/70 mesh	85	Slick water	450	491	5.10%	2	37,800	8.5%	344,295	5.77
40/70 mesh	85	Slick water	500	551	5.67%	2.25	47,250	10.6%	391,545	6.48
40/70 mesh	85	Slick water	500	557	5.67%	2.5	52,500	11.8%	444,045	6.55
Flush	85	Slick water	350	350	3.97%	0	0	0.0%	444,045	4.12

Continued

Table 25.14 Slick water schedule for 2000 lb/ft of sand and 40% 100 mesh and 60% 40/70—cont'd

Total clean volume	8820	BBLs
Stage length	222	ft
Sand/ft	2000	lb/ft
Water/ft	40	BBLs/ft
SWR	1.20	lb/gal
Total slurry volume excluding acid and ball	8939	BBLs
Pad%	3.92%	
100 mesh	178920	lbs	40%
40/70 mesh	265125	lbs	60%

After requesting the slick water schedule for all sand/ft cases for sensitivity analysis, the wellbore configuration is imported in FracPro using deviation survey, casing, and tubing information, available at http://mseel.org/Data/Wells_Datasets/MIP_3H/Completions/. The layer properties of all five layers including Hamilton, Upper Marcellus, Cherry Valley, Lower Marcellus, and Onondaga formations are imported to FracPro from our previous well log analysis. The values of Young's modulus and Poisson's ratios for each layer are obtained from geomechanical log while the pore pressure gradient is assumed to be 0.64 psi/ft. For fracture toughness and fracture gradient typical values of 2050 $psi.in^{0.5}$ and 1.16 psi/ft are assumed and later iterated on to history match the pressure response of the performed DFIT and main hydraulic fracture treatments. Before performing the simulation, it is essential to history match the net pressure obtained during the Mini-Frac job or actual hydraulic fracturing of the stages. For this purpose, we used the measured parameters and adjusted the unknown parameters and model setting to match the measured pressure data. For initial values, the closure pressure of 7402 psi (obtained from DFIT) and near wellbore friction pressure of 700 psi is used to history match the pressure response. The leak-off coefficient has been changed iteratively until the history-matched pressure profile matches the pressure readings as presented in Fig. 25.25. It's imperative to match the pressure decline following the pumping.

Table 25.15 shows the summary of fracture dimensions and properties obtained for different sand schedules using FracPro. These values are imported in our Marcellus Shale history-matched model to adjust the hydraulic fracture properties. The cumulative gas production (~2 years and 2 months) corresponding to each hydraulic fracture scenario is obtained and presented in Table 25.16.

Cumulative gas production obtained using history-matched Marcellus Shale gas model for different hydraulic fracture properties obtained from different sand schedule scenarios indicates that the higher the sand per foot injected during the stimulation, the higher the cumulative gas production. However, to obtain the optimum sand schedule, we need to run the economic analysis using information provided for this project.

Economics analysis to obtain the optimum sand schedule

BTAX and ATAX economic analysis following the step-by-step workflow in Chapter 18 is performed for various sand schedules of 1000, 1500, 2000,

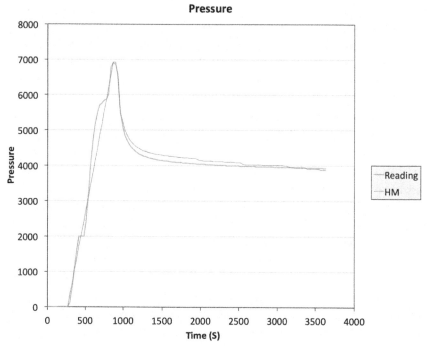

Fig. 25.25 Pressure history match of Mini-frac test.

Table 25.15 Summary of hydraulic fracture dimensions and properties obtained for different sand/ft schedules using FracPro

Sand/ft (lb/ft)	Fracture half-length (ft)	Propped half-length (ft)	Average fracture width (in)	Fracture permeability (md)	Dimensionless conductivity
1000	321	196	0.16	3390	27,452
1500	352	209	0.20	4071	38,460
2000	362	218	0.22	4455	44,900
2500	407	261	0.27	5247	53,689
3000	471	279	0.27	5592	56,144
3500	476	283	0.28	5660	58,119
4000	481	285	0.28	5726	59,480

2500, 3000, 3500, and 4000 lb/ft, to obtain an optimum sand schedule based on the highest NPV of all sand designs. The economic analysis is performed using different total CAPEX presented in Table 25.2. The net gas production per-month is used as a basis of our calculation. For gas pricing,

Table 25.16 Cumulative gas production obtained for different sand schedules

Sand/ft	Cumulative gas production (MMSCF)
1000	2453
1500	2582
2000	2717
2500	3216
3000	3893
3500	4195
4000	4254

first, the constant base values of $2.5, $3, $3.5, and $4 per MMBTU with a 3% annual increase applied on a monthly stair-stepped wise. Next, the NYMEX, that is, estimated monthly gas price for two years based on Henry Hub pricing, is performed. After 2 years of NYMEX predications, the same 3% annual increase on a monthly stair-stepped wise is applied. Table 25.17 presents the NPV calculated for different sand schedules based on different gas pricing. In all the gas pricing scenarios, the 3000 lb/ft sand schedule is found to be the optimum design based on the assumptions used in this study and NPV values except for high gas pricing of $3.5 and $4.0/MMBTU in which 4000 lb/ft of sand appeared to be the optimum case. Fig. 25.26 shows the optimum sand schedule based on NPV values for different pricing scenarios.

Table 25.17 NPV calculated for each sand schedule and pricing

Sand lb/ft	NPV ($1M)			
	$2.5/MMBTU	$3.0/MMBTU	$3.5/MMBTU	$4.0/MMBTU
1000	2.04	4.77	7.14	9.50
1500	**2.05**	4.39	6.73	9.07
2000	2.01	4.65	7.29	9.93
2500	1.98	4.94	7.90	10.86
3000	2.02	**5.24**	8.46	11.69
3500	1.83	5.20	8.56	11.92
4000	1.61	5.12	**8.62**	**12.12**

Bold values iterate on the optimum economic case.

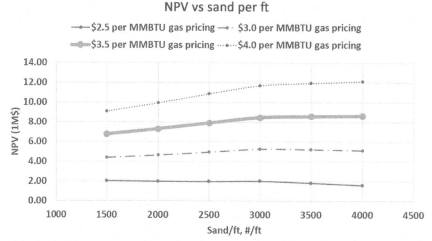

Fig. 25.26 Calculated NPV for each sand schedule at different gas pricing.

Phase VI: Modified hyperbolic decline curve

In Hyperbolic decline the decline rate "D" is a function of production rate and can be defined as follows:

$$D = K q^b$$

where b is a constant and K is defined as follows:

$$K = \frac{D_i}{q_i^b}$$

Here D_i and q_i are initial rates and decline rates. The application of hyperbolic decline in unconventional reservoirs tends to overestimate the cumulative production. Therefore, the transition from hyperbolic to exponential decline curve at some terminal decline rate is assumed in unconventional reservoirs. For this example, terminal decline rate of 5% is assumed. That is when annual effective decline rate "D_e" is reached to 5% the hyperbolic decline will be a switch to exponential decline (for more details please refer to Chapter 17 of this book). The modified hyperbolic decline is fitted through the optimum sand schedule production volume vs time that was obtained numerically to obtain the production rate vs time for 50 years. For this purpose, commercial software IHS harmony is used and monthly production forecast is used for the economic analysis.

Table 25.18 10 years and 50 years NPV comparison of the optimum sand schedule

Sand lb/ft	NPV ($1M)			
	$2.5/MMBTU	$3.0/MMBTU	$3.5/MMBTU	$4.0/MMBTU
3000 (10 years)	2.02	5.24	8.46	11.69
3000 (50 years)	3.13	7.33	11.34	15.3

Table 25.18 summarizes the comparison of the NPV obtained for 10 and 50 years of production with optimum sand schedule of 3000 lb/ft.

Conclusion

Using the actual MIP-3H well data including reservoir, completion, stimulation, and operation, a comprehensive study has been performed and a history-matched shale gas reservoir model developed that can resemble the actual well performance in terms of the cumulative gas production, gas rate, and bottom hole pressure. The well performance analysis studies have been performed and the results are compared with optimum parameters obtained through history-matching of Marcellus Shale reservoir model. The history-matched model then used to forecast the cumulative gas production of different hydraulic fracture properties obtained from different sand schedule scenarios using the commercial software FracPro. The economic analysis has been run for different cases and the optimum sand schedule design is obtained to be 3000 lb/ft. Next, the modified hyperbolic decline curve analysis is performed, and the 50-year production forecast obtained for optimum sand schedule design of 3000 lb/ft. Finally, economic analysis is performed and the ATAX NPV of the project is obtained and compared with 10 years forecast using the history-matched model.

References

Abe, H., Mura, T., Keer, L., 1976. Growth-rate of a penny-shaped crack in hydraulic fracturing of rocks. J. Geophys. Res. 81 (29), 5335–5340.

Adachi, J., Detournay, E., 2008. Plane strain propagation of a hydraulic fracture in a permeable rock. Eng. Fract. Mech. 75, 4666–4694.

Adesida, A., Akkutlu, I.Y., Resasco, D.E., Rai, C.S., 2011. Characterization of Barnett Shale Pore Size Distribution using DFT Analysis and Monte Carlo Simulations. SPE 147397.

Agarwal, R.G., Gardner, D.C., Kleinsteiber, S.W., Fussell, D.D., 1999. Analyzing Well Production Data Using Combined-Type-Curve and Decline-Curve Analysis Concepts. Society of Petroleum Engineers.

Akkutlu, I.Y., Fathi, E., 2012. Multiscale gas transport in shales with local Kerogen heterogeneities. SPE J. 17(4).

Ambrose, R.J., Hartman, R.C., Diaz-Campos, M., Akkutlu, I.Y., Sondergeld, C.H., 2010. New pore-scale considerations in shale gas in-place calculations. SPE-131772.

Ambrose, R.J., Hartman, R.C., Diaz-Campos, M., Akkutlu, I.Y., 2012. Shale gas in-place calculations. Part I—new pore-scale considerations. SPE J. 17 (1), 219–229.

Anderson, M.A., 1988. Predicting reservoir condition, pv compressibility from hydrostatic stress laboratory data. SPE J. 3, 1078–1082.

Anderson, J.R., Pratt, K.C., 1985. Introduction to Characterization and Testing of Catalysts. Academic Press, Sydney.

Aria, A., Gharib, M., 2011. Carbon nanotube arrays with tunable wettability and their applications. NSTI-Nanotech. 1.

Arnold, R., Anderson, R., 1908. Preliminary report on Coalinga oil district. U.S. Geol. Survey Bull. 357, 79.

Arps, J.J., 1944. Analysis of Decline Curves. Petroleum Engineering. A.I.M.E. Houston Meeting.

Aziz, K., Settari, A., 1979. Petroleum Reservoir Simulation. Applied Science Publishers Ltd., London.

Bailey, S., 2009. Closure and Compressibility Corrections to Capillary Pressure Data in Shales. Colorado School of Mines. October 19.

Bao, J.Q., Fathi, E., Ameri, S., 2014. A coupled method for the numerical simulation of hydraulic fracturing with a condensation technique. Eng. Fract. Mech. 131, 269–281.

Bao, J.Q., Fathi, E., Ameri, S., 2015. Uniform investigation of hydraulic fracturing propagation regimes in the plane strain model. Int. J. Numer. Anal. Methods Geomech. 39, 507–523.

Bao, J.Q., Fathi, E., Ameri, S., 2016. A unified finite element method for the simulation of hydraulic fracturing with and without fluid lag. Eng. Fract. Mech. 162, 164–178.

Barree, B., 2013. Overview of Current DFIT Analysis Methodology.

Barree, R.D., Baree, V.L., Craig, D., 2007. Holistic fracture diagnostics. In: Rocky Mountain Oil & Gas Technology Symposium, April 16–18, Denver, Colorado. SPE-107877.

Belyadi, F., 2014. Impact of gas desorption on production behavior of shale gas. PhD dissertation submitted to West Virginia University.

Belyadi, H., Smith, M., 2018. A Fast-Paced Workflow for Well Spacing and Completions Design Optimization in Unconventional Reservoirs. In: SPE Eastern Regional Meeting. SPE-191779.

Belyadi, H., Yuyi, S., Junca-Laplace, J., 2015. Production analysis using rate transient analysis. In: SPE Eastern Regional Meeting, 13-15 October, Morgantown, West Virginia, USA. SPE-177293.

Belyadi, H., Fathi, E., Belyadi, F., 2016a. Managed Pressure Drawdown in Utica/Point Pleasant with Case Studies. SPE Eastern Regional Meeting. SPE-184054.

Belyadi, H., Yuyi, J., Ahmad, M., Wyatt, J., 2016b. Deep Dry Utica Well Spacing Analysis with Case Study. SPE Eastern Regional Meeting. SPE-184045.

Besler, M., Steele, J., Egan, T., Wagner, J., 2007. Improving well productivity and profitability in the Bakken—a summary of our experiences drilling, stimulating, and operating horizontal wells. In: SPE Annual Technical Conference and Exhibition held in Anaheim. SPE-110679.

Binder, R.C., 1973. Fluid Mechanics. Prentice-Hall Inc, Englewood Cliffs, NJ.

Blanton, T.L., 1982. An experimental study of interaction between hydraulically induce and pre-existing fractures. In: SPE 10847. SPE/DOE Unconventional Gas Recovery Symposium, Pittsburgh, PA.

Brace, W.W., 1968. Permeability of granite under high pressure. J. Geophys. Res. 73, 2225.

Britt, L.K., 2011. Keys to Successful Multi-Fractured Horizontal Wells in Tight and Unconventional Reservoirs. NSI Fracturing & Britt Rock Mechanics Laboratory presentation.

Bui, K., Akkutlu, I.Y., 2015. Nanopore wall effect on surface tension of methane. J. Mol. Phys.

Bunger, A.P., Detournay, E., Garagash, D.I., 2005. Toughness-dominated hydraulic fracture with leak-off. Int. J. Fract. 134, 175–190.

Cheng, Y., 2010. Impacts of the number of perforation clusters and cluster spacing on production performance of horizontal shale gas wells. 138843-MS SPE Conference Paper.

Cheng, Y., 2012. Mechanical interaction of multiple fractures-exploring impacts of the selection of the spacing number of perforation clusters on horizontal shale-gas wells. SPE J. 17 (4), 992–1001.

Cinco-Ley, H., Samaniego, V.F., 1981. Transient pressure analysis for fractured wells. J. Petrol. Technol. 33(9).

Coleman, S.B., Clay, H.B., McCurdy, D.G., Norris III, L.H., 1991. A new look at predicting gas-well load-up. J. Pet. Technol. 43 (3), 329–333. Trans., AIME, 291. SPE-20280-PA, https://doi.org/10.2118/20280-PA.

Cramer, D.D., 1987. The application of limited-entry techniques in massive hydraulic fracturing treatments. In: SPE Production Operations Symposium, March 8–10, Oklahoma City, OK.

Cronquist, C., 2001. Estimation and Classification of Reserves of Crude Oil, Natural Gas and Condensates. SPE Book Series, Houston, TX, pp. 157–160.

Culter, W.W., 1924. Estimation of underground oil reserves by well production curves. U.S. Bur. Mines Bull. 228.

Culter, W.W., Johnson, H.R., 1940. Estimating Recoverable Oil of Curtailed Wells. Oil Weekly.

Curtis, M.E., Ambrose, R.J., Sondergeld, C.H., Rai, C.S., 2010. Structural characterization of gas shales on the micro- and nano-scales. In: CUSG/SPE 137693.

Curtis, M.E., Sondergeld, C.H., Rai, C.S., 2013. Investigation of the microstructure of shales in the oil window. In: Unconventional Resources Technology Conference, August 12–14, Denver, CO.

Dahi, A., Olson, J., 2011. Numerical modeling of multistranded-hydraulic-fracture propagation: accounting for the interaction between induced and natural fractures. SPE J. 16(3).

Daneshy, A.A., 1974. Hydraulic fracture propagation in the presence of planes of weakness. In: SPE 4852, SPE-European Spring Meeting, Amsterdam.

Dontsov, E.V., Peirce, A.P., 2015. An enhanced pseudo-3D model for hydraulic fracturing accounting for viscous height growth, non-local elasticity, and lateral toughness. Eng. Fract. Mech. 42, 116–139.

Doublet, L.E., Pande, P.K., McCollum, T.J., Blasingame, T.A., 1994. Decline curve analysis using type curves—analysis of oil well production data using material balance time:

application to field cases. In: Paper SPE 28688 presented at the 1994 Petroleum Conference and Exhibition of Mexico held in Veracruz, MEXICO, 10–13 October.

Drake, L.C., Ritter, H.L., 1945. Macropore-size distributions in some typical porous substances. Ind. Eng. Chem. Anal. Ed. 17 (12), 787–791.

Dubinin, M.M., 1960. The potential theory of adsorption of gases and vapors for adsorbents with energetically nonuniform surfaces. Chem. Rev. 60 (2), 235–241.

Dubinin, M.M., 1966. Chemistry and Physics of Carbon. Marcel Dekker, New York.

Duong, A.N., 2011. Rate-decline analysis for fracture-dominated shale reservoirs. SPE Reserv. Eval. Eng. 14(3).

Economides, M., Martin, T., 2007. Modern Fracturing. ET Publishing, Houston, TX.

Economides, M.J., Hill, A.D., Whlig-Economides, C., 1994. Petroleum Production Systems. Prentice Hall PTR, New Jersey, pp. 74–75.

Ellsworth, W.L., 2013. Injection-induced earthquakes. Science 341(6142).

Ely, J., 2012. Proppant Agents. Waynesburg, PA.

Fathi, E., Akkutlu, I.Y., 2009. Nonlinear sorption kinetics and surface diffusion effects on gas transport in low permeability formations. In: SPE Annual Technical Conference held in New Orleans, LA, October 4–7.

Fathi, E., Akkutlu, I.Y., 2011. Gas transport in shales with local Kerogen heterogeneities. In: SPE Annual Technical Conference and Exhibition (ATCE) 2011 in Denver, CO. SPE-146422-PP.

Fathi, E., Akkutlu, I.Y., 2013. Lattice Boltzmann method for simulation of shale gas transport in Kerogen. SPE J. 18(1).

Fathi, E., Akkutlu, I.Y., 2014. Multi-component gas transport and adsorption effects during CO injection and enhanced shale gas recovery. Int. J. Coal Geol. 123, 52–61.

Fathi, E., Tinni, A., Akkutlu, I.Y., 2012. Correction to Klinkenberg slip theory for gas dynamics in nano-capillaries. Int. J. Coal Geol. 103, 51–59.

Feng, F., Akkutlu, I.Y., 2015. Flow of hydrocarbons in nanocapillary: a non-equilibrium molecular dynamics study. In: SPE Asia Pacific Unconventional Resources Conference and Exhibition, November 9–11, Brisbane, Australia.

Finsterle, S., Persoff, P., 1997. Determining permeability of tight rock samples using inverse modeling. Water Resour. Res. 33(8).

Fisher, M.K., Heinze, J.R., Harris, C.D., Davidson, B.M., Wright, C.A., Dunn, K.P., 2004. Optimizing horizontal completion technologies in the Barnett shale using microseismic fracture mapping. In: Annual Technical Conference and Exhibition, Houston, TX. SPE 90051.

Gadde, P., Yajun, L., Jay, N., Roger, B., Sharma, M., 2004. Modeling proppant settling in water-fracs. In: Proc. SPE-89875-MS, SPE Annu. Tech. Conf. Exhib, September 26–29, Houston, TX.

Gan, H., Nandie, S.P., Walker, P.L., 1972. Nature of porosity in American coals. Fuel 51, 272–277.

Gao, Q., Cheng, Y., Fathi, E., Ameri, S., 2015. Analysis of stress-field variations expected on subsurface faults and discontinuities in the vicinity of hydraulic fracturing. SPE-168761SPE Reserv. Eval. Eng. J.

Garagash, D., 2006. Propagation of a plane-strain hydraulic fracture with a fluid lag: early-time solution. Int. J. Solids Struct. 43 (43), 5811–5835.

Garagash, D., 2007. Plane-strain propagation of a fluid-driven fracture during injection and shut-in: asymptotics of large toughness. Eng. Fract. Mech. 74, 456–481.

Gijtenbeek, K., Shaoul, J., Pater, H., 2012. Overdisplacing propped fracture treatments-good practice or asking for trouble? In: SPE EAGE Annual Conference and Exhibition. SPE-154397.

Goodway, B., Perez, M., Varsek, J., Abaco, C., 2010. Seismic petrophysics and isotropic–anisotropic AVO methods for unconventional gas exploration. Lead. Edge 29 (12), 1500–1508.

Gudmundsson, J.S., Hveding, F., Børrehaug, A., 1995. Transport or natural gas as frozen hydrate. In: The Fifth International Offshore and Polar Engineering Conference, June 11–16, The Hague, the Netherlands.

Haimson, B., Fairhurst, C., 1967. Initiation and extension of hydraulic fractures in rocks. Soc. Petrol. Eng. J. 7, 310.

Hartman, R.C., Ambrose, R.J., Akkutlu, I.Y., Clarkson, C.R., 2011. Shale gas in place calculations. Part II: multicomponent gas adsorption effects. In: SPE 144097, SPE Unconventional Gas Conference, Woodland, TX.

Hayashi, K., Haimson, B.C., 1991. Characteristics of shut-in curves in hydraulic fracturing stress measurements and determination of in situ minimum compressive stress. J. Geophys. Res. 96 (B11), 18311–18321.

Heidbach, O., 2008. Helmholtz Centre Potsdam GFZ German Research Centre for Geosciences.

Homfray, I.F., Physik, Z., 1910. Chemistry 74, 129.

Ilk, D., Rushing, J.A., Perego, A.D., Blasingame, T.A., 2008. Exponential vs. hyperbolic decline in tight gas sands—understanding the origin and implications for reserve estimates using Arps' decline curves. In: SPE Annual Technical Conference and Exhibition held in Denver, CO, September 21–24.

Jones, S., 1997. A technique for faster pulse-decay permeability measurements in tight rocks. SPE Form. Eval., 19–25.

Kanfar, M.S., Wattenbarger, R.A., 2012. Comparison of empirical decline curve methods for shale wells. In: Proceedings of the SPE Canadian Unconventional Resources Conferences, Calgary, AB, Canada, 30 October–1 November.

Kang, S.M., Fathi, E., Ambrose, R.J., Akkutlu, I.Y., Sigal, R.F., 2010. CO_2 applications. Carbon dioxide storage capacity of organic-rich shales. SPE J. 16 (4), 842–855.

Katz, D.L., et al., 1959. Handbook of Natural Gas Engineering. McGraw-Hill Publishing Co., New York City.

Kim, Y.I., Amadei, B., Pan, E., 1999. Modeling the effect of water, excavation sequence and rock reinforcement with discontinuous deformation analysis. Int. J. Rock. Mech. Min. 36, 949–970.

Kim, B.H., Kum, G.H., Seo, Y.G., 2003. Adsorption of methane and ethane into single-walled carbon nanotubes and slit-shaped carbonaceous pores. Korean J. Chem. Eng. 20, 104–109.

King, G., 2010. 30 Years of gas shale fracturing: what have we learnt? SPE 133456. In: SPE Annual Technical Conference and Exhibition, September 19–22, Florence, Italy.

Klenner, R., Liu, G., Stephenson, H., Murrell, G., Iyer, N., Virani, N., Anveshi, C., 2018. Characterization of fracture-driven interference and the application of machine learning to improve operational efficiency. In: SPE Liquids-Rich Basins Conference. SPE-191407-MS.

Kong, B., Fathi, E., Ameri, S., 2015. Coupled 3-D numerical simulation of proppant distribution and hydraulic fracturing performance optimization in Marcellus shale reservoirs. Int J. Coal Geol. 147–148, 35–45.

Krumbein, W.C., Sloss, L.L., 1963. Stratigraphy and Sedimentation. W.H. Freeman, San Francisco.

Lamont, N., Jessen, F., 1963. The effects of existing fractures in rocks on the extension of hydraulic fractures. J. Pet. Technol., 203–209. February.

Langmuir, I., 1916. The constitution and fundamental properties of solids and liquids. Part I. Solids. Am. Chem. Soc., 2221–2295.

Larkey, C.S., 1925. Mathematical determination of production decline curves. Trans. AIME 71, 1315.

Legarth, B., Huenges, E., Zimmermann, G., 2005. Hydraulic fracturing in sedimentary geothermal reservoir: results and implications. Int. J. Rock Mech. Min. Sci. 42 (7–8), 1028–1041.

Levasseur, S., Charlier, R., Frieg, B., Collin, F., 2010. Hydro-mechanical modell ing of the excavation damaged zone around an underground excavation at Mont Terri Rock Laboratory. Int. J. Rock Mech. Min. Sci. 47 (3), 414–425.

Li, D., Beckner, B., 2015. A practical and efficient uplayering method for scale-up of multimillion-cell geologic model. In: SPE-57273, Asia Pacific Improved Oil Recovery Conference, 25–26 October 1999, Kuala Lumpur, Malaysia.

Luffel, D.H., 1993. Matrix permeability measurement of gas productive shales. In: Paper SPE 26633, SPE Annual Technical Conference and Exhibition. SPE, Houston, TX.

Mallet, J.L., 2002. Geomodeling. Oxford University Press Inc, New York, NY.

Massaras, L., Dragomir, A., 2007. Enhanced fracture entry friction analysis of the rate step down test. In: SPE Hydraulic Fracture Technology Conference. SPE-106058.

Mavor, M.J., Owen, L.B., Pratt, T.J., 1990. Measurement and evaluation of coal sorption isotherm data. In: SPE-20728, Paper presented during the Annual Technical Conference and Exhibition of the SPE held in New Orleans, LA, September 23–26.

McClure, B., 2010. Investors Need A Good WACC. Investopedia Newsletter.

McNeil, R., Jeje, O., Renaud, A., 2009. Application of the power law loss-ratio method of decline analysis. In: Canadian International Petroleum Conference, June 16–18, Calgary, Alberta.

Miller, M., 2010. Gas shale evaluation techniques-things to think about. In: OGS Workshop presented July 28, 2010, Norman, OK.

Mobbs, A.T., Hammond, P.S., 2001. Computer simulations of proppant transport in a hydraulic fracture. SPE Prod. Facil. 16(2).

Morrill, J., Miskimins, J., 2012. Optimizing hydraulic fracture spacing in unconventional shales. In: Paper SPE 152595 presented at Hydraulic Fracturing Technology Conference, Woodlands, February 6–8.

Murdoch, L.C., 2002. Mechanical analysis of idealized shallow hydraulic fracture. J. Geotech. Geoenviron. Eng. 128 (6), 289–313.

Mutalik, P.N., Gibson, B., 2008. Case history of sequential and simultaneous fracturing of the Barnett shale in Parker county. In: Presented at the 2008 SPE Annual Technical Conference and Exhibition held in Denver, September 21–24.

Myers, A.L., Prausnitz, J.M., 1965. Thermodynamics of mixed-gas adsorption. AICHE J. 11 (1), 121–127.

Newman, G.H., 1973. Pore-volume compressibility of consolidated, friable and unconsolidated reservoir rock under hydrostatic loading. J. Pet. Technol. 25(2).

Ning, X., 1992. The measurement of matrix and fracture properties in naturally fractured low permeability cores using a pressure pulse method. PhD thesis, Texas A&M University.

Nolte, K.G., 1997. Background for After-Closure Analysis of Fracture Calibration Tests. Society of Petroleum Engineers.

Olson, J., Dahi, A., 2009. Modeling simultaneous growth of multiple hydraulic fractures and their interaction with natural fractures. In: Paper SPE 119739 presented at Hydraulic Fracturing Technology Conference, Woodlands, January 19–21.

Osholake, T.A., 2010. Factors Affecting Hydraulically Fractured Well Performance in the Marcellus Shale Gas Reservoirs. The Pennsylvania State University.

Ousina, E., Sondergeld, C., Rai, C., 2011. An NMR study on shale wettability. In: Canadian Unconventional Resources Conference Calgary, Alberta, November.

Ozkan, E., Brown, M., Raghavan, R., Kazemi, H., 2009. Comparison of fractured horizontal-well performance in conventional and unconventional reservoirs. In: Paper SPE 121290 presented at the SPE Western Regional Meeting, San Jose, March 24–26.

Passey, Q.R., Bohacs, K.M., Esch, W.L., Klimentidis, R., Sinha, S., 2010. From oil-prone source rock to gas-producing shale reservoir-geologic and petrophysical characterization of unconventional shale-gas reservoirs. In: SPE 131350 presented at the CPS/SPE International Oil & Gas Conference and Exhibition, Beijing, China, June 8–10.

Phani, B.G., Liu, Y., Norman, J., Bonnecaze, R., Sharma, M., 2004. Modeling proppant settling in Water-Fracs. In: 89875-MS SPE Conference Paper.

Pirson, S.J., 1935. Production decline curve of oil well may be extrapolated by loss-ratio. Oil Gas J.

Potluri, N., Zhu, D., Hill, A.D., 2005. Effect of natural fractures on hydraulic fracture propagation. In: SPE 94568, SPE European Formation Damage Conference, The Netherlands, 25–27 May.

Rafiee, M., Soliman, M.Y., Pirayesh, E., 2012. Hydraulic Fracturing Design and Optimization: A Modification to Zipper Frac. SPE 159786.

Rahmani Didar, B., Akkutlu, I.Y., 2013. Pore-size dependence of fluid phase behavior and properties in organic-rich shale reservoirs. In: SPE-164099, Paper Prepared for Presentation at the SPE Int. Symposium on Oilfield Chemistry held in Woodlands, TX, USA, April 8–10.

Rawlins, E.L., Schellhardt, M.A., 1935. Backpressure Data on Natural Gas Wells and Their Application to Production Practices. 7 Monograph Series, U.S. Bureau of Mines.

Rickman, V.R., Mullen, M.J., Petre, J.E., Erik, J., Grieser, W.V., Vincent, W., Kundert, D., 2008. A practical use of shale petrophysics for stimulation design optimization: all shale plays are not clones of the Barnett shale. In: SPE Annual Technical Conference and Exhibition, September 21–24, Denver, CO.

Rodvelt, G., Ahmad, M., Blake, A., 2015. Refracturing Early Marcellus Producers Accesses Additional Gas. In: SPE Eastern Regional Meeting. SPE-177295.

R.N.P. Roussel, M. Sharma, Strategies to minimize frac spacing and stimulate. Natural fractures in horizontal completions. (2011).

Ruthven, D.M., 1984. Principles of Adsorption and Adsorption Processes. John Wiley & Sons Inc, New York.

Rylander, E., Singer, P., Jiang, T., Lewis, R., McLin, R., 2013. NMR T2 distributions in the Eagle Ford shale: reflections on pore size. In: SPE 164554 presented at the Unconventional Resources Conference, Woodlands, TX, April 10–12.

Santos, J.M., Akkutlu, I.Y., 2012. Laboratory measurement of sorption isotherm under confining stress with pore volume effects. In: SPE 162595, Calgary, Canada.

Saunders, J.T., Tsai, B.M.C., Yang, R.T., 1985. Adsorption of gases on coals and heattreated coals at elevated temperature and pressure: 2. Adsorption from hydrogen-methane mixtures. Fuel 64 (5), 621–626.

Schlebaum, W., Scharaa, G., Vanriemsdijk, W.H., 1999. Influence of nonlinear sorption kinetics on the slow desorbing organic contaminant fraction in soil. Environ. Sci. Technol. 33, 1413–1417.

Seshadri, J., Mattar, L., 2010. Comparison of power law and modified hyperbolic decline methods. In: Paper SPE 137320 presented at the Canadian Unconventional Resources & International Petroleum Conference, Calgary, Alberta, Canada, October 19–21.

Shen, Y., 2014. A variational inequality formulation to incorporate the fluid lag in fluid-driven fracture propagation. Comput. Methods Appl. Mech. Eng. 272 (HF-69), 17–33.

Shi, J.Q., Durucan, S., 2003. A bidisperse pore diffusion model for methane displacement desorption in coal by CO_2 injection. Fuel 82, 1219–1229.

Siebrits, E., Peirce, A.P., 2002. An efficient multi-layer planar 3D fracture growth algorithm using a fixed mesh approach. Int. J. Numer. Methods Eng. 53, 691–717.

Sing, K.S., 1985. Reporting physisorption data for gas/solid systems with special reference to the determination of surface area and porosity. Pure Appl. Chem.

Singh, S.K., Sinha, A., Deo, G., Singh, J.K., 2009. Vapor — liquid phase coexistence, critical properties, and surface tension of confined alkanes. J. Phys. Chem. C 113 (17), 7170–7180.

Singha, P., Van Swaaijb, W.P.M., (Wim) Brilmanb, D.W.F., 2013. Energy efficient solvents for CO_2 absorption from flue gas: vapor liquid equilibrium and pilot plant study. Energy Procedia 37, 2021–2046.

Soliman, M.Y., East, L., Adams, D., 2004. Geomechanics aspects of multiple fracturing of horizontal and vertical wells. In: SPE International Thermal Operations and Heavy Oil Symposium and Western Regional Meeting, March 16–18, Bakersfield, CA.

Soliman, M.Y., Craig, D., Barko, K., Rahim, Z., Ansah, J., Adams, D., 2005. New Method for Determination of Formation Permeability, Reservoir Pressure, and Fracture Properties from a Minifrac Test. Paper ARMA/USRMS 05-658.

Soltanzadeh, H., Hawkes, C.D., 2009. Induced poroelastic and thermoelastic stress changes within reservoirs during fluid injection and production. In: Porous Media: Heat and Mass Transfer, Transport and Mechanics. Nova Science Publishers Inc, Hauppauge, NY (Chapter 2).

Srinivasan, R., Auvil, S.R., Schork, J.M., 1995. Mass transfer in carbon molecular sieves—an interpretation of Langmuir kinetics. Chem. Eng. J. 57, 137–144.

Stevenson, M.D., Pinczewski, W.V., Somers, M.L., Bagio, S.E., 1991. Adsorption/desorption of multi-component gas mixtures at in-seam conditions. In: SPE23026, SPE Asia-Specific Conference, Perth, Western Australia, November 4–7.

Tadmor, R., 2004. Line energy and the relation between advancing, receding, and young contact angles. Langmuir 20 (18), 7659–7664.

Taghichian, A., 2013. On the geomechanical optimization of hydraulic fracturing in unconventional shales. A thesis submitted to University of Oklahoma.

Thompson, J., Franciose, N., Schutt, M., Hartig, K., McKenna, J., 2018. Tank development in the Midland Basin, Texas: a case study of super-charging a reservoir to optimize production and increase horizontal well densities. In: Unconventional Resources Technology Conference. URteC: 2902895.

Tiab, D., Donaldson, E.C., 2015. Petrophysics. In: Theory and Practice of Measuring Reservoir Rock and Fluid Transport Properties, fourth ed. Gulf Professional Publishing.

Tinni, A., Fathi, E., Agrawal, R., Sondergeld, C., Akkutlu, I.Y., Rai, C., 2012. Shale permeability measurements on plugs and crushed samples. In: SPE-162235 Selected for presentation at the SPE Canadian Resources Conference held in Calgary, Alberta, Canada, 30 October–1 November.

Todd Hoffman, B., 2012. Comparison of various gases for enhanced recovery from shale oil reservoirs. In: 154329-MS SPE C SPE Improved Oil Recovery Symposium, April 14–18, Tulsa, OK.

Turner, R.G., Hubbard, M.G., Dukler, A.E., 1969. Analysis and prediction of minimum flowrate for the continuous removal of liquids from gas wells. J. Pet. Technol. 21 (11), 1475–1482. Trans., AIME, 246.SPE-2198-PA, https://doi.org/10.2118/2198-PA.

Valkó, P.P., 2009. Assigning value to stimulation in the Barnett shale: a simultaneous analysis of 7000 plus production histories and well completion records. In: Paper SPE 119369 presented at the SPE Hydraulic Fracturing Technology Conference, The Woodlands, TX, January 19–21.

Virk, P.S., 1975. Drag reduction fundamentals. AICHE J. 21 (4), 625–656.

Waples Douglas, W., 1985. Geochemistry in Petroleum Exploration. Brown and Ruth laboratories Inc, Denver, CO.

Warpinski, N.R., Teufel, L.W., 1987. Influence of geologic discontinuities on hydraulic fracture propagation. J. Pet. Technol. 39 (2), 209–220.

Warren, J.E., Root, P.J., 1963. The behavior of naturally fractured reservoirs. SPE 426-PA, SPE J. 3 (3), 245–255.

Washburn, E.W., 1921. Note on the method of determining the distribution of pore sizes in a porous material. Proc. Natl. Acad. Sci. U. S. A. 7, 115–116.

Waters, G., Dean, B., Downie, R., Kerrihard, K., Austbo, L., McPherson, B., 2009. Simultaneous Hydraulic Fracturing of Adjacent Horizontal Wells in the Woodford Shale. SPE 119635.

Weichun, C., Scott, K., Flumerfelt, R., 2018. A new technique for quantifying pressure interference in fractured horizontal shale wells. In: SPE Annul Technical Conference and Exhibition. SPE-191407-MS.

Xu, W., Tran, T.T., Srivastava, R.M., Journel, A.G., 1992. Integrating Seismic Data in Reservoir Modeling: The Collocated Cokriging Alternative. Society Petroleum Engineers.

Yamada, S.A., 1980. A review of a pulse technique for permeability measurements. SPE J., 357–358.

Yamamoto, K., Shimamoto, T., Maezumi, S., 1999. Development of a true 3D hydraulic fracturing simulator. In: SPE Asia Pacific Oil, pp. 1–10.

Yang, J.T., 1987. Gas Separation by Adsorption Processes. Butterworth Publishers.

Yang, Y., Aplin, A.C., 1998. Influence of lithology and compaction on the pore size distribution and modelled permeability of some mudstones from the Norwegian margin. Mar. Pet. Geol. 15, 163–175.

Yee, D., Seidle, J.P., Hanson, W.B., 1993. Gas sorption on coal measurements and gas content. In: Law, B.E., Rice, D.D. (Eds.), Hydrocarbons from Coal. AAPG Studies in Geology, pp. 203–218 (Chapter 9).

Yuyi, J., Belyadi, H., Blake, A., Wyatt, J., Dalton, J., Vangilder, C., Roth, B., 2016. Dry utica proppant and frac fluid design optimization. In: SPE Eastern Regional Meeting. SPE-184078.

Zamirian, M., Aminian, K., Ameri, S., Fathi, E., 2014a. New steady-state technique for measuring shale core plug permeability. In: SPE-171613-MS, SPE/CSUR Unconventional Resources Conference, September–2 October Calgary, Canada.

Zamirian, M., Aminian, K., Fathi, E., Ameri, S., 2014b. A fast and robust technique for accurate measurement of the organic-rich shales characteristics under steady-state conditions. In: SPE 171018, SPE Eastern Regional Meeting, October 21–23, 2014, Charleston, WV.

Zimmerman, R.W., 1991. Compressibility of Sandstones. Elsevier Science Publishing Company Inc, New York, NY.

Index

Note: Page numbers followed by *f* indicate figures and *t* indicate tables.

A

Aboveground storage tanks (ASTs), 274, 276*f*
Absolute pressure, 18–19
Absolute roughness of pipe, 112
Absolute volume factor (AVF), 149–150
ACA. *See* After-closure analysis (ACA)
Acidic buffer, 118
Acidization stage, 61–63
Acid-producing bacteria, 115
Acid solubility, 80
Acquisition, 362
Adaptive moment estimation (ADAM) neural networks, 449–450
Adjustable choke, 203
Adsorbed gas density, 27–29, 31–32
Adsorption isotherm types, 18*f*
Ad valorem tax, 355–356, 356*b*
Advanced shale reservoir characterization, 13
   pore compressibility measurements of shale, 21–22
   pore-size distribution measurement of shale, 13–15
   shale permeability measurement techniques, 22–26
   shale porosity measurements, 20–21
   shale sorption measurement techniques, 15–20
After-closure analysis (ACA), 252–256
   Horner plot, 252–254, 254*f*
   linear flow-time function *vs.* bottom-hole pressure, 254–255, 255*f*
   radial flow-time function *vs.* BHP, 255–256, 256*f*
After federal income tax (ATAX) monthly discounted net cash flow, 376–377
Agarwal-Gardner type curve, for Fractured gas well, 566–567, 567*f*

AI&ML. *See* Artificial intelligence and machine learning (AI&ML)
Amazon, 502–503
American Petroleum Industry unit (GAPI), 11
American Petroleum Institute (API)- recommended limit for sand, 79, 81–82
Amott wettability index, 44–45
Anisotropic closure stress, 549–550, 551*f*
Annular flow, liquid loading, 491
Aquifer interaction, hydraulic fracturing and, 101–102
Archie's equation, 548
Arps decline curve equations, for DCA, 319–329
   exponential decline equations, 320
   hyperbolic decline equations, 320–329, 321*f*, 324*t*, 328*t*
Arps models, 318
Arrows up system, 280–282, 282*f*
Artificial intelligence (AI), 499–500
   application of, 499
   definitions of, 499–500
   technique of, 449–450
Artificial intelligence and machine learning (AI&ML), 177–178, 200
Artificial neural network (ANN), 513–516*b*
   activation functions and characteristics, 511, 512*t*
   elements, 508
   supervised learning, 509–510
   unsupervised learning, 509
   weight and bias, 511, 513–514*t*, 513*f*
Assignment, 344
Associated gas, 4
ATAX monthly undiscounted NCF, 391–404, 391–404*b*

Automated high-pressure, high-temperature (HPHT) pulse-decay permeameter, 24f
AVF. *See* Absolute volume factor (AVF)

B

Back pressure model, 479–484, 481–482b, 483t, 483f
Back pressure regulator (BPR), 207, 209f
Bacteria, 114–115
Bakken Shale, 10–11, 67–68, 177–178, 184, 210–212
Barium sulfate, 116
Barnett Shale, 36, 49, 51–53, 119
Base case scenario, 456–458
Basic buffer, 118
Basis differential, 360
Basis price, 360–361, 360–361b
BCA. *See* Before-closure analysis (BCA)
Bean choke, 203
Before-closure analysis (BCA), 235–252, 576–577
  G-function analysis, 240–252, 246f
  log-log plot (log (BH ISIP-BHP) *vs.* log (time)), 237–240, 240–242f
  square root plot, 236–237, 237–238f
Before federal income tax (BTAX) monthly undiscounted net cash flow, 366
Bernoulli's principle, 277
Beta coefficient, 370–371
BHFP. *See* Bottom-hole frac pressure (BHFP)
BHP. *See* Bottom-hole pressure (BHP)
BHST. *See* Bottom-hole static temperature (BHST)
BHTP. *See* Bottom-hole treating pressure (BHTP)
Big data, 499
Biocide, 114–115
Biot's constant (poroelastic constant), 225–226
Biot's poroelastic coefficient, 35–36
Bi-wing fracture system, 52–53, 54f
Blasingame type curve, 568, 568f
Blender, 279–280, 284f
Blender sand concentration, 163
Blender tub, 287, 288f

Blending gas, 415–416, 423–441, 424t, 429t
Blowing sand, 280
Boost pump, 290
Bottom-hole frac pressure (BHFP), 125
Bottom-hole pressure (BHP), 90–91, 292, 563, 565
  vs. square root of time, 134, 134t, 237–238f
Bottom-hole static temperature (BHST), 161
Bottom-hole treating pressure (BHTP), 124–125, 132, 144, 161
Boundary-dominated flow, 315
Bounded *vs.* unbounded (inner vs outer), 196–198, 197f
Boyle's law, 20
BPR. *See* Back pressure regulator (BPR)
Brady sand, 83
Break even analysis, 415–416
Breakeven BTU content, 416–417, 438–440t
Breakeven ethane pricing, 441, 442t
Bretton Woods Agreement, 411
British thermal units (BTUs), 1–4, 346, 415–416
  of natural gas component, 2t
Brittleness and fracability ratios, 220–222, 222t
Brown sand, 83
B section. *See* Tubing head
BTAX monthly discounted net cash flow, 376–377
BTUs. *See* British thermal units (BTUs)
Bubble, liquid loading, 492
Buffer, 118
Bulk density, 20, 80
Burgeoning technologies, 5–6

C

Calcium carbonate, 116
CAPEX, 444, 456–458
Capex, 362, 364–365
Capillary pressure, 39–44
Capillary rise method, 40
Capital asset pricing model (CAPM), 370
Capital budgeting, 372
Capital expenditure cost, 362–364, 364b

Cap rock, 8–9
Carter's leak-off model, 265
Cash inflow, 341–342
Cash outflow, 341–342
Casing design, 121
Casing selection, 98
Casing size, 140
Casing string, 101f
Centralized impoundment, 274
Centralized tanks, 275, 276f
Centrifugal pumps, 279–280, 279f
Ceramic proppant, 74, 75f
Chaw Pressure Group (CPG), 447–448
Check valves, 297–299
Chemical chart, 293, 294–295f
Chemical coordination, 300–301
Chemical injection ports, 288–289, 289f
Chemical selection and design, hydraulic fracturing, 107
  biocide, 114–115
  buffer, 118
  cross-linker, 118–119
  FR breaker, 114
  FR flow loop test, 108–111
  friction reducer (FR), 108
  gel breaker, 117–118
  iron control, 120
  linear gel, 116–117, 117f
  pipe friction pressure, 111–112
  relative roughness of pipe, 112–114, 113t
  Reynolds number, 112
  scale inhibitor, 115–116
  surfactant, 119–120
Choke manifold, 203, 204f
"Choking" effect, 129
Clays, 10–11
Clean rate, 150–151
  no proppant, 164
  with proppant, 165
Clean volume, 150–151
Closure pressure, 132–135, 135f
Clustering, 82
Cluster spacing, 186, 193
CMG CMOST, 558–559
CNG. See Compressed natural gas (CNG)
Coal, 7–8
Coalbed methane, 7–10

Coalification, 7–8
Codevelopment analysis, step-by-step workflow for, 468–474, 470–471f
Cohesive zone models (CZMs), 262
Coleman rate, 493–498, 494–498b, 496–498t, 573–574
Commodity exchange (COMEX), 359
Commodity pricing, 443–444
Common stock, 368
Completed slick water schedule, 159–160t
Completion and stimulation design (CSD), 517–530
Completion (frac) methods, 177–178
Completions and flowback design, relation to production, 191–192
  bounded vs unbounded (inner vs. outer), 196–198
  cluster spacing, 193
  entry-hole diameter (EHD), 194
  flowback design, 201–203
  flowback equipment (see Flowback equipment)
  flowback equipment spacing guidelines, 212–213
  landing zone, 192
  proppant size and type, 195–196
  sand and water per foot, 194–195
  stage spacing, 193
  tubing analysis, 213
  up dip vs. down dip, 198, 198f
  water quality, 200–201
  well spacing, 199–200
Completions CAPEX/ft, 444–445
Completions design optimization. See Well spacing and completions optimization
Complex fracture system, 52f
Composite bridge (frac) plugs, 178–182, 179f
  simultaneous frac, 182
  stack fracing, 180–181
  zipper fracing, 182
Compositional reservoir simulation, 33–34
Compressed natural gas (CNG), 4
Compression, 349
Computer modeling, 155
Conceptual shale matrix model, 36

Condensate pricing, 415–416
Conductivity, 85
Conductor casing, 98–99
Confinement effects, 32
Constant-rate decline, 315
Constitutive law of linear elasticity, 266
Contact angle, 44
Conventional enhanced oil recovery technique, 6–7
Conventional methods, 257, 259
  of sampling and measuring, 13
Conventional plug-and-perf method, 177–180, 184–185, 201–202
Conventional pulse-decay setup, 25
Conventional resources, 47–48
Core plug pulse-decay permeameter, 24f
Corporation tax, 391
Cost-effective casing, 140
Cost function, 509–510
Cost of capital. *See* Discount rate
Cost of debt, 369, 369b
Cost of money. *See* Discount rate
Crack-tip open displacement (CTOD), 262
Crack-tip plasticity (CTP), 259–260
CRCS. *See* Curable resin-coated sand (CRCS)
Critical velocity, 494–498b
Cross-linked gel fluid system, 52–54, 119f
Cross-linked jobs, 89–90
Cross-linker, 118–119
Crushed samples, 20
Crush resistance, 79–80
CTOD. *See* Crack-tip open displacement (CTOD)
CTP. *See* Crack-tip plasticity (CTP)
"Cubic sugar" models, 9–10
Cumulative gas production, 581
Cumulative time relationship, 318
Curable resin-coated sand (CRCS), 72–73, 73f
Current tax payment act, 411
Curtailment, 444–445, 482–484
Cushing Hub, 361–362
C value, 480
Cyclic stress, 90–91
CZMs. *See* Cohesive zone models (CZMs)

D

Darcy friction factor, 112–114
Darcy's equation, 22
Darcy's law, 22, 23f, 88
Darcy's unit, 22
Data mining (DM), 500
Data partitioning, 526
Data preprocessing, 504–505, 519
Data science, 504–505
DBSCAN clustering, 537–539
DCA. *See* Decline curve analysis (DCA)
DD method. *See* Discontinuous displacement (DD) method
Debt, 367
Decline curve analysis (DCA), 191–192, 311–312, 565–566
  anatomy of
    effective decline, 312–313
    hyperbolic exponent, 313, 314t
    instantaneous production, 312
    nominal decline, 312
    pseudosteady state, 315
    shape of, 313, 313f
    unsteady state period, 314
  estimating future volumes, Arps decline curve equations for, 319–329
    exponential decline equations, 320
    hyperbolic decline equations, 320–329, 321f, 324t, 328t
    multisegment decline, 330–333, 332f, 332–333t
    pressure normalized rate, 333–340, 338–340t
  primary types
    Duong decline, 319
    exponential decline, 315
    harmonic decline, 317
    hyperbolic decline, 315–316, 316f
    modified hyperbolic decline curve, 317–318
    PLE, 318
    stretched exponential, 318
Decline factor, 312–315
De-ethanizer, 417–418
Delayed cross-linked, 118
Delta-T compressional logs (DTCO), 545–547

Index

De-methanizer, 417–418
Density-logging tool, 224
Densometer, 68f, 289
Depreciation, 388–389, 388t, 389b
Design exploration controlled evolution (DECE), 563
Desorption isotherm, 15–16, 18–19
DFITs. *See* Diagnostic fracture injection tests (DFITs)
Diagnostic fracture injection tests (DFITs), 12, 123, 233–234, 543, 574–577, 576–577f
   after-closure analysis (ACA), 252–256
      Horner plot, 252–254, 254f
      linear flow-time function *vs.* bottom-hole pressure, 254–255, 255f
      radial flow-time function *vs.* BHP, 255–256, 256f
   before-closure analysis (BCA), 235–252
      G-function analysis, 240–252, 246f
      log-log plot (log (BH ISIP-BHP) *vs.* log (time)), 237–240, 240–242f
      square root plot, 236–237, 237–238f
   data recording and reporting, 234–235
   typical DFIT procedure, 234, 235f
Digenesis, 8–9
Dimensionless fracture conductivity, 86
   vs. effective drainage radius, 93f
Dirty rate, 150–151
Discharge coefficient, 127
Discharge side centrifugal pumps, 280
Discontinuous displacement (DD) method, 257
Discounted cash-flow rate of return (DCFROR). *See* Internal rate of return (IRR)
Discount rate, 366–368, 410
Discrete models, 36
Domain expertise, 504–505, 530–532
Dot-com Crash, 412
Double-cell Boyle's law porosimeter, 25–26, 26f
Draw type curve boundaries, 537–539
Drilling, 273
Drilling CAPEX/ft, 444–445
Dual-porosity single-permeability models, 35

Dump valve, 207
du Noüy ring, 40
Duong decline, 319
Dynamic contact angle measurement, 41f
Dynamic Young's modulus, 216–217

E

Economic analysis, 311, 417–418
Economic evaluation
   ad valorem tax, 355–356, 356b
   BTAX and ATAX monthly discounted net cash flow, 366, 376–377
   BTU content, 346
   capital budgeting, 372
   capital expenditure cost, 362–364, 364b
   cost of debt, 369, 369b
   cost of equity, 370–371, 371b
   Cushing Hub and WTI, 361–362
   discount rate, 366–368
   Henry Hub and basis price, 360–361, 360–361b
   IRR, 377–382, 378–381b
   MIRR, 382–384, 383–384b
   Mont Belvieu and OPIS, 362
   NCF model, 341–342, 342f
   Net Opex, 356–357, 357b
   NPV, 372–376, 374t, 374–376b, 382
   NRI, 345–346, 346b
   NYMEX, 359
   operating expense, 348–351
   Opex, Capex and pricing escalations, 364–365
   payback method, 384–386, 384–386b
   profit, 365–366
   profitability index, 386–387, 387b
   revenue, 357–359, 358–359b
   royalty, 342–343
   severance tax, 354–355, 354–355b
   shrinkage factor, 346–348, 347–348b
   tax model
      ATAX monthly undiscounted NCF, 391–404, 391–404b
      corporation tax, 391
      depreciation, 388–389, 388t, 389b
      taxable income, 389–390, 389–390b
   total Opex per month, 351–353, 352–353b

Economic evaluation *(Continued)*
  weight of debt and equity, 368–369, 369b
  working interest, 343–345
Economics analysis, frac job sand schedule optimization, 581–583, 583t, 584f
Effective decline (De), 312–313
Effective pore volume, 20
Effective stress, 35
Entry-hole diameter (EHD), 127, 130–131, 141–143, 194
Environmental Protection Agency (EPA), 99, 101
Equilibrium condition, 266
Erosional velocity calculation, 488–490, 489–490b
Estimated ultimate recovery (EUR), 185, 191–192, 317–318
Excel solver, 147t, 148f
Exchange, 344
Exchange rate. *See* Discount rate
Expected ultimate recovery (EUR), 565–566
Exploration and production (E&P) companies, 107, 273
  design well spacing, 199–200
Exponential decline, 315, 320
Extended finite element method (XFEM), 270–271
Extraction methods, 5–6

F
Facebook, 503
Fancyimpute, 520–521
Fanning friction factor, 112–114
Fanny friction factor, 485
Fault plain, geometry of, 105f
Fault reactivation, 97–98, 102
  hydraulic fracturing and, 102–105
Federal funds rate, 408–410, 409f
Federal Open Market Committee (FOMC), 405
The Federal Reserve (The FED), 405
  discount rate, 410
  federal funds rate, 410
  financial and political events, 410–413
  interest rate management, 405–408

money supply management, 408–409, 409f
  prime rate, 410
FG. *See* Fracture gradient (FG)
Field ATAX NPV, 454, 457t, 459t, 462–463t, 465t
Field completion design optimization, 529–530, 531f
Filter cake forms, 89–90
Fines migration, 91–92
Finite *vs.* infinite conductivity, 93–95
Firm transportation (FT) cost, 350
Five-stage adsorption measurement technique, 20
Fixed BTU content, 432
Fixed realized gas pricing, 435
Flapper/dart check valves, 297–299
Flare stack, 210–211, 211f
Flowback design, 201
Flowback equipment, 203–212
  choke manifold, 203, 204f
  flare stack, 210–211, 211f
  high-stage separator, 205–209
  low-stage separator, 210
  oil tanks (upright tanks), 212, 212f
  sand trap, 204–205, 205f
  spacing guidelines, 212–213
Flow cross, 306, 306–307f
Flow loop apparatus, 109f
Flow loop test results, 110f
Fluid efficiency, 64
Fluid flow in hydraulic fractures, 265–266
Fluid leak-off regimes on G-function plot, 243–252
Fluid rate, 208–209
Fluid-rock molecular collision, 15
Flush stage, 66–68
Foam design schedule, 175t
Foam frac jobs, 163
Foam frac schedule and calculations, 161
Foam-fracturing fluid system, 55–57
Foam quality, 57–58
  vs. lb of proppant per gallon of foam
Foam stability, 58–59
Foam viscosity, 58
Foam volume, 161–162

Focused ion beam scanning electron microscopy (FIB/SEM), 13–15
Formation X codeveloped type curve, 470
Formation X stand-alone type curve, 470
Formation Y codeveloped type curve, 470
40/70 mesh sand size, 76–77
Four-phase horizontal separator, 208f
Four-phase separator, 205–206
Four-way entry frac head, 308, 308f
FR. *See* Friction reducer (FR)
Frac ball inside of composite bridge plug for frac stage isolation, 181f
Frac barriers, 192
Frac design basis, 61f
Frac fluid selection, 69
Frac gradient. *See* Fracture gradient (FG)
Frac head, 308, 308–309f
Frac hit, 447–449, 448f
  detection and mitigation strategies, 449–450
Fracing. *See* Hydraulic fracturing fluid systems
Frac job formation, 76, 78
Frac job sand schedule optimization, 577–583
  economics analysis, 581–583, 583t, 584f
  hydraulic fracture treatment design, 578–581, 579–580t, 582f, 582–583t
Frac manifold, 291, 291–292f
Frac microseismic data, 101–102
Frac packing, 48
FracPro, 543, 574, 578
Frac release valve (FRV), 295
Frac stage spacing (plug-to-plug spacing), 184
Fracture communication, 468
Fracture conductivity testing, 85, 87f, 93
Fractured gas well, Agarwal-Gardner type curve for, 566–567, 567f
Fracture extension pressure, 132, 133f
Fracture geometry, 315
Fracture gradient (FG), 124–125
Fracture growth limitation, 103f
Fracture orientation, 228–229
Fracture pressure analysis and perforation design, 121
  bottom-hole treating pressure (BHTP), 125
  closure pressure, 132–135, 135f
  fracture extension pressure, 132
  fracture gradient (FG), 124–125
  hydrostatic pressure, 122
  hydrostatic pressure gradient, 122
  instantaneous shut-in pressure (ISIP), 123–124, 124f
  near-wellbore friction pressure (NWBFP), 131–132
  net pressure, 135–137, 136f
  numbers of holes (perfs) and limited entry technique, 130–131
  open perforations, 128
  perforation design, 129–130
  perforation diameter and penetration, 131
  perforation efficiency, 128
  perforation erosion, 131
  perforation friction pressure, 127
  pipe friction pressure, 126–127
  pressure, 121
  production casing design, 139–140
  rate step-down test workflow, 141–148
  surface-treating pressure (STP), 137–139
  total friction pressure, 126
Fracture propagation models, 257
Fracture-tip extension, 248–250, 250f
Fracture-tip pressure, 136
Fracture toughness, 220
Fracture treatment design, 149
  absolute volume factor (AVF), 149–150
  blender sand concentration, 163
  clean rate (no proppant), 164
  clean rate (with proppant), 165
  foam frac schedule and calculations, 161
  foam volume, 161–162
  nitrogen rate (with and without proppant), 166–174
  nitrogen volume, 162
  sand per foot, 154
  sand-to-water ratio (SWR), 155
  slick water frac schedule, 155–160
  slurry density, 151–152
  slurry factor (SF), 163–164
  slurry rate (with proppant), 165–166
  stage fluid clean volume, 152

Fracture treatment design *(Continued)*
  stage fluid slurry volume, 153
  stage proppant, 153–154
  water per foot, 154
Fracture width, 86*f*
Frac valve. *See* Manual valve
Frac van
  chemical chart, 293, 294–295*f*
  NBHP chart, 292
  surface-treating pressure chart, 292, 293*f*
Frac wellhead, 304–308
  flow cross, 306, 306–307*f*
  frac head, 308, 308–309*f*
  hydraulic valve, 305, 305*f*
  lower master valve, 304–305
  manual valve, 306–307, 307*f*
  tubing head, 304
FR breaker, 114
FR concentration, 140
Free Fluid Volume (CMFF), 545–547
Free gas, 27–28
Free gas mass balance, 38
F-regression tests, 521–523
Freundlich isotherm, 17
FR flow loop test, 108–111
Friction factor, 102–104
Friction reducer (FR), 108
Full factorial design (FFD), 562
Full-field ATAX NPV, 453
Full-field development, 446–447, 452

G

Gamma ray, 545–547
Gamma ray log, 11, 11*f*
Gas chromatograph, 1–4, 3*f*
Gas pricing, 199–200, 444
Gas processing, 417–418
Gas production operation issues, 491–492
Gas pseudo-pressure, 476–479, 477–479*b*, 478–479*t*
Gas Research Institute (GRI) technique, 25–26
Gas resources pyramid, 7*f*
Gas sorption kinetics, 39
Gas transport in organic-rich shale reservoirs, 35

Gas well deliverability, 475
Gathering and compression cost (G&C), 349
Gel breaker, 117–118
Gel damage, 89–90
Gelling agents, 58–59
General and administrative (G&A) cost, 350
General heterotrophic bacteria, 115
Geological structure model, 258–259
Geomechanical logs, 549–554, 551–553*f*
G-function analysis, 240–252, 246*f*
G-function plot
  effective permeability, 250–252
  fluid leak-off regimes on, 243–252
  with height recession signature, 249*f*
Gibbs theory model, 16
Gini importance, 523
Glutaraldehyde, 114–115
Goat head. *See* Frac head
Google, 502–503
Gradient descent technique, 509–510
Grain density of sample, 20
Gravel packing, 48
Green's function, 269
GRG nonlinear (generalized reduced gradient) method, 143
Grid search, 527
GRI technique. *See* Gas Research Institute (GRI) technique
Gross shrunk gas, 430
Guar, 116

H

Harmonic decline, 317
Height recession behavior, 247*f*
Height recession leak-off, 247–248, 248*f*
Helium, 20
Henry Hub, 360–361, 360–361*b*
Henry's law isotherm, 16
HHP. *See* Hydraulic horsepower (HHP)
Hickory formation outcrops, 83
Hidden layer, ANN elements, 508, 511
High-pressure frac pumps, 291
High-pressure mercury injection, 20–21
High-stage separator, 205–209
High-strength proppant, 74–75
Histogram distribution, 528–529

Index

History-matching, 556–565, 560–561t, 561–564f, 565t
  algorithms, 23–24
Hopper screws, 282, 284f
Horizontal separators, 205–206, 208f
Horizontal well multistage completion techniques, 177–178
  cluster spacing, 186
  composite bridge (frac) plug, 179–182, 179f
    simultaneous frac, 182
    stack fracing, 180–181
    zipper fracing, 182
  conventional plug and perf, 178–179
  frac stage spacing (plug-to-plug spacing), 184
  refrac overview, 186–189
  shorter stage length (SSL), 184–186
  sliding sleeve, 182
    advantages, 182–183
    disadvantages, 183–184
Horner plot, 252–254, 254f
100 mesh sand size, 76, 77f
Hurdle rate. *See* Discount rate
Hybrid design, 184
Hybrid fluid system, 54–55
Hydraulically fractured (frac order), 196–198
Hydraulic and natural fracture interactions, 270–271
Hydraulic coupling, in multicontinuum approach, 38f
Hydraulic ESD, 306, 307f
Hydraulic frac job, design and fracture modeling of, 273
Hydraulic fracture, 9–10, 72, 88–89
  and aquifer interaction, 101–102
  chemical selection and design, 107
    biocide, 114–115
    buffer, 118
    cross-linker, 118–119
    FR breaker, 114
    FR flow loop test, 108–111
    friction reducer, 108
    gel breaker, 117–118
    iron control, 120
    linear gel, 116–117, 117f
    pipe friction pressure, 111–112
    relative roughness of pipe, 112–114, 113t
    Reynolds number, 112
    scale inhibitor, 115–116
    surfactant, 119–120
  design, 129
  environmental impacts of, 98
  and fault reactivation, 102–105
  geometry, 262–263
  and low-magnitude earthquakes, 106
  propagation, 549–550
  proppant transport and distribution in, 82–84
  treatment design, 578–581, 579–580t, 582f, 582–583t
Hydraulic fracture optimization
  ANN (*see* Artificial neural network (ANN))
  artificial intelligence, 499–500
    application of, 499
    definitions of, 499–500
  data mining, 500
  field completion design optimization, 529–530, 531f
  K-means clustering, 530–537
    application, 535
    data set division, 534, 534f
    distance equation calculation, 534
    for liquid-loading detection, 537, 538f
    process, 532–537
  machine learning, 499–502
    case usages, various industries, 502–504
    knowledge of data science and domain expertise, 504–505
    reinforcement learning, 501–502
    supervised learning, 501
    unsupervised ML, 501
    workflow for, 505–507, 507f
  neural network training, validation and testing, 526–527, 526–528f
  supervised ML, completion and stimulation optimization using, 500, 518f, 520t, 520f, 522f, 524–525f
  type curves, 528–529
  unsupervised ML algorithms, application of, 537–539

Hydraulic fracturing fluid systems, 47–48
  acidization stage, 61–63
  cross-linked gel fluid system, 52–54
  flush stage, 66–68
  foam fracturing, 55–57
  foam quality, 57–58
  foam stability, 58–59
  frac fluid selection, 69
  hybrid fluid system, 54–55
  pad stage, 63–65
  proppant stage, 65–66
  slick water fluid system, 49–52
  tortuosity, 59–61, 60f
  typical slick water frac steps, 61
Hydraulic fracturing propagation, numerical simulation of, 257–258
  fluid flow in hydraulic fractures, 265–266
  hydraulic and natural fracture interactions, 270–271
  pseudo-3D hydraulic fracturing models, 266–268, 267f
  simulators, development of, 259–262
  solid elastic response, 266
  stage merging and stress shadow effects, 271–272
  stratigraphic and geological structure modeling, 258–259
  3D hydraulic fracturing models, 268–270
  2D hydraulic fracturing models, 262–265
Hydraulic fracturing simulation, 121, 502–503
Hydraulic horsepower (HHP), 101–102, 138
Hydraulic valve, 305, 305f
Hydrostatic pressure, 122
Hydrostatic pressure gradient, 122
Hyperbolic decline equations, 315–316, 316–317f, 320–329, 321f, 324t, 328t
Hyperbolic exponent (b), 313, 314t
Hysteresis types, 19f

I

Ideal adsorption solution (IAS) theory, 32
Idiosyncratic risk, 370–371
IGIP. *See* Initial gas in place (IGIP)
Infinite conductivity, finite *vs.*, 93–95

Inflation, 407–408
Inflow performance relationship (IPR), 475
In-ground pits, 274, 275f
Initial gas in place (IGIP), 52
Input layer, ANN elements, 508
In situ stresses, 215
  Biot's constant (poroelastic constant), 225–226
  brittleness and fracability ratios, 220–222, 222t
  fracture orientation, 228–229
  fracture toughness, 220
  longitudinal fractures, 229–231, 230f
  maximum horizontal stress, 226–228
  minimum horizontal stress, 224–225
  Poisson's ratio, 217–219, 220f
  transverse fractures, 229, 230–231f
  various stress states, 228
  vertical stress, 223–224
  Young's modulus, 215–217, 217f
Instantaneous production (IP), 312
Instantaneous shut-in pressure (ISIP), 123–124, 124f
Interest rate. *See* Discount rate
Interest rate management, 405–408
Interfacial tension (IFT), 39–44
Inter-lateral spacing, 199–200
Intermediate casing, 100
Intermediate-strength ceramic proppant, 74
Internal rate of return (IRR), 377–382, 378–381b, 468–473
International Organization for Standardization (ISO) conductivity testing, 86–89, 94
International Union of Pure and Applied Chemistry (IUPAC), 13–14, 18–19
Inventory difference, 461–464
Inventory influence, 445
Iron control, 120
Iron sulfide, 116
ISIP. *See* Instantaneous shut-in pressure (ISIP)
Isolation manifold. *See* Frac manifold
IUPAC. *See* International Union of Pure and Applied Chemistry (IUPAC)

J

Joint operating agreement (JOA), 344–345
Jones' technique, 25

K

Kerogen, 8–9, 9t, 13–14
Khristianovic-Geertsma de Klerk (KGD) model, 262–264, 263f
Kickoff point (KOP), 100
Klun system, 280–282, 283f
K-means clustering, 530–537
   application, 535
   data set division, 534, 534f
   distance equation calculation, 534
   for liquid-loading detection, 537, 538f
   process, 532–537
K-nearest neighbor (KNN), 505, 519–521
Knudsen diffusion, 14
KOP. *See* Kickoff point (KOP)
K-value testing, 79–80

L

Landing zone, 192
Langmuir equation, 29
Langmuir equilibrium equation, 16–17
Langmuir model, 16, 17f
Langmuir-type adsorption, 18–19
Langmuir volume, 33–34, 34f
Laplace's law, 39–40
Last resort valve. *See* Lower master valve
Late transient period, 314
Latin hypercube sampling (LHS), 526
LCP. *See* Lightweight ceramic proppant (LCP)
Leak-off, 64
Leak-off coefficient, 581
Learning rate, 510
LEFM. *See* Linear elastic fracture mechanics (LEFM)
Lifting cost, 348
Lightweight ceramic proppant (LCP), 74
Limited entry, 129–131, 130t
Linear elastic fracture mechanics (LEFM), 259–262
Linear flow-time function *vs.* bottom-hole pressure, 254–255, 255f
Linear gel, 116–117, 117f

Liquefied natural gas (LNG), 4–5
Liquefied petroleum gas (LPG), 4
Liquid additive (LA) pumps, 277–279
Liquid capacity, 208–209
   of separator, 205–206
Liquid level controller (LLC), 207, 209f
Liquid loading, 491–492
   well performance analysis, 573–574, 575f
Liquid-loading detection, k-means clustering for, 537, 538f
Liquid shrinkage, 347
LLC. *See* Liquid level controller (LLC)
LNG. *See* Liquefied natural gas (LNG)
Local outlier factor (LOF), 530–532
Log-log plot (log (BH ISIP-BHP) *vs.* log (time)), 237–240, 240–242f
Longitudinal fractures, 228–231, 230f
Lower master valve, 304–305
Low-magnitude earthquakes, hydraulic fracturing and, 106
Low-magnitude seismic events, 97–98
Low-permeability reservoirs, 49
Low Poisson's ratio, 50–51, 60–61, 69
Low-stage separator, 210
Low-temperature nitrogen adsorption technique, 19–20
LPG. *See* Liquefied petroleum gas (LPG)

M

Machine learning (ML), 499–502
   case usages, various industries, 502–504
   knowledge of data science and domain expertise, 504–505
   reinforcement learning, 501–502
   supervised learning, 501
   unsupervised ML, 501
   workflow for, 505–507, 507f
Manual valves, 297–299, 306–307, 307f
Marcellus Shale, 49, 53–55, 101–102, 119
   operations, 205–206
   reservoir, 84
Market basket analysis, 503
Matrix and hydraulic fracture interactions, 87f
Maximum horizontal stress, 226–228
Mean decrease in impurity (MDI), 523
Measuring, conventional methods of, 13

Mechanical pop-off, 207, 210f, 295, 297f
Mercury injection technique, 42–44
Metamorphism. See Coalification
Meter, 208
Microemulsion, 119
Micropores, 15
Minimum horizontal stress, 224–225
Missile, 290–291, 290f
Mist flow, liquid loading, 491
MLPRegressor, 526–527
Modern well performance analysis, 565–566
Modified hyperbolic decline curve, 317–318, 317f, 584–585, 585t
Modified internal rate of return (MIRR), 382–384, 383–384b
Modified pulse-decay permeameter, 21–22
Molecular dynamic technique, 29
Money supply management, 408–409, 409f
Mont Belvieu, 362
Monte-Carlo analysis, 529–530, 531f
Monthly hyperbolic cumulative volume, 322
Monthly nominal decline, 324–325
Multicontinuum modelling of shale reservoirs, 36–39, 37f
Multientry-point frac sleeve, 183
Multiphase flow, 88–89
Multiport technology, 182
Multiscale fluid flow, 35–36
   interfacial tension (IFT) and capillary pressure, 39–44
   multicontinuum modelling of shale reservoirs, 36–39, 37f
   wettability effects on shale recovery, 44–45
Multisegment decline, 330–333, 332f, 332–333t
Multistage hydraulic fracturing, 177
   of horizontal wells, 259

N

National Seismic Hazard Map, 97–98
Natural fractures, 9–10
Natural gas, 1–4
   components of, 2t
   hydrates, 7
   transportation of, 5
   types of, 4
Natural gas liquid (NGL), 4, 415
   yield calculation, 418–423, 420–423b, 421t, 425–426t, 428–429t, 433–434t, 436–440t
Natural gasoline, 4
Natural language processing (NLP), 503
Near-wellbore friction pressure (NWBFP), 131–132, 144–148
Near-wellbore pressure differential, 142
Net bottom-hole pressure (NBHP) chart, 136–137, 292
Net Capex, 363–364
Net cash-flow (NCF) model, 341–342, 342f, 365–366
Net Opex, 356–357, 357b
Net present value (NPV), 341–342, 372–376, 374t, 374–376b, 382, 468–473
Net pressure, 135–137, 136f
   charts, 136–137
   vs. time, 136–137
Net revenue interest (NRI), 345–346, 346b, 468
Net sold condensate, 426
Neural network training, 526–527, 526–528f
Neutrally wet, 42
Newtonian fluid, 112
New York mercantile exchange (NYMEX), 359
NGL. See Natural gas liquid (NGL)
Nitrogen foam frac, 55–57
Nitrogen rate (with and without proppant), 166–174
Nitrogen volume, 162
NMR. See Nuclear magnetic resonance (NMR)
Nodal analysis, 475–476
Nolty chart. See Net bottom-hole pressure (NBHP) chart
Nominal decline (Di), 312
Nonassociated gas, 4
Nonconductive mud, 11
Non-Darcy effect, 480
Non-darcy flow, 88
Nondestructive techniques, 45

Index

Nonemulsifiers, 120
Nonequilibrium molecular dynamics, 44
Nonlinear adsorption kinetics, 39
Nonlinear history-matching algorithm, 25
Non-Newtonian fluid, 89–90
Nonoxidizing biocide, 115
Non productive time (NPT), 273
Nonwetting phase, 41–42
Normal fault environment, 228
Normal stress, 260
North American shale plays, TOC of, 10t
Nuclear magnetic resonance (NMR), 14, 45, 548–549
Numbers of holes (perfs) and limited entry technique, 130–131
Numerical simulation of hydraulic fracturing propagation, 257–258
   fluid flow in hydraulic fractures, 265–266
   hydraulic and natural fracture interactions, 270–271
   pseudo-3D hydraulic fracturing models, 266–268, 267f
   simulators, development of, 259–262
   solid elastic response, 266
   stage merging and stress shadow effects, 271–272
   stratigraphic and geological structure modeling, 258–259
   3D hydraulic fracturing models, 268–270
   2D hydraulic fracturing models, 262–265
NWBFP. *See* Near-wellbore friction pressure (NWBFP)

O

OGIP. *See* Original gas-in-place (OGIP)
Oil, 1
   and gas industry, 499, 517
   and gas inventory, 444
Oil mud, 11
Oil Price Information Services (OPIS), 362
Oil tanks (upright tanks), 212, 212f
Open perforations, 128
Operating expense (Opex), 348–351, 364–365
Operations and execution
   blender, 279–280
   chemical coordination, 300–301
   flowback, tips for, 301–303
   frac van
      chemical chart, 293, 294–295f
      NBHP chart, 292
      surface-treating pressure chart, 292, 293f
   hydration unit, 277–278
   overpressuring safety devices, 294–297
      check valves and manual valves, 297–299
      pressure transducers, 297, 298f
      PRVs, 295, 297f
      pump trips, 294, 296f
   postscreen-out injection test, 303
   sand coordination, 300
   sand master, 280–282, 284f
   stage treatment, 301
   T-belt, 280–289, 284f
      chemical injection ports, 288–289, 289f
      densometer, 289
      frac manifold, 291, 291–292f
      missile, 290–291, 290f
      sand screws, 283–287, 285f, 288f
   water coordination, 299
   water sources, 273–277
      ASTs, 274, 276f
      centralized impoundment, 274, 275f
      centralized tanks, 275, 276f
      delivery, 276–277
Opportunity cost, 467, 473
Opportunity cost of capital. *See* Discount rate
Optimized chemical package, 107
Optimum economic rate, 91–93
Organic-rich shale
   samples, 20
   transport in, 35–36
      interfacial tension (IFT) and capillary pressure, 39–44
      multicontinuum modelling of shale reservoirs, 36–39, 37f
      wettability effects on shale recovery, 44–45
Original gas-in-place (OGIP), 13, 27–28, 31, 33–34, 570–571
Original oil-in-place (OOIP), 13
Ottawa sand, 83

Outer wells, 446–447
Output layer, ANN elements, 508
Overlying shale layer, 227
Overpressuring safety devices, 294–297
  check valves and manual valves, 297–299
  pressure transducers, 297, 298f
  PRVs, 295, 297f
  pump trips, 294, 296f
Oxidizing biocide, 115

P

Pad stage, 63–65
Payback method, 384–386, 384–386b
PayPal, 502–503
PDL. *See* Pressure-dependent leak-off (PDL)
Pearson correlation coefficient, 523–525
Penny-shaped model, 264
Perforation design, 129–130, 194
Perforation diameter, 127
  and penetration, 131
Perforation efficiency, 128
Perforation erosion, 131
Perforation friction-pressure, 127–128, 142, 145
Perforation guns, 180f
Perforation phasing, 194f
Perkins and Kern (PKN) model, 262–264, 264f
Petra, 545–547
Petrophysics, 545–554, 546f, 547t
  geomechanical logs, 549–554, 551–553f
Photoelectric absorption log (PEFZ), 545–547
Physical sorption, 15–16
Pipe, relative roughness of, 112–114, 113t
Pipe friction pressure, 111–112, 126–127
Pipeline water delivery, 276
PKN model. *See* Perkins and Kern (PKN) model
Plackett-Burman "PB" design, 559, 560t
Plant inlet yields liquid shrinkage, 418–423
Plant outlet yields liquid shrinkage, 419
Plotted *vs.* relative pressure, 18–19
Plug-and-cluster spacing, 180–181
Plug and perf, 177–179, 182
Pneumatic ESD, 306, 307f

Poisson's ratio, 10–11, 217–219, 220f, 224–226, 543, 549–554
Polyacrylamide, 108
Polyvinyl chloride (PVC), 299
POOH. *See* Pulled out of the hole (POOH)
Pop-offs. *See* Pressure-relief valves (PRVs)
Pore compressibility measurements of shale, 21–22
Pore pressure. *See* Reservoir pressure
Pore-size distribution measurement of shale, 13–15
Poroelastic constant, 225
Porosity, 20
Porosity correlation, 223
Post extraction residue gas, 419
Postfracturing flowback fluids, 108–109
Postscreen-out injection test, 303
Potential theory model, 16
Pounds per revolution (PPR), 282, 287
Power law exponential decline model (PLE), 318
PPAL. *See* Precision Petrophysical Analysis Laboratory (PPAL)
PRCS. *See* Precured resin-coated sand (PRCS)
Precision Petrophysical Analysis Laboratory (PPAL), 26
Precured resin-coated sand (PRCS), 72–73
Preferred stock, 367
Pre job water testing, 114–115
Pressure, 121
Pressure-dependent leak-off (PDL), 243–246
Pressure normalized rate, 333–340, 338–340t
Pressure pulse-decay techniques, 25–26
Pressure-relief valves (PRVs), 295, 297f
Pressure squared approach, 476
Pressure transducers, 297, 298f
Pressure-volume-temperature (PVT) measurements, 32
Pricing escalations, 364–365
Prime borrowers, 410
Prime rate, 410
Principal stress, 261
Principal stress reversal, 271–272
Processing cost, 350

Index

Processing gas, 423–441, 424t
Processing plant, 417–418
Production analysis
   back pressure model, 479–484, 481–482b, 483t, 483f
   Coleman rate, 493–498, 494–498b, 496–498t
   erosional velocity calculation, 488–490, 489–490b
   gas production operation issues, 491–492
   nodal analysis, 475–476
   pseudo-pressure concept and calculation, 476–479, 477–479b, 478–479t
   tubing performance relationship, 485–488, 485–486b, 487t, 487f
   Turner rate, 492–493
Production casing, 100
   design, 139–140
Profit, 365–366
Profitability index (PI), 386–387, 387b
Proppant, 280–282, 287
   size and type, 195–196
Proppant characteristics and application design, 71, 79–82
   curable resin-coated sand (CRCS), 72–73, 73f
   cyclic stress, 90–91
   dimensionless fracture conductivity, 86
   fines migration, 91–92
   finite vs. infinite conductivity, 93–95
   fracture conductivity, 85
   gel damage, 89–90
   high-strength proppant, 74–75
   intermediate-strength ceramic proppant, 74
   International Organization for Standardization conductivity test, 86–87
   lightweight ceramic proppant (LCP), 74
   multiphase flow, 88–89
   non-darcy flow, 88
   precured resin-coated sand (PRCS), 72
   proppant particle-size distributions, 82
   proppant size
      20/40 mesh, 78–79
      30/50 mesh, 77–78
      40/70 mesh, 76–77
      100 mesh, 76, 77f
   proppant transport and distribution in hydraulic fracture, 82–84
   reduced proppant concentration, 89
   sand, 71–72
   time degradation, 93
Proppant crushing embedment, 91f
Proppant particle-size distributions, 82
Proppant permeability, 85
Proppant size
   20/40 mesh, 78–79
   30/50 mesh, 77–78
   40/70 mesh, 76–77
   100 mesh, 76
Proppant stage, 65–66
Proppant transport and distribution in hydraulic fracture, 82–84
Pseudo-3D hydraulic fracturing models, 266–268, 267f
Pseudolinear flow, 237–239
Pseudo-pressure, 476–479, 477–479b, 478–479t
Pseudosteady state, 315
Pseudo-steady-state flow, 476
Pulled out of the hole (POOH), 178–179
Pulse-decay permeameters, 23–24
Pumped proppants, 82–83
Pump trips, 294, 296f
Put option strategy, 412
PVT measurements. See Pressure-volume-temperature (PVT) measurements
Python, 526–527
Python "sklearn" package, 521

Q

Quantitative easing (QE), 408–409

R

Radial flow-time function vs. BHP, 255–256, 256f
Radial fracture geometry, 264f
Radial fracture model, 264
Rate of return (ROR). See Internal rate of return (IRR)
Rate step-down test workflow, 141–148
Rate transient analysis (RTA), 191–192, 565–566

Rawlins and Schellhardt equation, 479–484, 481–482*b*, 483*t*, 483*f*
Real field Marcellus Shale reservoir development and stimulation, 541
  frac job sand schedule optimization, 577–583
    economics analysis, 581–583, 583*t*, 584*f*
    hydraulic fracture treatment design, 578–581, 579–580*t*, 582*f*, 582–583*t*
  MIP-3H well available data, 541, 542*t*, 550
  modified hyperbolic decline curve, 584–585, 585*t*
  petrophysics and well log analysis, 545–554, 546*f*, 547*t*
    geomechanical logs, 549–554, 551–553*f*
  problem statement, 541–545
  sensitivity analysis and history-matching, 556–565, 560–561*t*, 561–562*f*
  Shale gas reservoir base model development
    petrophysical properties of layer, 554–555
    process of, 554
    3D numerical model, 555–556, 556*f*
  well performance analysis, 565–566
    Blasingame type curve, 568, 568*f*
    DFIT, 574–577, 576–577*f*
    fractured gas well, Agarwal-Gardner type curve for, 566–567, 567*f*
    liquid loading, 573–574, 575*f*
    transient type curves, 569, 570*f*
    variable flowing pressure, 570–573, 572*f*
    Wattenbarger type curve analysis, 568–569, 569*f*
Reason surface-treating rate, 116–117
Recovery, 408–409
Recovery factor, 33–34
Reduced proppant concentration, 89
Refrac, 186–189
Regularization techniques, 521–523
Reinforcement learning, 501–502
Relative permeability curve, 88*f*
Relative roughness of pipe, 112–114, 113*t*

Reservoir pressure, 12
Residual gel effect, 89–90
Residual optimization algorithm, 451
Resin-coated sand, 72, 83–84
Revenue, 357–359, 358–359*b*
The Revenue Act, 412
Reverse (thrust) fault environment, 228
Reynolds number, 82–83, 112, 114
Ridge regression, 521–523
Risk-free rate (Rf), 370
Risk premium (R_m-R_f), 370–371
Rock brittleness, 10–11
Rock mechanical properties, 10–11, 215
  Biot's constant (poroelastic constant), 225
  Biot's constant estimation, 225–226
  brittleness and fracability ratios, 220–222, 222*t*
  fracture orientation, 228–229
  fracture toughness, 220
  longitudinal fractures, 229–231, 230*f*
  maximum horizontal stress, 226–228
  minimum horizontal stress, 224–225
  Poisson's ratio, 217–219, 220*f*
  transverse fractures, 229, 230–231*f*
  various stress states, 228
  vertical stress, 223–224
  Young's modulus, 215–217, 217*f*
Roundness, 79, 80*f*
Royalty, 342–343
RTA. *See* Rate transient analysis (RTA)
Rule of thumb, 194

S

Safety factor, 139
Salt-mud, 11
Sample bulk volume, 20
Sampling, conventional methods of, 13
Sand, 71–72
  coordination, 300
  and water per foot design, 194–195
Sand box system, 280–282, 288*f*
Sanding off, 64–65
Sand master, 280–282, 284*f*
Sand per foot, 154
Sand screws, 283–287, 285*f*
Sandstone layer, 227
Sand storm system, 280–282, 285*f*

Index

Sand-to-water ratio (SWR), 155
Sand trap, 204–205, 205f
Saturation pressure, 18–19
Scale inhibitor, 115–116
Scanning transmission electron microscopy imaging (STEM), 15
SCF. *See* Standard cubic feet (SCF)
Scrubber pot, 207
Seat, 203
Selective coupling, 36
Sensitivity analysis, 556–565, 560–561t, 561–564f, 565t
Series coupling, 36
Settling velocity, 82–83
7402 psi, 576–577
Severance tax, 354–355, 354–355b
SF. *See* Slurry factor (SF)
Shale gas, 7–9
Shale gas reservoir base model development
   petrophysical properties of layer, 554–555
   process of, 554
   3D numerical model, 555–556, 556f
Shale gas reservoirs, 22, 27–28
   multicontinuum modelling of, 36–39, 37f
Shale initial gas-in-place calculation, 27
   adsorbed gas density, 31–32
   recovery factor, 33–34
   total gas-in-place calculation, 27–31
Shale matrix bulk volume, 28f
Shale permeability measurement techniques, 22–26
Shale porosity measurements, 20–21
Shale recovery, wettability effects on, 44–45
Shale sorption measurement techniques, 15–20
Shape of sorption isotherms, 18–19
Shareholder's value, 471, 473–474
Shear environment, 228
Shorter stage length (SSL), 184–186
Shrinkage factor, 346–348, 347–348b
Sieve analysis, 80–81
Silt and fine particles, 81–82
Simultaneous frac, 182
Simultaneous optimization, 443–464
   well spacing and completions optimization, 445–464
   dynamic workflow, 450–464, 452f, 455t, 459t, 460f, 462t
   frac hit and influence, 447–449, 448f
   fracture hit detection and mitigation strategies, 449–450
   outer and standalone wells' influence on, 446–447
Single-component monolayer Langmuir isotherm, 39
Single-entry-point frac sleeve, 183
Single hydraulic fracture, geometry of, 105f
Single-well applications, 204–205
Sintered bauxite, 74–75, 75t
Slick water fluid system, 49–52, 112
Slick water frac fluid system, 47–48, 65–66, 107, 141
Slick water frac jobs, 108
Slick water frac schedule, 155–160
Slick water multistage horizontal stimulation, 47–48
Slick water schedule, 157t
Sliding sleeve, 177–178, 182
   advantages, 182–183
   disadvantages, 183–184
Slip tendency, 104, 105f
Slugging, liquid loading, 491
Slurry density, 151–152
Slurry factor (SF), 163–164
Slurry fluid, 290
Slurry rate, 292
   with proppant, 165–166
Social Security Act, 411
Society of Petroleum Engineers (SPE) in the Journal of Petroleum Technology, 6
Sodium chloride, 116
Solid elastic response, 266
Sonic porosity (SPHI), 545–547
Sorage in organic-rich shale reservoirs, 35
Sorbed gas quantity, 20
Sorption isotherms, 15–16
Source rock, 8–9
Specific gravity (SG), 80
   of HCl acid, 62t
Sphericity, 79, 80f
Square root plot, 236–237, 237–238f
SSL. *See* Shorter stage length (SSL)

Stacked pay codevelopment, 467–474, 470–471f
Stack fracing, 180–181
Stage fluid clean volume, 152
Stage fluid slurry volume, 153
Stage proppant, 153–154
Stage spacing, 193
Stage treatment, 301
Stand-alone development, 470, 470f
Standalone wells, 446–447
Standard API crush-testing procedure, 79–80
Standard cubic feet (SCF), 161
Standard sieve openings, 81t
STEM. See Scanning transmission electron microscopy imaging (STEM)
Stokes' law, 82–83
STP. See Surface-treating pressure (STP)
Strain, 261
Stratigraphic structure model, 258–259
Stress, 260
Stress states, 228
Stretched exponential decline, 318
Strike-slip (shear) environment, 228
Subprime Crash, 412
Suction side centrifugal pumps, 280
Sulfate-reducing bacteria, 115
Supervised learning, 501
    completion and stimulation optimization using, 500, 518f, 520t, 520f, 522f, 524–525f
Surface casing, 99
Surface-treating pressure (STP), 121, 137–139, 292, 293f
Surfactant, 119–120
Swab valve. See Manual valve
Sweep, 65
SWR. See Sand-to-water ratio (SWR)
Symmetric geometry, 264
Synthetic-based mud, 11
System back offs, 444–445

T

Tank batteries, 275, 276f
Target zones, 192
Taxable income, 389–390, 389–390b
Tax model
    economic evaluation
        ATAX monthly undiscounted NCF, 391–404, 391–404b
        corporation tax, 391
        depreciation, 388–389, 388t, 389b
        taxable income, 389–390, 389–390b
T-belt, 280–289, 284f
    chemical injection ports, 288–289, 289f
    densometer, 289
    frac manifold, 291, 291–292f
    missile, 290–291, 290f
    sand screws, 283–287, 285f, 288f
TDS. See Total dissolved solids (TDS)
Tensile strength, 261–262
Terminal decline, 317–318
Test Sieve shaker, 81f
Thermal maturity (TM), 8–9
30/50 mesh sand size, 77–78
Three-dimensional (3D) geological models, 258
Three-dimensional (3D) hydraulic fracturing models, 268–270
Three-leg frac manifolds, 291, 291f
Three-phase separator, 205–206
Thrust fault environment, 228
Tighter stage spacing, 193
Tight gas sands, 7–8
Time degradation, 93
Time zero, 341–342, 365–366
Tip extension leak-off, 248–252, 250f
TM. See Thermal maturity (TM)
TOCs. See Total organic contents (TOCs)
Toe sleeve/valve, 183
Top valve. See Manual valve
Tortuosity, 59–61, 60f
Tortuosity friction pressure, 142, 145
Total dissolved solids (TDS), 200
Total friction pressure, 126, 141–142
Total gas-in-place calculation, 27–31
Total Magnetic Resonance Porosity (TCMR), 545–547
Total Opex per month, 351–353, 352–353b
Total organic contents (TOCs), 7–9, 11, 192, 545–547
    of North American shale plays, 10t
Totes, 277–279
Traditional volumetric technique, 19–20

Transient flow, 314
Transient type curves, 569, 570f
Transmissibility, 255–256
Transport in organic-rich shale, 35–36
  interfacial tension (IFT) and capillary pressure, 39–44
  multicontinuum modelling of shale reservoirs, 36–39, 37f
  wettability effects on shale recovery, 44–45
Transverse fractures, 228–229, 230–231f
Trapezoidal method, 477
Trucking water delivery, 277
Tub agitator, 279
Tubing analysis, 213
Tubing head, 304, 304f
Tubing performance curve (TPC), 475
Tubing performance relationship (TPR), 485–488, 485–486b, 487t, 487f
Turned in line (TIL), 341–342, 467–468
Turner and Coleman techniques, 537
Turner rates, 492–493, 573–574
20/40 mesh sand size, 78–79
20lb linear gel system, 56f
2D hydraulic fracturing models, 262–265
2D plane-strain model, 263
Two-phase separator, 205–206
Type I adsorption isotherms, 18–19
Type II adsorption isotherms, 18–19
Type III adsorption isotherms, 18–19
Type IV adsorption isotherms, 18–19
Type V adsorption isotherms, 18–19
Type VI adsorption isotherms, 18–19
Type I kerogen, 8–9
Type III kerogen, 8–9
Type IV kerogen, 8–9
Typical DFIT procedure, 234, 235f
Typical frac wellhead, 308, 309f
Typical slick water frac steps, 61

U

Ultra-low permeability formations, 84
Ultra-low-permeability shale reservoirs, 97–98
Unconventional gas resources, 7
Unconventional reservoir development footprints, 97–98
  casing selection, 98
  conductor casing, 98–99
  hydraulic fracturing
    and aquifer interaction, 101–102
    and fault reactivation, 102–105
    and low-magnitude earthquakes, 106
  intermediate casing, 100
  production casing, 100
  surface casing, 99
Unconventional reservoirs, 5–12
Unconventional shale reservoir, 47
  characterization, 13
Uncurtailed volume, 453–454
Underlying shale layer, 227
Unsteady (transient flow) state period, 314
Unsupervised machine learning, 501, 537–539
Up dip vs. down dip, 198, 198f
Upper master valve. See Manual valve
Uranium Concentration (HURA), 545–547
US Bureau of Mines (USBM) wettability index, 44
US Energy Information Administration (EIA), 6
US Geological Survey (USGS), 6, 97–98, 106
Utica/Point Pleasant, 91–93
Utica Shale, 120, 210–212

V

Valve, 203
Van der Waals equation of state, 29, 31–32
Variable flowing pressure, 570–573, 572f
Vertical stress, 223–224
Vitrinite reflectance values vs. reservoir relationship, 9t
Volumetric method, 27
Volumetric sorption measurement techniques, 19–20

W

Washburn equation, 14
Wastewater deposition, 97–98
Water coordination, 299
Water disposal cost, 350
Water frac, 50–51
Water per foot, 154

Water quality, 200–201
Water saturation of Marcellus Shale, 10–11
Water sources, 273–277
　delivery, 276–277
　water storage
　　ASTs, 274, 276f
　　centralized impoundment, 274, 275f
　　centralized tanks, 275, 276f
Wattenbarger type curve analysis, 568–569, 569f
Weighted average BTU factor, 4t
Weighted average cost of capital (WACC). See Discount rate
Weight of debt and equity, 368–369, 369b
Wellbore integrity, 98
Wellhead design, 475
　production analysis and (see Production analysis)
Wellhead treating pressure (WHTP). See Surface-treating pressure (STP)
Well log analysis, 545–554, 546f, 547t
Well performance analysis, 565–566
　Blasingame type curve, 568, 568f
　DFIT, 574–577, 576–577f
　fractured gas well, Agarwal-Gardner type curve for, 566–567, 567f
　liquid loading, 573–574, 575f
　transient type curves, 569, 570f
　variable flowing pressure, 570–573, 572f
　Wattenbarger type curve analysis, 568–569, 569f
Well spacing, 199–200
Well spacing and completions optimization, 443–444
　gas pricing and CAPEX influence, 444
　inventory influence, 445
　lateral length influence, 444–445
　simultaneous optimization, 445–464
　　dynamic workflow, 450–464, 452f, 455t, 459t, 460f, 462t
　　frac hit and influence, 447–449, 448f
　　fracture hit detection and mitigation strategies, 449–450
　　outer and standalone wells' influence on, 446–447
West Texas Intermediate (WTI), 361–362
Wettability effects on shale recovery, 44–45
Wetting phase, 41–42
Wild Well Control Technical Data Book (WWCTDB), 139–140
Wireline process, 178–179
Wobbe Index (WI), 415
Working interest (WI), 343–345
World Shale Map, 6
World stress map, 229
WWCTDB. See Wild Well Control Technical Data Book (WWCTDB)

X

XFEM. See Extended finite element method (XFEM)

Y

Young-Dupré equation, 41–42
Young-Laplace equation, 42
Young's modulus, 10–11, 50–51, 60–61, 69, 135–136, 215–217, 217f, 220–221, 261, 549–554, 581

Z

Zipper fracing, 182
Z-score test, 519